P9-DCG-064

THE ADVANCED THEORY OF STATISTICS

Volume 1

DISTRIBUTION THEORY

Other books on theoretical and applied statistics

Scientific truth and statistical method — M. Boldrini (trans. Ruth Kendall)
Multivariate analysis — Sir Maurice Kendall
Time-series — Sir Maurice Kendall
Rank correlation methods — Sir Maurice Kendall
An introduction to the theory of statistics — G. U. Yule and Sir Maurice Kendall
Statistical papers of George Udny Yule — ed. A. Stuart and Sir Maurice Kendall
Studies in the history of statistics and probability
 Volume I — ed. E. S. Pearson and Sir Maurice Kendall
 Volume II — ed. Sir Maurice Kendall and R. L. Plackett
Exercises in theoretical statistics — Sir Maurice Kendall
Exercises in probability and statistics (for mathematics undergraduates) — N. A. Rahman
Practical exercises in probability and statistics — N. A. Rahman
A course in theoretical statistics — N. A. Rahman
Rapid statistical calculations — M. H. Quenouille
Characteristic functions — E. Lukacs
Experimental design: selected papers — F. Yates
Sampling methods for censuses and surveys — F. Yates
Biomathematics — Cedric A. B. Smith
Combinatorial chance — F. N. David and D. E. Barton
Exercises in mathematical economics and econometrics — J. E. Spencer and R. C. Geary
Problems in linear and nonlinear programming — S. Vajda
Distribution management: mathematical modelling and practical analysis — S. Eilon, C. D. T. Watson-Gandy and N. Christofides
Mathematical model building in economics and industry (1st and 2nd series) — ed. Sir Maurice Kendall
Computer simulation models — John Smith
The mathematical theory of infectious diseases — N. T. J. Bailey
Estimation of animal abundance and related parameters — G. A. F. Seber
Statistical method in biological assay — D. J. Finney
Statistical communication and detection: with special reference to digital data processing of radar and seismic signals — E. A. Robinson

THE ADVANCED THEORY OF STATISTICS

Sir Maurice Kendall, Sc.D., F.B.A.

and

Alan Stuart, D.Sc. (Econ.)

Professor of Statistics in the University of London

IN THREE VOLUMES

Volume 1

DISTRIBUTION THEORY

Fourth edition

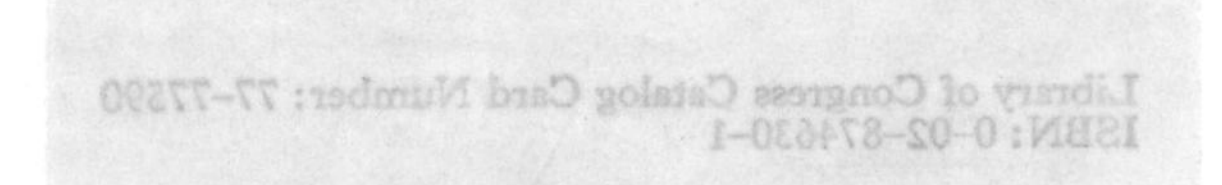

MACMILLAN PUBLISHING CO., INC.

NEW YORK

Published in USA by
Macmillan Publishing Co., Inc.
866 Third Avenue, New York, N.Y. 10022

Distributed in Canada by
Collier Macmillan Canada, Ltd.

By arrangement with the originating publisher
CHARLES GRIFFIN & COMPANY LIMITED
London and High Wycombe

Library of Congress Catalog Card Number: 77–77590
ISBN: 0–02–847630–1

Made and printed in Great Britain by
Butler & Tanner Ltd, Frome and London

"Let us sit on this log at the roadside," says I, "and forget the inhumanity and ribaldry of the poets. It is in the glorious columns of ascertained facts and legalized measures that beauty is to be found. In this very log we sit upon, Mrs. Sampson," says I, "is statistics more wonderful than any poem. The rings show it was sixty years old. At the depth of two thousand feet it would become coal in three thousand years. The deepest coal mine in the world is at Killingworth, near Newcastle. A box four feet long, three feet wide, and two feet eight inches deep will hold one ton of coal. If an artery is cut, compress it above the wound. A man's leg contains thirty bones. The Tower of London was burned in 1841."

"Go on, Mr. Pratt," says Mrs. Sampson. "Them ideas is so original and soothing. I think statistics are just as lovely as they can be."

O. HENRY, *The Handbook of Hymen*

PREFACE TO FIRST EDITION

This book, in its original form, was planned before the outbreak of the Second World War and written mostly during the early years of that troubled period, the first volume being published in 1943 and the second in 1946. Since that time there have been many advances in statistical theory and an expansion in the use of statistical methods which is striking even by comparison with the development of scientific work generally. Five editions of the original Volume 1, and three of Volume 2, have appeared with little more than the correction of errors and the addition of references. But the time has clearly arrived for a complete revision, of which this is the first volume. Owing to the growth of the subject, two further volumes are projected. I regret the size of the new version, but nothing smaller would have covered the field after the manner of my original design.

The revised work falls into three main divisions. The theory of distributions, including sampling distributions, is covered by the present new Volume 1. The second volume will deal with estimation, the general theory of statistical inference, and with statistical relationship. Volume 3 will cover the theory of the design and analysis of sample surveys and experiments, multivariate analysis, and time-series.

A good deal of the material in the original Volume 1 has survived in this volume. Substantial additions, however, have been necessary, notably on standard distributions, on the theory of approximations to sampling distributions, on order-statistics, on multivariate normal distributions, and on distributions associated with the normal distribution. Three of the chapters are new and much new material has been incorporated in the rest of the text. Very considerable additions and modifications will be necessary for the new Volumes 2 and 3, but as it may be several years before the complete work appears I hope that the original Volume 2 will be useful in the meantime.

Many new examples and exercises have been added. There are now over 300 exercises in this volume. Any student who remains unsated after working through them may like to proceed to my *Exercises in Theoretical Statistics*, after which he should be in a position to compose his own.

The references at the end of this volume cover only those articles which are specifically mentioned in the text. Further references will be given in Volumes 2 and 3.

The heavy labour of revision was something I felt unable to face alone, and I have been very fortunate in enlisting the co-operation of my colleague, Mr. Alan Stuart, who has thrown himself into the work with characteristic energy and ability. This revision is to be regarded as of joint authorship ; for all of it we both accept responsibility.

For permission to quote some of the tables at the end of the book we are indebted to Dr. F. N. David, Professor Sir Ronald Fisher, the late Dr. John Wishart, Dr. Frank Yates, Messrs. Oliver & Boyd, and the editors of *Biometrika*. To these and to all the authors on whose work we have drawn we express our grateful thanks. Our thanks are also due to Mr. James Durbin, who read the galley proofs with great care and made some valuable suggestions for improvement. We are equally indebted to Mr. E. V. Burke, of Charles Griffin and Company, Limited, who took very great pains with the production of the book in all its stages.

Some readers of previous editions went to considerable trouble to draw attention to misprints, imperfections in the presentation, and points where amplification was desirable. We shall continue to be grateful to any reader who will perform the same service for this reset edition.

M. G. K.

LONDON,
May, 1958

PREFACE TO FOURTH EDITION

There have been major revisions of the text for this edition, and many minor improvements and extensions, including new exercises, based on recent research. We are, as always, grateful to the readers who have helped us to remove misprints and to clarify difficult points.

M. G. K.
A. S.

LONDON,
February, 1976

CONTENTS

INTRODUCTORY NOTE

0.1 The chapter-sections in this work are numbered serially. The serial numbers are prefixed by the number of the chapter in which they occur and are separated therefrom by a period, e.g. **14.13** refers to the thirteenth section of Chapter 14. A similar procedure is followed for tables, examples (in the text) and exercises (at the ends of chapters). Similarly, (7.15) refers to the fifteenth equation of Chapter 7. Thus chapter-sections are denoted by bold-face type, equations by ordinary type in parentheses.

0.2 References are given, by author's name and date of publication, to the list of references at the end of the book. Occasionally (e.g. in table-headings), a fuller reference is given in the text; this is always done when no individual author is named, as in extensive sets of statistical tables published by official bodies. Where journal articles are referred to, the number of the volume is given in bold-face type and the number of the first page of the article in ordinary type, e.g. *Ann. Math. Statist.*, **10**, 275, refers to the article beginning on page 275 of volume 10 of the *Annals of Mathematical Statistics*. Where an exercise is followed by an author's name and a date, the result given in the exercise appears in the article listed in the references under those particulars.

0.3 The mathematical notation is that in current use, but a few symbols may be explained.

(1) The exclamation mark ! written after an integer denotes the factorial of that integer. Some writers give the symbol a more extended use for non-integral numbers x by writing

$$x! = \Gamma(x+1) = \int_0^\infty e^{-t} t^x \, dt.$$

This, of course, accords with the factorial notation, but will not be used in this book.

(2) The use of horizontal lines over expressions to indicate bracketed terms has been avoided; instead, parentheses or brackets are used where necessary.

(3) The combinatorial sign $\binom{n}{r} = \dfrac{n!}{r!\,(n-r)!}$ is used.

(4) The summation sign is written as Σ, e.g. $\sum_{j=1}^{j=n} x_j = x_1 + x_2 + \ldots + x_n$.

The symbol $\sum_{j=1}^{j=n}$ can as a rule be shortened to $\sum_{j=1}^{n}$ and in many cases to $\sum_j$ or merely to Σ, the extent of the summation being clear from the context.

(5) The ordinary notation for the Γ-function (given above), the B-function, and the hypergeometric function is used, i.e.

$$B(p, q) = \int_0^1 x^{p-1}(1-x)^{q-1} \, dx = \frac{\Gamma(p)\,\Gamma(q)}{\Gamma(p+q)}$$

and

$$F(\alpha,\beta,\gamma,x) = 1+\frac{\alpha.\beta}{1.\gamma}x+\frac{\alpha(\alpha+1).\beta(\beta+1)}{1.2.\gamma(\gamma+1)}x^2+\frac{\alpha(\alpha+1)(\alpha+2)\beta(\beta+1)(\beta+2)}{1.2.3.\gamma(\gamma+1)(\gamma+2)}x^3+\dots$$

(6) Where the exponent is concise, the exponential function is written as a power of e, for example $e^{\frac{1}{2}x^2}$. But where it is lengthy we use the notation exemplified by $\exp\{-\frac{1}{2}(x^2-2\rho xy+y^2)\}$ instead of $e^{-\frac{1}{2}(x^2-2\rho xy+y^2)}$.

(7) The notation $O(f(n))$ signifies that an expression is of order not exceeding that of $f(n)$, i.e. that the limit of the ratio of the expression to $f(n)$ is a bounded constant C. If $C = 1$, we say that the expression is asymptotically equal to $f(n)$ and use the symbol $\sim$ to express this. If $C = 0$, we write $o(f(n))$ instead of $O(f(n))$ to signify that the order is less than that of $f(n)$.

0.4 It is necessary to preserve a distinction between a quantity in a population and that quantity in a sample. Where possible, the former will be denoted by a Greek letter and the latter by a Roman letter, e.g. the product-moment correlation coefficient of a population is denoted by ρ and that of a sample by r. It is not, however, always desirable to follow this rule; for the multiple correlation coefficient R, e.g., a Greek capital could be confused with the Roman P, so a different convention is followed. Complete notational consistency can only be achieved at the expense of jettisoning a great deal of accepted statistical usage, and even then would probably result in some cumbrous symbols.

0.5 In order to enable the reader to follow the worked examples and illustrative material, a few tables of functions commonly required are given at the end of this volume. These tables are in no way a substitute for the comprehensive sets that have been published, which are a necessary adjunct to most practical and a good deal of theoretical work. Frequent reference will be made to the following :—

Biometrika Tables for Statisticians, Volumes I and II, edited by E. S. Pearson and H. O. Hartley, Cambridge University Press. Unless otherwise stated, reference is made to Vol. I.

Statistical Tables for use in Biological, Agricultural and Medical Research, by Sir Ronald Fisher and F. Yates. Oliver and Boyd, Edinburgh.

Tracts for Computers is a very useful series published by Cambridge Univ. Press.

Some useful specialized tables are contained in

Selected Tables in Mathematical Statistics, Vols. I, II and III, edited for the Institute of Mathematical Statistics by H. L. Harter and D. B. Owen, with J. M. Davenport in Vol. III; published by The American Mathematical Society, Providence, R.I., U.S.A.

There is also a comprehensive

Guide to Tables in Mathematical Statistics, by J. A. Greenwood and H. O. Hartley, Princeton U.P. and Oxford U.P., 1962,

which covers fairly completely all tables published up to 1954 and less completely covers later tables.

Useful tables are described at many points in the text. References to the scope of tables are given in what is now fairly standard form; e.g. $x = 0(0{\cdot}1)10(1)50(2)100$ means that the function is tabulated for an argument x which proceeds by steps of 0·1 from 0 to 10, thence by steps of 1 from 10 to 50 and thence by steps of 2 from 50 to 100. In describing tables, " decimal places " is sometimes abbreviated to " d.p."

GLOSSARY OF ABBREVIATIONS

The following abbreviations are sometimes used:

ARE	Asymptotic Relative Efficiency
ASN	Average Sample Number
AV	Analysis of Variance
BAN	Best Asymptotically Normal
BCR	Best Critical Region
BIB	Balanced Incomplete Blocks
c.f.	characteristic function
c.g.f.	cumulant-generating function
d.f.	distribution function
d.fr.	degrees of freedom
d.p.	decimal places
f.f.	frequency function
f.m.g.f.	factorial moment-generating function
g.f.	generating function
LF	Likelihood Function
LR	Likelihood Ratio
LS	Least Squares
MCS	Minimum Chi-Square
m.g.f.	moment-generating function
ML	Maximum Likelihood
MS	Mean Square
m.s.e.	mean-square-error
MV	Minimum Variance
MVB	Minimum Variance Bound
$N(a, b)$	(multi-)normal with mean (-vector) a and variance (dispersion matrix) b
OC	Operating Characteristic
PBIB	Partially Balanced Incomplete Blocks
p.p.s.	probability proportional to size
s.e.	standard error
SPR	Sequential Probability Ratio
SS	Sum of Squares
UMP	Uniformly Most Powerful
UMPU	Uniformly Most Powerful Unbiased
USF	Uniform Sampling Fraction

CHAPTER 1

FREQUENCY DISTRIBUTIONS

1.1 The fundamental notion in statistical theory is that of the group or aggregate, a concept for which statisticians use a special word—" population ". This term will be generally employed to denote any collection of objects under consideration, whether animate or inanimate ; for example, we shall consider populations of men, of plants, of mistakes in reading a scale, of barometric heights on different days, and even populations of ideas, such as that of the possible ways in which a hand of cards might be dealt. The notion common to all these things is that of aggregation.

The science of Statistics deals with the properties of populations. In considering a population of men we are not interested, statistically speaking, in whether some particular individual has brown eyes or is a forger, but rather in how many of the individuals have brown eyes or are forgers, and whether the possession of brown eyes goes with a propensity to forgery in the population. We are, so to speak, concerned with the properties of the population itself. Such a standpoint can occur in physics as well as in demographic sciences. For instance, in discussing the properties of a gas we are, as a rule, not so much interested in the behaviour of particular molecules, as in that of the aggregate of molecules which go to compose the gas. The statistician, like Nature, is mainly concerned with the species and is careless of the individual.

1.2 We may therefore begin an approach to a definition of our subject by the following : Statistics is the branch of scientific method which deals with the properties of populations. This, however, is rather too general. Statistics deals only with the *numerical* properties. A dictionary, for example, sets out a population of words, and among the properties of that population which are a suitable subject for scientific inquiry is that of word-derivation. It is not of statistical concern, however, to know that some words are derived from Latin, some from Anglo-Saxon and some from Hindustani. The subject would only assume a statistical aspect if we were to inquire *how many* words were derived from the different sources.

1.3 As a second approximation to our definition we may then try the following : Statistics is the branch of scientific method which deals with the data obtained by counting or measuring the properties of populations.

This again is a little too general. A set of logarithm tables is a population of numerals, but it is hardly a subject for statistical inquiry, for every numeral is determined according to mathematical laws. The statistician is rather concerned with populations which occur in Nature and are thus subject to the multitudinous influences at work in the world at large. His populations rarely, if ever, conform exactly to simple mathematical rules, and in fact it is in the departure from such rules that he often finds topics of the greatest statistical interest. To allow for this factor we may then formulate our definition as follows :—

Statistics is the branch of scientific method which deals with the data obtained by counting or measuring the properties of populations of natural phenomena. In this definition " natural phenomena " includes all the happenings of the external world, whether human or not.

1.4 For the avoidance of misunderstandings in the interpretation of this definition it may be as well to point out that "Statistics", the name of the scientific method, is a collective noun and takes the singular. The same word " statistics " is also applied to the numerical material with which the method operates, and in such a case takes the plural. Later in this book we shall meet the singular form " statistic", which is defined as a function of the observations in a sample from some population. " Statistic " in this sense takes the plural "statistics".

Frequency distributions

1.5 Consider a population of members each of which bears some numerical value of a variable, e.g. of men measured according to height or of flowers classified according to numbers of petals. This variable we shall call a variate. If it can assume only a number of isolated values it will be called discrete or discontinuous, and if it can assume any value of a continuous range, continuous. The population of members will then correspond to a population of variate-values, and it is the properties of this latter population which we have to consider.

If the population consists of only a few members we can without much difficulty consider the population of variate-values exhibited by them ; but if, as usually happens, the aggregate is large (or, in a sense defined later, infinite), the set of variate-values has to be reduced in some way before the mind can grasp their significance. This is done by classification of the individuals into intervals of the variate. So far as possible the intervals should be equal, so that the numbers falling into different intervals are comparable. The interval is called the class-interval (or simply the interval) and the number of members bearing a variate-value falling into a given class-interval is the class-frequency (or simply the frequency). The manner in which the class-frequencies are distributed over the class-intervals is called the frequency distribution (or simply the distribution).

1.6 Tables 1.1 and 1.2 give two frequency distributions of observed populations classified according to a single variate. Table 1.1 shows the distribution of 1475 local districts in England and Wales for 1953, classified according to birth-rate (number of births per 1000 of the population). The distribution is shown in this table in a way which would be quite impossible if each of the 1475 districts were shown separately. The greatest number of districts fall within the range 15·5–16·5 per thousand and the frequencies tail off on either side of this value. Table 1.2 shows the number of persons subject to surtax in the United Kingdom in 1950 classified according to the variate " income ". The class-intervals here are unequal—a typical feature of official figures—and in the last column of the table is a reduction of the class-frequencies to comparability, namely, to frequency per £500 within the class-interval concerned.

Table 1.1—Showing local government areas in England and Wales, 1953, with specified birth-rates per 1000 of population

(Data from the Registrar-General's Statistical Review of England and Wales for 1953)

Birth-rate	Number of districts	Birth-rate	Number of districts
2·5–	1	15·5–	250
3·5–	0	16·5–	178
4·5–	0	17·5–	129
5·5–	1	18·5–	64
6·5–	0	19·5–	39
7·5–	3	20·5–	20
8·5–	7	21·5–	12
9·5–	19	22·5–	7
10·5–	39	23·5–	1
11·5–	96	24·5–	2
12·5–	151	25·5–	3
13·5–	231	26·5–	0
14·5–	221	27·5–	1
		TOTAL	1475

Note—2·5– means " 2·5 and less than 3·5 " etc.

Table 1.2—Number of persons in the United Kingdom liable to surtax in the year beginning 6 April, 1950, classified according to the magnitude of their annual income (*from Annual Abstract of Statistics*, 1950)

Annual income (£000)	Number of persons	Estimated frequency per £500 interval
2 and not exceeding 2·5	60,336	60,336
2·5 " " 3	41,033	41,033
3 " " 4	45,532	22,766
4 " " 5	23,263	11,632
5 " " 6	13,475	6,737
6 " " 8	13,456	3,364
8 " " 10	6,419	1,605
10 " " 12	3,551	888
12 " " 15	2,926	488
15 " " 20	2,007	201
20 " " 25	820	82
25 " " 30	399	40
30 " " 40	376	19
40 " " 50	134	6
50 " " 75	128	2
75 " " 100	45	1
100 and over	38	?
TOTAL NUMBER OF PERSONS	213,938	—

Looking at this column we see that the maximum frequency per £500 in this case is at the beginning of the frequency distribution.

1.7 The frequency distribution may be represented graphically. Measuring the variate-value along the x-axis and frequency per class-interval along the y-axis, we erect at the abscissa corresponding to the centre of each class-interval an ordinate equal to the frequency per unit interval in that interval. The ends of these ordinates

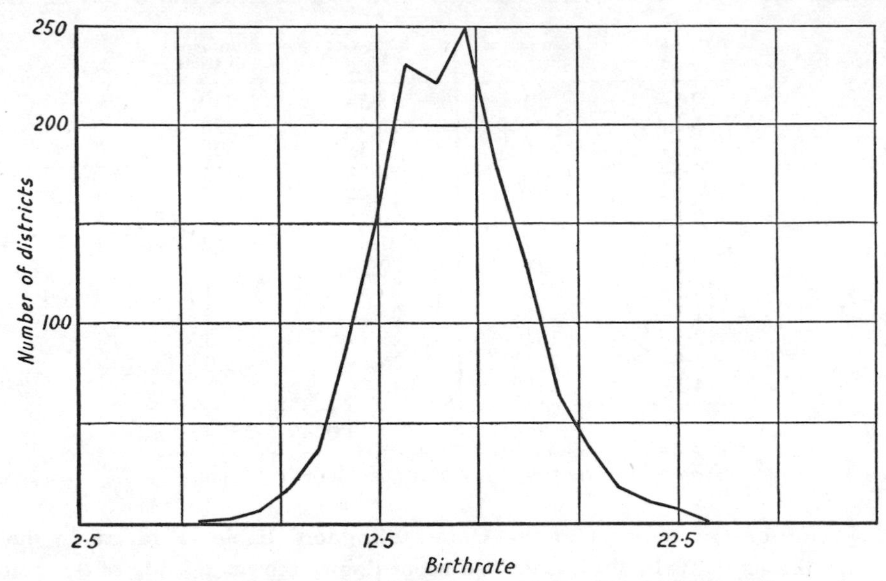

Fig. 1.1—Frequency polygon of the data of Table 1.1

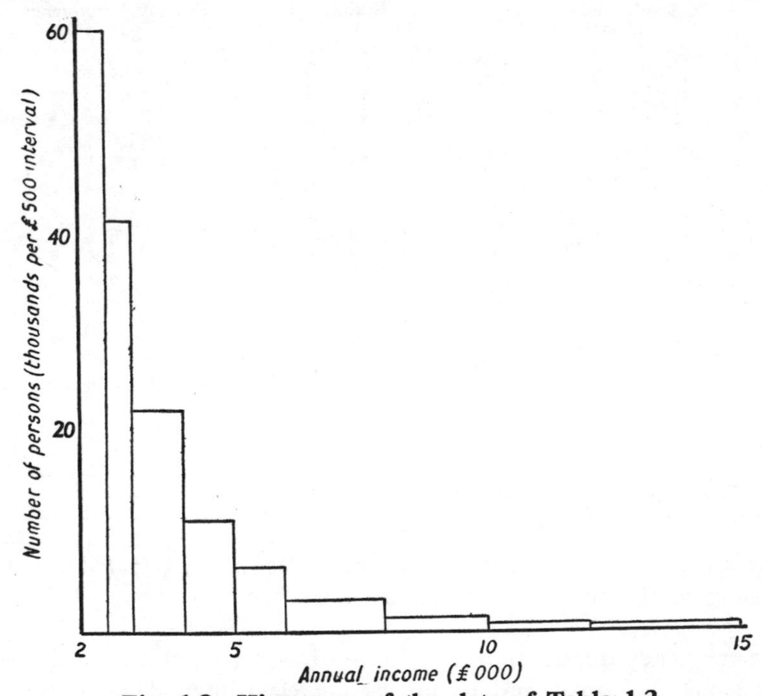

Fig. 1.2—Histogram of the data of Table 1.2

are joined by straight lines, one to the next. The diagram so obtained is called a frequency polygon. Fig. 1.1 shows the frequency polygon for the data of Table 1.1.

As a variant of this procedure we may erect on the abscissa range corresponding to each class-interval a rectangle whose area is proportional to the frequency in that interval. A diagram constructed in this way is called a histogram. Fig. 1.2 shows such a histogram for the data of Table 1.2. It is evident that the histogram is a more suitable form of representation when the class-intervals are unequal.

1.8 A few practical points in the tabulation of observed frequency distributions may be noted.

(1) It has been remarked that wherever possible the class-intervals should be equal. The importance of this will be more appreciated in subsequent chapters; but it is already evident that comparability is difficult to carry out by inspection when there exist inequalities in class-intervals. On running the eye down the second column of Table 1.2, for example, we note that the frequency in the interval 3–4 is greater than in the immediately preceding interval; but this is merely due to a change in the width of the intervals at that point and, as is seen from the third column, the frequency per unit interval decreases steadily.

(2) It is important to specify the class-interval with precision. We not infrequently meet with such classifications as "0–10, 10–20, 20–30", etc. To which interval is a member with variate-value 10 assigned? Obviously the classification is ambiguous if such values can in fact arise. We must either take the intervals "greater than or equal to 0 and less than 10, greater than or equal to 10 and less than 20," or make it clear what convention we use to allot a variate-value falling on the border between two neighbouring intervals, e.g. it might be decided to allot one-half of the member to each. There are various ways of indicating the class-interval in practical tables, e.g. "10–, 20–, 30–" means "greater than or equal to 10 and less than 20", and so forth. Sometimes, where a continuous variate is concerned, there is an element of imprecision in the specification of the fineness to which the measurements are made; for example, if we are measuring lengths in centimetres to the nearest centimetre, an interval shown as "greater than 15 and less than 18" means an interval of "greater than 15·5 and less than 17·5". When the precision of the measurements is known we can specify an interval by its middle point, for example, in this case, 16·5.

(3) Remark (1) about the importance of equality of class-intervals should not be held to preclude the specification of frequencies in finer intervals where the frequency is changing very rapidly. Table 1.3, for instance, shows the number of deaths from meningitis in England and Wales in 1953 according to the variate "age at death." If the frequencies in the interval "0 and less than 5" were not subdivided and were thus shown as a total 208 for the interval, we might be doubtful whether the greatest number of deaths occurred in the first year of life. This is so; the individual frequencies in the first five years bring out the relatively heavy mortality in the first year.

(4) Perhaps it is hardly necessary to add that the histogram is not a suitable method of representing data classified according to discrete variates. It shows the class-frequency uniformly dispersed over the whole interval, whereas if the variate is discrete, frequencies must necessarily be concentrated at certain points.

Table 1.3—Deaths from meningitis (except meningo-coccal and tuberculous) in England and Wales in 1953

(From the Registrar-General's Statistical Review for England and Wales for 1953)

Age, years	Number of deaths		Age, years	Number of deaths
0–	162	208	35–	5
1–	29		40–	11
2–	11		45–	17
3–	6		50–	21
4–	0		55–	32
5–	10		60–	23
10–	8		65–	22
15–	6		70–	15
20–	4		75–	11
25–	3		80–	10
30–	5		85 and over	3
			TOTAL	414

Frequency distributions: discrete (discontinuous) variates

1.9 It will be useful at this stage to give some further examples of the frequency distributions which occur in practice.

Table 1.4 shows the distribution of digits in numbers taken from a four-figure telephone directory. The numbers were chosen by opening the directory haphazardly

Table 1.4—Showing numbers of digits from a London Telephone Directory

(M. G. Kendall and B. Babington Smith (1938))

Digit	0	1	2	3	4	5	6	7	8	9	TOTAL
Frequency	1026	1107	997	966	1075	933	1107	972	964	853	10,000

and taking the last two digits of all the numbers on the page except those in heavy type. The distribution is irregular, but from a cursory inspection of the table we are inclined to suppose that the digits occur approximately equally frequently in the larger population from which these 10,000 members were chosen. We shall see later that the divergences from the average frequency per digit, 1000, are not accidental sampling effects ; but at this stage it is sufficient to note that the data suggest for consideration a population of equally frequent members.

Table 1.5 shows a number of unit samples of foliage on orange trees infested by black-scale insect. The greatest frequency is at zero and the frequencies for higher values of the variate tail off more or less regularly.

Table 1.5—Number of black-scale insects on 821 unit samples (ten adjacent leaves and their stem) of orange trees

(W. M. Upholt and R. Craig, 1940, *Jour. Ecol. Entom.*, **33**, 113)

Number of scales	0	1	2	3	4	5	6	7	8	over 8	TOTAL
Frequency	199	124	106	65	42	46	36	30	14	159	821

In Table 1.6, on the other hand, showing suicides among women in some German states in certain years according to the variate " number of suicides per year ", the distribution reaches its maximum frequency in the region 1–3 suicides and then tails off rather slowly.

Table 1.6—Showing suicides of women in eight German States in fourteen years
(Von Bortkiewicz, *Das Gesetz der kleinen Zahlen*, 1898)

Number of suicides per year	0	1	2	3	4	5	6	7	8	9	10 and over	TOTAL
Frequency . . .	9	19	17	20	15	11	8	2	3	5	3	112

Frequency distributions : continuous variates

1.10 Table 1.7 shows a number of adult males in the United Kingdom (including, at the time of the collection of the data, the whole of Ireland), distributed according to the variate " height in inches." The frequency polygon is shown in Fig. 1.3. It

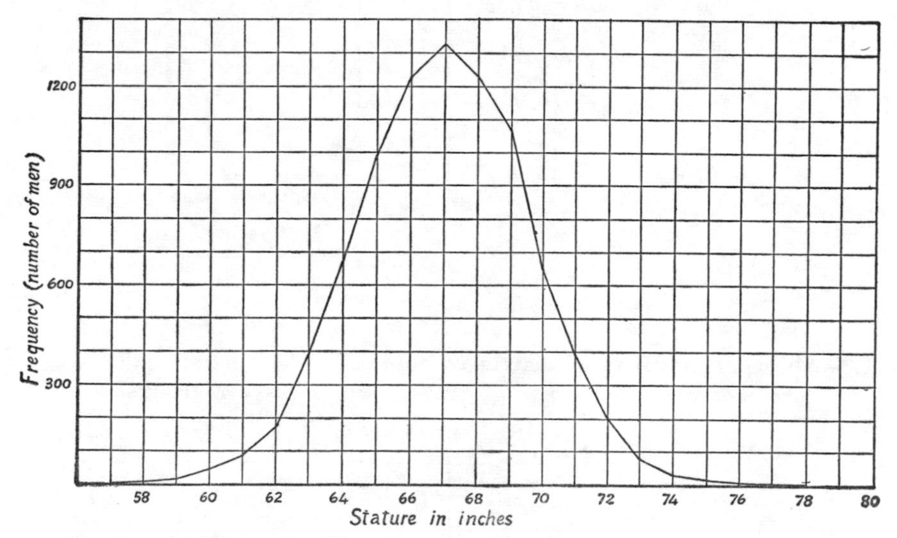

Fig. 1.3—Frequency distribution of the data of Table 1.7. Values of the abscissa correspond to the beginning of class-intervals

will be seen that the distribution is almost symmetrical, there being a maximum ordinate at 67– inches and a steady decrease in frequency on either side of the maximum.

This more-or-less uniform " tailing off " of frequencies is very common in observed distributions, but the symmetrical property is comparatively rare. Table 1.1 is roughly symmetrical, but Tables 1.8 and 1.9, showing respectively a number of Australian marriages distributed according to bridegroom's age, and a number of dairy herds distributed according to costs of production of milk, illustrate that various degrees of asymmetry can occur. An extreme form is shown in Table 1.3. In this connection Table 1.10, showing a number of men distributed according to weight, is of interest for comparison with the height data of Table 1.7. The latter is symmetrical but the former is not.

Table 1.7—Showing the frequency distribution of statures for adult males born in the United Kingdom (including the whole of Ireland)

(Final Report of the Anthropometric Committee to the British Association, 1883, p. 256)

As measurements are stated to have been taken to the nearest $\frac{1}{8}$th of an inch, the class-intervals are here presumably $56\frac{15}{16}$–$57\frac{15}{16}$, $57\frac{15}{16}$–$58\frac{15}{16}$, and so on

Height without shoes (inches)	Number of men within said limits of height	Height without shoes (inches)	Number of men within said limits of height
57–	2	68–	1230
58–	4	69–	1063
59–	14	70–	646
60–	41	71–	392
61–	83	72–	202
62–	169	73–	79
63–	394	74–	32
64–	669	75–	16
65–	990	76–	5
66–	1223	77–	2
67–	1329		
		TOTAL	8585

Table 1.8—Showing numbers of marriages contracted in Australia, 1907–14, arranged according to the age of bridegroom in 3-year groups

(From S. J. Pretorius (1930))

Age of bridegroom (central value of 3-year range, in years)	Number of marriages	Age of bridegroom (central value of 3-year range, in years)	Number of marriages
16·5	294	55·5	1,655
19·5	10,995	58·5	1,100
22·5	61,001	61·5	810
25·5	73,054	64·5	649
28·5	56,501	67·5	487
31·5	33,478	70·5	326
34·5	20,569	73·5	211
37·5	14,281	76·5	119
40·5	9,320	79·5	73
43·5	6,236	82·5	27
46·5	4,770	85·5	14
49·5	3,620	88·5	5
52·5	2,190		
		TOTAL	301,785

Table 1.9—Number of dairy herds in a sample of herds from England and Wales, 1948/9 according to cost of production of milk

(From National Investigation into the Economics of Milk Production—Milk Marketing Board, 1950)

Cost in pence per gallon	Number of herds	Cost in pence per gallon	Number of herds
less than 12	3	24–	70
12–	19	26–	49
14–	52	28–	31
16–	96	30–	16
18–	121	32–	6
20–	115	34–	8
22–	86	36 and over	7
		TOTAL	679

Table 1.10—Frequency distribution of weights for adult males born in the United Kingdom

(Source: as in Table 1.7. Weights were taken to the nearest pound, consequently the true class-intervals are 89·5–99·5, 99·5–109·5, etc.)

Weight in lb	Frequency	Weight in lb	Frequency
90–	2	190–	263
100–	34	200–	107
110–	152	210–	85
120–	390	220–	41
130–	867	230–	16
140–	1623	240–	11
150–	1559	250–	8
160–	1326	260–	1
170–	787	270–	—
180–	476	280–	1
		TOTAL	7749

1.11 When the asymmetry of a distribution such as that of Table 1.3 becomes extreme we may be unable to determine whether, near the maximum ordinate, there is a fall on either side, or whether the maximum occurs right at the start of the distribution. This would have been the case in Table 1.3 if we had not the finer grouping for the first five years of life; and it is the case in Table 1.2, in which the maximum frequency apparently occurs at or very close to an income of £2000 per annum. Asymmetrical distributions are sometimes called " skew "; and those such as Table 1.2 are called " J-shaped ".

1.12 In rare cases the distribution may have maxima at both ends, as in Table 1.11,

Table 1.11—Showing the frequencies of estimated intensities of cloudiness at Greenwich during the years 1890–1904 (excluding 1901) for the month of July

(Data from Gertrude E. Pearse, 1928, *Biometrika*, 20A, 336)

Degrees of cloudiness	Frequency	Degrees of cloudiness	Frequency
10	676	4	45
9	148	3	68
8	90	2	74
7	65	1	129
6	55	0	320
5	45		
		TOTAL	1715

showing a number of days distributed according to degree of cloudiness. This is known as a U-shaped distribution.

1.13 Distributions also occur which in general appearance resemble sections of the types already mentioned. A J-shaped distribution, for example, resembles the "tail" of the symmetrical distribution of Table 1.7. The suicide data of Table 1.6 may be regarded as a symmetrical distribution truncated just below the maximum ordinate by the impossibility of the occurrence of negative values of the variate. This sort of conception is sometimes useful in fitting curves to observed data—a given analytical curve may fit the data quite well in a certain variate-range, but may also extend into regions where the data cannot, so to speak, follow it.

1.14 The distributions considered up to this point have one thing in common—they have only one maximum or, in the case of the U-shaped curve, only one minimum. Distributions also occur showing several maxima, Tables 1.12 and 1.13 being instances in point. The first, showing a number of deaths according to age at death, is typical of death distributions. Near the start of the distribution there is a maximum and a rapid fall in the frequency, which then rises to a pronounced maximum about the age 70–75, the frequencies beyond that point tailing off to zero. It is natural to wonder whether such a distribution can be usefully considered as two or more superposed distributions, e.g. a J-shaped distribution indicative of infantile mortality, and a skew distribution with a maximum at 70–75, the ordinary death curve of senescence.

A similar dissection of a complex distribution could be undertaken for the data of Table 1.13, showing a number of trypanosomes from the tsetse fly, *Glossina morsitans*, classified according to length. We are led to suspect here that the distribution is composed of the addition of several others (and this, by the way, has led to a suggestion that the trypanosomes are a mixture of distinct types).

Table 1.12—Showing the number of male deaths in England and Wales for 1953, classified by ages at death (*from reference to Table* 1.1)

Age at death	Number of deaths	Age at death	Number of deaths
0–	12,244	45–	9,016
5–	1,043	50–	14,507
10–	665	55–	19,204
15–	1,104	60–	26,802
20–	1,640	65–	34,292
25–	1,932	70–	39,644
30–	2,449	75–	40,162
35–	3,068	80–	29,061
40–	5,104	80 and over	17,553
		TOTAL	259,490

Table 1.13—Showing number of Trypanosomes from *Glossina morsitans* classified according to length in microns

(From K. Pearson, 1914–15, *Biometrika*, **10**, 112. Length presumably to nearest micron)

Length (microns)	Frequency	Length (microns)	Frequency
15	7	26	110
16	31	27	127
17	148	28	133
18	230	29	113
19	326	30	96
20	252	31	54
21	237	32	44
22	184	33	11
23	143	34	7
24	115	35	2
25	130		
		TOTAL	2500

Frequency functions and distribution functions

1.15 The examples given above illustrate the remarkable fact that the majority of the frequency distributions encountered in practice possess a high degree of regularity. The form of the frequency polygons and histograms above suggests, almost inevitably, that our data are approximations to distributions which can be specified by smooth curves and simple mathematical expressions. This approach to the concept of the frequency *function*, however, requires some care, particularly for continuous distributions.

Consider in the first place a discrete distribution such as that of Table 1.4. Let us represent our variate by ξ. Then we may say that ξ can take any of the ten values 0, 1, . . ., 9 and that the frequency of a value of ξ equal to x, say $f(x)$, is given by the table, that is to say, $f(0) = 1026$, $f(1) = 1107$, $f(2) = 997$, and so on. The frequency table, in fact, defines the frequency function. Furthermore, most of the frequencies in the table are approximately 1000, and we may then consider the observed distribution as approximating to that defined by

$$f(x) = 1000, \qquad x = 0, 1, \ldots 9, \tag{1.1}$$

or, more generally, to the distribution

$$f(x) = k, \qquad x = 0, 1, \ldots 9. \tag{1.2}$$

This is perhaps the simplest form of discrete frequency function, $f(x)$ being a constant for all permissible values of x.

In Table 1.5 we have a discrete variate which can, theoretically, take an infinite number of values, namely, any one of the positive integers. In practice, of course, there must be a limit to the number, but since we do not know that limit we may imagine our variate as infinite in range. The frequency function for the table itself is again simply defined by the frequencies therein; but if we wish to proceed to a conceptual generalization of such a table we must admit a discrete function $f(x)$ defined for all positive integral values of x. This occasions no difficulty provided that we are able to attach some meaning to the total frequency, i.e. that $\sum_{j=1}^{\infty} f(x_j)$ converges.

1.16 Consider now the case of a continuous variate. In the ordinary data of experience our distributions are invariably discontinuous, because our measurements can only attain a certain degree of accuracy. For instance, we are accustomed to suppose that the height of a man may in reality be any real number of inches in a certain range, say 50 to 80, such as 20π. In fact, we can measure heights only to a certain accuracy, say to the nearest thousandth of an inch. Our measurements thus consist of whole numbers (of thousandths) from 50,000 to 80,000, and such a number as 62,831·85 ($= 20{,}000\pi$ approximately) cannot appear. All physical measurements are subject to this limitation, but we accept it and nevertheless speak of our variables as "continuous," the underlying supposition being that the measurements are approximations to numbers which can fall anywhere in the arithmetic continuum.

1.17 With this understanding we can consider the distribution of grouped frequencies as leading to the concept of a frequency function for a continuous variate. If, in one of the distributions above, say that of Table 1.7, we were to subdivide the intervals, we should probably find that *up to a point* the resulting frequencies were smoother and smoother. The reader can verify the appearance of this effect for himself by grouping the data of Table 1.7 in intervals of 8, 4, and 2 inches. We cannot, however, take the process too far, because, with a finite population, continued subdivision of the interval would sooner or later result in irregular frequencies, there being only a few members in each interval. But we may suppose that for ranges Δx, not too small, the distribution may be specified by a function $f(x)\Delta x$, expressing that in the range $\pm\frac{1}{2}\Delta x$ centred at x the frequency is $f(x)\Delta x$, *wherever x may be in the permissible range of the variate.* We may suppose further that as Δx tends to zero the population is perpetually replenished so as to prevent the occurrence of small and irregular frequencies; and in this way we arrive at the concept of the frequency function for a continuous variable. We write

$$dF = f(x)\,dx \tag{1.3}$$

expressing that the element of frequency dF between $x-\frac{1}{2}dx$ and $x+\frac{1}{2}dx$ is $f(x)\,dx$, for all x and for dx however small.

1.18 This admittedly somewhat intuitive approach to the concept of the continuous frequency distribution appears to be the best for statistical purposes, and is certainly the way in which the concept was originally reached. In formulating the axioms and postulates of a rigorous mathematical theory, however, the mathematician considers a rather more general function.

Let $F(x)$ be a function of x which is defined at every point in a range and is continuous except perhaps at a denumerable number of points. In our convention $F(x)$ is always continuous on the right. We require that F shall be zero at the lower point of the range (which may be $-\infty$) and a constant N at the upper point (which may be $+\infty$) and that it shall not decrease at any point. Such a function is called a *distribution function.* It corresponds to the cumulated frequency of a frequency distribution, N being the total frequency; for example, in Table 1.4, $F(x) = 0$ for $x<0$, $F(x) = 1026$ for $0\leqslant x<1$, $F(x) = 2133$ $(= 1026+1107)$ for $1\leqslant x<2$, and so on. Here there are ten points of discontinuity for $F(x)$. These points are called " saltuses " (jumps) and $F(x)$ in this case is called a step function.

If there is no saltus in the range, $F(x)$ is everywhere continuous and monotonically non-decreasing. If it possesses a derivative everywhere we have the equation in differentials

$$dF = F'(x)\,dx = f(x)\,dx \tag{1.4}$$

corresponding to (1.3). $f(x)$ is called the *frequency function.* The mathematics of this branch of the subject is then that of the study of functions of the class $F(x)$ and $f(x)$.

1.19 The functions as thus defined are more general than those arrived at from the statistical approach in two ways: (i) $F(x)$ can be continuous throughout part of the range and elsewhere possess saltuses, i.e. the variate may be continuous up to a point and then discrete—in statistical practice a variate is usually continuous or discrete throughout its range (although there are important exceptions—cf. Exercise 5.22 below); (ii) where no saltus exists $F(x)$ can exist without there existing a frequency function, since a continuous function need not necessarily possess a derivative. In all the cases we shall consider, the existence of a continuous variate will be accompanied by the existence of a continuous frequency function.

The function $F(x)$ is sometimes called the cumulative distribution function, but we shall use the term " distribution function " only. $f(x)$ is often called the probability density (for reasons which will become evident in Chapter 7 when we consider the theory of probability) or simply the density, but we shall always refer to it as the frequency function, a usage which has the merit of applying also to discrete variates.

Fig. 1.4 shows the distribution function of the data of Table 1.2 in graphical form, $F(x)$ being given as ordinate against x as abscissa. In Fig. 1.5 we show the distribution function of the data of Table 1.7. In both cases we have " smoothed " the functions to some extent by joining the ends of the ordinates instead of erecting narrow rectangles. For a continuous variate such a process is clearly often justified; it approximates to the continuous distribution which we imagine to underlie the data.

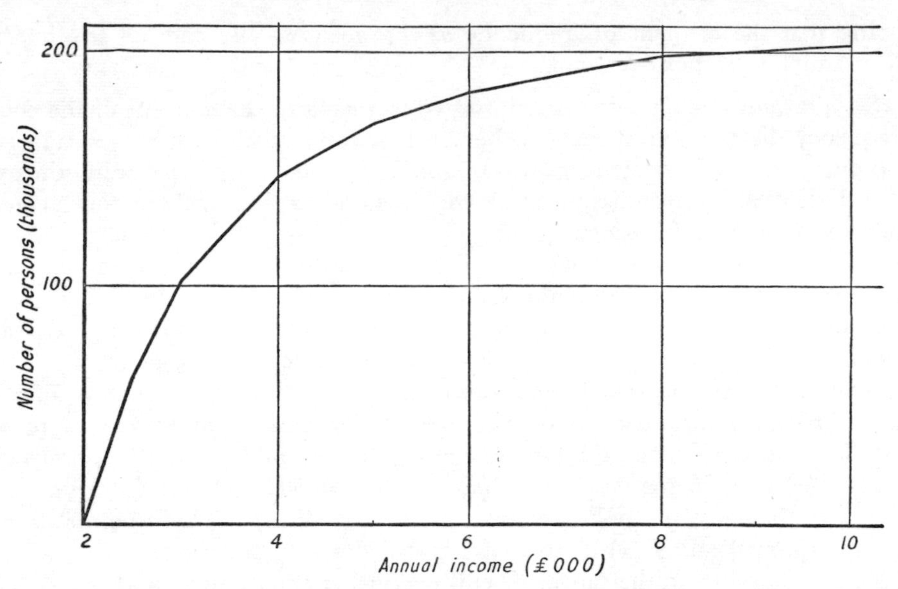

Fig. 1.4—Distribution function of the data of Table 1.2

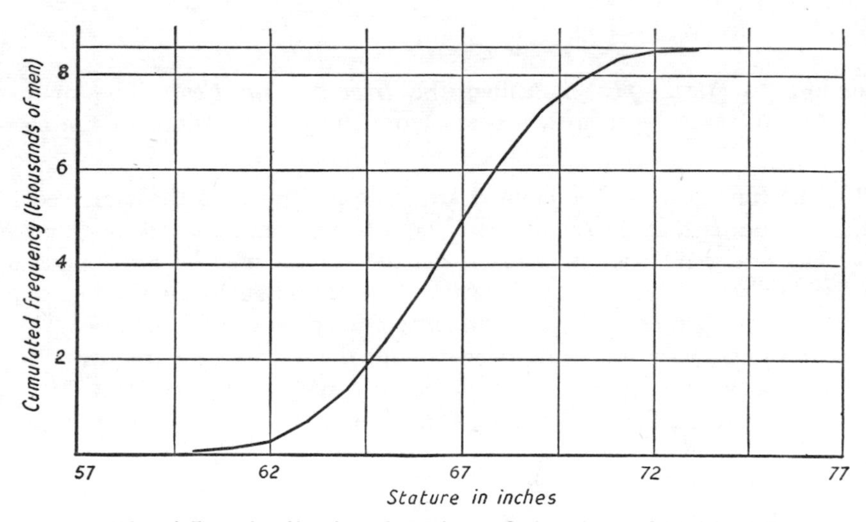

Fig. 1.5—Distribution function of the data of Table 1.7

1.20 If the discrete frequency function is $f(x)$, and $F(x)$ is taken to be the total frequency less than *or equal to* x, we have

$$F(x_r) = \sum_{j=1}^{r} f(x_j). \tag{1.5}$$

In the continuous case

$$F(x) = \int_a^x dF(u)$$

$$= \int_a^x f(u)\,du, \tag{1.6}$$

where the range is from a to b. We now introduce two conventions which simplify these expressions to some extent. We shall suppose, unless the contrary is specified, that in these mathematical expressions our frequencies are always expressed as proportions of the total frequency N, so that the total frequency is unity and the sum or integral over the whole range of the frequency function is also unity, i.e. $F(b) = 1$. Secondly, to avoid the constant specification of the limits a and b we may, without loss of generality, suppose that $F(x)$ and $f(x)$ are zero for any x less than a, and that $F(x) = 1$ and $f(x) = 0$ for any x greater than b. With this convention we may write

$$\left.\begin{aligned} F(x_r) &= \sum_{j=-\infty} f(x_j) \\ F(x) &= \int_{-\infty}^{x} f(u)\,du \end{aligned}\right\} \tag{1.7}$$

and

$$\left.\begin{aligned} \sum_{j=-\infty}^{\infty} f(x_j) &= F(\infty)-F(-\infty) = 1 \\ \int_{-\infty}^{\infty} f(u)\,du &= F(\infty)-F(-\infty) = 1 \end{aligned}\right\} \tag{1.8}$$

Where it is necessary to take account of the total frequency we may do so by multiplying by N the frequencies given by the frequency function.

1.21 We shall often find it useful to abbreviate the expression " frequency function " to f.f., and similarly " distribution function " to d.f. (In Chapter 4 we shall encounter the " characteristic function ", abbreviated to c.f.)

One other convention will occasionally help us to avoid cumbrous notation. Strictly speaking we should use different symbols, say ξ to denote the variate and x to denote the range of values which it may take. Thus $F(x)$ denotes, as a function of x, the totality of values for which $\xi \leqslant x$. It is often convenient, however, to use the same symbol for both variate ξ and variable x. We may then speak of a *variate* x with f.f. $f(x)$ and d.f. $F(x)$. Wherever this is likely to lead to confusion we shall revert to the stricter notation.

Excursus on integrals

1.22 We have noted above that the functions which arise for consideration in statistical practice are less complicated than the more general type of distribution function which the mathematician wishes to discuss. This illustrates a point which causes constant difficulty in developing a systematic course on theoretical statistics. Modern mathematics has devised rigorous theories of great generality, but largely at the expense of added difficulty and abstractness. For most statistical purposes this elaborate apparatus is not required, and statistical functions can be adequately investigated by simpler methods without loss of rigour. In this book we shall, as a rule, prefer these simpler methods. But this course will not overcome all our difficulties.

Some results which we require are not obtainable by simple methods; for others, the so-called elementary proofs are so long and tedious as to be worse than the more sophisticated versions; and for others again the simpler treatment may obscure the essential statistical point.

1.23 The modern theory of integration is a case in point. For many purposes it would be enough to consider the integral as defined in introductory treatments, or in the manner of Cauchy. Thus if $f(x)$ is defined in the interval (a, b) we divide the range at points $x_1, x_2, \ldots x_n$ and consider the sum

$$s_n = f(a)(x_1-a)+f(x_1)(x_2-x_1)+ \ldots +f(x_n)(b-x_n). \tag{1.9}$$

It may be shown that under certain conditions (such as the continuity of $f(x)$ in the range) this sum tends to a limit as the length of the intervals tends to zero, independently of where the dividing-points are drawn or the way in which the tendency to zero proceeds. This limit is the Cauchy integral: $\int_a^b f(x)\,dx$.

A more general type of integration is due to Riemann and Stieltjes. Let $\psi(x)$ be a continuous function of x. Choose ξ_1 in the range a to x_1, ξ_2 in the range x_1 to x_2 and so on. Consider the sum

$$s_n' = \psi(\xi_1)\{F(x_1)-F(a)\}+\psi(\xi_2)\{F(x_2)-F(x_1)\}+ \ldots +\psi(\xi_n)\{F(b)-F(x_n)\}. \tag{1.10}$$

It may be shown that as the length of the intervals tends to zero uniformly s tends to a limit which is independent of the location of the points ξ or the boundary points of the intervals. We then write this limit as

$$\int_a^b \psi(x)\,dF. \tag{1.11}$$

It is known as the Stieltjes integral of $\psi(x)$ with respect to $F(x)$. Riemann considered the case when $F(x) = x$.

1.24 The advantage of the Stieltjes integral is that it reduces to the Cauchy integral if $f(x)$ is continuous and to an ordinary summation if $f(x)$ is discrete. To save writing all our formulae down twice, once for the continuous f.f. and once for the discrete f.f., we shall usually employ an integral of this type, replacing it in special cases by an ordinary integral or a sum as the circumstances require.

Many of the theorems of ordinary integration are true of the Stieltjes integral. We shall frequently require the following:

$$\left|\int_a^b \psi\,dF\right| \leqslant \int_a^b |\psi|\,dF \tag{1.12}$$

$$\leqslant M\int_a^b dF$$

$$\leqslant M, \tag{1.13}$$

where M is the upper bound of $\psi(x)$ in the range (a, b).

$$\int_a^b \psi\,dF = \psi(\xi)\int_a^b dF, \tag{1.14}$$

where ξ is a value of x in the range (a, b).

If a and b are finite,

$$\int_a^b \sum_{j=1}^{\infty} f_j(x)\,dF = \sum_{j=1}^{\infty} \int_a^b f_j(x)\,dF, \tag{1.15}$$

provided that $\Sigma f_j(x)$ converges uniformly in the range. The theorem is not necessarily true if a or b is infinite.

The ordinary rules of partial integration are also applicable to Stieltjes integrals.

1.25 A more complicated integral is due to Lebesgue. It is equal to the Riemann-Stieltjes integral when the latter exists, but also exists in cases where the Riemann-Stieltjes integral does not. We do not normally require it in the ordinary theory of distributions, but it seems essential for a rigorous treatment of Fourier transforms which we need for certain parts of the theory of time-series.(*)

Variate-transformations

1.26 Suppose that we have a variable y related functionally to a variable x by the one-to-one relation

$$x = x(y), \qquad y = y(x), \tag{1.16}$$

y being continuous and differentiable in x and vice versa. We then have

$$dx = \frac{dx}{dy}dy. \tag{1.17}$$

Consequently, for a continuous distribution with d.f. $F(x)$,

$$F(x) = \int_{-\infty}^{x} dF = \int_{-\infty}^{x} f(u)\,du$$
$$= \int_{-\infty}^{x(y)} f(u)\frac{du}{dy}dy.$$

We may thus write the distribution as

$$dF = f\{x(y)\}\frac{dx}{dy}dy, \tag{1.18}$$

expressing the fact that an element of frequency between $y-\frac{1}{2}dy$ and $y+\frac{1}{2}dy$ is $f\{x(y)\}\,dx/dy$. If ξ is the variate corresponding to x we may regard this expression as defining a variate η corresponding to y. The equation determining the frequency function is then transformable as if it were an equation in differentials. Such variate-transformations are important in the theory of continuous distributions. By their means many mathematically specified distributions may be reduced to a known form, either exactly or approximately.

(*) The reader who feels that his mathematical equipment needs strengthening or revision may care to refer to C. A. B. Smith's *Biomathematics* (Griffin, London) which covers most of the mathematics required for statistical work at the intermediate level. More advanced material is covered by Sir Harold and Lady Jeffreys' *Methods of Mathematical Physics* (Oxford U.P.).

For example, a distribution which we shall have to study in the theory of sampling is

$$dF = \frac{1}{2^{\frac{1}{2}\nu-1}\Gamma(\frac{1}{2}\nu)} e^{-\frac{1}{2}\chi^2} \chi^{\nu-1}\, d\chi, \qquad 0 \leqslant \chi \leqslant \infty. \tag{1.19}$$

It is readily verified that $F(\infty) = 1$.

By the transformation $t = \frac{1}{2}\chi^2$ we reduce this to

$$dF = \frac{1}{\Gamma(\frac{1}{2}\nu)} e^{-t}\, t^{\frac{1}{2}\nu-1}\, dt, \qquad 0 \leqslant t \leqslant \infty. \tag{1.20}$$

This is a well-known form in analysis, the distribution function being the incomplete Γ-function:

$$F(t) = \Gamma_t(\tfrac{1}{2}\nu)/\Gamma(\tfrac{1}{2}\nu). \tag{1.21}$$

Again, the distribution

$$dF = \frac{y_0}{(1+t^2/\nu)^{\frac{1}{2}(\nu+1)}}\, dt, \qquad -\infty \leqslant t \leqslant \infty \tag{1.22}$$

(y_0 being chosen so that $F(\infty) = 1$), a symmetrical peaked distribution of infinite range rather like that of Fig. 1.3, may, by the substitution of $t = \sqrt{\nu}\tan\theta$, be transformed into

$$\begin{aligned} dF &= \frac{y_0\sqrt{\nu}\sec^2\theta\, d\theta}{\sec^{\nu+1}\theta}, \qquad -\tfrac{1}{2}\pi \leqslant \theta \leqslant \tfrac{1}{2}\pi, \\ &= y_0\sqrt{\nu}\cos^{\nu-1}\theta\, d\theta, \end{aligned} \tag{1.23}$$

a distribution of finite range $-\frac{1}{2}\pi$ to $+\frac{1}{2}\pi$, but still symmetrical. Putting now $\sin\theta = u$, we have

$$dF = y_0\sqrt{\nu}(1-u^2)^{\frac{1}{2}\nu-1}\, du, \qquad -1 \leqslant u \leqslant 1, \tag{1.24}$$

and again with $u^2 = x$,

$$dF = y_0\sqrt{\nu}\, x^{-\frac{1}{2}}(1-x)^{\frac{1}{2}\nu-1}\, dx, \qquad 0 \leqslant x \leqslant 1. \tag{1.25}$$

The last substitution is to be noted. u ranges from -1 to $+1$, and as it does so x ranges from $+1$ to 0 and back to $+1$, the relation being two-to-one. (1.18) is applied separately for $-1 \leqslant u < 0$ and $0 \leqslant u \leqslant 1$, with $\left|\frac{du}{dx}\right| = \frac{1}{2}x^{-\frac{1}{2}}$ in each interval. Bringing these two contributions to $f(x)$ together, we obtain (1.25). Whenever substitutions are made under which there is not a (1, 1) relation between the variates, points such as this require some watching.

1.27 There is one variate-transformation which is worth special attention. In the distribution

$$dF = f(x)\, dx$$

put

$$y = \int_{-\infty}^{x} f(u)\, du = F(x).$$

Then

$$\begin{aligned} dF &= f(x)\frac{dx}{dy}\, dy \\ &= \frac{f(x)}{f(x)}\, dy \\ &= dy, \qquad 0 \leqslant y \leqslant 1, \end{aligned} \tag{1.26}$$

so that the distribution is transformed into the very simple "rectangular" or "uniform" distribution, in which all values of the variate from 0 to 1 are equally frequent. This is called the *probability integral transformation.* Any continuous distribution can be transformed into the rectangular form; and it follows that there exists at least one transformation which will transform any continuous frequency distribution into any other continuous frequency distribution, viz. the transformation which transforms one into the rectangular form coupled with the inverse of that which transforms the other into the rectangular form.

The genesis of frequency distributions

1.28 Up to this point we have not inquired into the origin of the various observed frequency distributions which have been adduced in illustration. Certain of them may be considered apart from any question of origination from a larger population. The death distribution of Table 1.12 is an example; if we are interested only in the distribution of male deaths in England and Wales in 1953 the whole of the population under consideration is before us.

But in the great majority of cases the population which we are able to examine is only part of a larger population on which our main interest is centred. The height distribution of Table 1.7 is only a part of the population of men in the United Kingdom living at the time of the inquiry, and it is mainly of importance in the light of the information which it gives us about that population. Similarly the distribution of dairy herds of Table 1.9 is mainly of interest in the information it gives about costs of milk production for the whole country.

1.29 In the two cases just mentioned, height and costs of milk production, we have information about a certain sample of individuals chosen from an existing population. Only lack of time and opportunity prevents us from examining the whole population. It sometimes happens, however, that we have data which do not emanate from a finite existent population in this way. Table 1.14 is an example. It shows the distribution of throws with dice.

Table 1.14—Showing the number of successes (throws of 4, 5 or 6) with throws of 12 dice

(Weldon's data, cited by F. Y. Edgeworth, *Encyclopædia Britannica*, 11th edn, **22**, 39)

Number of successes	Frequency	Number of successes	Frequency
0	0	7	847
1	7	8	536
2	60	9	257
3	198	10	71
4	430	11	11
5	731	12	0
6	948		
		TOTAL	4096

Now it is clear that, in a sense, we have not in these data got a complete population, for we can add to them by further casting of the dice. But these further throws do not exist in the sense that the unexamined men of the United Kingdom or the unexamined dairy herds of England and Wales exist. They have a kind of hypothetical existence conferred on them by our notion of the throwing of the dice.

Even distributions which appear at first sight to be existent may be considered in this light. The trypanosome distribution of Table 1.13, for instance, was obtained from certain tsetse flies. We may consider it as a sample of all the tsetse flies in existence, whether harbouring trypanosomes or not—an existent population ; but we may also consider it as a sample of what the distribution would be if all the tsetse flies were infected with trypanosomes—a hypothetical population.

The population conceived of as parental to an observed distribution is fundamental to statistical inference. We shall take up this matter again in later chapters when we consider the sampling problem. The point is mentioned here because it will occasionally arise before we reach those chapters. It must be emphasized that the distinction between existent and hypothetical universes is not merely a matter of ontological speculation—if it were we could safely ignore it—but one of practical importance when inferences are drawn about a population from a sample generated from it.

Multivariate distributions

1.30 In the foregoing sections we have considered the members of a population according to a single variate, and the frequency distributions may thus be called univariate. The work may be readily generalized to include populations of members considered according to two or more variates, yielding bivariate, trivariate . . . multivariate frequency distributions. Table 1.15, for example, shows the distribution of a number of beans according to both length and breadth. The border frequencies show the univariate distributions of the beans according to length and breadth separately, and the body of the table shows how the two qualities vary together.

Table 1.15—Showing frequencies of beans with specified lengths and breadths

(Johannsen's data, cited by S. J. Pretorius (1930))

Lengths in millimetres (central values)

Breadth in millimetres (central values)	17	16·5	16	15·5	15	14·5	14	13·5	13	12·5	12	11·5	11	10·5	10	9·5	TOTALS
9·125	—	2	—	—	3	—	—	—	—	—	—	—	—	—	—	—	5
8·875	4	8	17	19	—	—	—	—	—	—	—	—	—	—	—	—	48
8·625	2	23	101	156	93	23	2	—	—	—	—	—	—	—	—	—	400
8·375	—	18	105	494	574	227	56	9	—	—	—	—	—	—	—	—	1483
8·125	—	4	44	375	956	913	362	73	12	3	—	—	—	—	—	—	2742
7·875	—	—	7	81	385	871	794	330	89	19	3	—	—	—	—	—	2579
7·625	—	—	1	4	65	236	469	361	175	55	27	4	—	—	—	—	1397
7·375	—	—	—	—	6	23	91	137	124	78	37	22	11	—	1	—	530
7·125	—	—	—	—	—	1	13	18	28	35	25	32	11	6	1	—	170
6·875	—	—	—	—	—	—	—	1	9	8	21	12	13	7	1	—	72
6·625	—	—	—	—	—	—	—	—	—	—	2	—	1	4	3	—	10
6·375	—	—	—	—	—	—	—	—	—	1	—	—	—	1	1	1	4
TOTALS	6	55	275	1129	2082	2294	1787	929	437	199	115	70	36	18	7	1	9440

As for the univariate case, the variates may be discontinuous or continuous and we sometimes meet cases in which one variate is of one kind and one of the other.

1.31 In generalization of the frequency polygon and the histogram we may construct three-dimensional figures to represent the bivariate distribution. Imagine a horizontal plane containing a pair of perpendicular axes and ruled like a chessboard into cells, the ruled lines being drawn at points corresponding to the terminal points of class-intervals. At the centre of each interval we erect a vertical line proportional in length to the frequency in that interval. The summits of these verticals are joined, each to the four summits of verticals in the neighbouring cells possessing the same values of one or the other variate. The resulting figure is the bivariate frequency polygon or stereogram.

Similarly we may erect on each cell a pillar proportional in volume to the frequency in that cell and thus obtain a bivariate histogram. Fig. 1.6 shows such a figure for the bean data of Table 1.15.

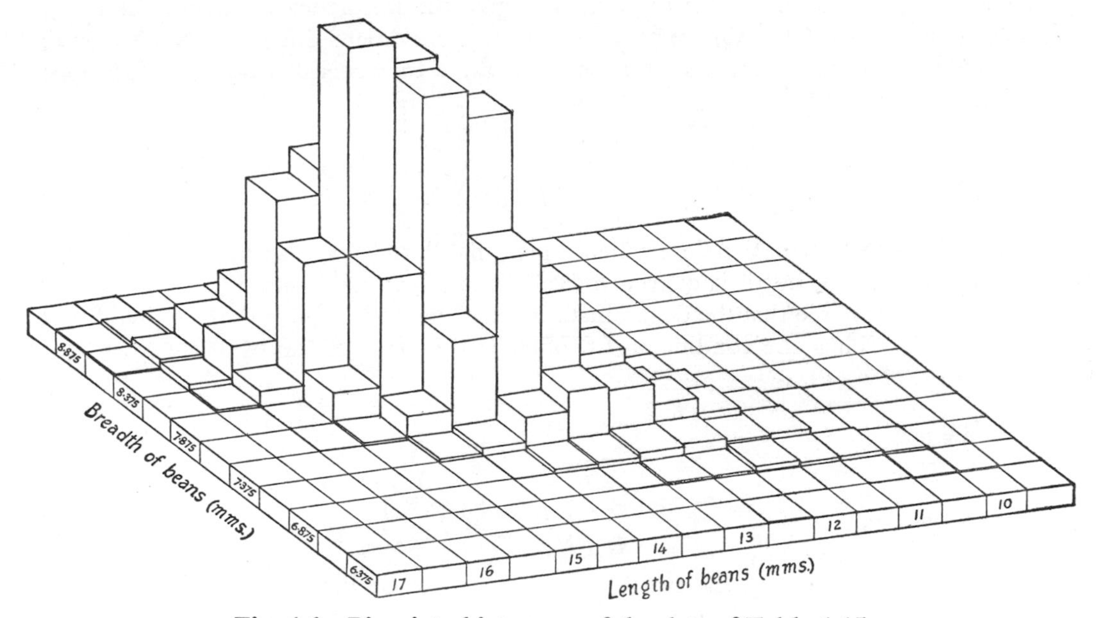

Fig. 1.6—Bivariate histogram of the data of Table 1.15

1.32 We may write the bivariate frequency function with variates x_1, x_2, as

$$dF = f(x_1, x_2)\, dx_1\, dx_2. \tag{1.27}$$

We then define the bivariate distribution function

$$F(x_1, x_2) = \int_{-\infty}^{x_1} \int_{-\infty}^{x_2} f(u_1, u_2)\, du_1\, du_2 \tag{1.28}$$

this integral also being understood in the Stieltjes sense, reducing to ordinary integration if $f(x_1, x_2)$ is continuous and to ordinary summation if it is discrete. Clearly $F(x_1, -\infty) = F(-\infty, x_2) = 0$ and by convention $F(+\infty, +\infty) = 1$ as in **1.20** for the univariate case, while F is a non-decreasing function of each of its arguments since $f(x_1, x_2) \geqslant 0$.

For any fixed value of one variate, say $x_2 = k$, the other variate will, of course, have a univariate distribution. If as usual we follow the convention that the total frequency should be unity this will be given by

$$p(x_1 \mid x_2 = k) = \frac{f(x_1, k)}{\int_{-\infty}^{\infty} f(x_1, k)\, dx_1}, \tag{1.29}$$

where the vertical bar on the left is read as " given ". (1.29) is called the conditional f.f. of x_1 for $x_2 = k$.

Marginal distributions

1.33 The sums in the margins of tables such as Table 1.15 give us the frequency-distributions of each variate by itself. For any value of x_2 the total frequency is obtained by summing over all values of x_1 to give the frequency for that value of x_2. Thus, if the d.f. is $F(x_1, x_2)$ we may express the univariate d.f. of x_1 as $F(x_1, \infty)$; and similarly the univariate d.f. of x_2 is $F(\infty, x_2)$. In terms of frequency functions

$$dF_1(x_1) = \left\{\int_{-\infty}^{\infty} f(x_1, x_2)\, dx_2\right\} dx_1, \tag{1.30}$$

$$dF_2(x_2) = \left\{\int_{-\infty}^{\infty} f(x_1, x_2)\, dx_1\right\} dx_2. \tag{1.31}$$

These univariate distributions are called " marginal ", for obvious reasons. The term is strictly redundant, but it is convenient and almost universally used.

From (1.31) we see that the denominator on the right of (1.29) is the marginal f.f. of x_2 at $x_2 = k$. Thus the conditional f.f. of x_1 is the ratio of the bivariate f.f. to the marginal f.f. of x_2.

Independence

1.34 If and only if

$$F(x_1, x_2) = F(x_1, \infty) F(\infty, x_2) \tag{1.32}$$
$$= F_1(x_1) F_2(x_2), \quad \text{say,}$$

for all values of x_1 and x_2, the two variates are said to be independent. Where f.f.'s exist we have

$$\int_{-\infty}^{x_1}\int_{-\infty}^{x_2} f(u_1, u_2)\, du_1\, du_2 = \int_{-\infty}^{x_1} f_1(u_1)\, du_1 \int_{-\infty}^{x_2} f_2(u_2)\, du_2$$

and hence

$$f(x_1, x_2) = f_1(x_1) f_2(x_2) \tag{1.33}$$

for all x_1, x_2. Conversely, (1.33) leads back to (1.32). Thus the d.f. and the f.f. factorize into two parts, one for each variate alone. It is readily seen that this definition of statistical independence conforms to the colloquial use of the word and also to its mathematical use. The conditional distribution of x_2 for any fixed x_1 (e.g. the distribution in a row or column of the bivariate frequency table) is the same whatever the fixed value of x_1, that is to say, the distribution of x_2 is independent of x_1. (1.33) must hold for all values of x_1 and x_2 if it is to be sufficient for their independence—cf. Exercise 1.23.

Two variates which are not independent are said to be dependent. Evidently those of Table 1.15 are dependent, for the distributions in rows or in columns are far from similar.

Example 1.1

Consider the distribution

$$dF = z_0 \exp\left\{-\frac{1}{2(1-\rho^2)}\left(\frac{x_1^2}{\sigma_1^2}-\frac{2\rho x_1 x_2}{\sigma_1\sigma_2}+\frac{x_2^2}{\sigma_2^2}\right)\right\} dx_1\, dx_2, \qquad -\infty \leqslant x_1, x_2 \leqslant \infty. \quad (1.34)$$

Here z_0 is to be chosen so that the total frequency is unity. Put

$$y_1 = \frac{1}{(1-\rho^2)^{\frac{1}{2}}}\left(\frac{x_1}{\sigma_1}-\frac{\rho x_2}{\sigma_2}\right)$$

$$y_2 = \frac{x_2}{\sigma_2}.$$

We have for the Jacobian of the transformation (cf. **1.35** below)

$$\frac{\partial(y_1, y_2)}{\partial(x_1, x_2)} = \begin{vmatrix} \dfrac{1}{\sigma_1(1-\rho^2)^{\frac{1}{2}}} & -\dfrac{\rho}{\sigma_2(1-\rho^2)^{\frac{1}{2}}} \\ 0 & \dfrac{1}{\sigma_2} \end{vmatrix}$$

$$= \frac{1}{\sigma_1\sigma_2(1-\rho^2)^{\frac{1}{2}}}.$$

Also
$$y_1^2+y_2^2 = \frac{1}{1-\rho^2}\left\{\frac{x_1^2}{\sigma}-\frac{2\rho x_1 x_2}{\sigma_1\sigma_2}+\frac{x_2^2}{\sigma_2^2}\right\}.$$

The distribution then becomes

$$dF = z_0\sigma_1\sigma_2(1-\rho^2)^{\frac{1}{2}} \exp\{-\tfrac{1}{2}(y_1^2+y_2^2)\}\, dy_1\, dy_2$$

$$= z_0\sigma_1\sigma_2(1-\rho^2)^{\frac{1}{2}} e^{-\frac{1}{2}y_1^2} dy_1\, e^{-\frac{1}{2}y_2^2} dy_2. \qquad (1.35)$$

The original variables are clearly dependent, for (1.34) cannot be factorized into a product of a function of x_1 times a function of x_2. From (1.35) we see that the transformed variables are independent.

Incidentally, this gives us a method of evaluating z_0. For

$$\int_{-\infty}^{\infty} e^{-\frac{1}{2}u^2}\, du = \sqrt{(2\pi)}$$

and on integrating the right-hand side of (1.35) with respect to y_1 and y_2 and equating the result to unity we find

$$z_0 = 1/\{2\pi\sigma_1\sigma_2(1-\rho^2)^{\frac{1}{2}}\}.$$

1.35 In this example we have used a Jacobian to transform the integral from one pair of variables to another. The process is evidently quite general. If

$$dF = f(x_1, x_2)\, dx_1\, dx_2$$

and
$$x_1 = x_1(y_1, y_2), \qquad x_2 = x_2(y_1, y_2),$$

the frequency function for y_1 and y_2 will be given by

$$g(y_1, y_2) = f\{x_1(y_1, y_2),\ x_2(y_1, y_2)\}J, \qquad (1.36)$$

where

$$J = \frac{\partial(x_1, x_2)}{\partial(y_1, y_2)} = \begin{vmatrix} \frac{\partial x_1}{\partial y_1} & \frac{\partial x_2}{\partial y_1} \\ \frac{\partial x_1}{\partial y_2} & \frac{\partial x_2}{\partial y_2} \end{vmatrix}$$

and is to be taken with positive sign in (1.36). If J changes sign in the domain of integration the variate-transformation requires special attention. We have already met this difficulty in its simplest form in passing from (1.24) to (1.25), and we shall discuss it generally in Chapter 11.

1.36 An extension of the bivariate results to n-variate results is immediate. For instance, if and only if a d.f. $F(x_1, x_2, \ldots x_n)$ can be expressed in the form

$$F(x_1, x_2, .., x_n) = F(x_1, \infty, .., \infty)F(\infty, x_2, .., \infty) \ldots F(\infty, \infty, .., x_n) \tag{1.37}$$

the corresponding variates are *completely* independent. It is not sufficient for this that each variate should be independent of each of the others. That is to say, if

$$F(x_1, x_2, \infty, .., \infty) = F(x_1, \infty, \infty, .., \infty)F(\infty, x_2, \infty, .., \infty) \tag{1.38}$$

and similarly for every pair of x's, equation (1.37) does not necessarily follow, for there are as many marginal distributions of the multivariate distribution as there are subsets of its n variates, i.e. $2^n - 2$ in all. (1.38) states only that each of the $\binom{n}{2}$ *bivariate* marginal distributions has independent constituents, while (1.37) ensures this for all the higher-dimensional marginal distributions also. (1.38) will be described as *pairwise* independence. An interesting example of pairwise independence without complete independence will occur in Exercise 16.11 below.

Unless otherwise stated, we shall use " independence " to mean " complete independence ".

Frequency-generating functions

1.37 For distributions which are mathematically specified it is often convenient, especially for discontinuous functions, to summarize the specification in a generating function. Suppose that the frequency of a variate x, which can take values 0, 1, 2,... is f_r for $x = r$. This frequency is then the coefficient of t^r in

$$P(t) \equiv \sum_{r=0}^{\infty} f_r t^r \tag{1.39}$$

which is thus a frequency-generating function.

Example 1.2

For the binomial distribution, which we shall study in some detail in Chapter 5, the frequency at $x = r$ $(r = 0, 1, .., n)$ is given by

$$f_r = \binom{n}{r} p^r (1-p)^{n-r}, \qquad 0 \leqslant p \leqslant 1. \tag{1.40}$$

Hence a generating function of type (1.39) is given by

$$P(t) = \sum_{j=0}^{n} \binom{n}{j} p^j (1-p)^{n-j} t^j$$
$$= \{pt+(1-p)\}^n. \tag{1.41}$$

EXERCISES

1.1 Draw frequency polygons or histograms of the following distributions :—

Table 1.16—Frequency distribution of successes in twelve dice thrown 4096 times, a throw of 6 points reckoned as a success (*Weldon's data ; source as in Table* 1.14)

Number of successes . . .	0	1	2	3	4	5	6	7 and over	TOTAL
Number of throws . . .	447	1145	1181	796	380	115	24	8	4096

Table 1.17—Showing numbers of sentences of given lengths in passages from Macaulay's essays on Bacon and on Chatham

(From G. Udny Yule, 1939, *Biometrika*, **30**, 363)

Length of sentence in words	Number of sentences	Length of sentence in words	Number of sentences
1–5	46	66–	2
6–	204	71–	4
11–	252	76–	8
16–	200	81–	2
21–	186	86–	2
26–	108	91–	1
31–	61	96–	2
36–	68	101–	1
41–	38	106–	—
46–	24	111–	1
51–	20	116–	—
56–	12	121–	1
61–	8		
		TOTAL	1251

Table 1.18—Showing the numbers of old Egyptian skulls with specified lengths of the left occipital bone in millimetres

(From T. L. Woo, 1930, *Biometrika*, **22**, 324)

Length (central values)	Frequency	Length (central values)	Frequency
84·5	12	102·5	74
86·5	12	104·5	68
88·5	32	106·5	36
90·5	48	108·5	18
92·5	79	110·5	7
94·5	116	112·5	4
96·5	104	114·5	4
98·5	126	116·5	—
100·5	123	118·5	1
		TOTAL	864

Table 1.19—Showing the frequency distribution of fecundity, i.e. the ratio of the number of yearling foals produced to the number of coverings, for brood-mares (racehorses) covered eight times at least

(Pearson, Lee and Moore, 1899, *Phil. Trans.*, A, **192**, 303. Where a case fell on the border between two intervals, one-half was assigned to each)

Fecundity	Number of mares with fecundity between the given limits	Fecundity	Number of mares with fecundity between the given limits
1/30– 3/30	2·0	17/30–19/30	315·0
3/30– 5/30	7·5	19/30–21/30	337·0
5/30– 7/30	11·5	21/30–23/30	293·5
7/30– 9/30	21·5	23/30–25/30	204·0
9/30–11/30	55·0	25/30–27/30	127·0
11/30–13/30	104·5	27/30–29/30	49·0
13/30–15/30	182·0	29/30–1	19·0
15/30–17/30	271·5		
		TOTAL	2000·0

Table 1.20—Frequency distribution of plots according to yield of grain in pounds from plots of $\frac{1}{500}$th acre in a wheat-field

(Mercer and Hall, 1911, *Jour. Agr. Science*, 4, 107)

Yield of grain in pounds per $\frac{1}{500}$th acre (central value of range)	2·8	3·0	3·2	3·4	3·6	3·8	4·0	4·2	4·4	4·6	4·8	5·0	5·2	TOTAL
Number of plots . . .	4	15	20	47	63	78	88	69	59	35	10	8	4	500

Table 1.21—Frequency distribution of 166 London bus-drivers showing the number who met with 0, 1, 2, . . . accidents in an exposure of (*a*) one year, (*b*) five years

(Data of Farmer, E. and Chambers, E. G., 1939, Industrial Health Research Board Report 84. London, H.M. Stationery Office)

Number of accidents	Number of drivers meeting with said number of accidents	
	In one year	In five years
0	45	1
1	36	2
2	40	3
3	19	14
4	12	17
5	8	21
6	3	17
7	2	14
8	1	14
9	—	12
10	—	13
11	—	9
12	—	6
13	—	2
14	—	6
15 or more	—	15
TOTAL	166	166

Table 1.22—Showing the number of women aborting at specified term in weeks

(From T. V. Pearce, 1930, *Biometrika*, **22**, 250)

Term (weeks)	Frequency	Term (weeks)	Frequency
4	3	17	13
5	7	18	14
6	10	19	8
7	13	20	4
8	14	21	2
9	29	22	10
10	22	23	4
11	21	24	4
12	18	25	3
13	28	26	4
14	16	27	6
15	19	28	1
16	10		
		TOTAL	283

1.2 Sketch the following curves and compare their shapes with those of the distributions in the previous exercise :—

$$y = y_0 e^{-\frac{1}{2}x^2}, \qquad -\infty \leqslant x \leqslant \infty.$$
$$y = y_0 e^{-x}, \qquad 1 \leqslant x \leqslant \infty.$$
$$y = y_0 x^{-\gamma}, \qquad \gamma > 1, 0 \leqslant x \leqslant \infty.$$
$$y = \frac{y_0}{(1+x^2)^n}, \qquad n > 0, -\infty \leqslant x \leqslant \infty.$$
$$y = y_0(1-x)^a x^b, \qquad a, b > 1, 0 \leqslant x \leqslant 1.$$
$$y = y_0 e^{-x} x^{\gamma}, \qquad \gamma > 1, 0 \leqslant x \leqslant \infty.$$
$$y = y_0(1-x^2)^a, \qquad a < 0, -1 \leqslant x \leqslant 1.$$

1.3 Show that the following distributions can all be transformed into

$$dF = y_0 x^{p-1}(1-x)^{q-1}\,dx, \qquad 0 \leqslant x \leqslant 1,$$

and find the transformations :

$$dF = r_0(1-r^2)^{\frac{1}{2}(n-4)}\,dr, \qquad -1 \leqslant r \leqslant 1.$$

$$dF = \frac{t_0}{(1+t^2/\nu)^{\frac{1}{2}(\nu+1)}}\,dt, \qquad -\infty \leqslant t \leqslant \infty.$$

$$dF = \frac{z_0 e^{\nu_1 z}}{(\nu_1 e^{2z}+\nu_2)^{\frac{1}{2}(\nu_1+\nu_2)}}\,dz, \qquad -\infty \leqslant z \leqslant \infty.$$

(All these distributions are important in statistical theory. The distribution to which they are reduced is called the Type I distribution or Beta distribution of the first kind, to be studied in Chapter 6.)

1.4 Sketch the stereograms or bivariate histograms of the following distributions :—

Table 1.23—Sample of students in the University of London, 1955, classified by (1) the number of newspapers read and (2) the number of newspapers looked at only, on a particular day. (*H. S. Booker, unpublished data*)

(1) Number looked at only

(2) Number read	0	1	2	3	4	5 or more	TOTALS
0	77	75	19	10	1	2	184
1	179	136	65	20	15	2	417
2	86	70	45	18	1	3	223
3	17	21	13	3	4	—	58
4	4	2	2	2	—	—	10
5 or more	2	2	—	2	—	—	6
TOTALS	365	306	144	55	21	7	898

Table 1.24—Number of cows distributed according to (1) age in years and (2) yield of milk per week in 4912 Ayrshire cows

(Data from J. F. Tocher, 1928, *Biometrika*, 20B, 106)

(1) Age in years

(2) Yield of milk per week (gallons) (central value of intervals)	3	4	5	6	7	8	9	10	11	12	13	14	15	16	17	18	TOTALS
8	—	—	—	—	—	—	—	—	1	—	—	—	—	—	—	—	1
9	—	2	2	—	1	—	—	—	—	—	—	—	—	—	—	—	5
10	3	5	1	1	3	—	—	—	—	—	—	—	—	—	—	—	13
11	2	10	8	7	1	—	1	—	2	1	—	1	—	—	—	—	33
12	2	25	17	9	5	4	4	2	1	1	—	—	—	1	—	—	71
13	9	76	29	18	9	2	4	1	1	1	—	1	—	—	—	—	151
14	11	76	57	38	23	9	7	6	4	2	3	—	—	—	—	—	236
15	11	115	79	43	34	24	11	8	4	5	1	2	1	—	1	—	339
16	15	149	119	74	59	23	23	16	9	7	4	—	—	—	1	—	499
17	16	148	131	94	58	34	32	15	12	6	5	—	1	—	—	—	552
18	11	146	132	83	73	49	39	22	17	6	5	1	1	—	—	—	585
19	10	117	112	113	87	51	35	33	11	10	2	3	1	—	—	1	586
20	8	97	107	79	69	51	25	30	13	10	3	3	—	—	1	—	496
21	3	63	93	88	70	49	31	29	9	7	4	—	1	—	1	—	448
22	5	42	63	49	45	32	14	18	10	3	1	2	—	—	—	—	284
23	1	19	33	38	38	27	17	17	12	7	1	2	2	—	—	—	214
24	2	20	23	34	27	19	13	9	3	2	1	—	—	—	—	—	153
25	3	10	15	22	17	20	8	10	3	4	—	—	—	—	—	—	112
26	—	7	13	7	4	15	2	4	2	3	1	—	—	—	—	—	58
27	—	2	7	9	5	5	4	2	—	—	—	—	—	1	—	—	35
28	—	—	2	1	4	2	1	1	2	—	—	—	—	—	—	—	13
29	—	—	2	2	4	1	3	—	3	—	—	—	—	—	—	—	15
30	—	—	—	—	—	2	—	—	2	—	—	—	—	—	—	—	4
31	—	—	2	1	—	—	2	—	—	—	—	—	—	—	—	—	5
32	—	—	—	2	—	—	—	—	—	—	—	—	—	—	—	—	2
33	—	—	—	—	—	—	—	—	—	—	1	—	—	—	—	—	1
34	—	—	—	—	—	—	—	—	1	—	—	—	—	—	—	—	1
TOTALS	112	1129	1047	812	636	419	276	223	122	75	32	15	7	2	4	1	4912

1.5 Show that the conditions that the function

$$f(x_1, x_2) = z_0 \exp\{Ax_1^2 + 2Hx_1x_2 + Bx_2^2\}, \qquad -\infty \leqslant x_1, x_2 \leqslant \infty,$$

may represent a frequency function are

(*a*) $A \leqslant 0$,
(*b*) $B \leqslant 0$,
(*c*) $AB - H^2 \geqslant 0$.

Show further that if these conditions are satisfied and the integral of $f(x_1, x_2)$ between $-\infty$ and ∞ for both variates is unity, then

$$z_0 = \frac{1}{\pi}\begin{vmatrix} -A & H \\ H & -B \end{vmatrix}^{\frac{1}{2}}.$$

1.6 For the bivariate distribution

$$dF = \frac{ky}{(1+x)^4}\exp\left(-\frac{y}{1+x}\right)dx\,dy, \qquad 0 \leqslant x, y \leqslant \infty,$$

show by a transformation $u = y/(1+x)$, $v = 1/(1+x)$ that the distribution of u and v is

$$dF = kue^{-u}\,du\,dv, \qquad 0 \leqslant u \leqslant \infty, 0 \leqslant v \leqslant 1,$$

and hence that $k = 1$.

1.7 For the bivariate distribution

$$dF = \frac{k(1+x+y)}{(1+x)^4(1+y)^4}dx\,dy, \qquad 0 \leqslant x, y \leqslant \infty,$$

show that $k = 9/2$. Show also that the marginal distribution of x is

$$dF = \frac{3}{4}\frac{(2x+3)}{(1+x)^4}dx, \qquad 0 \leqslant x \leqslant \infty.$$

1.8 A distribution is given by

$$dF = k\,\text{sech}^n x\,dx, \qquad -\infty \leqslant x \leqslant \infty.$$

Sketch the form of the frequency curve and show that

$$k = \frac{\Gamma\{\frac{1}{2}(n+1)\}}{\Gamma(\frac{1}{2}n)\sqrt{\pi}}.$$

1.9 Three variates are independent and each is distributed in the form

$$dF = \frac{1}{\sqrt{(2\pi)}}e^{-\frac{1}{2}x^2}dx, \qquad -\infty \leqslant x \leqslant \infty.$$

Transforming to new variates by the equations

$$\begin{aligned} y_1 &= (x_1 - x_2)/\sqrt{2} \\ y_2 &= (x_1 + x_2 - 2x_3)/\sqrt{6} \\ y_3 &= (x_1 + x_2 + x_3)/\sqrt{3}, \end{aligned}$$

show that the y's are also completely independent and are distributed in the same form.

1.10 In the previous exercise, make a transformation of the polar type

$$\begin{aligned} x_1 &= r\cos\theta_1\cos\theta_2 \\ x_2 &= r\cos\theta_1\sin\theta_2 \\ x_3 &= r\sin\theta_1. \end{aligned}$$

Hence show that the variates corresponding to r, θ_1 and θ_2 are completely independent and that the distribution of the first is given by

$$dF = \sqrt{\left(\frac{2}{\pi}\right)}e^{-\frac{1}{2}r^2}r^2\,dr, \qquad 0 \leqslant r \leqslant \infty.$$

1.11 In the distribution

$$dF = \frac{1}{2\pi\sigma_1\sigma_2(1-\rho^2)^{\frac{1}{2}}}\exp\left\{\frac{-1}{2(1-\rho^2)}\left(\frac{x^2}{\sigma_1^2} - \frac{2\rho xy}{\sigma_1\sigma_2} + \frac{y^2}{\sigma_2^2}\right)\right\}dx\,dy, \qquad -\infty \leqslant x, y \leqslant \infty,$$

show that the distribution of x is

$$dF = \frac{1}{\sigma_1\sqrt{(2\pi)}}\exp\left(-\frac{x^2}{2\sigma_1^2}\right)dx, \qquad -\infty \leqslant x \leqslant \infty.$$

Sketch the distribution of x and y and consider the limiting case as $\rho \rightarrow 1$.

1.12 In the distribution

$$dF = \frac{1}{2\pi\,\sigma_1\sigma_2(1-\rho^2)^{\frac{1}{2}}}\,\frac{n-1}{n-2}\,\frac{1}{\left[1+\frac{1}{2(n-2)(1-\rho^2)}\left(\frac{x^2}{\sigma_1^2}-\frac{2\rho\,xy}{\sigma_1\,\sigma_2}+\frac{y^2}{\sigma_2^2}\right)\right]^n}\,dx\,dy \qquad -\infty \leqslant x, y \leqslant \infty,$$

show that the distribution of x is given by

$$dF = \frac{1}{\sigma_1\sqrt{(2\pi)}\,\sqrt{(n-2)}}\,\frac{\Gamma(n-\frac{1}{2})}{\Gamma(n-1)}\,\frac{1}{\left(1+\frac{x^2}{2(n-2)\sigma_1^2}\right)^{n-\frac{1}{2}}}\,dx, \qquad -\infty \leqslant x \leqslant \infty.$$

(K. Pearson, 1923)

1.13 The so-called Poisson distribution is such that the frequency at $x = r$ ($r = 0, 1, 2, \ldots$) is $e^{-\lambda}\lambda^r/r!$ Show that a frequency-generating function is

$$P(t) = \exp\{\lambda(t-1)\},$$

where the frequency at $x = r$ is the coefficient of t^r in $P(t)$.

1.14 A discrete distribution (the so-called negative binomial) has frequencies

$$f_x = \binom{x+r-1}{r-1}p^r q^x, \quad x = 0, 1, 2, \ldots; \; 0<p<1, \; q = 1-p, \; r \text{ a positive integer.}$$

Show that the frequency-generating function is

$$P(t) = p^r(1-qt)^{-r}.$$

1.15 A variate is distributed over the interval 0 to ∞ in the form $dF = f(x)\,dx$. Write down the distribution of the variate in the interval $x_0 \leqslant x \leqslant \infty$ ($x_0 > 0$). If this is of the same form as the original variate show that $f(x) = k\,e^{-kx}$. It may be assumed that $f(x)$ is differentiable.

1.16 A n-variate distribution has the form

$$dF = k\exp\left\{-\sum_{i,j=1}^{n} a_{ij}\,x_i\,x_j\right\}dx_1\ldots dx_n, \qquad -\infty \leqslant x_1, \ldots, x_n \leqslant \infty,$$

with $a_{ij} = a_{ji}$. By considering x_n show that the exponent can be put in the form

$$-a_{nn}\left(x_n+\frac{1}{a_{nn}}\sum_{i=1}^{n-1} a_{ni}\,x_i\right)^2 - \sum_{i,j=1}^{n-1} b_{ij}x_i\,x_j,$$

where $b_{ij} = a_{ij}-\frac{a_{in}a_{jn}}{a_{nn}}$, $\qquad i,j = 1, 2, \ldots, n-1$.

Hence show how to transform the original variates by linear transformation to a completely independent set each of which is distributed in a form of the type

$$dF = \frac{1}{\sigma\sqrt{(2\pi)}}e^{-u^2/2\sigma^2}\,du, \qquad -\infty \leqslant u \leqslant \infty.$$

1.17 For a bivariate distribution with a differentiable frequency function $f(x, y)$, show that there exists a linear transformation to new variates which are independent if and only if

$$\left(A\frac{\partial^2}{\partial x^2}+2H\frac{\partial^2}{\partial x\,\partial y}+B\frac{\partial^2}{\partial y^2}\right)\log f = 0$$

where A, H and B are constants. Hence show that for the distribution of Exercise 1.11 there exists an infinite number of linear transformations to independent variates.

1.18 Three non-negative variates represented by x, y and z have the distribution

$$dF = k x^{l-1} y^{m-1} z^{n-1} \, dx \, dy \, dz,$$

where $x+y+z \leqslant 1$. Show that

$$k = \frac{\Gamma(l+m+n+1)}{\Gamma(l)\,\Gamma(m)\,\Gamma(n)}.$$

1.19 n non-negative variates have the distribution

$$dF = k x_1^{l_1-1} x_2^{l_2-1} \ldots x_n^{l_n-1} g(x_1+x_2+\ldots+x_n) \, dx_1 \ldots dx_n,$$

where $\sum_{i=1}^{n} x_i \leqslant 1$. Show that

$$1/k = \frac{\Gamma(l_1)\Gamma(l_2)\ldots\Gamma(l_n)}{\Gamma(l_1+l_2+\ldots+l_n)} \int_0^1 v^{L-1} g(v) \, dv,$$

where

$$L = \sum_{i=1}^{n} l_i.$$

The particular case $g(v) = (1-v)^{l_{n+1}-1}$ is called the *Dirichlet distribution.*

1.20 Show that if a variate x is distributed symmetrically about a value θ, its distribution function satisfies

$$F(x) = 1 - F(2\theta - x)$$

and its frequency function correspondingly

$$f(x) = f(2\theta - x).$$

1.21 Two independent variates x_1, x_2 each have the distribution

$$dF = \frac{1}{\sqrt{(2\pi)}} \exp(-\tfrac{1}{2}x^2) \, dx, \qquad -\infty \leqslant x \leqslant \infty.$$

Show that the variates

$$y_1 = \exp\{-\tfrac{1}{2}(x_1^2 + x_2^2)\},$$

$$y_2 = \frac{1}{2\pi} \arctan(x_1/x_2)$$

are independently rectangularly distributed on the range (0, 1).

(Box and Muller, 1958)

1.22 If $F_1(x_1)$, $F_2(x_2)$ are univariate d.f.'s, show that

$$F(x_1, x_2) = F_1(x_1) F_2(x_2) [1 + \theta \{1 - F_1(x_1)\} \{1 - F_2(x_2)\}]$$

(where θ is a constant, $-1 \leqslant \theta \leqslant 1$) is a bivariate d.f. with marginal d.f.'s F_1 and F_2.

(Cf. Gumbel, 1960)

1.23 By considering the bivariate frequency function

$$f(x, y) = g(x)\,h(y) \quad \text{if} \quad \begin{Bmatrix} a < x < b \\ a < y < b \end{Bmatrix} \quad \text{or} \quad \begin{Bmatrix} b < x < c \\ b < y < c \end{Bmatrix}$$

$$= 0 \qquad \text{otherwise,}$$

show that factorization of $f(x, y)$ whenever it is non-zero is not sufficient for x to be independent of y.

CHAPTER 2

MEASURES OF LOCATION AND DISPERSION

2.1 It has been seen in Chapter 1 that the frequency distributions occurring in statistical practice vary considerably in general nature. Some are finite in range and some are not. Some are symmetrical and some markedly skew. Some present only a single maximum and others present several. Amid this variety we may, however, discern four general types : (*a*) the symmetrical distribution with a single maximum, such as that of Table 1.7 ; (*b*) the asymmetrical distribution, or skew distribution, with a single maximum, such as those of Tables 1.8 and 1.9 ; (*c*) the extremely skew, or J-shaped, distribution, such as that of Table 1.2 ; and (*d*) the U-shaped distribution, such as that of Table 1.11. To make this classification comprehensive we should have to add a fifth class comprising the miscellaneous distributions not falling into the other four.

The distributions with a single maximum will hereafter be called *unimodal.*

2.2 It frequently happens in statistical work that we have to compare two distributions. If one is unimodal and the other J-shaped or multimodal a concise comparison is clearly difficult to make, and in such a case it would probably be necessary to specify both distributions completely. But if both are of the same type (and it is in such cases that comparisons most frequently arise) we may be able to make a satisfactory comparison merely by examining their principal characteristics ; e.g. if both are unimodal it might be sufficient to compare (*a*) the whereabouts of some central value, such as the maximum—this, as it were, locates the distributions ; (*b*) the degree of scatter about this value—the dispersion ; and (*c*) the extent to which the distributions deviate from the symmetrical—the skewness.

The same point emerges when our distributions are specified by some mathematical function. If, for example, we have two distributions of the type

$$dF = k\exp\left\{-\frac{(x-\mu)^2}{2\sigma^2}\right\}dx, \qquad -\infty \leqslant x \leqslant \infty,$$

symmetrical about $x = \mu$, a complete comparison can be made by comparing the value of the constants μ and σ in the distributions. Such constants are called *parameters* of the distribution.(*)

Measures of location : the arithmetic mean

2.3 There are three groups of measures of location in common use : the means

(*) There has, in the past, been some confusion as to what is meant by a parameter of a distribution. We shall use the term only to denote quantities, such as μ and σ above, which appear explicitly in the specification of the distribution. Thus the mean and variance are not parameters in our sense ; although it so happens that in the frequency function given above μ is equal to the mean and σ^2 to the variance.

(arithmetic, geometric and harmonic), the median and the mode. We consider them in turn.

The arithmetic mean is perhaps the most generally used statistical measure, and in fact is far older than the science of statistics itself. If the proportional frequency of the values x of a distribution is $f(x)$, the arithmetic mean μ_1' about the point $x = a$ is defined by

$$\left.\begin{aligned}\mu_1' &= \int_{-\infty}^{\infty}(x-a)f(x)\,dx = \int_{-\infty}^{\infty}(x-a)\,dF\\ &= \int_{-\infty}^{\infty}x\,dF-a\int_{-\infty}^{\infty}dF = \int_{-\infty}^{\infty}x dF-a.\end{aligned}\right\} \tag{2.1}$$

This integral is to be understood in the Stieltjes sense and hence includes summation in the discontinuous case; e.g. the arithmetic mean about zero of a set of discrete values x is their sum divided by the number of values. In formula (2.1) the frequency, in accordance with our usual convention, is expressed as a proportion of the total frequency. If the *actual* frequencies are $g(x)$, totalling N, we have to divide the integral (or sum, in the discrete case) by N.

The value of the arithmetic mean thus depends on the value of a, the point from which it is measured. For a mathematically specified distribution the integral (2.1) need not necessarily converge, in which case no arithmetic mean exists.

If b is some other arbitrary variate-value, (2.1) implies that

$$\mu_1' \text{ (about } a) = \mu_1' \text{ (about } b)+b-a. \tag{2.2}$$

In other words, we can find the mean about any point very simply when we know the mean about any other. In calculating the arithmetic mean we can then take an arbitrary point as origin and transfer to any other desired point afterwards. It is convenient to choose this arbitrary point somewhere near the maximum frequency to reduce numerical computations. For mathematically specified distributions, μ_1' is usually calculated about the origin zero.

Example 2.1

To calculate the arithmetic mean of the population of males distributed according to height as in Table 1.7.

If there were relatively few values in the population, we should simply sum them and divide by N to obtain μ_1' about zero, but a further point arises in grouped data such as these. We do not know *exactly* the variate-values of the individuals within a certain class interval. We therefore assume them concentrated at the centre of the interval. Corrections for any distortion thus introduced will be considered in Chapter 3. In fact, no correction is required for the arithmetic mean in the case when the frequency "tails off" at both ends of the distribution.

In the particular case before us we take an arbitrary origin at the centre of the interval 67– inches, i.e. at the point $67\frac{7}{16}$ inches, and measure $\xi(=x-a)$ from that point. Column 2 in Table 2.1 shows the frequency, column 3 the value of ξ and column 4 the value of ξf. We find, having due regard to sign,

$$\Sigma(\xi f) = 8763-8584 = 179.$$

Hence the mean about $x = 0$ is $67\frac{7}{16}+\dfrac{179}{8585} = 67{\cdot}46$ inches.

Table 2.1—Calculation of the arithmetic mean for the distribution of Table 1.7

(1) Height, inches	(2) Frequency f	(3) Deviation from arbitrary value ξ	(4) Product ξf
57–	2	− 10	20
58–	4	− 9	36
59–	14	− 8	112
60–	41	− 7	287
61–	83	− 6	498
62–	169	− 5	845
63–	394	− 4	1576
64–	669	− 3	2007
65–	990	− 2	1980
66–	1223	− 1	1223
67–	1329	0	− 8584
68–	1230	+ 1	1230
69–	1063	+ 2	2126
70–	646	+ 3	1938
71–	392	+ 4	1568
72–	202	+ 5	1010
73–	79	+ 6	474
74–	32	+ 7	224
75–	16	+ 8	128
76–	5	+ 9	45
77–	2	+ 10	20
TOTALS	8585	—	+ 8763

Example 2.2

For a distribution specified by a mathematical function, the determination of the mean is a matter of evaluating the integral (2.1), when it exists. For instance, to find the mean of the distribution

$$dF = \frac{1}{B(p,q)}x^{p-1}(1-x)^{q-1}\,dx, \qquad 0\leqslant x\leqslant 1\,;\ p,q>0,$$

we have, about the origin zero,

$$\begin{aligned}\mu_1' &= \frac{1}{B(p,q)}\int_0^1 x^p(1-x)^{q-1}\,dx\\ &= \frac{B(p+1,q)}{B(p,q)} = \frac{\Gamma(p+1)\Gamma(q)}{\Gamma(p+q+1)}\cdot\frac{\Gamma(p+q)}{\Gamma(p)\Gamma(q)}\\ &= \frac{p}{p+q}.\end{aligned}$$

2.4 If a distribution is specified by a generating function the mean can be evaluated in the following manner.

Let

$$P(t) = \sum_{j=0}^{\infty} f_j t^j. \tag{2.3}$$

Then

$$\begin{aligned}\left[\frac{dP}{dt}\right]_{t=1} &= \left[\Sigma f_j t^{j-1} j\right]_{t=1} \\ &= \Sigma f_j j \\ &= \mu_1' \text{ (about zero).}\end{aligned} \tag{2.4}$$

Example 2.3

From Example 1.2, the binomial distribution is specified by

$$P(t) = \{pt+(1-p)\}^n$$

and we have

$$\begin{aligned}\left[\frac{dP}{dt}\right]_{t=1} &= np\,\{p+(1-p)\}^{n-1} \\ &= np,\end{aligned}$$

which is therefore the mean about zero.

The geometric mean and the harmonic mean

2.5 Two other types of mean are in use in elementary statistics, though they are of less importance in advanced theory.

The geometric mean of N variate-values is the Nth root of their product and is not used if any of the variate-values is zero or negative. For proportional frequencies $f(x)$ we have

$$\left.\begin{aligned} G &= \prod_{j=-\infty}^{\infty} (x_j^{f_j}) \\ \text{or}\qquad \log G &= \sum_{j=-\infty}^{\infty} f_j \log x_j \end{aligned}\right\} \tag{2.5}$$

and for actual frequencies $g(x)$, totalling N,

$$\left.\begin{aligned} G &= \Pi\,(x_j^{g_j})^{\frac{1}{N}} \\ \log G &= \frac{1}{N}\Sigma g_j \log x_j \end{aligned}\right\} \tag{2.6}$$

The harmonic mean of N variate-values is the reciprocal of the arithmetic mean of their reciprocals. In the usual notation

$$\frac{1}{H} = \int_{-\infty}^{\infty} \frac{dF}{x} = \int_{-\infty}^{\infty} \frac{f(x)\,dx}{x} \tag{2.7}$$

or, for actual frequencies,

$$\frac{1}{H} = \frac{1}{N}\int_{-\infty}^{\infty} \frac{g(x)\,dx}{x} \tag{2.8}$$

provided, of course, that the integral exists.

Example 2.4

To find the geometric and harmonic means of the distribution of Example 2.2,

$$dF = \frac{1}{B(p,q)} x^{p-1}(1-x)^{q-1}\,dx, \qquad 0 \leqslant x \leqslant 1\,;\; p,q>0,$$

we have
$$\log G = \frac{1}{B(p,q)}\int_0^1 x^{p-1}(1-x)^{q-1}\log x\,dx.$$

Now, since by definition

$$\int_0^1 x^{p-1}(1-x)^{q-1}\,dx = B(p,q)$$

we have, differentiating both sides with respect to p, an operation which is legitimate in virtue of the uniform convergence of the integral and the existence of the resulting expressions,

$$\int_0^1 x^{p-1}(1-x)^{q-1}\log x\,dx = \frac{\partial}{\partial p}B(p,q).$$

Thus
$$\begin{aligned}\log G &= \frac{1}{B(p,q)}\frac{\partial}{\partial p}B(p,q)\\ &= \frac{\partial}{\partial p}\log\frac{\Gamma(p)\Gamma(q)}{\Gamma(p+q)}\\ &= \frac{\partial}{\partial p}\{\log\Gamma(p)-\log\Gamma(p+q)\}.\end{aligned}$$

The harmonic mean is given by

$$\frac{1}{H} = \frac{1}{B(p,q)}\int_0^1 x^{p-2}(1-x)^{q-1}\,dx.$$

To keep this expression finite, however, we must require $p>1$. Then

$$\begin{aligned}\frac{1}{H} = \frac{B(p-1,q)}{B(p,q)} &= \frac{\Gamma(p-1)}{\Gamma(p+q-1)}\cdot\frac{\Gamma(p+q)}{\Gamma(p)}\\ &= \frac{p+q-1}{p-1},\end{aligned}$$

so that
$$H = \frac{p-1}{p+q-1}.$$

We may note that the arithmetic mean (Example 2.2) is greater than the harmonic mean, for

$$\mu_1' = \frac{p}{p+q} = 1-\frac{q}{p+q}, \quad \text{and } H = \frac{p-1}{p+q-1} = 1-\frac{q}{p+q-1}$$

so
$$\mu_1' > H.$$

2.6 In general it may be shown that for distributions in which the variate-values are not negative

$$H \leqslant G \leqslant \mu_1'. \tag{2.9}$$

D

Consider in fact the quantity

$$A(t) = \left|\frac{1}{N}(x_1^t + x_2^t + \ldots x_N^t)\right|^{\frac{1}{t}}$$

where the x's are real numbers. We shall show that this is an increasing function of t, i.e. $A(t_1) > A(t_2)$ if $t_1 > t_2$. As a trivial case these inequalities may be replaced by equalities, namely if all the x's are equal.

Note that if we put $t = 1$ in $A(t)$ we have the arithmetic mean; when $t = -1$ we have the harmonic mean; and as t tends to zero we have the geometric mean, for

$$\lim_{t \to 0} \log A = \lim\{\log(\Sigma x^t/N)/t\}$$

and by L'Hôpital's rule, this is

$$= \lim(\Sigma x^t \log x)/\Sigma x^t$$

$$= \Sigma \log x/N.$$

Put

$$\left.\begin{aligned} I &= \frac{1}{N}\Sigma x^t \\ F &= t^2 \frac{d}{dt}\log A. \end{aligned}\right\}$$

Then

$$F = t^2 \frac{d}{dt}\left(\frac{\log I}{t}\right) = \frac{t}{I}\frac{dI}{dt} - \log I,$$

and hence

$$\frac{dF}{dt} = \frac{t}{I^2}\left\{I\frac{d^2 I}{dt^2} - \left(\frac{dI}{dt}\right)^2\right\}. \tag{2.10}$$

The expression in braces in (2.10) is non-negative, in virtue of the Cauchy–Schwarz inequality, for it is equal to

$$\left(\frac{1}{N}\Sigma x^t\right)\left(\frac{1}{N}\Sigma x^t \log^2 x\right) - \left(\frac{1}{N}\Sigma x^t \log x\right)^2.$$

Hence dF/dt has the sign of t and F thus has a minimum at $t = 0$. But when $t = 0$. $F = 0$ and hence F is non-negative. Therefore $d\log A/dt$ is non-negative, and since A is non-negative dA/dt is non-negative and thus A is a non-decreasing function, It will, in fact, be increasing unless all the x's are equal. The inequality (2.9) follows.

For simplicity we have stated these results for a discontinuous variate. The analysis, however, is easily seen to remain true for Stieltjes integrals and hence is generally valid.

Hereafter when the " mean " is mentioned without qualification, the arithmetic mean is to be understood.

Exercise 2.3 shows that G is approximately the mean of μ_1' and H when dispersion is relatively small: Exercise 2.22 gives an upper bound for μ_1/H.

2.7 There are many statistical uses of the Cauchy–Schwarz inequality and it may be useful to interpolate a few comments on it. The inequality given by Cauchy in 1821 states that if a_i, b_i $(i = 1, 2, \ldots n)$ are real numbers $\left(\sum_{i=1}^{n} a_i^2\right)\left(\sum_{i=1}^{n} b_i^2\right) > \left(\sum_{i=1}^{n} a_i b_i\right)^2$ unless the a's and b's are proportional, in which case the inequality becomes an equality. The corresponding result for integrals was given by Buniakowsky in 1859 and Schwarz in 1885. Many other inequalities can be derived from it, e.g. Hölder's of 1889: if $a_i, b_i, \ldots, l_i$ $(i = 1, \ldots n)$ are non-negative, $\alpha, \beta, \ldots, \lambda$ are positive with $\alpha+\beta+\ldots+\lambda = 1$, then

$$\Sigma a^{\alpha} b^{\beta} \ldots l^{\lambda} < \Sigma a^{\alpha} \Sigma b^{\beta} \ldots \Sigma l^{\lambda},$$

unless one set of the (a) etc. are all zero or unless one set is proportional to all the others, in which cases the inequality becomes an equality. A very clear and comprehensive account is given in the book *Inequalities* by Hardy, Littlewood and Pólya, 1934, Cambridge University Press.

The median

2.8 The median(*) is that value of the variate which divides the total frequency into two equal halves, i.e. is the value x_m such that

$$\int_{-\infty}^{x_m} f(x)\,dx = \int_{x_m}^{\infty} f(x)\,dx = \tfrac{1}{2}. \tag{2.11}$$

There is some small indeterminacy in this definition when the distribution is discrete which may be removed by convention. If there are $(2N+1$ members of the population, we take the median to be the value of the $(N+1$ th member. If there are $2N$ we take it to be halfway between the values of the Nth and the $(N+1$ th. When the distribution is numerically specified in class-intervals there is the usual indeterminacy due to grouping, which may be dealt with in the manner of the following example.

Example 2.5

To find the median value of the distribution of heights considered in Example 2.1.

Half the total frequency of 8585 observations is 4292·5.
There are, up to and including the interval beginning at $66\frac{15}{16}$ inches, 3589
leaving 703·5.
The frequency in the next interval is 1329.
Hence we take the median to be

$$66\tfrac{15}{16} + \frac{703{\cdot}5}{1329} = 67{\cdot}47 \text{ inches.}$$

The mean (Example 2.1) is 67·46 inches, practically the same.

A graphical method of determining the median is given later in this chapter (**2.15**).

2.9 From the mathematical viewpoint the indeterminacy in the median can be removed. For a set of values $x_1, \ldots, x_N$ it may be shown that the sum $\Sigma|\xi - x_i|^p$, considered

(*) The name "median" was first used by Galton in 1883, but the concept was used by him, and by Fechner independently, around 1870.

as a function of ξ, has a unique minimum for some ξ_p if $p>1$; and further, that as p tends to unity ξ_p tends to some unique value, which may be defined as the median. See Jackson (1921).

An extension of the concept of median to the multivariate case is made if we use the result of Exercise 2.1 as the *definition* of the median, and generally define a multivariate median as that point which minimizes the sum of absolute deviations for the distribution. This is computationally more complicated than combining the medians of the univariate marginal distributions, but has theoretical attractions—cf. Haldane (1948).

The mode

2.10 If the frequency function has a local maximum at a value x, i.e. if $f(x)$ is greater at x than at neighbouring values below and above x, there is said to be a *mode* of the distribution at x. If there is only one such modal value, the distribution is called unimodal, as we have already seen; if there are several, the distribution is multimodal. If the frequency at an extreme permissible value of x is at a peak (e.g. if the distribution if J-shaped, as in Fig. 1.2), it is sometimes called a half-mode.

If the frequency function is everywhere continuous and twice-differentiable, and there is no terminal peak, a mode must be a solution of

$$f'(x) = \frac{d}{dx}f(x) = 0, \qquad f''(x) = \frac{d^2}{dx^2}f(x) < 0. \tag{2.12}$$

If $f'(x)$ vanishes and $f''(x)$ is greater than zero we have a local minimum, and such a point is sometimes called an antimode.

In numerically specified distributions and discrete distributions generally the mode is sometimes difficult to determine exactly. Where the number of observations is large enough to permit grouping, there will usually be an interval containing a maximum frequency, and we may regard the mode as lying in that interval. In the height distribution of Table 1.7, for instance, the mode may be considered as lying somewhere in the interval 67– inches. To estimate its position more accurately it is necessary to fit a continuous curve to the distribution and determine the mode of the curve. The process of fitting will be considered in Chapter 6.

2.11 In a symmetrical unimodal distribution the mean, the median and the mode coincide. For skew distributions they differ. There is an interesting empirical relationship between the three quantities which appears to hold for unimodal curves of moderate asymmetry, namely

$$\text{mean} - \text{mode} = 3\,(\text{mean} - \text{median}). \tag{2.13}$$

A mathematical explanation of this relationship is given in Exercise 6.20 below. Exercise 2.16 gives a particular distribution for which (2.13) holds approximately.

It is a useful mnemonic to observe that the mean, median and mode of a unimodal distribution occur in the same order (or the reverse order) as in the dictionary; and that the median is nearer to the mean than to the mode, just as the corresponding words are nearer together in the dictionary.

In elementary theory the median and the mode have considerable claims to use as measures of location for unimodal distributions. They are readily interpretable in terms of ordinary ideas—the median is the middle value and the mode is the most popular value—and the median is usually more easily determined than the mean in numerically specified distributions. What gives the arithmetic mean the greater importance in advanced theory is its superior mathematical tractability and certain sampling properties ; but the median has compensating advantages—it is, for instance, less sensitive to the configuration of the outlying parts of the frequency distribution than is the mean.

2.12 Mention should also be made of a little-used measure of location, the mid-range. This is the variate-value half-way between the terminal points of the distribution if it possesses such points. Its dependence on the terminal points and the use of mathematical distributions which extend to infinity in one or both directions make it unsuitable for many purposes but, like the range (**2.17**, below), it has some uses in sampling theory and is very easily calculated.

Quantiles

2.13 The concept of median value can be easily extended to locate the curve more accurately by the use of several measures. We may, for example, find the three variate-values which divide the total frequency into four equal parts. The middle one of these will be the median itself ; the other two are called the lower and upper *quartiles*(*) respectively. Similarly, we may find the nine variate-values which divide the total frequency into ten equal parts—the *deciles*—or into a hundred equal parts—the *percentiles*(*) or *percentage points*. Generally we may find the $(n-1)$ variate-values which divide the total frequency into n equal parts—the *quantiles*. Evidently the knowledge of the quantiles for some fairly high n, such as 10, gives a very good idea of the general form of the frequency distribution. Even the quartiles and the median are valuable general guides. In Chapter 14, we shall study statistics of this type systematically.

2.14 The determination of the quantiles of a numerically specified distribution proceeds as for the median, indeterminacies being resolved by similar conventions. That of the quantiles of a mathematically specified distribution, say the jth quantile, is a matter of solving for x the equation

$$j/n = \int_{-\infty}^{x} dF \tag{2.14}$$

which can be done without difficulty by interpolation when the integral of dF has been tabulated.

Example 2.6

To find the quartiles of the height distribution considered in Example 2.1.

One-quarter of the total frequency is 8585/4 =	2146·25	
Up to the interval 65– there are	1376	members
leaving	770·25	members
In the next interval there are	990	members
Thus the lower quartile is $64\frac{15}{16}+\frac{770{\cdot}25}{990}$ =	65·71	inches
The upper quartile will be found to be	69·21	inches
We have already found (Example 2.5) that the median x_m is	67·47	inches

(*) Percentiles were defined in 1885 by Galton, who also brought quartiles into general use a little earlier.

Denoting the quartiles by Q_1 and Q_3 we see that

$$Q_1 - x_m = -1{\cdot}76 \text{ inches}$$
$$Q_3 - x_m = 1{\cdot}74 \text{ inches}$$

so that the median is almost half-way between the quartiles, an indication of the symmetry of the distribution.

2.15 The quantiles can be easily determined graphically from the graph of the distribution function. Fig. 2.1 gives the curve for the data of Table 1.7. To find

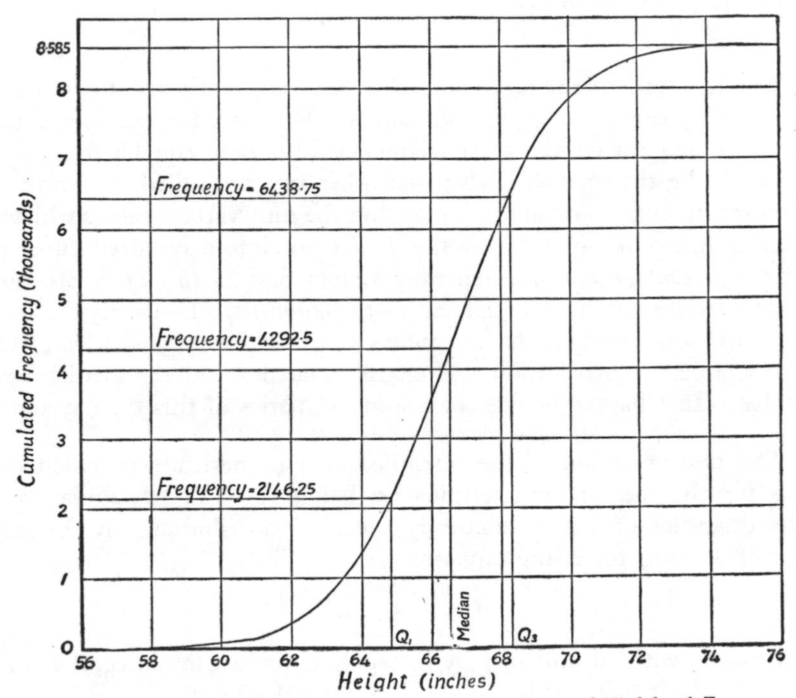

Fig. 2.1—Distribution curve of the data of Table 1.7

(Heights shown to correspond to entries in the Table, e.g. cumulated frequency at 64 inches is the frequency up to and including the range 64— and therefore up to $64\frac{15}{16}$ inches)

the median we determine the ordinate corresponding to the abscissa of $\frac{1}{2}N$ and so on. The positions of the quartiles and the median are shown in Fig. 2.1, and the reader may care to compare the values obtained by reading the graph by eye with those given in Example 2.6.

Measures of dispersion

2.16 We now proceed to consider the quantities which have been proposed to measure the dispersion of a distribution. They fall into three groups :—

(*a*) Measures of the distance (in terms of the variate) between certain representative values, such as the range, the interdecile range or the interquartile range.

(*b*) Measures compiled from the deviations of every member of the population from some central value, such as the mean deviation from the mean, the mean deviation from the median, and the standard deviation.

(*c*) Measures compiled from the deviations of all the members of the population among themselves, such as the mean difference.

In advanced theory the outstandingly important measure is the standard deviation; but they all require some mention.

Range and interquantile differences

2.17 The range of a distribution is the difference of the greatest and least variate-values borne by its members. As a descriptive measure of a population it has very little use. A knowledge of the whereabouts of the end values obviously tells little about the way the bulk of the distribution is condensed inside the range; and for distributions of infinite range it is obviously wholly inappropriate.

More useful rough-and-ready measures may be obtained from the quantiles, and there are two such in general use. The interquartile range is the distance between the upper and lower quartiles, and is thus a range which contains one-half the total frequency. The interdecile range (or perhaps, more accurately, the 1–9th interdecile range) is the distance between the first and the ninth decile. Both these measures evidently give some approximate idea of the "spread" of a distribution, and are easily calculable. For this reason they are fairly generally used in elementary descriptive statistics. In advanced theory they suffer from the disadvantage of being relatively difficult to handle mathematically in the theory of sampling.

Mean deviations

2.18 The amount of scatter in a population is evidently measured to some extent by the totality of deviations from the mean. It is easily seen from (2.1) that the sum of these deviations taken with appropriate sign is zero. We may however write

$$\delta_1 = \int_{-\infty}^{\infty} |x-\mu_1'|\,dF \tag{2.15}$$

where the deviations are now taken absolutely, and define δ_1 to be a coefficient of dispersion. We shall call it the mean deviation about the mean. It is clearly unaffected by change of origin for x, since μ_1' changes correspondingly by (2.1). We therefore may measure x from zero.

Because $\int_{-\infty}^{\infty} (x-\mu_1')\,dF \equiv 0$, where μ_1' is taken about zero, (2.15) may be expressed in the forms

$$\delta_1 = 2\int_{\mu_1'}^{\infty}(x-\mu_1')\,dF = -2\int_{-\infty}^{\mu_1'}(x-\mu_1')\,dF.$$

Similarly for the median x_m we may write

$$\delta_2 = \int_{-\infty}^{\infty} |x-x_m|\,dF \tag{2.16}$$

and call δ_2 the mean deviation about the median. The reader is asked to prove in Exercise 2.1 that δ_2 is no greater than the mean deviation measured from any other point.

In future the unqualified words "mean deviation" will be taken to refer to the mean deviation about the mean.

Both (2.15) and (2.16) have merits in elementary work, being fairly easily calculable. Once again, however, they are practically excluded from advanced work by their intractability in the theory of sampling.

Standard deviation

2.19 We have seen that the mean about an arbitrary point a is given by

$$\mu_1' = \int_{-\infty}^{\infty} (x-a)\,dF.$$

This is called the first moment(*) of the population. Similarly, the second moment is

$$\mu_2' = \int_{-\infty}^{\infty} (x-a)^2\,dF. \tag{2.17}$$

The second moment about the mean is written without the prime, thus:

$$\mu_2 = \int_{-\infty}^{\infty} (x-\mu_1')^2\,dF \tag{2.18}$$

and is called the variance(*). The positive square root of the variance is called the standard deviation(*), and usually denoted by σ, so that we have

$$\sigma = |\sqrt{\mu_2}|. \tag{2.19}$$

The variance is thus the mean of the squares of the deviations from the mean, and the standard deviation is their root-mean-square. The device of squaring and then taking the square root of the resultant sum in order to obtain the standard deviation may appear a little artificial, but it makes the mathematics of the sampling theory simpler than in the case of the mean deviation, where the absolute values complicate the theory.

The calculation of the variance and the standard deviation proceeds by an easy extension of the methods used for the mean. In particular, if b is some arbitrary value,

$$\begin{aligned}\mu_2' \text{ (about } a) &= \int_{-\infty}^{\infty} (x-a)^2\,dF \\ &= \int_{-\infty}^{\infty} \{(x-b)^2+2(b-a)(x-b)+(b-a)^2\}\,dF \\ &= \mu_2' \text{ (about } b)+2(b-a)\mu'_1 \text{ (about } b)+(b-a)^2.\end{aligned} \tag{2.20}$$

If now b is the mean we have

$$\mu_2' = \mu_2+(\mu_1'-a)^2 \quad \text{or} \quad \mu_2 = \mu_2'-(\mu_1'-a)^2. \tag{2.21}$$

Thus the variance can easily be found from the second moment about an arbitrary point, which can be selected to simplify the calculations.

Example 2.7

To find the mean deviation and the standard deviation for the distribution of men according to height considered in Example 2.1 (Table 1.7).

In the case of the mean deviation for a grouped distribution, the sum of deviations should first be calculated from the centre of the class-interval in which the mean lies

(*) The term " moment", used by analogy with its meaning in Statics, appears at least as long ago as the work of A. Quetelet (1796–1874) and has been regularly used since being adopted by K. Pearson, who first used the term " standard deviation " in the 1890's. " Variance " was not used until 1918, when R. A. Fisher defined it in a paper on genetics.

and then reduced to the mean as origin. It so happens that in Table 2.1 the mean fell in the interval taken as origin, so that the preliminary arithmetic already exists in the Table.

The sum of positive deviations is 8763 and that of negative deviations −8584. Hence the sum of deviations regardless of sign is 17,347, the unit being the class-interval and the origin the centre of the interval.

To reduce to the mean as origin, we note that if the number of observations below the mean is N_1 and the number above the mean is N_2, and $d = \mu_1' - a$, we have to add $N_1 d$ to the sum of deviations about the centre of the interval and subtract $N_2 d$. In this case $d = 0{\cdot}02$ (Example 2.1), $N_1 = 4918$, $N_2 = 3667$. Hence we add $(4918-3667)0{\cdot}02 = 25$. Hence the mean deviation

$$\delta_1 = \frac{17{,}347+25}{8585} = 2{\cdot}02 \text{ inches.}^{(*)}$$

For the standard deviation some further calculation is required, as shown in Table 2.2.

Table 2.2—Calculation of the standard deviation for the distribution of Table 1.7 *some) preliminary calculation already carried out in Table 2.1)*

(1) Height, inches	(2) Frequency f	(3) Deviation ξ	(4) $\xi^2 f$
57–	2	−10	200
58–	4	−9	324
59–	14	−8	896
60–	41	−7	2,009
61–	83	−6	2,988
62–	169	−5	4,225
63–	394	−4	6,304
64–	669	−3	6,021
65–	990	−2	3,960
66–	1223	−1	1,223
67–	1329	0	0
68–	1230	1	1,230
69–	1063	2	4,252
70–	646	3	5,814
71–	392	4	6,272
72–	202	5	5,050
73–	79	6	2,844
74–	32	7	1,568
75–	16	8	1,024
76–	5	9	405
77–	2	10	200
TOTALS	8585	—	56,809

Column (4) shows the sum $\Sigma\xi^2 f$, where f is the actual frequency. We then have, for the second moment about the arbitrary origin,

$$\mu_2' = \frac{56{,}809}{8585} = 6{\cdot}6172.$$

(*) This calculation can be refined slightly—cf. Exercise 2.21.

We have already found in Example 2.1 that

$$\mu_1' - a = \frac{179}{8585} = 0{\cdot}0209.$$

Hence, in virtue of (2.21) $\mu_2 = 6{\cdot}6172 - (0{\cdot}0209)^2$

$$= 6{\cdot}6168$$

$$\sigma = \sqrt{\mu_2} = 2{\cdot}57 \text{ inches.}$$

It may be noted that in Example 2.7 the mean deviation is about 80 per cent. of the standard deviation. This relationship often holds approximately for unimodal curves approaching symmetry. The reason will become apparent when we study the so-called " normal " distribution in Chapter 5.

Chakrabarti (1948) has shown that, in general, for a set of $n > 2$ values,

$$\left.\begin{aligned}(2/n)^{\frac{1}{2}} \leqslant \delta_1/\sigma &\leqslant (1-n^{-2})^{\frac{1}{2}}, && n \text{ odd} \\ &\leqslant 1, && n \text{ even,}\end{aligned}\right\} \tag{2.22}$$

the equalities being attainable. Exercise 2.4 is to show that $\delta_1/\sigma \leqslant 1$. Exercise 2.25 gives other relations between δ_1, μ_2 and the d.f.

Example 2.8

To find the variance of the distribution

$$dF = \frac{1}{B(p,q)} x^{p-1}(1-x)^{q-1}\,dx, \qquad 0 \leqslant x \leqslant 1\,;\ p, q > 0.$$

We have, about the origin,

$$\mu_2' = \frac{1}{B(p,q)} \int_0^1 x^{p+1}(1-x)^{q-1}\,dx$$

$$= \frac{B(p+2,q)}{B(p,q)} = \frac{(p+1)p}{(p+q+1)(p+q)}.$$

We have already found (Example 2.2) that

$$\mu_1' = \frac{p}{p+q}.$$

Thus $\mu_2 = \mu_2' - \mu_1'^2$

$$= \frac{(p+1)p}{(p+q+1)(p+q)} - \frac{p^2}{(p+q)^2}$$

$$= \frac{pq}{(p+q+1)(p+q)^2}.$$

Example 2.9

We may find the variance, like the mean, by differentiating the frequency-generating function. Reverting to the binomial distribution considered in Example 2.3 we have

$$\frac{d^2P}{dt^2} = \Sigma\,\{j(j-1)t^{j-2}f_j\}$$

and on putting $t = 1$ we have, with moments about zero,

$$\left[\frac{d^2P}{dt^2}\right]_{t=1} = \Sigma\,\{j(j-1)f_j\} = \mu_2' - \mu_1.$$

In this particular case

$$\left[\frac{d^2}{dt^2}\{pt+(1-p)\}^n\right]_{t=1} = n(n-1)p^2.$$

Thus $$\mu_2' = n(n-1)p^2+np,$$

and $$\mu_2 = n(n-1)p^2+np-n^2p^2$$
$$= np(1-p). \tag{2.23}$$

Sheppard's corrections

2.20 The treatment of the values of a grouped frequency distribution as if they were concentrated at the mid-points of intervals is an approximation, and in certain circumstances it is possible to make corrections for any distortion introduced thereby. These so-called " Sheppard's corrections " will be discussed at length in the next chapter, but at this stage we may indicate without proof the appropriate correction for the second moment.

If the distribution is continuous and has high order contact with the variate-axis at its extremities, i.e. if it " tails off " smoothly, the crude second moment calculated from grouped frequencies should be corrected by subtracting from it $h^2/12$, where h is the width of the interval. For example, in the height data of Example 2.7, we have $h = 1$, and the corrected second moment is

$$6{\cdot}6168-0{\cdot}0833 = 6{\cdot}5335.$$

The corrected value of σ is $\sqrt{6{\cdot}5335} = 2{\cdot}56$, as against an uncorrected value of 2·57. Exercise 2.8 provides a general guide to the magnitude of the correction.

Mean difference

2.21 The coefficient of mean difference (not to be confused with mean deviation) is defined by

$$\Delta = \int_{-\infty}^{\infty}\int_{-\infty}^{\infty}|x-y|\,dF(x)\,dF(y)$$
$$= \int_{-\infty}^{\infty}\int_{-\infty}^{\infty}|x-y|f(x)f(y)\,dx\,dy. \tag{2.24}$$

In the discrete case two formulae arise. We have either

$$\Delta = \frac{1}{N(N-1)}\sum_{j=-\infty}^{\infty}\sum_{k=-\infty}^{\infty}|x_j-x_k|f(x_j)f(x_k), \qquad j\neq k, \tag{2.25}$$

the mean difference without repetition, or

$$\Delta = \frac{1}{N^2}\sum_{j=-\infty}^{\infty}\sum_{k=-\infty}^{\infty}|x_j-x_k|f(x_j)f(x_k), \tag{2.26}$$

the mean difference with repetition. The distinction lies only in the divisor and is unimportant if N is large.

The mean difference is the average of the differences of all the possible pairs of variate-values, taken regardless of sign. In the coefficient with repetition each value is taken with itself, adding of course nothing to the sum of deviations, but resulting in the total number of pairs being N^2. In the coefficient without repetition only distinct values are taken, so that the number of pairs is $N(N-1)$. Hence the divisors in (2.25) and (2.26).

2.22 The mean difference was proposed by Gini (1912), after whom it is usually named, but was discussed by Helmert and other German writers in the 1870's—cf. H. A. David (1968). It has a certain theoretical attraction, being dependent on the spread of the variate-values among themselves and not on the deviations from some central value. Further, its defining integral (2.24) may converge when that of the variance, (2.18), does not converge. Since it is essentially a linear rather than a quadratic measure of dispersion, it converges whenever the mean exists—cf. Exercises 2.9 and 2.19. It is, however, more difficult to compute than the standard deviation, and the appearance of the absolute values in the defining equations indicates, as for the mean deviation, the appearance of difficulties in the theory of sampling. It might be thought that this inconvenience could be overcome by the definition of a coefficient

$$E^2 = \int_{-\infty}^{\infty}\int_{-\infty}^{\infty}(x-y)^2\,dF(x)\,dF(y).$$

This, however, is nothing but twice the variance.

For
$$\begin{aligned} E^2 &= \int_{-\infty}^{\infty}\int_{-\infty}^{\infty}\{x^2-2xy+y^2\}\,dF(x)\,dF(y) \\ &= \int_{-\infty}^{\infty}x^2\,dF(x)\int_{-\infty}^{\infty}dF(y)-2\int_{-\infty}^{\infty}x\,dF(x)\int_{-\infty}^{\infty}y\,dF(y) \\ &\qquad\qquad +\int_{-\infty}^{\infty}dF(x)\int_{-\infty}^{\infty}y^2\,dF(y) \\ &= 2\mu_2'-2\mu_1'^2 = 2\mu_2. \end{aligned} \tag{2.27}$$

This interesting relation shows that the variance may in fact be defined as half the mean square of all possible variate-differences, that is to say, without reference to deviations from a central value, the mean.

Another unusual formula for the variance is given in Exercise 2.24.

Coefficients of variation: standardization

2.23 The foregoing measures of dispersion have all been expressed in terms of units of the variate. It is thus difficult to compare dispersions in different populations unless the units happen to be identical; and this has led to a search for measures which shall be independent of the variate scale, that is to say, shall be pure numbers.

Several coefficients of this kind may be constructed, such as the mean deviation divided by the mean or by the median. Only two have been used at all extensively in practice, Karl Pearson's coefficient of variation, defined by

$$V = \frac{\sigma}{\mu_1'}, \tag{2.28}$$

and Gini's coefficient of concentration, defined by

$$G = \Delta/(2\mu_1'). \tag{2.29}$$

These coefficients both suffer from the disadvantage of being affected very much by μ_1', the value of the mean measured from some arbitrary origin, and are not usually employed unless there is a natural origin of measurement or comparisons are being made between distributions with similar origins.

We shall see at the end of **2.25** that $0 \leqslant G \leqslant 1$. There is no upper bound for V in general, but Exercise 2.23 gives one for a finite set of non-negative observations.

2.24 For our purposes, comparability may be attained in a somewhat different way. Let us take σ itself as a new unit and express the frequency function in terms of a new variable y related to x by

$$y = (x - \mu_1')/\sigma. \tag{2.30}$$

Any distribution expressed in this way has zero mean and unit variance. It is then said to be expressed in standard measure, or standardized. Two distributions in standard measure can be readily compared in regard to form, skewness, and other qualities, though not of course in regard to mean and variance.

Concentration

2.25 Gini's coefficient of concentration (2.29) arises in a natural way from the following approach:—

Writing, as usual

let us define

$$\left.\begin{aligned} F(x) &= \int_{-\infty}^{x} f(u)\,du \\ \Phi(x) &= \frac{1}{\mu_1'}\int_{-\infty}^{x} u f(u)\,du \end{aligned}\right\} \tag{2.31}$$

$\Phi(x)$ exists, of course, only if μ_1' exists. Just as $F(x)$ varies from 0 to 1, $\Phi(x)$ varies from 0 to 1 provided that the origin is taken to the left of the start of the frequency distribution, assumed to be finite. $\Phi(x)$ may be called the *incomplete* first moment.

Now (2.31) may be regarded as defining a relationship between the variables F and Φ in terms of parametric equations in x. The curve whose ordinate and abscissa are Φ and F is called the concentration curve or Lorenz curve (see Fig. 2.2).

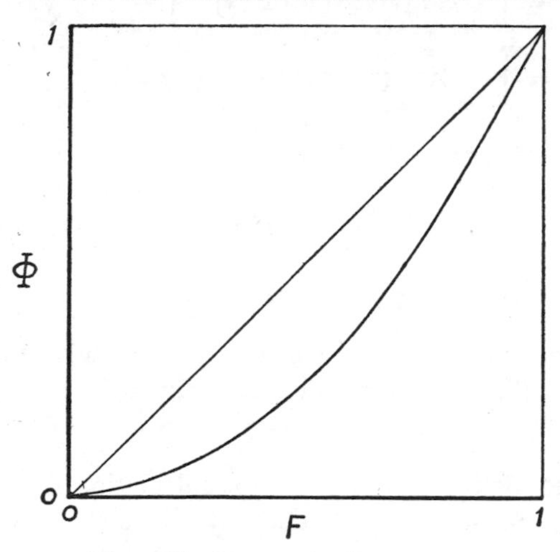

Fig. 2.2—Concentration curve

The concentration curve must be convex to the F-axis, for we have

$$\mu_1' \frac{d\Phi}{dF} = \mu_1' \frac{d\Phi}{dx} \Big/ \frac{dF}{dx} = \frac{xf(x)}{f(x)} = x,$$

which is positive since our origin is taken to the left of the start of the distribution. Also

$$\mu_1' \frac{d^2\Phi}{dF^2} = \frac{dx}{dF} = \frac{1}{f(x)} > 0.$$

Thus the tangent to the curve makes a positive acute angle with the F-axis, and the angle increases as F increases; in other words, the curve is convex to the F-axis.

The distance between the line $\Phi = F$ and the concentration curve is $F - \Phi(F)$. It has a turning point where $\frac{d}{dF}\{F-\Phi\} = 0$, i.e. $d\Phi/dF = 1$. From above, this is at $x = \mu_1'$; at the mean, the concentration curve is parallel to $\Phi = F$ and most distant from it. This maximum distance is

$$F(\mu_1') - \Phi(\mu_1') = -\frac{1}{\mu_1'} \int_{-\infty}^{\mu_1'} (u - \mu_1')\, dF(u) = \tfrac{1}{2}\frac{\delta_1}{\mu_1'}$$

from **2.18**, where δ_1 is the mean deviation from the mean (2.15).

The area between the concentration curve and the line $\Phi = F$ is called the area of concentration. We proceed to show that it is equal to one-half the coefficient of concentration.

In fact, we have from Fig. 2.2

$$2\,(\text{area of concentration}) = \int_0^1 F\, d\Phi - \int_0^1 \Phi\, dF$$

and thus from above

$$\begin{aligned} 2\mu_1'\,(\text{area}) &= \int_{-\infty}^{\infty} F(x)\, x\, dF(x) - \mu_1' \int_{-\infty}^{\infty} \Phi(x)\, dF(x) \\ &= \int_{-\infty}^{\infty} x \int_{-\infty}^{x} dF(y)\, dF(x) - \int_{-\infty}^{\infty}\int_{-\infty}^{x} y\, dF(y)\, dF(x) \\ &= \int_{-\infty}^{\infty}\int_{-\infty}^{x} (x-y)\, dF(y)\, dF(x). \end{aligned}$$

Now $\int_{-\infty}^{\infty}\int_{-\infty}^{\infty} (x-y)\, dF(y)\, dF(x) = 0$, and hence (as in **2.18** for the mean deviation)

$$\begin{aligned} 2\mu_1'\,(\text{area}) &= \tfrac{1}{2}\left[\int_{-\infty}^{\infty}\int_{-\infty}^{x} (x-y)\, dF(y)\, dF(x) + \int_{-\infty}^{\infty}\int_{x}^{\infty} (y-x)\, dF(y)\, dF(x)\right] \\ &= \tfrac{1}{2}\int_{-\infty}^{\infty}\int_{-\infty}^{\infty} |x-y|\, dF(x)\, dF(y) = \tfrac{1}{2}\Delta. \end{aligned}$$

Thus the area of concentration is equal to $\frac{1}{4}\Delta/\mu_1'$, namely one-half of the coefficient of concentration G at (2.29). From Fig. 2.2, the area lies between 0 and $\frac{1}{2}$, so $0 \leqslant G \leqslant 1$.

2.26 Various methods have been given for calculating the mean difference. The following is probably the simplest, particularly for distributions specified in equal class-intervals.

Let us, without loss of generality, take an origin at the start of the distribution. We may then write

$$\sum_{j=1}^{N} \sum_{k=1}^{N} |x_j - x_k| = 2\Sigma'(x_j - x_k),$$

the summation Σ' being taken over values such that $j \geqslant k$. We have also

$$x_j - x_k = (x_j - x_{j-1}) + (x_{j-1} - x_{j-2}) + \ldots + (x_{k+1} - x_k).$$

Thus
$$\Sigma'(x_j - x_k) = \sum_{h=1}^{N-1} C_h(x_{h+1} - x_h),$$

where C_h is the number of terms of type $(x_j - x_k)$ in Σ' containing $x_{h+1} - x_h$. Since h is the number of values of j less than or equal to h (the origin being at the start of the distribution) and $N-h$ the number greater than or equal to $h+1$, we have $C_h = h(N-h)$, and thus (2.26) is

$$\begin{aligned} \Delta_1 &= \frac{2}{N^2}\Sigma'(x_j - x_k) \\ &= \frac{2}{N^2}\sum_{h=1}^{N-1} h(N-h)(x_{h+1} - x_h). \end{aligned} \tag{2.32}$$

This form is particularly useful if all the intervals are equal. F_h being the distribution function of x_h we then have

$$\Delta = \frac{2}{N^2}\sum_{h=1}^{N-1}(NF_h)(N - NF_h) = 2\sum_{h=1}^{N-1} F_h(1 - F_h). \tag{2.33}$$

If the actual cumulated frequency for x_h is G_h we have

$$\Delta = \frac{2}{N^2}\sum_{h=1}^{N-1} G_h(N - G_h), \tag{2.34}$$

the most convenient form in practice.

Exercise 2.10 gives another form for the calculation.

Example 2.10

Returning once more to the height distribution considered in previous examples, we may calculate $\Sigma G_h(N - G_h)$ as in Table 2.3.

We have, from (2.34), for the mean difference with repetition,

$$\begin{aligned} \Delta &= \frac{2 \times 105{,}990{,}850}{8585^2} \\ &= 2{\cdot}88 \text{ inches} \end{aligned}$$

as against a mean deviation of 2·02 inches and a standard deviation of 2·57 inches (Example 2.7).

The mean difference is here about one-eighth larger than the standard deviation. Like the mean deviation's relation to the standard deviation remarked below Example 2.7, this relation has its roots in the properties of the " normal " distribution—cf. **10.14** below. Exercise 2.4 is to show that $\Delta \leqslant \sigma\sqrt{2}$. The sharper inequality $\Delta \leqslant 2\sigma/\sqrt{3}$ follows from Exercise 31.13, Vol. 2.

Table 2.3—Calculation of the mean difference for the height distribution of Table 1.7

Height, inches	Frequency	G_h	$N - G_h$	$G_h(N - G_h)$
57–	2	2	8583	17,166
58–	4	6	8579	51,474
59–	14	20	8565	171,300
60–	41	61	8524	519,964
61–	83	144	8441	1,215,504
62–	169	313	8272	2,589,136
63–	394	707	7878	5,569,746
64–	669	1376	7209	9,919,584
65–	990	2366	6219	14,714,154
66–	1223	3589	4996	17,930,644
67–	1329	4918	3667	18,034,306
68–	1230	6148	2437	14,982,676
69–	1063	7211	1374	9,907,914
70–	646	7857	728	5,719,896
71–	392	8249	336	2,771,664
72–	202	8451	134	1,132,434
73–	79	8530	55	469,150
74–	32	8562	23	196,926
75–	16	8578	7	60,046
76–	5	8583	2	17,166
77–	2	8585	—	—
TOTALS	8585	—	—	105,990,850

Mean values in general

2.27 The concept of the arithmetic mean of a variate can be easily extended to that of the mean of a function, say $h(x)$. In fact we define

$$E(h) = \int_{-\infty}^{\infty} h(x)\, dF, \tag{2.35}$$

subject, of course, to the existence of the sum or integral on the right. Mean values such as this play a fundamental role in the theory of statistics and of probability. The notation E derives from the fact that in probability theory such values are often known as " expected values " or " expectations ". It is a remarkable and fruitful fact that in the theory of sampling we are often able to determine certain mean values, especially those of powers of the variate, more easily than the corresponding frequency functions.

Generally, for a function $h(x_1, x_2, \ldots, x_n)$ we have

$$E(h) = \int_{-\infty}^{\infty} \cdots \int_{-\infty}^{\infty} h(x_1, \ldots, x_n)\, dF(x_1, \ldots, x_n) \tag{2.36}$$

In particular if h is a function of x_1 only, we have, for the multivariate distribution, on integrating with respect to $x_2, \ldots, x_n$,

$$E(h) = \int_{-\infty}^{\infty} h(x_1)\, dF(x_1) \tag{2.37}$$

where $F(x_1)$ is the (marginal) distribution function of x_1. This is required for consistence with (2.35) and shows us that the mean value of a variate (or some function of it) is the same whether the variate forms part of a multivariate complex or not.

2.28 Two simple but important properties of mean values are to be noted:

(*a*) The mean of a sum is the sum of the means. If h_1 and h_2 are two functions,

$$E(h_1+h_2) = E(h_1)+E(h_2). \tag{2.38}$$

This follows at once from the definitions.

(*b*) The mean value of a product of two functions is the product of their mean values if they are independent. For example, let x_1 and x_2 represent independent variates. Their distribution function then factorizes into two components. Any two functions $h_1(x_1)$, $h_2(x_2)$ will be independent and

$$\begin{aligned} E(h_1 h_2) &= \int_{-\infty}^{\infty}\int_{-\infty}^{\infty} h_1(x_1)\,h_2(x_2)\,dF(x_1, x_2) \\ &= \int_{-\infty}^{\infty} h_1(x_1)\,dF(x_1, \infty) \int_{-\infty}^{\infty} h_2(x_2)\,dF(\infty, x_2) \\ &= E(h_1)\,E(h_2). \end{aligned} \tag{2.39}$$

Equation (2.38) is true even if the variates are dependent; (2.39) will not be true in general for dependent variates. When it is true, the functions h_1 and h_2 are said to be *uncorrelated*, a weaker property than independence. We shall study correlation in detail in Chapter 26, Vol. 2.

EXERCISES

2.1 Show that the mean deviation about an arbitrary point is least when that point is the median.

2.2 Show that the mean (about the origin) of the discrete distribution whose frequencies at $0, 1, 2, \ldots, r, \ldots$ are

$$e^{-\lambda}\left(1, \lambda, \frac{\lambda^2}{2!}, \ldots, \frac{\lambda^r}{r!}, \ldots\right)$$

is λ, and that the variance is also λ.

2.3 Show that, if deviations are small compared with the value of the mean, we have approximately, for the geometric and harmonic means,

$$G = \mu_1'\left(1 - \frac{1}{2}\frac{\sigma^2}{\mu_1'^2}\right) \qquad H = \mu_1'\left(1 - \frac{\sigma^2}{\mu_1'^2}\right)$$

and hence that

$$\mu_1' - 2G + H = 0.$$

2.4 Show that the mean deviation about the mean is not greater than the standard deviation; and that the mean difference cannot exceed $\sqrt{2}$ times the standard deviation.

2.5 Show that for the " rectangular " distribution

$$dF = dx, \qquad 0 \leqslant x \leqslant 1,$$
$$\mu_1' \text{ (about the origin)} = \tfrac{1}{2}$$
$$\mu_2 = \tfrac{1}{12}$$
$$\delta_1 = \tfrac{1}{4}$$
$$\Delta = \tfrac{1}{3}.$$

2.6 Show that for the distribution (the exponential)

$$dF = k e^{-x/\sigma} dx, \qquad 0 \leqslant x \leqslant \infty, \; \sigma > 0,$$

the mean, standard deviation and mean difference are all equal to σ; and that the interquartile range is $\sigma \log_e 3$.

2.7 Show that for the distribution

$$dF = k e^{-\frac{1}{2}\chi^2} \chi^{\nu-1} d\chi, \qquad 0 \leqslant \chi \leqslant \infty,$$
$$\mu_1 \text{ (about the origin)} = \sqrt{2}\, \Gamma\{\tfrac{1}{2}(\nu+1)\}/\Gamma(\tfrac{1}{2}\nu)$$
$$\mu_2' = \nu.$$

2.8 Show that if a range of six times the standard deviation contains at least 18 class-intervals, Sheppard's correction will make a difference of less than 0·5 per cent. in the uncorrected value of the standard deviation.

2.9 Show that for a continuous distribution whose mean exists

$$\Delta = 2\int_{-\infty}^{\infty} F(x)\{1 - F(x)\} dx = 4\int_{-\infty}^{\infty} x\{F(x) - \tfrac{1}{2}\}\, dF.$$

Hence confirm that $0 \leqslant G \equiv \Delta/(2\mu_1') \leqslant 1$. (Cf. (2.33) in the discrete case.)

2.10 If the variate-values of a distribution are $x_1, \ldots, x_N$ in ascending order of magnitude and

$$s_r = \sum_{j=1}^{r} x_j \qquad U = \sum_{r=1}^{N} s_r$$

$$t_r = \sum_{j=1}^{r} x_{N-j+1} \qquad V = \sum_{r=1}^{N} t_r$$

show that (2.26) is
$$\Delta = \frac{2}{N^2}(V-U) = \frac{2}{N^2}\{2V-N(N+1)\mu_1'\}$$
$$= \frac{2}{N^2}\{N(N+1)\mu_1'-2U\}.$$

(Cf. **2.25** in the continuous case.)

2.11 Referring to Exercise 1.14, show that the mean of the negative binomial distribution is rq/p and that its variance is rq/p^2.

2.12 Show that the variance of the distribution
$$dF = \frac{k}{1+x^4}dx, \qquad -\infty \leqslant x \leqslant \infty,$$
is unity.

2.13 With reference to Exercise 1.10 find the mean value of $x_1^2+x_2^2+x_3^2$ (a) directly and (b) from the distribution of r^2.

2.14 For the distribution of Exercise 1.11 show that the mean value of x^2 is σ_1^2 and that of y^2 is σ_2^2. By differentiating $\int dF = 1$ with respect to ρ show that $E(xy) = \rho\sigma_1\sigma_2$.

2.15 A variate takes only non-negative values and has mean μ_1'. Show that for any positive t
$$F(t) > 1-\mu_1'/t.$$

2.16 Show that if a distribution
$$dF = \frac{1}{\sqrt{(2\pi)}}e^{-\frac{1}{2}\xi^2}d\xi, \qquad -\infty \leqslant \xi \leqslant \infty,$$
is transformed by $\xi = \gamma+\delta\log(x-\mu)$, the distribution of x has
$$\frac{\text{mean}-\text{mode}}{\text{mean}-\text{median}} = \frac{\exp\left(\frac{1}{2\delta^2}\right)-\exp\left(-\frac{1}{\delta^2}\right)}{\exp\left(\frac{1}{2\delta^2}\right)-1},$$
and that this ratio tends to the value 3 as $\delta \to \infty$, in accordance with (2.13).

2.17 A non-increasing continuous frequency function $f(x)$ on the range 0 to ∞ has mean μ and median m. Show that $\mu \geqslant m$ and hence that $\mu \geqslant 1/\{2f(0)\}$.

2.18 As in Exercise 2.17, show that for a continuous, differentiable unimodal distribution $f(x)$, the second moment μ_2' about the mode $\tilde{\mu}$ satisfies $\mu_2' \geqslant [3\{f(\tilde{\mu})\}^2]^{-1}$, and hence $f(\tilde{\mu}) \geqslant (3\mu_2')^{-\frac{1}{2}}$.

2.19 Show that for a distribution with finite mean the mean difference Δ also exists. Illustrate by reference to the so-called Pareto distribution
$$dF = \frac{k}{x^\alpha}dx, \qquad 0<a\leqslant x\leqslant \infty,\ \alpha>1,$$
where the mean may exist but the variance does not exist for $\alpha \leqslant 3$.

2.20 For the distribution of Exercise 1.11 show that if $E(x^2)E(y^2) = E(x^2y^2)$ then the variates are independent.

2.21 In the mean deviation calculation for grouped data in Example 2.7, the interval in which the mean fell made contribution $f\,|d|$ to the sum of deviations. Show that if the frequency f is split into frequencies above and below μ_1' proportionately to the fractions of the interval-width h above and below μ_1', and these frequencies are taken to be concentrated at the centre of their intervals, the sum of deviations is increased by $fh\left(\frac{|d|}{h}-\frac{1}{2}\right)^2$. Confirm that in Example 2.7, this correction increases δ_1 by 0·04.

2.22 Show that for any variate distributed over the range from l (>0) to u,

$$\frac{\mu_1'}{H} \leqslant \frac{(l+u)^2}{4lu},$$

the equality being attainable if and only if half of the frequency lies at each extreme variate-value.

Show similarly that $\sigma^2 \leqslant \frac{1}{4}(u-l)^2$.

(Jacobson (1964) shows that for unimodal distributions $\sigma^2 \leqslant \frac{1}{9}(u-l)^2$.)

2.23 Show that for a set of n ($\geqslant 2$) non-negative values x_i, not all equal, the coefficient of variation V defined at (2.28) cannot exceed $(n-1)^{\frac{1}{2}}$, attaining this value if and only if all but one of the x_i are zero.

(Katsnelson and Kotz, 1957)

2.24 Show that for any distribution function $F(x)$ with $F(a) = 0$, $F(b) = 1$,

$$\int_a^b (x-a)\,dF = \int_a^b \{1-F(x)\}\,dx$$

and hence that the means of distributions in the same interval may be compared graphically from cumulative values of their distribution functions. Show further that the variance of x is

$$\sigma^2 = \int_a^b \left[\int_x^b \{1-F(y)\}\,dy\right]^2 \{1-F(x)\}^{-2} dF(x).$$

(Cf. Pyke (1965))

2.25 If a f.f. may be written in the form

$$f(x, \theta) = a(x)\,e^{\theta x}/g(\theta)$$

and $F(x, \theta)$ is the corresponding d.f., show that the mean and variance are

$$\mu = \frac{\partial \log g(\theta)}{\partial \theta}, \quad \mu_2 = \frac{\partial \mu}{\partial \theta}$$

and that

$$\frac{\partial f}{\partial \theta} = (x-\mu)f.$$

Hence show that if x varies continuously, the mean deviation is

$$\delta_1 = -2\int_{-\infty}^{\mu} \frac{\partial f}{\partial \theta}\,dx$$

$$= 2\mu_2 f(\mu, \theta) - 2\frac{\partial F(\mu, \theta)}{\partial \theta},$$

so that if $F(\mu, \theta)$ does not depend upon θ,

$$\delta_1 = 2\mu_2 f(\mu, \theta). \quad \text{(A)}$$

In particular show that this holds (a) if f is symmetric about μ, or (b) if θ is a scale parameter not functionally related to μ.

If x takes only non-negative integer values and $m = [\mu]$, show that

$$\delta_1 = -2\sum_{x=0}^{m} \frac{\partial f(x, \theta)}{\delta \theta} = -2\frac{d}{d\theta}F(m, \theta).$$

$$= -2\mu_2 \frac{d}{d\mu} F(m, \theta). \quad \text{(B)}$$

(Cf. Kamat (1965))

CHAPTER 3

MOMENTS AND CUMULANTS

Definition of moments

3.1 In the previous chapter we defined the first moment (arithmetic mean) about an arbitrary point a by the Stieltjes integral

$$\mu_1' = \int_{-\infty}^{\infty} (x-a)\,dF \tag{3.1}$$

and the second moment about the point by

$$\mu_2' = \int_{-\infty}^{\infty} (x-a)^2\,dF. \tag{3.2}$$

In generalization of these equations we may define a series of coefficients μ_r', $r = 1, 2 \ldots$, by the relation

$$\mu_r' = \int_{-\infty}^{\infty} (x-a)^r\,dF. \tag{3.3}$$

μ_r' is called the moment of order r about the point a. When a is the mean μ_1' we write the moment without the prime,

$$\mu_r = \int_{-\infty}^{\infty} (x-\mu_1')^r\,dF, \tag{3.4}$$

and call it a central moment or a moment about the mean. In particular $\mu_1 = 0$, and we may also define a moment of zero order

$$\mu_0' = \mu_0 = \int_{-\infty}^{\infty} dF = 1.$$

It is assumed that when reference is made to the rth moment of a particular distribution, the appropriate integral (3.3) converges for that distribution. As will be seen later, some of the theoretical distributions encountered in statistics do not possess moments of all orders; some, in fact, possess only a few moments of low order, and one or two do not possess any, except of course the moment of order zero. It is easily seen by considering the orders of magnitude of the integrands that if μ_r' exists, so does μ_s' for $s<r$; and that if μ_r' does not exist, nor does μ_s' for $s>r$.

3.2 If a and b are two variate-values, let $b-a = c$ and denote the moments about a and b by $\mu'(a)$ and $\mu'(b)$ respectively. Then we have, by the binomial theorem,

$$(x-a)^r = (x-b+b-a)^r = (x-b+c)^r$$

$$= \sum_{j=0}^{r} \binom{r}{j} (x-b)^{r-j} c^j.$$

Hence

$$\begin{aligned}\mu'_r(a) &= \int_{-\infty}^{\infty} (x-a)^r\, dF \\ &= \int_{-\infty}^{\infty} \sum_{j=0}^{r} \binom{r}{j} (x-b)^{r-j} c^j\, dF \\ &= \sum_{j=0}^{r} \binom{r}{j} c^j \int_{-\infty}^{\infty} (x-b)^{r-j}\, dF \\ &= \sum_{j=0}^{r} \binom{r}{j} \mu'_{r-j}(b)\, c^j.\end{aligned}$$

This equation gives the rth moment about a in terms of the rth and lower moments about b. It may be written in a symbolic form which will be found to provide a useful mnemonic, namely

$$\mu'_r(a) = \{\mu'(b)+c\}^r \tag{3.5}$$

with the convention that the expression on the right is to be expanded binomially and the form $\{\mu'(b)\}^j$ replaced by $\mu'_j(b)$.

The equation (3.5) is of particular importance if one of the values a or b is the mean of the distribution. In this case we have

$$\mu'_r = \sum_{j=0}^{r} \binom{r}{j} \mu_{r-j}\mu_1'^j. \tag{3.6}$$

$$\mu_r = \sum_{j=0}^{r} \binom{r}{j} \mu'_{r-j}(-\mu'_1)^j. \tag{3.7}$$

In particular

$$\left.\begin{aligned}\mu'_2 &= \mu_2+\mu_1'^2 \\ \mu'_3 &= \mu_3+3\mu'_1\mu_2+\mu_1'^3 \\ \mu'_4 &= \mu_4+4\mu'_1\mu_3+6\mu_1'^2\mu_2+\mu_1'^4\end{aligned}\right\} \tag{3.8}$$

and

$$\left.\begin{aligned}\mu_2 &= \mu'_2-\mu_1'^2 \\ \mu_3 &= \mu'_3-3\mu'_1\mu'_2+2\mu_1'^3 \\ \mu_4 &= \mu'_4-4\mu'_1\mu'_3+6\mu_1'^2\mu'_2-3\mu_1'^4\end{aligned}\right\} \tag{3.9}$$

Calculation of moments

3.3 For a distribution specified numerically in a frequency table the calculation of moments of third and higher orders is akin to that of the first and second moments. For grouped data (high order moments are hardly ever required for ungrouped data) the observations are regarded as concentrated at the mid-points of intervals ; a convenient arbitrary origin a is chosen, the moments about a calculated, and then if necessary the moments about the mean are ascertained from (3.6) or (3.7). The effect of grouping may be corrected for in certain cases.

In practice numerical moments of order higher than the fourth are rarely required, being so sensitive to sampling fluctuations that values computed from moderate numbers of observations are subject to a large margin of error.

There are two methods in general use for arriving at the moments about an arbitrary origin. The first is an immediate generalization of the methods used in Chapter 2 for the first two moments. The second will be considered in **3.10** in connection with factorial moments.

Example 3.1

To find the first four moments about the mean of the distribution of Australian marriages of Table 1.8.

Until the last stage we work in units of three years, the variate interval. A working mean is taken at 28·5 years. To check the arithmetic we use an identity of type

$$(x+1)^3 = x^3+3x^2+3x+1,$$
$$(x+1)^4 = x^4+4x^3+6x^2+4x+1.$$

Thus, for instance, the value of $g(x)(x+1)^r$ is found in addition to that of $g(x)x^r$ and the two checked by identities such as

$$\Sigma g(x)(x+1)^3 = \Sigma g(x)x^3+3\Sigma g(x)x^2+3\Sigma g(x)x+\Sigma g(x),$$

$g(x)$ being the actual frequencies. The arithmetical work is shown in Table 3.1.

Table 3.1—Calculation of the first four moments of the distribution of marriages of Table 1.8

Mid-value of inter-vals, years		x	xg	$(x+1)g$	x^2g	$(x+1)^2g$	x^3g	$(x+1)^3g$	x^4g	$(x+1)^4g$
16·5	294	−4	− 1,176	− 882	4,704	2,646	− 18,816	− 7,938	75,264	23,814
19·5	10,995	−3	− 32,985	−21,990	98,955	43,980	−296,865	− 87 960	890,595	175,920
22·5	61,001	−2	−122,002	−61,001	244,004	61,001	−488,008	− 61,001	976,016	61,001
25·5	73,054	−1	− 73,054	−83,873	73,054	—	− 73,054	−156,899	73,054	—
28·5	56,501	0	−229,217	56,501	—	56,501	−876,743	56,501	—	56,501
31·5	33,478	1	33,478	66,956	33,478	133,912	33,478	267,824	33,478	535,648
34·5	20,569	2	41,138	61,707	82,276	185,121	164,552	555,363	329,104	1,666,089
37·5	14,281	3	42,843	57,124	128,529	228,496	385,587	913,984	1,156,761	3,655,936
40·5	9,320	4	37,280	46,600	149,120	233,000	596,480	1,165,000	2,385,920	5,825,000
43·5	6,236	5	31,180	37,416	155,900	224,496	779,500	1,346,976	3,897,500	8,081,856
46·5	4,770	6	28,620	33,390	171,720	233,730	1,030,320	1,636,110	6,181,920	11,452,770
49·5	3,620	7	25,340	28,960	177,380	231,680	1,241,660	1,853,440	8,691,620	14,827,520
52·5	2,190	8	17,520	19,710	140,160	177,390	1,121,280	1,596,510	8,970,240	14,368,590
55·5	1,655	9	14,895	16,550	134,055	165,500	1,206,495	1,655,000	10,858,455	16,550,000
58·5	1,100	10	11,000	12,100	110,000	133,100	1,100,000	1,464,100	11,000,000	16,105,100
61·5	810	11	8,910	9,720	98,010	116,640	1,078,110	1,399,680	11,859,210	16,796,160
64·5	649	12	7,788	8,437	93,456	109,681	1,121,472	1,425,853	13,457,664	18,536,089
67·5	487	13	6,331	6,818	82,303	95,452	1,069,939	1,336,328	13,909,207	18,708,592
70·5	326	14	4,564	4,890	63,896	73,350	894,544	1,100,250	12,523,616	16,503,750
73·5	211	15	3,165	3,376	47,475	54,016	712,125	864,256	10,681,875	13,828,096
76·5	119	16	1,904	2,023	30,464	34,391	487,424	584,647	7,798,784	9,938,999
79·5	73	17	1,241	1,314	21,097	23,652	358,649	425,736	6,097,033	7,663,248
82·5	27	18	486	513	8,748	9,747	157,464	185,193	2,834,352	3,518,667
85·5	14	19	266	280	5,054	5,600	96,026	112,000	1,824,494	2,240,000
88·5	5	20	100	105	2,000	2,205	40,000	46,305	800,000	972,405
TOTALS OF +VE TERMS	301,785	—	318,049	474,490	2,155,838	2,635,287	13,675,105	19,991,056	137,306,162	202,091,751

From this table we find

$$\begin{aligned}\Sigma(xg) &= 88{,}832\\ \Sigma(x^2g) &= 2{,}155{,}838\\ \Sigma(x^3g) &= 12{,}798{,}362\\ \Sigma(x^4g) &= 137{,}306{,}162.\end{aligned}$$

The values will be found to check and we have, about the working mean, on dividing by the total frequency 301,785,

$$\begin{aligned}\mu_1' &= 0{\cdot}294{,}355{,}253\\ \mu_2' &= 7{\cdot}143{,}622{,}115\\ \mu_3' &= 42{\cdot}408{,}873{,}867\\ \mu_4' &= 454{\cdot}980{,}075{,}219.\end{aligned}$$

For the moments about the mean, substitution in equations (3.9) gives

$$\begin{aligned}\mu_2 &= 7{\cdot}056{,}977\\ \mu_3 &= 36{\cdot}151{,}595\\ \mu_4 &= 408{\cdot}738{,}210.\end{aligned}$$

These are expressed in class-intervals, which are units of three years. To express the results in units of one year we multiply the rth moment by 3^r, e.g.

$$\mu_2 = 7{\cdot}056{,}977 \times 9 = 63{\cdot}512{,}79.$$

3.4 If a distribution is specified mathematically the determination of moments is equivalent to the evaluation of certain sums or integrals. It is usually necessary to consider whether the moments exist. Some examples will illustrate the general principles involved.

Example 3.2

Consider the binomial distribution of Examples 2.3 and 2.9. We will write $q = 1-p$ so that the generating function of the distribution is $P(t) = (pt+q)^n$. In the manner of those examples we have

$$\begin{aligned}\left[\frac{d^3P}{dt^3}\right]_{t=1} &= \sum_{j=0}^{n} j(j-1)(j-2)f_j\\ &= \Sigma(j^3f_j)-3\Sigma(j^2f_j)+2\Sigma(jf_j)\\ &= \mu_3'-3\mu_2'+2\mu_1',\end{aligned}$$

the moments being measured from zero origin. This gives us

$$\mu_3' = 3\mu_2'-2\mu_1'+n(n-1)(n-2)p^3.$$

We have already found that

$$\mu_2' = n(n-1)p^2+np, \qquad \mu_1' = np,$$

and hence we find, after substitution and reduction,

$$\mu_3' = n(n-1)(n-2)p^3+3n(n-1)p^2+np.$$

We are usually more interested in the moments about the mean, and from (3.9) we have

$$\mu_3 = \mu_3'-3\mu_1'\mu_2'+2\mu_1'^3$$

which, on substitution on the right, reduces to

$$\mu_3 = np(p-1)(2p-1) \tag{3.10}$$

$$= npq(q-p). \tag{3.11}$$

In a similar manner we find

$$\mu_4 = 3n^2p^2(1-p)^2+np(1-p)(1-6p+6p^2) \tag{3.12}$$

$$= 3n^2p^2q^2+npq(1-6pq). \tag{3.13}$$

The process may evidently be extended to give as many moments as are desired. But we shall see later that there are easier ways of finding the moments of the binomial distribution.

Example 3.3

Consider the distribution

$$dF = \frac{k}{(1+x^2)^m}dx, \qquad -\infty \leqslant x \leqslant \infty;\ m \geqslant 1.$$

This is a unimodal distribution symmetrical about $x = 0$. All finite moments of odd order about the origin therefore vanish. The constant k is given by the equation

$$1 = k\int_{-\infty}^{\infty} \frac{dx}{(1+x^2)^m}$$

$$= k\frac{\Gamma(\frac{1}{2})\Gamma(m-\frac{1}{2})}{\Gamma(m)}.$$

The moment about the mean of order r, if it exists, is given by

$$\mu_r = k\int_{-\infty}^{\infty} \frac{x^r}{(1+x^2)^m}dx,$$

and this integral converges if and only if

$$2m-r > 1.$$

Thus μ_r exists if and only if $r < 2m-1$.

If $m = 1$, the integral defining the mean, $\lim_{n\to\infty,\, n'\to\infty} \int_{-n}^{n'} \frac{kx\,dx}{(1+x^2)}$, does not converge, although the principal value

$$I = \lim_{n\to\infty} \int_{-n}^{n} \frac{kx\,dx}{(1+x^2)}$$

does and is equal to zero. Thus even the mean does not exist in this case.(*) For $m > 1$ the mean exists and is located at the origin.

Making the substitution $z = \frac{1}{1+x^2}$ in the formula for μ_{2r}, we find

---

(*) A disadvantage of the convention that the distribution has a mean for $m = 1$ because I is finite is that it is no longer necessarily true that the mean of a sum is the sum of the means. Cf. Fréchet (1937), page 45, and Exercise 7.19 below.

$$\mu_{2r} = k\int_0^1 (1-z)^{r-\frac{1}{2}} z^{m-r-\frac{3}{2}} dz$$

$$= k\frac{\Gamma(r+\frac{1}{2})\Gamma(m-r-\frac{1}{2})}{\Gamma(m)}$$

and on substituting for k,

$$\mu_{2r} = \frac{\Gamma(r+\frac{1}{2})\Gamma(m-r-\frac{1}{2})}{\Gamma(\frac{1}{2})\Gamma(m-\frac{1}{2})} \qquad 2r < 2m-1.$$

Example 3.4

Consider the " normal " distribution

$$dF = \frac{1}{\sigma\sqrt{(2\pi)}} e^{-\frac{x^2}{2\sigma^2}} dx, \qquad -\infty \leqslant x \leqslant \infty. \tag{3.14}$$

This is symmetrical about the origin. All moments exist, those of odd order vanishing. Thus

$$\mu_{2r} = \frac{1}{\sigma\sqrt{(2\pi)}}\int_{-\infty}^{\infty} x^{2r} e^{-\frac{x^2}{2\sigma^2}} dx.$$

This may be evaluated by partial integration, but a more direct method is as follows.

Consider the integral

$$M(t) = \frac{1}{\sigma\sqrt{(2\pi)}}\int_{-\infty}^{\infty} e^{tx} e^{-\frac{x^2}{2\sigma^2}} dx$$

$$= \exp(\tfrac{1}{2}\sigma^2 t^2). \tag{3.15}$$

We have, for all real values of t,

$$e^{tx} e^{-\frac{x^2}{2\sigma^2}} = \sum_{r=0}^{\infty}\left(\frac{t^r}{r!} x^r e^{-\frac{x^2}{2\sigma^2}}\right).$$

The series on the right is uniformly convergent in x and may be integrated term by term if the resulting series is uniformly convergent. We then have

$$M(t) = \sum_{r=0}^{\infty}\left(\frac{t^r}{r!}\mu_r\right).$$

In other words, μ_r is the coefficient of $t^r/r!$ in $\exp(\frac{1}{2}\sigma^2 t^2)$ and hence

$$\mu_{2r} = \frac{\sigma^{2r}}{2^r}\frac{(2r)!}{r!}. \tag{3.16}$$

Evidently, we may obtain μ_r from $M(t)$ by differentiating r times with respect to t and putting $t = 0$.

Moment-generating functions and characteristic functions

3.5 The previous example shows that in some cases we can derive from the distribution- or the frequency-function a function $M(t)$ which, when expanded in powers of t, will yield the moments of the distribution as the coefficients of those powers. Such a function is accordingly called a moment-generating function. If the frequency-generating function is $P(t)$, the moment-generating function is simply $P(e^t)$. The name is conveniently abbreviated to m.g.f.

For many frequency functions the integral $\int_{-\infty}^{\infty} e^{tx}\, dF$ or the sum $\{\Sigma e^{tx_j} f(x_j)\}$ does not exist for some or all real values of t. This is, for example, true of the function $dF = k(1+x^2)^{-m}\, dx$ for finite positive values of m. A more serviceable auxiliary function is

$$\phi(t) = \int_{-\infty}^{\infty} e^{itx}\, dF \quad (t \text{ real}) \tag{3.17}$$

where i is the complex operator ($i^2 = -1$). This is known as the characteristic function (often abbreviated to c.f.) and is of great theoretical importance. It will be seen in Chapter 4 that under certain general conditions the characteristic function determines and is completely determined by the distribution function. It also yields many valuable results in the theory of sampling.

Since by the nature of the distribution function the integral $\int_{-\infty}^{\infty} dF$ converges,

$$|\phi(t)| \leqslant \int_{-\infty}^{\infty} |e^{itx}|\, dF = 1$$

and hence the Stieltjes integral (3.17) converges absolutely and uniformly in t. It may therefore be integrated under the summation signs with respect to t, and may be differentiated provided that the resulting expressions exist and are uniformly convergent. We have, for example, writing D_t for $\frac{d}{dt}$,

$$D_t^r \phi(t) = i^r \int_{-\infty}^{\infty} e^{itx} x^r\, dF,$$

and hence, putting $t = 0$,

$$\mu_r = (-i)^r [D_t^r \phi(t)]_{t=0} \tag{3.18}$$

provided that μ_r' exists. If $\phi(t)$ be expanded in powers of t, μ_r' must thus be equal to the coefficient of $(it)^r/r!$ in the expansion. Thus the characteristic function is also a moment-generating function. For many formal purposes it is sufficient to write, say, θ instead of it and to ignore the fact that it is imaginary.

Our earlier m.g.f., $M(t)$, may similarly be differentiated to obtain the moments if it exists in an open interval including the origin.

Example 3.5

Consider again the binomial distribution of Example 3.2. The frequency-generating function is $P(t) = (pt + q)^n$ and hence a moment-generating function is

$$M(t) = P(e^t) = (pe^t + q)^n.$$

Hence

$$\mu_1' = \left[\frac{d}{dt} M(t)\right]_{t=0} = np.$$

$$\mu_2' = \left[\frac{d^2}{dt^2} M(t)\right]_{t=0} = np + n(n-1)p^2$$

and so on.

Example 3.6

Consider the distribution

$$dF = \frac{a^\gamma}{\Gamma(\gamma)} e^{-ax} x^{\gamma-1} dx, \qquad 0 \leqslant x \leqslant \infty; \ a > 0, \ \gamma > 0,$$

which is called a Gamma distribution (cf. **6.9** below). The distribution may have a variety of shapes, depending on the value of γ, but moments of all orders exist in virtue of the convergence of the integral $\int_0^\infty e^{-ax} x^r dx$, the Γ-function integral, for $r > -1$. We have then, for the characteristic function,

$$\phi(t) = \frac{a^\gamma}{\Gamma(\gamma)} \int_0^\infty e^{x(-a+it)} x^{\gamma-1} dx.$$

By the substitution $z = x(a-it)$ this becomes

$$\begin{aligned} \phi(t) &= \frac{a^\gamma}{\Gamma(\gamma)(a-it)^\gamma} \int_0^\infty e^{-z} z^{\gamma-1} dz \\ &= \frac{1}{\left(1-\frac{it}{a}\right)^\gamma}, \end{aligned} \tag{3.19}$$

since $\int_0^\infty e^{-z} z^{\gamma-1} dz = \Gamma(\gamma)$ whether z is real or complex.

Hence

$$\phi(t) = 1 + \gamma\frac{it}{a} + \frac{\gamma(\gamma+1)}{2!}\left(\frac{it}{a}\right)^2 + \dots$$

and thus

$$\mu_1' = \frac{\gamma}{a},$$

$$\mu_2' = \frac{\gamma(\gamma+1)}{a^2},$$

$$\mu_3' = \frac{\gamma(\gamma+1)(\gamma+2)}{a^3},$$

and so on. In particular,

$$\left.\begin{aligned} \mu_2 &= \frac{\gamma}{a^2}, \\ \mu_3 &= \frac{2\gamma}{a^3}. \end{aligned}\right\} \tag{3.20}$$

Absolute moments

3.6 The quantity

$$\nu_r' = \int_{-\infty}^{\infty} |x-a|^r dF \tag{3.21}$$

is called the absolute moment of order r about a. The absolute moments about the mean are written without primes. Clearly if the moment of order r exists, the absolute moment exists for any order less than or equal to r.

If r is even, the absolute moment is equal to the ordinary moment, and if the range of the distribution is positive the absolute moments about any point to the left of the start of the distribution are equal to the ordinary moments of corresponding order.

There are some interesting inequalities concerning the absolute moments. Referring to the function $A(t)$ of **2.6** and remembering that it is a non-decreasing function of t, we find, on putting $t = 1, 2, \ldots$, that

$$(\nu_1')^{1/1} \leqslant (\nu_2')^{\frac{1}{2}} \leqslant (\nu_3')^{\frac{1}{3}} \ldots \leqslant (\nu_r')^{1/r}.$$

A more general inequality, due to Liapounoff (1901), is

$$(\nu_b')^{a-c} \leqslant (\nu_c')^{a-b} \, (\nu_a')^{b-c}, \qquad a \geqslant b \geqslant c \geqslant 0. \tag{3.22}$$

A proof of this result is given in Exercise 3.15.

Factorial moments

3.7 The factorial expression

$$x(x-h)(x-2h)\ldots\{x-(r-1)h\}$$

may conveniently be written $x^{[r]}$, a notation which brings out an analogy with the power x^r. Taking first differences with respect to x and with unit h, we have

$$\begin{aligned}\Delta x^{[r]} &= (x+h)^{[r]} - x^{[r]} \\ &= (x+h)x(x-h)\ldots\{x-(r-2)h\} - x(x-h)\ldots\{x-(r-1)h\} \\ &= rx^{[r-1]}h,\end{aligned}$$

which may be compared with the equation in differentials

$$dx^r = rx^{r-1}\,dx.$$

Conversely

$$\sum_{x=0}^{x} x^{[r]} = \frac{1}{(r+1)h}(x+h)^{[r+1]}$$

corresponding to

$$\int_0^x x^r dx = \frac{1}{r+1} x^{r+1}.$$

The rth factorial moment about an arbitrary origin may then be defined by the equation

$$\mu'_{[r]} = \sum_{j=-\infty}^{\infty} (x_j - a)^{[r]} f(x_j) \tag{3.23}$$

where we have chosen the summation sign Σ rather than the Stieltjes integral because it is almost entirely for discontinuous distributions, or continuous distributions grouped in intervals of width h, that the factorial moments are used. In statistical theory they are not very prominent, but they provide very concise formulae for the moments of certain discontinuous distributions of the binomial type.

As usual, when it is necessary to distinguish between factorial moments about the mean and those about an arbitrary point we may write the former without the prime.

3.8 The factorial moments obey laws of transformation similar to those of equation (3.5) governing ordinary moments. In fact we have the expansion (*)

$$(a+b)^{[r]} = \sum_{j=0}^{r} \binom{r}{j} a^{[r-j]} b^{[j]}$$

and hence $$(x-a)^{[r]} = (x-b+c)^{[r]}, \text{ where } c = b-a,$$

$$= \sum_{j=0}^{r} \binom{r}{j} (x-b)^{[r-j]} c^{[j]}$$

and hence $$\mu'_{[r]}(a) = \sum_{j=0}^{r} \binom{r}{j} \mu'_{[r-j]}(b)\, c^{[j]} \qquad (3.24)$$

which may be written symbolically

$$\mu'_{[r]}(a) = \{\mu'(b)+c\}^{[r]}.$$

3.9 By direct expansion of (3.23) it is seen that

$$\left.\begin{aligned} \mu'_{[1]} &= \mu'_1 \\ \mu'_{[2]} &= \mu'_2 - h\mu'_1 \\ \mu'_{[3]} &= \mu'_3 - 3h\mu'_2 + 2h^2\mu'_1 \\ \mu'_{[4]} &= \mu'_4 - 6h\mu'_3 + 11h^2\mu'_2 - 6h^3\mu'_1 \end{aligned}\right\} \qquad (3.25)$$

and conversely that

$$\left.\begin{aligned} \mu'_2 &= \mu'_{[2]} + h\mu'_{[1]} \\ \mu'_3 &= \mu'_{[3]} + 3h\mu'_{[2]} + h^2\mu'_{[1]} \\ \mu'_4 &= \mu'_{[4]} + 6h\mu'_{[3]} + 7h^2\mu'_{[2]} + h^3\mu'_{[1]} \end{aligned}\right\} \qquad (3.26)$$

Since the first moments are equal the equations remain true when the primes are dropped and terms in first moments omitted.

Frisch (1926) gives general formulae showing the factorial moments about one point in terms of the ordinary moments about another, and vice versa. In fact

$$\mu'_r(a) = \sum_{j=0}^{r} \left\{ \binom{r}{j} B_{r-j}^{(-j)}(c) h^{r-j} \mu'_{[j]}(b) \right\} \qquad (3.27)$$

$$\mu'_{[r]}(a) = \sum_{j=0}^{r} \left\{ \binom{r}{j} B_{r-j}^{(r+1)}(c+h) h^{r-j} \mu'_j(b) \right\} \qquad (3.28)$$

where $B_r^{(n)}(x)$ is the Bernoulli polynomial of order n and degree r in x (see **3.25** below).

Calculation of factorial moments

3.10 The calculation of factorial moments for grouped data may be effected by a process of progressive summation which is illustrated in Table 3.2.

(*) It is clear that $(a+b)^{[r]}$ will be a polynomial of degree r in a, and may therefore be equated to $\sum_{j=0}^{r} k_j a^{[r-j]}$, where the k's are polynomials in b and h but do not contain a. Putting $a = 0$ we obtain $b^{[r]} = k_r$. Taking first differences with respect to a and putting $a = 0$ we obtain $rb^{[r-1]} = k_{r-1}$. Successive differences give the k's and the above result follows.

Table 3.2

(1) Frequency	(2) First summation	(3) Second summation	(4) Third summation
f_1	$f_1 + \ldots + f_n$		
f_2	$f_2 + \ldots + f_n$	$f_2 + 2f_3 + \ldots + (n-1)f_n$	
f_3	$f_3 + \ldots + f_n$	$f_3 + 2f_4 + \ldots + (n-2)f_n$	$f_3+3f_4+\ldots+\frac{(n-1)(n-2)}{2}f_n$
f_4	$f_4 + f_5 + \ldots + f_n$	$+ \ldots + (n-3)f_n$	$f_4 + 3f_5+\ldots+\frac{(n-2)(n-3)}{2}f_n$
.	.	.	
.	.	.	
.	.	.	.
f_{n-2}	$f_{n-2} + f_{n-1} + f_n$	$f_{n-2} + 2f_{n-1} + 3f_n$	$f_{n-2} + 3f_{n-1} + 6f_n$
f_{n-1}	$f_{n-1} + f_n$	$f_{n-1} + 2f_n$	$f_{n-1} + 3f_n$
f_n	f_n	f_n	f_n
TOTALS	$\Sigma(jf_j) = \mu'_{[1]}$	$\Sigma\left\{\frac{j(j-1)}{2}f_j\right\} = \frac{1}{2!}\mu'_{[2]}$	$\Sigma\left\{\frac{j(j-1)(j-2)}{3!}f_j\right\} = \frac{1}{3!}\mu'_{[3]}$

For simplicity, we put $a = 0$ and take the class-interval h as unit.

Writing the proportional frequencies in the successive n intervals as $f_1 \ldots f_n$, as shown in the left-hand column, we construct column 2 by adding frequencies from the bottom. In the nth row we write f_n, in the $(n-1)$th row the sum f_n+f_{n-1}, in the $(n-2)$th row the sum $f_n+f_{n-1}+f_{n-2}$, and so on, the first row containing the sum $f_n+f_{n-1}+\ldots+f_1$.

In column 3 the process is repeated with the rows of column 2, stopping at the second row, e.g. the nth row contains f_n, the $(n-1)$th row $(f_n+f_{n-1})+f_n = 2f_n+f_{n-1}$, and so on, the second row containing the sum $(n-1)f_n+(n-2)f_{n-1}+\ldots+2f_3+f_2$.

Column 4 repeats the process with the entries of column 3, but stopping at the third row; and so on.

Consider now the sum of the entries in column 2. In that sum f_1 appears once, f_2 twice, . . . , f_n n times. Hence the sum is equal to $\sum_{j=1}^{n} jf_j = \mu'_{[1]}$.

In column 3, f_2 appears once, f_3 appears 3 times, . . . , f_n appears $\frac{1}{2}n(n-1)$ times. Hence

$$\text{Sum} = \sum_{j=1}^{n}\left(\frac{j(j-1)}{2}f_j\right)$$
$$= \tfrac{1}{2}\mu'_{[2]}.$$

In general, the sum of the $(r+1)$th column will be given by

$$\text{Sum} = \frac{1}{r!}\mu'_{[r]}.$$

If the actual frequencies are used instead of the proportional frequencies the sums have to be divided by the total frequency N.

Thus the process of summation gives the factorial moments directly.

Example 3.7

Consider again the data of Table 1.7, showing the distribution of 8585 men according to height in inches. The columns on the right in Table 3.3 show the successive sums.

At the top of the last three columns there has been placed within brackets the number which would have been obtained if the summation were continued up the column one place further than is required for the sum at the foot. These bracketed figures are useful to have as a check since each must equal the sum at the foot of the preceding column.

Table 3.3—Calculation of the factorial moments of a distribution of men according to height in inches (*Table 1.7*)

Height	Frequency	First sum	Second sum	Third sum	Fourth sum
57–	2	8,585	(94,614)	—	—
58–	4	8,583	86,029	(502,459)	—
59–	14	8,579	77,446	416,430	(1,709,785)
60–	41	8,565	68,867	338,984	1,293,355
61–	83	8,524	60,302	270,117	954,371
62–	169	8,441	51,778	209,815	684,254
63–	394	8,272	43,337	158,037	474,439
64–	669	7,878	35,065	114,700	316,402
65–	990	7,209	27,187	79,635	201,702
66–	1,223	6,219	19,978	52,448	122,067
67–	1,329	4,996	13,759	32,470	69,619
68–	1,230	3,667	8,763	18,711	37,419
69–	1,063	2,437	5,096	9,948	18,438
70–	646	1,374	2,659	4,852	8,490
71–	392	728	1,285	2,193	3,638
72–	202	336	557	908	1,445
73–	79	134	221	351	537
74–	32	55	87	130	186
75–	16	23	32	43	56
76–	5	7	9	11	13
77–	2	2	2	2	2
TOTALS	8,585	94,614	502,459	1,709,785	4,186,163

From this table we find

$$\begin{aligned} \mu'_{[1]} &= 11{\cdot}020{,}850{,}320{,}33 \\ \mu'_{[2]} &= 117{\cdot}055{,}096{,}097{,}84 \\ \mu'_{[3]} &= 1{,}194{\cdot}957{,}483{,}983{,}69 \\ \mu'_{[4]} &= 11{,}702{\cdot}727{,}082{,}119{,}98. \end{aligned}$$

From these values we may derive the ordinary moments, using equations (3.26) and find

$$\begin{aligned} \mu'_1 &= 11{\cdot}020{,}850{,}320{,}33 \\ \mu'_2 &= 128{\cdot}075{,}946{,}418{,}2 \\ \mu'_3 &= 1{,}557{\cdot}143{,}622{,}597{,}5 \\ \mu'_4 &= 19{,}702{\cdot}878{,}509{,}027{,}3, \end{aligned}$$

from which we find, for the moments about the mean,

$$\begin{aligned} \mu_2 &= 6{\cdot}616{,}805 \\ \mu_3 &= -0{\cdot}207{,}840 \\ \mu_4 &= 137{\cdot}689{,}185, \end{aligned}$$

the units being one inch.

Factorial moment-generating functions

3.11 If the frequency-generating function is given by

$$P(t) = \sum_{j=0}^{\infty} f_j t^j$$

we have, replacing t by $1+t$,

$$P(1+t) = \Sigma f_j(1+t)^j = \sum_{j=0}^{\infty} f_j \sum_{i=1}^{j} \binom{j}{i} t^i$$

$$= \sum_{i=0}^{\infty} \frac{t^i}{i!} \sum_{j=0}^{\infty} [f_j j(j-1)\ldots\{j-i+1\}]$$

$$= \sum_{i=0}^{\infty} (t^i \mu'_{[i]}/i!). \tag{3.29}$$

Hence $P(1+t)$ is a factorial moment-generating function (f.m.g.f.), the factorial moment of order r being the coefficient of $t^r/r!$ The moments here are taken with a unit interval between the frequencies.

The results obtained by differentiating $P(t)$ and putting $t = 1$, in **2.4** and Example 2.9, are also obtained by differentiating $P(1+t)$ and putting $t = 0$. They are, as we now see, the factorial moments.

Example 3.8

Consider again the binomial distribution of Example 3.5, with $P(t) = \{pt+(1-p)\}^n$. We have

$$P(1+t) = (1+pt)^n = \sum_{j=0}^{n} p^j t^j \binom{n}{j}$$

and hence, putting $t = 0$,

$$\mu'_{[r]} = p^r n^{[r]}, \quad r \leqslant n,$$
$$= 0, \quad r > n.$$

We can now convert to the ordinary moments, if desired, by equations (3.26). For instance

$$\mu'_1 = \mu_{[1]} = pn,$$
$$\mu'_2 = p^2 n(n-1)+pn, \text{ etc.}$$

Cumulants

3.12 The moments are a set of descriptive constants of a distribution which are useful for measuring its properties and, in certain circumstances, for specifying it. Their use in these connections will be considered in later chapters. They are not, however, the only set of constants for the purpose, or even the best set. Another series of constants, the so-called cumulants, have properties which are more useful from the theoretical standpoint.

Formally, the cumulants $\kappa_1, \kappa_2, \ldots, \kappa_r$ are defined by the identity in t

$$\exp\left\{\kappa_1 t+\frac{\kappa_2 t^2}{2!}+\ldots+\frac{\kappa_r t^r}{r!}+\ldots\right\} = 1+\mu'_1 t+\frac{\mu'_2 t^2}{2!}+\ldots+\frac{\mu'_r t^r}{r!}+\ldots \tag{3.30}$$

It is sometimes more convenient to write the same equation with it for t, thus:

$$\exp\left\{\kappa_1(it)+\kappa_2\frac{(it)^2}{2!}+\ldots+\kappa_r\frac{(it)^r}{r!}+\ldots\right\} = 1+\mu'_1\frac{(it)}{1!}+\ldots+\mu'_r\frac{(it)^r}{r!}+\ldots$$
$$= \int_{-\infty}^{\infty} e^{itx}\, dF$$
$$= \phi(t). \tag{3.31}$$

Thus, whereas μ'_r is the coefficient of $(it)^r/r!$ in $\phi(t)$, the characteristic function, κ_r is the coefficient of $(it)^r/r!$ in $\log \phi(t)$, if an expansion in power series exists. Log $\phi(t)$

may then be called a cumulant-generating function and denoted by c.g.f. "Cumulative function" is an older synonym for c.g.f.

3.13 If in equation (3.31) the origin is changed from a to b, where as usual $b-a = c$, the effect on $\phi(t)$ is to multiply it by e^{-itc}, for $\int e^{itx}\,dF$ becomes $\int e^{it(x-c)}\,e^{itc}\,dF$. Hence the effect on $\log \phi(t)$ is merely to add the term $-itc$, and consequently the coefficients in $\log \phi(t)$ are unchanged, except the first, which is decreased by c.

Hence the cumulants are invariant under change of origin, except the first. In this they stand in sharp contrast to the moments about an arbitrary point.

Both cumulants and moments have another property of an invariantive kind, namely, that if the variate-values are multiplied by a constant a, μ_r' and κ_r are multiplied by a^r. This is at once evident from their definitions. Thus any linear transformation of the kind

$$\xi = lx+m \tag{3.32}$$

leaves the cumulants unchanged so far as the constant m is concerned and multiplies κ_r by l^r. The sole exception is the first cumulant, which is equal to the mean. In particular, if we transform a distribution to standard measure, the only effect is to multiply κ_r by σ^{-r}, σ being the standard deviation and, as we shall see in a moment, being equal to $\kappa_2^{\frac{1}{2}}$.

The invariantive properties of the cumulants were the origin of their original name of semi-invariants, seminvariants or half-invariants (Thiele, 1903). In accordance with the theory of algebraic invariants, however, it seems best to reserve the word "seminvariant" for any constant λ_r which, under the transformation (3.32), is multiplied by l^r. The cumulants and the moments *about the mean* are thus particular cases of seminvariants.

Relations between moments and cumulants

3.14 Subject to conditions of existence, we have, from (3.30),

$$\begin{aligned}
1+\mu_1'\frac{t}{1!}+\ldots+\mu_r'\frac{t^r}{r!}+\ldots &= \exp\left\{\kappa_1\frac{t}{1!}+\kappa_2\frac{t^2}{2!}+\ldots\kappa_r\frac{t^r}{r!}\ldots\right\}\\
&= \exp\left(\frac{\kappa_1 t}{1!}\right)\exp\left(\frac{\kappa_2 t^2}{2!}\right)\ldots\exp\left(\frac{\kappa_r t^r}{r!}\right)\ldots\\
&= \left\{1+\frac{\kappa_1 t}{1!}+\frac{\kappa_1^2 t^2}{2!}+\ldots\right\}\left\{1+\frac{\kappa_2 t^2}{2!}+\frac{1}{2!}\left(\frac{\kappa_2 t^2}{2!}\right)^2+\ldots\right\}\\
&\qquad\ldots\left\{1+\left(\frac{\kappa_r t^r}{r!}\right)+\frac{1}{2!}\left(\frac{\kappa_r t^r}{r!}\right)^2+\ldots\right\}\ldots
\end{aligned}$$

Picking out the terms in the exponential expansions which, when multiplied together, give a power of t^r, we have

$$\mu_r' = \sum_{m=0}^{r}\sum\left(\frac{\kappa_{p_1}}{p_1!}\right)^{\pi_1}\left(\frac{\kappa_{p_2}}{p_2!}\right)^{\pi_2}\ldots\left(\frac{\kappa_{p_m}}{p_m!}\right)^{\pi_m}\frac{r!}{\pi_1!\pi_2!\ldots\pi_m!}, \tag{3.33}$$

where the second summation extends over all non-negative values of the π's such that

$$p_1\pi_1+p_2\pi_2+\ldots p_m\pi_m = r. \tag{3.34}$$

It is worth noting that the rather tedious process of writing down the explicit relations for particular values of r may be shortened considerably. In fact, differentiating (3.30) with respect to κ_j we have

$$\frac{t^j}{j!}\left(1+\mu'_1 t+\ldots+\frac{\mu'_r t^r}{r!}+\ldots\right) = \frac{\partial \mu'_1}{\partial \kappa_j}t+\ldots+\frac{t^r}{r!}\frac{\partial \mu'_r}{\partial \kappa_j}+\ldots$$

and hence, identifying powers of t,

$$\frac{\partial \mu'_r}{\partial \kappa_j} = \binom{r}{j}\mu'_{r-j}. \tag{3.35}$$

In particular

$$\frac{\partial \mu'_r}{\partial \kappa_1} = r\mu'_{r-1} \tag{3.36}$$

and thus, given any μ'_r in terms of the κ's, we can write down successively those of lower orders by a differentiation.(*)

The first ten of these expressions are, for moments about an arbitrary point :—

$$\left.\begin{aligned}
\mu'_1 &= \kappa_1, \\
\mu'_2 &= \kappa_2 + \kappa_1^2, \\
\mu'_3 &= \kappa_3 + 3\kappa_2\kappa_1 + \kappa_1^3, \\
\mu'_4 &= \kappa_4 + 4\kappa_3\kappa_1 + 3\kappa_2^2 + 6\kappa_2\kappa_1^2 + \kappa_1^4, \\
\mu'_5 &= \kappa_5 + 5\kappa_4\kappa_1 + 10\kappa_3\kappa_2 + 10\kappa_3\kappa_1^2 + 15\kappa_2^2\kappa_1 + 10\kappa_2\kappa_1^3 + \kappa_1^5, \\
\mu'_6 &= \kappa_6 + 6\kappa_5\kappa_1 + 15\kappa_4\kappa_2 + 15\kappa_4\kappa_1^2 + 10\kappa_3^2 + 60\kappa_3\kappa_2\kappa_1 + 20\kappa_3\kappa_1^3 \\
&\quad + 15\kappa_2^3 + 45\kappa_2^2\kappa_1^2 + 15\kappa_2\kappa_1^4 + \kappa_1^6, \\
\mu'_7 &= \kappa_7 + 7\kappa_6\kappa_1 + 21\kappa_5\kappa_2 + 21\kappa_5\kappa_1^2 + 35\kappa_4\kappa_3 + 105\kappa_4\kappa_2\kappa_1 \\
&\quad + 35\kappa_4\kappa_1^3 + 70\kappa_3^2\kappa_1 + 105\kappa_3\kappa_2^2 + 210\kappa_3\kappa_2\kappa_1^2 + 35\kappa_3\kappa_1^4 \\
&\quad + 105\kappa_2^3\kappa_1 + 105\kappa_2^2\kappa_1^3 + 21\kappa_2\kappa_1^5 + \kappa_1^7, \\
\mu'_8 &= \kappa_8 + 8\kappa_7\kappa_1 + 28\kappa_6\kappa_2 + 28\kappa_6\kappa_1^2 + 56\kappa_5\kappa_3 + 168\kappa_5\kappa_2\kappa_1 + 56\kappa_5\kappa_1^3 \\
&\quad + 35\kappa_4^2 + 280\kappa_4\kappa_3\kappa_1 + 210\kappa_4\kappa_2^2 + 420\kappa_4\kappa_2\kappa_1^2 + 70\kappa_4\kappa_1^4 \\
&\quad + 280\kappa_3^2\kappa_2 + 280\kappa_3^2\kappa_1^2 + 840\kappa_3\kappa_2^2\kappa_1 + 560\kappa_3\kappa_2\kappa_1^3 + 56\kappa_3\kappa_1^5 \\
&\quad + 105\kappa_2^4 + 420\kappa_2^3\kappa_1^2 + 210\kappa_2^2\kappa_1^4 + 28\kappa_2\kappa_1^6 + \kappa_1^8, \\
\mu'_9 &= \kappa_9 + 9\kappa_8\kappa_1 + 36\kappa_7\kappa_2 + 36\kappa_7\kappa_1^2 + 84\kappa_6\kappa_3 + 252\kappa_6\kappa_2\kappa_1 \\
&\quad + 84\kappa_6\kappa_1^3 + 126\kappa_5\kappa_4 + 504\kappa_5\kappa_3\kappa_1 + 378\kappa_5\kappa_2^2 + 756\kappa_5\kappa_2\kappa_1^2 \\
&\quad + 126\kappa_5\kappa_1^4 + 315\kappa_4^2\kappa_1 + 1260\kappa_4\kappa_3\kappa_2 + 1260\kappa_4\kappa_3\kappa_1^2 + 1890\kappa_4\kappa_2^2\kappa_1 \\
&\quad + 1260\kappa_4\kappa_2\kappa_1^3 + 126\kappa_4\kappa_1^5 + 280\kappa_3^3 + 2520\kappa_3^2\kappa_2\kappa_1 + 840\kappa_3^2\kappa_1^3 \\
&\quad + 1260\kappa_3\kappa_2^3 + 3780\kappa_3\kappa_2^2\kappa_1^2 + 1260\kappa_3\kappa_2\kappa_1^4 + 84\kappa_3\kappa_1^6 + 945\kappa_2^4\kappa_1 \\
&\quad + 1260\kappa_2^3\kappa_1^3 + 378\kappa_2^2\kappa_1^5 + 36\kappa_2\kappa_1^7 + \kappa_1^9, \\
\mu'_{10} &= \kappa_{10} + 10\kappa_9\kappa_1 + 45\kappa_8\kappa_2 + 45\kappa_8\kappa_1^2 + 120\kappa_7\kappa_3 + 360\kappa_7\kappa_2\kappa_1 \\
&\quad + 120\kappa_7\kappa_1^3 + 210\kappa_6\kappa_4 + 840\kappa_6\kappa_3\kappa_1 + 630\kappa_6\kappa_2^2 + 1260\kappa_6\kappa_2\kappa_1^2 \\
&\quad + 210\kappa_6\kappa_1^4 + 126\kappa_5^2 + 1260\kappa_5\kappa_4\kappa_1 + 2520\kappa_5\kappa_3\kappa_2 + 2520\kappa_5\kappa_3\kappa_1^2 \\
&\quad + 3780\kappa_5\kappa_2^2\kappa_1 + 2520\kappa_5\kappa_2\kappa_1^3 + 252\kappa_5\kappa_1^5 + 1575\kappa_4^2\kappa_2 + 1575\kappa_4^2\kappa_1^2 \\
&\quad + 2100\kappa_4\kappa_3^2 + 12600\kappa_4\kappa_3\kappa_2\kappa_1 + 4200\kappa_4\kappa_3\kappa_1^3 + 3150\kappa_4\kappa_2^3 \\
&\quad + 9450\kappa_4\kappa_2^2\kappa_1^2 + 3150\kappa_4\kappa_2\kappa_1^4 + 210\kappa_4\kappa_1^6 + 2800\kappa_3^3\kappa_1 \\
&\quad + 6300\kappa_3^2\kappa_2^2 + 12600\kappa_3^2\kappa_2\kappa_1^2 + 2100\kappa_3^2\kappa_1^4 + 12600\kappa_3\kappa_2^3\kappa_1 \\
&\quad + 12600\kappa_3\kappa_2^2\kappa_1^3 + 2520\kappa_3\kappa_2\kappa_1^5 + 120\kappa_3\kappa_1^7 + 945\kappa_2^5 + 4725\kappa_2^4\kappa_1^2 \\
&\quad + 3150\kappa_2^3\kappa_1^4 + 630\kappa_2^2\kappa_1^6 + 45\kappa_2\kappa_1^8 + \kappa_1^{10};
\end{aligned}\right\} \tag{3.37}$$

(*) The coefficients of μ's in terms of κ's are those of the unitary symmetric functions (1^r) in terms of the augmented symmetric functions and can be read from the tables of David and Kendall (1949) as far as order 12. The inverse relations may also be obtained from the tables on multiplication by appropriate factorials. See **12.5** below.

or, for moments about the mean ($\kappa_1 = 0$),

$$\left.\begin{aligned}
\mu_2 &= \kappa_2,\\
\mu_3 &= \kappa_3,\\
\mu_4 &= \kappa_4 + 3\kappa_2^2,\\
\mu_5 &= \kappa_5 + 10\kappa_3\kappa_2,\\
\mu_6 &= \kappa_6 + 15\kappa_4\kappa_2 + 10\kappa_3^2 + 15\kappa_2^3,\\
\mu_7 &= \kappa_7 + 21\kappa_5\kappa_2 + 35\kappa_4\kappa_3 + 105\kappa_3\kappa_2^2,\\
\mu_8 &= \kappa_8 + 28\kappa_6\kappa_2 + 56\kappa_5\kappa_3 + 35\kappa_4^2 + 210\kappa_4\kappa_2^2 + 280\kappa_3^2\kappa_2 + 105\kappa_2^4,\\
\mu_9 &= \kappa_9 + 36\kappa_7\kappa_2 + 84\kappa_6\kappa_3 + 126\kappa_5\kappa_4 + 378\kappa_5\kappa_2^2 + 1260\kappa_4\kappa_3\kappa_2 + 280\kappa_3^3\\
&\quad + 1260\kappa_3\kappa_2^3,\\
\mu_{10} &= \kappa_{10} + 45\kappa_8\kappa_2 + 120\kappa_7\kappa_3 + 210\kappa_6\kappa_4 + 630\kappa_6\kappa_2^2 + 126\kappa_5^2\\
&\quad + 2520\kappa_5\kappa_3\kappa_2 + 1575\kappa_4^2\kappa_2 + 2100\kappa_4\kappa_3^2 + 3150\kappa_4\kappa_2^3\\
&\quad + 6300\kappa_3^2\kappa_2^2 + 945\kappa_2^5.
\end{aligned}\right\} \qquad (3.38)$$

Conversely we have

$$\frac{\kappa_1 t}{1!} + \ldots + \frac{\kappa_r t^r}{r!} + \ldots = \log\left(1 + \frac{\mu_1' t}{1!} + \ldots + \frac{\mu_r' t^r}{r!} + \ldots\right).$$

Expanding the logarithm and picking out powers of t^r as before, we have

$$\kappa_r = r! \sum_{m=0}^{r} \Sigma \left(\frac{\mu_{p_1}'}{p_1!}\right)^{\pi_1} \ldots \left(\frac{\mu_{p_m}'}{p_m!}\right)^{\pi_m} \frac{(-1)^{\rho-1}(\rho-1)!}{\pi_1!\ldots\pi_m!}, \qquad (3.39)$$

the second summation extending over all non-negative π's and ρ's, subject to (3.34) and the further condition

$$\pi_1 + \pi_2 + \ldots + \pi_m = \rho. \qquad (3.40)$$

The first ten formulae are, in terms of moments about an arbitrary point :—

$$\left.\begin{aligned}
\kappa_1 &= \mu_1',\\
\kappa_2 &= \mu_2' - \mu_1'^2,\\
\kappa_3 &= \mu_3' - 3\mu_2'\mu_1' + 2\mu_1'^3,\\
\kappa_4 &= \mu_4' - 4\mu_3'\mu_1' - 3\mu_2'^2 + 12\mu_2'\mu_1'^2 - 6\mu_1'^4,\\
\kappa_5 &= \mu_5' - 5\mu_4'\mu_1' - 10\mu_3'\mu_2' + 20\mu_3'\mu_1'^2 + 30\mu_2'^2\mu_1' - 60\mu_2'\mu_1'^3 + 24\mu_1'^5,\\
\kappa_6 &= \mu_6' - 6\mu_5'\mu_1' - 15\mu_4'\mu_2' + 30\mu_4'\mu_1'^2 - 10\mu_3'^2 + 120\mu_3'\mu_2'\mu_1' - 120\mu_3'\mu_1'^3\\
&\quad + 30\mu_2'^3 - 270\mu_2'^2\mu_1'^2 + 360\mu_2'\mu_1'^4 - 120\mu_1'^6,\\
\kappa_7 &= \mu_7' - 7\mu_6'\mu_1' - 21\mu_5'\mu_2' + 42\mu_5'\mu_1'^2 - 35\mu_4'\mu_3' + 210\mu_4'\mu_2'\mu_1'\\
&\quad - 210\mu_4'\mu_1'^3 + 140\mu_3'^2\mu_1' + 210\mu_3'\mu_2'^2 - 1260\mu_3'\mu_2'\mu_1'^2 + 840\mu_3'\mu_1^4\\
&\quad - 630\mu_2'^3\mu_1' + 2520\mu_2'^2\mu_1'^3 - 2520\mu_2'\mu_1'^5 + 720\mu_1'^7,\\
\kappa_8 &= \mu_8' - 8\mu_7'\mu_1' - 28\mu_6'\mu_2' + 56\mu_6'\mu_1'^2 - 56\mu_5'\mu_3' + 336\mu_5'\mu_2'\mu_1'\\
&\quad - 336\mu_5'\mu_1'^3 - 35\mu_4'^2 + 560\mu_4'\mu_3'\mu_1' + 420\mu_4'\mu_2'^2 - 2520\mu_4'\mu_2'\mu_1'^2\\
&\quad + 1680\mu_4'\mu_1'^4 + 560\mu_3'^2\mu_2' - 1680\mu_3'^2\mu_1'^2 - 5040\mu_3'\mu_2'^2\mu_1'\\
&\quad + 13440\mu_3'\mu_2'\mu_1'^3 - 6720\mu_3'\mu_1'^5 - 630\mu_2'^4 + 10080\mu_2'^3\mu_1'^2\\
&\quad - 25200\mu_2'^2\mu_1'^4 + 20160\mu_2'\mu_1'^6 - 5040\mu_1'^8,\\
\kappa_9 &= \mu_9' - 9\mu_8'\mu_1' - 36\mu_7'\mu_2' + 72\mu_7'\mu_1'^2 - 84\mu_6'\mu_3' + 504\mu_6'\mu_2'\mu_1'\\
&\quad - 504\mu_6'\mu_1'^3 - 126\mu_5'\mu_4' + 1008\mu_5'\mu_3'\mu_1' + 756\mu_5'\mu_2'^2 - 4536\mu_5'\mu_2'\mu_1'^2\\
&\quad + 3024\mu_5'\mu_1'^4 + 630\mu_4'^2\mu_1' + 2520\mu_4'\mu_3'\mu_2' - 7560\mu_4'\mu_3'\mu_1'^2\\
&\quad - 11340\mu_4'\mu_2'^2\mu_1' + 30240\mu_4'\mu_2'\mu_1'^3 - 15120\mu_4'\mu_1'^5 + 560\mu_3'^3
\end{aligned}\right\} \qquad (3.41)$$

$$\left.\begin{aligned}
&-15120\mu_3'^2\mu_2'\mu_1'+20160\mu_3'^2\mu_1'^3-7560\mu_3'\mu_2'^3+90720\mu_3'\mu_2'^2\mu_1'^2\\
&-151200\mu_3'\mu_2'\mu_1'^4+60480\mu_3'\mu_1'^6+22680\mu_2'^4\mu_1'-151200\mu_2'^3\mu_1'^3\\
&+272160\mu_2'^2\mu_1'^5-181440\mu_2'\mu_1'^7+40320\mu_1'^9,\\
\kappa_{10} = {} & \mu_{10}'-10\mu_9'\mu_1'-45\mu_8'\mu_2'+90\mu_8'\mu_1'^2-120\mu_7'\mu_3'+720\mu_7'\mu_2'\mu_1'\\
&-720\mu_7'\mu_1'^3-210\mu_6'\mu_4'+1680\mu_6'\mu_3'\mu_1'+1260\mu_6'\mu_2'^2\\
&-7560\mu_6'\mu_2'\mu_1'^2+5040\mu_6'\mu_1'^4-126\mu_5'^2+2520\mu_5'\mu_4'\mu_1'\\
&+5040\mu_5'\mu_3'\mu_2'-15120\mu_5'\mu_3'\mu_1'^2-22680\mu_5'\mu_2'^2\mu_1'+60480\mu_5'\mu_2'\mu_1'^3\\
&-30240\mu_5'\mu_1'^5+3150\mu_4'^2\mu_2'-9450\mu_4'^2\mu_1'^2+4200\mu_4'\mu_3'^2\\
&-75600\mu_4'\mu_3'\mu_2'\mu_1'+100800\mu_4'\mu_3'\mu_1'^3-18900\mu_4'\mu_2'^3\\
&+226800\mu_4'\mu_2'^2\mu_1'^2-378000\mu_4'\mu_2'\mu_1'^4+151200\mu_4'\mu_1'^6-16800\mu_3'^3\mu_1'\\
&-37800\mu_3'^2\mu_2'^2+302400\mu_3'^2\mu_2'\mu_1'^2-252000\mu_3'^2\mu_1'^4+302400\mu_3'\mu_2'^3\mu_1'\\
&-1512000\mu_3'\mu_2'^2\mu_1'^3+1814400\mu_3'\mu_2'\mu_1'^5-604800\mu_3'\mu_1'^7\\
&+22680\mu_2'^5-567000\mu_2'^4\mu_1'^2+2268000\mu_2'^3\mu_1'^4-3175200\mu_2'^2\mu_1'^6\\
&+1814400\mu_2'\mu_1'^8-362880\mu_1'^{10};
\end{aligned}\right\} \quad (3.42)$$

or, for moments about the mean,

$$\left.\begin{aligned}
\kappa_2 &= \mu_2,\\
\kappa_3 &= \mu_3,\\
\kappa_4 &= \mu_4-3\mu_2^2,\\
\kappa_5 &= \mu_5-10\mu_3\mu_2,\\
\kappa_6 &= \mu_6-15\mu_4\mu_2-10\mu_3^2+30\mu_2^3,\\
\kappa_7 &= \mu_7-21\mu_5\mu_2-35\mu_4\mu_3+210\mu_3\mu_2^2,\\
\kappa_8 &= \mu_8-28\mu_6\mu_2-56\mu_5\mu_3-35\mu_4^2+420\mu_4\mu_2^2+560\mu_3^2\mu_2-630\mu_2^4,\\
\kappa_9 &= \mu_9-36\mu_7\mu_2-84\mu_6\mu_3-126\mu_5\mu_4+756\mu_5\mu_2^2+2520\mu_4\mu_3\mu_2\\
&\quad +560\mu_3^3-7560\mu_3\mu_2^3,\\
\kappa_{10} &= \mu_{10}-45\mu_8\mu_2-120\mu_7\mu_3-210\mu_6\mu_4+1260\mu_6\mu_2^2-126\mu_5^2\\
&\quad +5040\mu_5\mu_3\mu_2+3150\mu_4^2\mu_2+4200\mu_4\mu_3^2-18900\mu_4\mu_2^3\\
&\quad -37800\mu_3^2\mu_2^2+22680\mu_2^5.
\end{aligned}\right\} \quad (3.43)$$

Existence of cumulants

3.15 The formal expression (3.30) may be regarded as defining the cumulants in terms of the moments, and it is thus evident that the cumulant of order r exists if the moments of orders r and lower exist. If, however, we look to the equation

$$\exp\left(\Sigma\,\kappa_r\frac{(it)^r}{r!}\right) = \phi(t)$$

as defining the cumulants, it is not quite so easy to show that κ_r exists if μ_r and lower μ's exist.

Since e^{itx} is an analytic function we can expand it in the expression for the characteristic function and obtain

$$\begin{aligned}
\phi(t) &= \int_{-\infty}^{\infty}\left\{\sum_0^r\frac{(itx)^j}{j!}+O\{|tx|^{r+1}/(r+1)!\}\right\}dF\\
&= \sum_{j=0}^{r}\mu_j'(it)^j/j!+R_r
\end{aligned} \quad (3.44)$$

where R_r is a remainder term not greater than $\rho\nu_{r+1}|t|^{r+1}/(r+1)!$ with $|\rho|\leqslant 1$.

We may write this as

$$\phi(t) = \sum_{j=0}^{r} \mu_j'(it)^j/j!+o(t^r).$$

For some small t we may then take logarithms and expand to obtain

$$\log\phi(t) = \sum_{j=1}^{r} \kappa_j(it)^j/j!+o(t^r), \tag{3.45}$$

the coefficients κ being the cumulants by definition. Thus if ν_{r+1} exists κ_r and lower cumulants exist. It is possible to prove stronger results, but this is sufficient for many purposes.

Calculation of cumulants

3.16 The cumulants are not, like the moments, directly ascertainable by summatory or integrative processes, and to find them it is necessary either to find the moments and employ equations (3.43) or to derive them from the characteristic function. The following examples will illustrate the processes involved.

Example 3.9

In Example 3.7 we found the following values for the moments about the mean of the height data of Table 1.7 :—

$$\begin{aligned} \mu_1' &= 11{\cdot}020{,}850 \\ \mu_2 &= 6{\cdot}616{,}805 \\ \mu_3 &= -0{\cdot}207{,}840 \\ \mu_4 &= 137{\cdot}689{,}185, \end{aligned}$$

whence, from (3.43), κ_2 and κ_3 have the same values as μ_2 and μ_3 and

$$\begin{aligned} \kappa_4 &= \mu_4-3\mu_2^2 \\ &= 6{\cdot}342{,}86. \end{aligned}$$

κ_1 is the same as μ_1', in this case measured from the centre of the interval 56– inches.

The same results would, of course, have been obtained if we had used equations (3.42) and moments about the origin.

Example 3.10

Consider the discrete (so-called Poisson) distribution whose frequencies at $0, 1, \ldots, j, \ldots$ are $e^{-\lambda}\left(1, \frac{\lambda}{1!}, \ldots, \frac{\lambda^j}{j!}, \ldots\right)$. The characteristic function is given by

$$\begin{aligned} \phi(t) &= e^{-\lambda} \sum_{j=0}^{\infty} \frac{\lambda^j}{j!} e^{itj} \\ &= e^{-\lambda}\exp(\lambda e^{it}) \\ &= \exp\{\lambda(e^{it}-1)\}. \end{aligned}$$

Since the variate is non-negative, for any r the absolute moment is the same as the ordinary moment, and we have

$$\mu_r' = e^{-\lambda} \sum_{j=0}^{\infty} \frac{\lambda^j j^r}{j!};$$

and since this converges (*) cumulants of all orders exist. They are therefore given by the expansion of $\log \phi(t)$ as a power series in t. But

$$\log \phi(t) = \lambda(e^{it}-1)$$

$$= \lambda \sum_{j=1}^{\infty} (it)^j/j!$$

and hence

$$\kappa_r = \lambda$$

for all r. Thus all cumulants of the distribution are equal to λ.

Example 3.11

In Example 3.4 we found, in effect, for the characteristic function of the distribution

$$dF = \frac{1}{\sigma\sqrt{(2\pi)}} \exp\left(-\frac{x^2}{2\sigma^2}\right) dx, \qquad -\infty \leqslant x \leqslant \infty,$$

$$\phi(t) = \exp(-\tfrac{1}{2}\sigma^2 t^2).$$

Thus

$$\log \phi(t) = -\tfrac{1}{2}\sigma^2 t^2.$$

It is easily seen that the absolute moments and hence the cumulants of all orders exist. Thus

$$\kappa_2 = \sigma^2$$

and

$$\kappa_r = 0, \qquad r > 2.$$

Example 3.12

In Example 3.6 it was found that for the distribution

$$dF = \frac{a^\gamma}{\Gamma(\gamma)} e^{-ax} x^{\gamma-1} dx, \qquad 0 \leqslant x \leqslant \infty\,;\ a, \gamma > 0,$$

the characteristic function is given by

$$\phi(t) = \frac{1}{\left(1-\dfrac{it}{a}\right)^\gamma}.$$

It is readily verified that cumulants of all orders exist and hence

$$\kappa_r = \text{coeff. of } \frac{(it)^r}{r!} \text{ in } -\gamma \log\left(1-\frac{it}{a}\right)$$

$$= \gamma(r-1)!\, a^{-r}.$$

Example 3.13

Consider again the distribution of Example 3.3,

$$dF = \frac{k}{(1+x^2)^m} dx, \qquad -\infty \leqslant x \leqslant \infty\,;\ m \geqslant 1.$$

(*) For the ratio of the $(n+1)$th term of the series to the nth is

$$\frac{\lambda^{n+1}(n+1)^r}{(n+1)!} \bigg/ \frac{\lambda^n n^r}{n!} = \frac{\lambda}{n+1}\left(1+\frac{1}{n}\right)^r = O\left(\frac{\lambda}{n}\right),$$

and thus the series converges for all finite values of λ.

The characteristic function is given by

$$\phi(t) = k\int_{-\infty}^{\infty} \frac{e^{ixt}}{(1+x^2)^m}dx,$$

which, since $\sin xt$ is an odd function, reduces to

$$k\int_{-\infty}^{\infty} \frac{\cos xt}{(1+x^2)^m}dx.$$

This integral may be evaluated by complex integration round a contour consisting of the x-axis, the infinite semicircle above the x-axis and the infinitely small circle round the point $x = i$. It is found that

$$\phi(t) = \frac{k\pi}{2^{2m-2}(m-1)!}\, e^{-|t|}\Big\{(2|t|)^{m-1}+m(m-1)(2|t|)^{m-2} + \frac{(m+1)(m)(m-1)(m-2)}{2!}(2|t|)^{m-3}+\dots+\frac{(2m-2)!}{(m-1)!}\Big\}$$

For $m = 1$, $k = \pi^{-1}$ by Example 3.3, and $\phi(t) = \exp\{-|t|\}$.

If $r<2m-1$ the absolute moment of order r

$$\nu_r = k\int_{-\infty}^{\infty} \frac{|x|^r\,dx}{(1+x^2)^m}$$

exists and hence so does the cumulant of order r. But in this case we cannot expand $\log\phi(t)$ in an *infinite* series of powers of t, though this might perhaps be thought possible from the form of $\phi(t)$. In fact, we can only expand $\log\phi(t)$ in powers of t up to the point at which the derivatives of $\phi(t)$ exist, for $t = 0$.

To simplify the discussion, consider the case when $m = 2$. We have then, since $k = 2/\pi$ in this case,

$$\phi(t) = e^{-|t|}\{|t|+1\}$$

$$\log\phi(t) = -|t|+\log\{1+|t|\}.$$

If t is positive this equals

$$-\tfrac{1}{2}t^2+\tfrac{1}{3}t^3-\dots$$

but if t is negative it equals

$$-\tfrac{1}{2}t^2-\tfrac{1}{3}t^3-\dots$$

the two expressions differing in the sign of the term in t^3 and every second term thereafter. There is thus no unique expansion of $\log\phi(t)$ in powers of t about the point $t = 0$. There are two forms of the function expressing $\log\phi(t)$ according as t is positive or negative.

However, these expressions coincide as far as their terms in t and t^2, and the first and second derivatives of $\log\phi(t)$ are uniquely defined when $t = 0$. Thus the first and second cumulants exist and are given by

$$\kappa_1 = 0 \qquad \kappa_2 = 1.$$

Cumulants of higher orders do not exist.

Factorial cumulants

3.17 Analogously to the relation between cumulants and moments we may define a factorial cumulant $\kappa_{[r]}$ as the coefficient of $t^r/r!$ in the expansion of the logarithm of the factorial moment-generating function. Thus if $P(t)$ is the frequency-generating function the factorial cumulant-generating function is

$$\omega(t) = \log P(1+t). \tag{3.46}$$

Like the factorial moment-generating function, $\omega(t)$ is mainly of use for certain classes of discontinuous distribution.

The relations between cumulants and factorial cumulants are formally similar to those between moments and factorial moments. Taking a unit interval between frequencies we have, analogously to (3.25) and (3.26),

$$\left.\begin{aligned} \kappa_{[1]} &= \kappa_1 \\ \kappa_{[2]} &= \kappa_2 - \kappa_1 \\ \kappa_{[3]} &= \kappa_3 - 3\kappa_2 + 2\kappa_1 \\ \kappa_{[4]} &= \kappa_4 - 6\kappa_3 + 11\kappa_2 - 6\kappa_1 \end{aligned}\right\} \tag{3.47}$$

Conversely

$$\left.\begin{aligned} \kappa_2 &= \kappa_{[2]} + \kappa_{[1]} \\ \kappa_3 &= \kappa_{[3]} + 3\kappa_{[2]} + \kappa_{[1]} \\ \kappa_4 &= \kappa_{[4]} + 6\kappa_{[3]} + 7\kappa_{[2]} + \kappa_{[1]} \end{aligned}\right\} \tag{3.48}$$

Example 3.14

Reverting once more to the binomial distribution we have

$$\omega(t) = n\log(1+pt)$$

and hence

$$\kappa_{[r]} = (-1)^{r+1}n(r-1)!p^r.$$

Corrections for grouping (Sheppard's corrections)

3.18 When moments are calculated from a numerically specified distribution which is grouped, there is present a certain amount of approximation owing to the fact that the frequencies are assumed to be concentrated at the mid-points of intervals. It is possible to correct for this effect under certain conditions.

Let $f(x)$ be continuous and of finite range a to b; and let the range be divided into n intervals of width h. The frequency in the jth interval, centred at $x_j = a+(j-\frac{1}{2})h$, is then given by

$$f_j = \int_{-\frac{1}{2}h}^{\frac{1}{2}h} f(x_j+\xi)\,d\xi. \tag{3.49}$$

The moments calculated from the grouped frequencies, which we will denote by a bar, are then given by

$$\bar{\mu}'_r = \sum_{j=1}^{n} x_j^r \int_{-\frac{1}{2}h}^{\frac{1}{2}h} f(x_j+\xi)\,d\xi. \tag{3.50}$$

Now from one version of the Euler–Maclaurin sum formula (*) we have, for a function $y(x)$ with $2m$ derivatives,

$$\frac{1}{h}\int_a^b y(x)\,dx = [y(a+\tfrac{1}{2}h)+y(a+\tfrac{3}{2}h)+\ldots+y\{a+(n-\tfrac{1}{2})h\}] \\ -\sum_{j=1}^{m}\frac{h^{2j-1}}{(2j)!}B_{2j}(\tfrac{1}{2})\left[y^{(2j-1)}(x)\right]_a^b - S_{2m} \tag{3.51}$$

(*) See, for example, Milne-Thomson, *The Calculus of Finite Differences*, section 7.5.

where $y^{(r)}$ is the rth derivative of y, $B_{2m}(\frac{1}{2})$ is the value of the $2m$'th Bernoulli polynomial for argument $\frac{1}{2}$ and S_{2m} is a remainder term which can be expressed as

$$S_{2m} = \frac{nh^{2m}}{(2m)!}B_{2m}(\tfrac{1}{2})y^{(2m)}(a+nh\theta), \qquad 0<\theta<1. \tag{3.52}$$

Putting

$$y(x) = x^r\int_{-\frac{1}{2}h}^{\frac{1}{2}h} f(x+\xi)\,d\xi$$

in (3.51) and neglecting S_{2m} we have, if the first $2m-3$ derivatives of y vanish at the terminals,

$$\frac{1}{h}\int_a^b x^r\int_{-\frac{1}{2}h}^{\frac{1}{2}h} f(x+\xi)\,d\xi\,dx = \bar{\mu}_r'. \tag{3.53}$$

Hence

$$\begin{aligned}\bar{\mu}_r' &= \frac{1}{h}\int_a^b\int_{-\frac{1}{2}h}^{\frac{1}{2}h}(u-\xi)^r f(u)\,du\,d\xi \\ &= \frac{1}{h}\int_a^b\frac{(u+\frac{1}{2}h)^{r+1}-(u-\frac{1}{2}h)^{r+1}}{r+1}f(u)\,du \\ &= \sum_{j=0}^{[\frac{1}{2}r]}\binom{r}{2j}(\tfrac{1}{2}h)^{2j}\frac{1}{2j+1}\mu_{r-2j}'\end{aligned} \tag{3.54}$$

where $[\frac{1}{2}r]$ is the integral part of $\frac{1}{2}r$. This gives the raw (i.e., uncorrected) moments in terms of the actual moments. In practice we require the latter in terms of the former, and it is easy to find from (3.54) the following expressions:

$$\left.\begin{aligned}\mu_1' &= \bar{\mu}_1' \\ \mu_2' &= \bar{\mu}_2'-\frac{1}{12}h^2 \\ \mu_3' &= \bar{\mu}_3'-\frac{1}{4}\bar{\mu}_1'h^2 \\ \mu_4' &= \bar{\mu}_4'-\frac{1}{2}\bar{\mu}_2'h^2+\frac{7}{240}h^4 \\ \mu_5' &= \bar{\mu}_5'-\frac{5}{6}\bar{\mu}_3'h^2+\frac{7}{48}\bar{\mu}_1'h^4 \\ \mu_6' &= \bar{\mu}_6'-\frac{5}{4}\bar{\mu}_4'h^2+\frac{7}{16}\bar{\mu}_2'h^4-\frac{31}{1344}h_6\end{aligned}\right\} \tag{3.55}$$

The general expression for these formulae is

$$\mu_r' = \sum_{j=0}^{r}\left\{\binom{r}{j}(2^{1-j}-1)B_j h^j\bar{\mu}_{r-j}'\right\} \tag{3.56}$$

where B_j is the Bernoulli number of order j (Wold, 1934a). (See also Exercise 3.17.)

3.19 We have made various assumptions in arriving at these formulae: that there is high-order contact at the terminals of the range, that the range itself is finite and that the remainder term may be neglected. In absolute magnitude $B_{2m}(\frac{1}{2})/(2m)!$

is less than $4/(2\pi)^{2m}$ and thus $|S_{2m}|$ in (3.52) is less than $4nh^{2m}/(2\pi)^{2m}$ multiplied by some value of $y^{(2m)}(x)$ in the range a to b. The extent to which the remainder term is negligible then depends on the behaviour of $f^{(2m-1)}(x)$ in the range a to b.

In practice the corrections result in improvements if there is reasonably high-order contact, but may break down if there is not. For distributions of infinite range the frequency tails off near the ends and we may still apply the corrections with some confidence to those distributions (in practice, of course, necessarily of finite range) where the major part of the distribution is concentrated and the frequencies at the extremes are small.

Example 3.15

Consider the distribution

$$dF = \frac{1}{B(12,6)} x^{11}(1-x)^5\,dx, \qquad 0 \leqslant x \leqslant 1,$$

a case of the so-called Type I distribution. The exact frequencies for intervals of 0·1 may be obtained from the Tables of the Incomplete B-Function, and are as follows :—

Centre of interval	Frequency
0·05	0·000,000,0
0·15	0·000,009,2
0·25	0·000,646,8
0·35	0·009,938,2
0·45	0·061,137,4
0·55	0·192,199,6
0·65	0·332,887,7
0·75	0·297,479,9
0·85	0·101,033,7
0·95	0·004,667,5
TOTAL	1·000,000,0

The raw moments about $x = 0$ are shown in the following table :—

Moment	Raw	Exact	Corrected
μ_1'	0·666,662,8	0·666,666,7	0·666,662,8
μ_2'	0·456,965,5	0·456,140,4	0·456,132,2
μ_3'	0·320,952,3	0·319,298,2	0·319,285,7
μ_4'	0·230,335,1	0·228,070,2	0·228,053,2
μ_5'	0·168,512,9	0·165,869,2	0·165,848,0
μ_6'	0·125,433,2	0·122,599,0	0·122,574,0

The exact values are calculable by direct integration. It is to be noted that the corrected second and third moments are more accurate than the order of the terms, $h^2/12$ and $h^2/4$, used in making the corrections. The corrected fourth moment is in error by a term of order 2×10^{-5}, which is of the same magnitude as one of the correcting terms, $7h^4/240$, used in arriving at it. Similarly the errors in the corrected fifth and sixth moments are of the same order as or higher order than some of the correcting terms.

Example 3.16

As an illustration of the way in which Sheppard's corrections break down when the condition for high-order contact is violated, an example is taken from a paper by Pairman and Pearson (1919). The following table shows the frequencies in a certain range of the normal distribution

$$dF = \frac{1}{\sqrt{(2\pi)}} e^{-\frac{1}{2}x^2} dx,$$

with intervals of width 0·5.

Interval centred at	Frequency
1·5	0·655,91
2·0	0·278,34
2·5	0·092,45
3·0	0·024,02
3·5	0·004,89
4·0	0·000,78
4·5	0·000,10
5·0	0·000,01
TOTAL	1·056,50

The distribution has high-order contact at one end but not at the start of the curve, being in fact J-shaped and very abrupt at that point.

The following table shows the raw moments about the mean up to the fourth order, the moments with Sheppard's corrections, and the true moments calculated from the continuous normal distribution :—

Moment	Raw	Exact	Corrected
μ_2	0·158,524	0·172,222	0·137,691
μ_3	0·104,226	0·098,612	0·104,226
μ_4	0·149,090	0·156,405	0·131,097

It will be noted that in the two cases where the corrections are made they operate in the wrong direction. For the fourth moment they increase the difference between calculated and true values from about 4 per cent. to about 16 per cent. It is clear that, at least for the fairly coarse grouping of this example, Sheppard's corrections may fail completely.

3.20 Equations (3.55) were written in terms of moments about an arbitrary point. This point can, in particular, be the mean of the distribution, and accordingly we may drop the dashes and put μ_1' equal to zero in (3.55), to get the corrections appropriate for moments about the mean.

Average corrections

3.21 There is a distinct type of problem which also leads to the Sheppard corrections. Suppose there is given a distribution of unknown range and the frequencies falling into specified intervals. One may ask what are the corrections to be applied to the raw moments so as to bring them *on the average* into closer relation with the real moments. In other words, supposing that the interval-mesh is located at random on the distribution, what are the average values of the raw moments?

Let X_j be a fixed set of values of x, j varying from $-\infty$ to ∞ by integral values. As x_j varies from $X_j-\frac{1}{2}h$ to $X_j+\frac{1}{2}h$, x_k varies from $X_k-\frac{1}{2}h$ to $X_k+\frac{1}{2}h$.

By definition

$$\bar{\mu}'_r = \sum_{j=-\infty}^{\infty}\left\{x_j^r \int_{-\frac{1}{2}h}^{\frac{1}{2}h} f(x_j+\xi)\,d\xi\right\}.$$

Denoting by $\mathrm{E}(\bar{\mu}'_r)$ the average as x_j varies from $X_j-\frac{1}{2}h$ to $X_j+\frac{1}{2}h$, we have

$$\begin{aligned}\mathrm{E}(\bar{\mu}'_r) &= \frac{1}{h}\int_{X_j-\frac{1}{2}h}^{X_j+\frac{1}{2}h} \Sigma\left\{x_j^r\int_{-\frac{1}{2}h}^{\frac{1}{2}h} f(x_j+\xi)\,d\xi\right\}dx_j \\ &= \frac{1}{h}\sum_{j=-\infty}^{\infty}\int_{X_j-\frac{1}{2}h}^{X_j+\frac{1}{2}h} x_j^r \int_{-\frac{1}{2}h}^{\frac{1}{2}h} f(x_j+\xi)\,d\xi\,dx_j \\ &= \frac{1}{h}\int_{-\infty}^{\infty} x^r \int_{-\frac{1}{2}h}^{\frac{1}{2}h} f(x+\xi)\,d\xi\,dx, \end{aligned} \tag{3.57}$$

which is the same as equation (3.53) with the substitution of $\mathrm{E}(\bar{\mu}'_r)$ for $\bar{\mu}'_r$. Thus the Sheppard corrections apply for the average group-moments whatever the nature of the terminal contact.

They cannot, however, be applied indiscriminately on that ground. In place of the conditions about terminal contact and bounded derivatives, which ensure the applicability of Sheppard's corrections to any particular distribution, there is the condition that the grouping intervals are located at random on the range, which implies that although the corrections may be wrong in any given instance, the average effect in a large number of cases will be correct. In actual fact the condition about the random location of grouping does not operate very frequently for J- and U-shaped distributions, where the Sheppard corrections would not ordinarily apply; for instance, in a distribution of incomes or deaths at given ages it is almost inevitable to begin the grouping at zero.

3.22 It is also illegitimate to drop the dashes in order to obtain corrections for moments about the mean. If the mean of the grouped distribution is denoted by m, the average value of the rth moment about the mean is given by

$$\mathrm{E}(\bar{\mu}_r) = \frac{1}{h}\int_{-\infty}^{\infty}(x-m)^r\int_{-\frac{1}{2}h}^{\frac{1}{2}h} f(x+\xi)\,d\xi\,dx,$$

where m is a function of x and the transformation of the integral which has been used earlier in this chapter is no longer legitimate. Explicit expressions for average corrections to moments about the mean have not yet been obtained. From a consideration of some particular distributions, however, Kendall (1938) concluded that for all ordinary purposes it is sufficient to use equations (3.55) as if the mean were a fixed point.

3.23 Average corrections may also be applied to discrete data which have been grouped in wider intervals, but are different from those of the continuous case. Cf. Exercise 3.13 and C. C. Craig (1936b). For corrections when the Sheppard conditions are violated see Pairman and Pearson (1919), Sandon (1924), Martin (1934), Elderton (1938) and Exercise 3.10.

Sheppard's corrections for factorial moments

3.24 It has been shown by Wold (1934a) that for factorial moments the Sheppard corrections are as follows :—

$$\left.\begin{aligned}
\mu'_{[1]} &= \bar{\mu}'_{[1]} \\
\mu'_{[2]} &= \bar{\mu}'_{[2]} - \frac{h^2}{12} \\
\mu'_{[3]} &= \bar{\mu}'_{[3]} - \frac{h^2}{4}\bar{\mu}'_{[1]} + \frac{h^3}{4} \\
\mu'_{[4]} &= \bar{\mu}'_{[4]} - \frac{h^2}{2}\bar{\mu}'_{[2]} + h^3\bar{\mu}'_{[1]} - \frac{71}{80}h^4 \\
\mu'_{[5]} &= \bar{\mu}'_{[5]} - \frac{5}{6}\bar{\mu}'_{[3]}h^2 + \frac{5}{2}\bar{\mu}'_{[2]}h^3 - \frac{71}{16}\bar{\mu}'_{[1]}h^4 + \frac{31}{8}h^5 \\
\mu'_{[6]} &= \bar{\mu}'_{[6]} - \frac{5}{4}\bar{\mu}'_{[4]}h^2 + 5\,\bar{\mu}'_{[3]}h^3 - \frac{213}{16}\bar{\mu}'_{[2]}h^4 \\
&\qquad + \frac{93}{4}\bar{\mu}'_{[1]}h^5 - \frac{9129}{448}h^6
\end{aligned}\right\} \tag{3.58}$$

and in general are given by

$$\mu'_{[r]} = \sum_{j=0}^{r} \binom{r}{j} B_j^{(j+2)}(\tfrac{3}{2})\, h^j \bar{\mu}'_{[r-j]}, \tag{3.59}$$

where the Bernoulli polynomial $B_j^{(j+2)}(\frac{3}{2})$ is equal to

$$(-1)^{j+1}\frac{(2j)!}{2^{2j}(j+1)!}\left(\tfrac{1}{3}+\tfrac{1}{5}+\ldots+\frac{1}{2j-1}\right), \qquad j>1,$$

and

$$B_0^{(2)}(\tfrac{3}{2}) = 1, \quad B_1^{(3)}(\tfrac{3}{2}) = 0.$$

Note on Bernoulli numbers and polynomials

3.25 The Bernoulli numbers and (to a smaller extent) the Bernoulli polynomials have several important applications in statistics. The number of order j, B_j, is defined as the coefficient of $t^j/j!$ in $t/(e^t-1)$. Thus

$$\frac{t}{e^t-1} = \sum_{j=0}^{\infty} \frac{B_j t^j}{j!}. \tag{3.60}$$

Direct expansion gives us

$B_0 = 1$; $B_1 = -\frac{1}{2}$; odd order B's $= 0$ except B_1;
$B_2 = 1/6$; $B_4 = -1/30$; $B_6 = 1/42$; $B_8 = -1/30$; $B_{10} = 5/66$;
$B_{12} = -691/2730$; $B_{14} = 7/6$.

An expression we shall require (one of many expressible in terms of the Bernoulli numbers) is obtained by putting (3.60) in the form

$$\frac{1}{e^t-1} - \frac{1}{t} + \frac{1}{2} = \sum_{j=2}^{\infty} \frac{B_j t^{j-1}}{j!}$$

whence, on integration from 0 to t, we find

$$\log\{(\sinh \tfrac{1}{2}t)/\tfrac{1}{2}t\} = \sum_{j=2}^{\infty} (B_j t^j/j!\,j) = \sum_{j=1}^{\infty} B_{2j} t^{2j}/(2j)!\,2j. \tag{3.61}$$

An expansion of frequent use is the so-called Stirling's series :—

$$\log \Gamma(x) = (x-\tfrac{1}{2})\log x - x + \tfrac{1}{2}\log(2\pi) + \sum_{j=1}^{\infty} \frac{B_{2j}}{2j(2j-1)x^{2j-1}}. \qquad (3.62)$$

This is an asymptotic expansion in the sense that the true value of $\log\Gamma(x)$, for x real and positive, always lies between the sum of n and $n+1$ terms, whatever n may be. In particular

$$\log\Gamma(x)$$
$$= (x-\tfrac{1}{2})\log x - x + \tfrac{1}{2}\log(2\pi) + \frac{1}{12x} - \frac{1}{360x^3} + \frac{1}{1260x^5} - \frac{1}{1680x^7} + \frac{1}{1188x^9} - \dots \qquad (3.63)$$

and
$$\Gamma(x) = e^{-x}x^{x-\frac{1}{2}}\sqrt{(2\pi)}\left\{1 + \frac{1}{12x} + \frac{1}{288x^2} - \frac{139}{51840x^3} - \frac{571}{2488320x^4}\dots\right\}. \qquad (3.64)$$

The Bernoulli polynomial of order n and degree j in x, $B_j^{(n)}(x)$, is defined by

$$e^{xt}t^n/(e^t-1)^n = \sum_{j=0}^{\infty} t^j B_j^{(n)}(x)/j! \qquad (3.65)$$

Sheppard's corrections for cumulants

3.26 As in **3.18** we have, writing θ for it,

$$\Sigma\left\{e^{\theta x_j}\int_{-\frac{1}{2}h}^{\frac{1}{2}h} f(x_j+\xi)/d\xi\right\} = \frac{1}{h}\int_{-\infty}^{\infty} e^{\theta x}\,dx \int_{-\frac{1}{2}h}^{\frac{1}{2}h} f(x+\xi)\,d\xi$$
$$= \frac{1}{h}\int_{-\frac{1}{2}h}^{\frac{1}{2}h} e^{-\theta\xi}\,d\xi \int_{-\infty}^{\infty} e^{\theta x}f(x)\,dx$$
$$= \frac{\sinh\frac{1}{2}\theta h}{\frac{1}{2}\theta h}\int_{-\infty}^{\infty} e^{\theta x}f(x)\,dx. \qquad (3.66)$$

The expression on the left gives the characteristic function for the grouped data and the integral on the right the true c.f. Taking logarithms and recalling (3.61) we have, for the coefficient of $\theta^r/r!$,

$$\kappa_r = \bar{\kappa}_r - B_r h^r/r, \qquad r>1, \qquad (3.67)$$

an attractively simple result for the Sheppard corrections to cumulants. Since all B's of odd order are zero except B_1 and the first cumulant is equal to the mean, no cumulant of odd order needs any correction. For the others we have

$$\left.\begin{aligned} \kappa_2 &= \bar{\kappa}_2 - \frac{h^2}{12} \\ \kappa_4 &= \bar{\kappa}_4 + \frac{h^4}{120} \\ \kappa_6 &= \bar{\kappa}_6 - \frac{h^6}{252} \end{aligned}\right\} \qquad (3.68)$$

Multivariate moments and cumulants

3.27 The foregoing results in this chapter may be readily generalized to the multivariate case. To avoid complicating the algebraic expressions, we shall deal mainly with two variates x_1 and x_2 ; generalizations to more variates are straightforward.

The bivariate moment μ'_{rs} about an origin a_1 for x_1 and a_2 for x_2 is defined by

$$\mu'_{rs} = \int_{-\infty}^{\infty}\int_{-\infty}^{\infty} (x_1-a_1)^r (x_2-a_2)^s \, dF. \tag{3.69}$$

Evidently, μ'_{r0} and μ'_{0s} denote respectively the rth moment of (the marginal distribution of) x_1 and the sth moment of x_2. If $r \neq 0 \neq s$, μ'_{rs} is called a *product-moment*. By (2.39), $\mu'_{rs} = \mu'_{r0}\,\mu'_{0s}$ if x_1 and x_2 are independent.

As in the univariate case, bivariate moments about certain points can be expressed in terms of those about other points. If the x_1 origin is transferred from a_1 to b_1, where $c_1 = b_1-a_1$, and the x_2 origin from a_2 to b_2, where $c_2 = b_2-a_2$, we have

$$\mu'_{rs}(a_1, a_2) = (\mu'+c_1)^r (\mu'+c_2)^s \tag{3.70}$$

where the product $\mu'^j \mu'^k$ on the right is to be replaced by $\mu'_{jk}(b_1, b_2)$. This corresponds to (3.5) in the univariate case. If the variate means are taken as origins, μ_{rs} is written without the prime as before. μ_{11} is called the *covariance* of x_1 and x_2 and is important in the theory of correlation.

Methods of calculating the product-moments for numerically specified distributions will be considered in Volume 2 when we discuss the theory of correlation. The determination of bivariate moments for a mathematically specified population is a matter of evaluating double sums or double integrals, and no new statistical points call for comment.

Example 3.17

Consider the "bivariate normal" distribution

$$dF = \frac{1}{2\pi\sigma_1\sigma_2(1-\rho^2)^{\frac{1}{2}}}\exp\left\{-\frac{1}{2(1-\rho^2)}\left(\frac{x_1^2}{\sigma_1^2}-\frac{2\rho x_1 x_2}{\sigma_1\sigma_2}+\frac{x_2^2}{\sigma_2^2}\right)\right\} dx_1\,dx_2,$$

$$\rho^2<1\,;\ \sigma_1, \sigma_2>0\,;\ -\infty \leqslant x_1,\ x_2 \leqslant \infty.$$

Let us evaluate

$$M(t_1, t_2) = \int_{-\infty}^{\infty}\int_{-\infty}^{\infty} e^{x_1 t_1 + x_2 t_2}\, dF.$$

Making the substitution

$$\xi = x_1-\sigma_1^2 t_1-\rho\sigma_1\sigma_2 t_2$$
$$\eta = x_2-\rho\sigma_1\sigma_2 t_1-\sigma_2^2 t_2$$

we find

$$M(t_1, t_2) = \exp\left\{\tfrac{1}{2}(t_1^2\sigma_1^2+2t_1t_2\sigma_1\sigma_2\rho+t_2^2\sigma_2^2)\right\}\times$$
$$\frac{1}{2\pi\sigma_1\sigma_2(1-\rho^2)^{\frac{1}{2}}}\int_{-\infty}^{\infty}\int_{-\infty}^{\infty}\exp\left\{-\frac{1}{2(1-\rho^2)}\left(\frac{\xi^2}{\sigma_1^2}-\frac{2\rho\,\xi\eta}{\sigma_1\sigma_2}+\frac{\eta^2}{\sigma_2^2}\right)\right\} d\xi\, d\eta$$
$$= \exp\left\{\tfrac{1}{2}(t_1^2\sigma_1^2+2\rho\sigma_1\sigma_2 t_1 t_2+t_2^2\sigma_2^2)\right\}.$$

Now μ_{rs} is the coefficient of $\frac{t_1^r}{r!}\frac{t_2^s}{s!}$ in $M(t_1, t_2)$ and thus we find, for instance,

$$\mu_{20} = \sigma_1^2, \quad \mu_{11} = \rho\sigma_1\sigma_2, \quad \mu_{02} = \sigma_2^2;$$
$$\mu_{30} = \mu_{21} = \mu_{12} = \mu_{03} = 0;$$
$$\mu_{40} = 3\sigma_1^4, \quad \mu_{31} = 3\rho\sigma_1^3\sigma_2, \quad \mu_{22} = (1+2\rho^2)\sigma_1^2\sigma_2^2,$$
$$\mu_{13} = 3\rho\,\sigma_1\sigma_2^3, \quad \mu_{04} = 3\sigma_2^4.$$

3.28 The bivariate analogue of equation (3.30) may be written

$$\exp\left\{\frac{\kappa_{10}}{1!0!}t_1+\frac{\kappa_{01}}{0!1!}t_2+\ldots+\frac{\kappa_{rs}}{r!s!}t_1^r t_2^s+\ldots\right\}$$
$$= 1+\frac{\mu'_{10}}{1!0!}t_1+\frac{\mu'}{0!1!}t_2+\ldots+\frac{\mu'_{rs}}{r!\,s!}t_1^r t_2^s+\ldots$$

or symbolically,

$$\exp\left\{\Sigma\frac{1}{p!}\kappa(t_1+t_2)^p\right\} = \Sigma\frac{1}{p!}\mu'(t_1+t_2)^p \tag{3.71}$$

where
$$\kappa(t_1+t_2)^p = p!\left\{\frac{\kappa_{p0}}{p!0!}t_1^p+\frac{\kappa_{p-1,1}}{(p-1)!1!}t_1^{p-1}t_2+\ldots\right\}. \tag{3.72}$$

In terms of characteristic functions we may define

$$\phi(t_1, t_2) = \int_{-\infty}^{\infty}\int_{-\infty}^{\infty} e^{ix_1t_1+ix_2t_2}dF \tag{37.3}$$

and, as before, write

$$\phi(t_1, t_2) = \sum_{r,s=0}^{\infty}\mu'_{rs}\frac{(it_1)^r}{r!}\frac{(it^2)^s}{s!}$$
$$= \exp\left\{\sum_{r,s=0}^{\infty}\kappa_{rs}\frac{(it_1)^r}{r!}\frac{(it_2)^s}{s!}\right\}, \tag{37.4}$$

subject to conditions of existence, where κ_{00} is defined equal to zero. If $r \neq 0$, $s \neq 0$, κ_{rs} is called a *product-cumulant.*

From these equations the bivariate moments can be expressed in terms of bivariate cumulants and vice versa. It is also possible to derive bivariate equations from the univariate equations by the following simple symbolic process.

3.29 Consider the equation (one of (3.37))

$$\mu'_4 = \kappa_4+4\kappa_3\kappa_1+3\kappa_2^2+6\kappa_2\kappa_1^2+\kappa_1^4. \tag{37.5}$$

Write this formally as

$$\mu'(r^4) = \kappa(r^4)+4\kappa(r^3)\kappa(r)+3\{\kappa(r^2)\}^2+6\kappa(r^2)\{\kappa(r)\}^2+\{\kappa(r)\}^4.$$

Treat this as a function of r and operate by $s\frac{\partial}{\partial r}$, giving

$$4\mu'(r^3s) = 4\kappa(r^3s)+12\kappa(r^2s)\kappa(r)+4\kappa(r^3)\kappa(s)+12\kappa(rs)\kappa(r^2)$$
$$+12\kappa(rs)\{\kappa(r)\}^2+12\kappa(r^2)\kappa(r)\kappa(s)+4\{\kappa(r)\}^3\kappa(s).$$

Divide through by a factor in 4 and replace r and s by suffixes relating to the first and second variate. We obtain

$$\mu'_{31} = \kappa_{31}+3\kappa_{21}\kappa_{10}+\kappa_{30}\kappa_{01}+3\kappa_{11}\kappa_{20}+3\kappa_{11}\kappa_{10}^2+3\kappa_{20}\kappa_{10}\kappa_{01}+\kappa_{10}^3\kappa_{01}.$$

This is the expression of μ'_{31} in terms of cumulants. The process is quite general and can be justified by reference to (3.71)—cf. Kendall (1940c). It also applies to the expression of cumulants in terms of moments and to formulae involving central moments. For example, from

$$\mu_4 = \kappa_4+3\kappa_2^2 \tag{3.76}$$

we derive, in the same manner,

$$\mu_{31} = \kappa_{31}+3\kappa_{11}\kappa_{20}. \tag{3.77}$$

A further operation of the same kind will yield

$$\mu_{22} = \kappa_{22}+\kappa_{20}\kappa_{02}+2\kappa_{11}^2. \tag{3.78}$$

Similarly, we may use the symbolic process to generate, e.g., the trivariate result

$$\mu_{211} = \kappa_{211}+\kappa_{200}\kappa_{011}+2\kappa_{110}\kappa_{101} \tag{3.79}$$

from either (3.77) or 3.78) by using the operator $t\dfrac{\partial}{\partial r}$ on $\mu(r^3 s)$ or $\mu(r^2 s^2)$.

The reverse process is much simpler, for suffixes to μ and κ need only be added correspondingly to produce the univariate from the bivariate formulae, or the bivariate from the trivariate. Thus, adding suffixes in (3.77) or (3.78) gives us (3.76), while addition of pairs of suffixes in (3.79) gives (3.77) and (3.78). For the reason, see **13.11–14** below.

The following formulae are sometimes required :—

$$\left.\begin{aligned}
\mu_{11} &= \kappa_{11}\,; \qquad \mu_{21} = \kappa_{21}\,; \qquad \mu_{31} = \kappa_{31}+3\kappa_{20}\kappa_{11}\,;\\
\mu_{22} &= \kappa_{22}+\kappa_{20}\kappa_{02}+2\kappa_{11}^2\,; \qquad \mu_{41} = \kappa_{41}+4\kappa_{30}\kappa_{11}+6\kappa_{21}\kappa_{20}\,;\\
\mu_{32} &= \kappa_{32}+\kappa_{30}\kappa_{02}+6\kappa_{21}\kappa_{11}+3\kappa_{20}\kappa_{12}\,;\\
\mu_{51} &= \kappa_{51}+5\kappa_{40}\kappa_{11}+10\kappa_{31}\kappa_{20}+10\kappa_{30}\kappa_{21}+15\kappa_{20}^2\kappa_{11}\,;\\
\mu_{42} &= \kappa_{42}+\kappa_{40}\kappa_{02}+8\kappa_{31}\kappa_{11}+4\kappa_{30}\kappa_{12}+6\kappa_{22}\kappa_{20}+6\kappa_{21}^2+3\kappa_{20}^2\kappa_{02}+12\kappa_{20}\kappa_{11}^2\,;\\
\mu_{33} &= \kappa_{33}+3\kappa_{31}\kappa_{02}+\kappa_{30}\kappa_{03}+9\kappa_{22}\kappa_{11}+9\kappa_{21}\kappa_{12}+3\kappa_{20}\kappa_{13}+9\kappa_{20}\kappa_{11}\kappa_{02}+6\kappa_{11}^3.
\end{aligned}\right\} \tag{3.80}$$

$$\left.\begin{aligned}
\kappa_{31} &= \mu_{31}-3\mu_{20}\mu_{11}\,; \qquad \kappa_{22} = \mu_{22}-\mu_{20}\mu_{02}-2\mu_{11}^2\,;\\
\kappa_{41} &= \mu_{41}-4\mu_{30}\mu_{11}-6\mu_{21}\mu_{20}\,;\\
\kappa_{32} &= \mu_{32}-\mu_{30}\mu_{02}-6\mu_{21}\mu_{11}-3\mu_{20}\mu_{12}\,;\\
\kappa_{51} &= \mu_{51}-5\mu_{40}\mu_{11}-10\mu_{31}\mu_{20}-10\mu_{30}\mu_{21}+30\mu_{20}\mu_{11}\,;\\
\kappa_{42} &= \mu_{42}-\mu_{40}\mu_{02}-8\mu_{31}\mu_{11}-4\mu_{30}\mu_{12}-6\mu_{22}\mu_{20}-6\mu_{21}^2+6\mu_{20}^2\mu_{02}+24\mu_{20}\mu_{11}^2\,;\\
\kappa_{33} &= \mu_{33}-3\mu_{31}\mu_{02}-\mu_{30}\mu_{03}-9\mu_{22}\mu_{11}-9\mu_{21}\mu_{12}-3\mu_{20}\mu_{13}+18\mu_{20}\mu_{11}\mu_{02}+12\mu_{11}^3
\end{aligned}\right\} \tag{3.81}$$

The formulae have been given for μ'_{rs} in terms of κ_{rs} and vice versa for orders $r+s \leqslant 6$ by Cook (1951). The symbolic process has now been computerized and used by Kratky *et al.* (1972) to give the moments and the central moments in terms of cumulants for all bivariate cases up to order 9 and 10 respectively, and for many other multivariate cases up to order 10.

3.30 Wold (1934b) has given the following expressions for Sheppard corrections to bivariate moments and cumulants, the variates being grouped in intervals h_1, h_2.

$$\mu'_{rs} = \sum_{j=0}^{r}\sum_{k=0}^{s} h_1^j h_2^k \binom{r}{j}\binom{s}{k}(2^{1-j}-1)(2^{1-k}-1)B_j B_k \bar{\mu}'_{r-j,\,s-k}. \tag{3.82}$$

In particular

$$\left.\begin{aligned} \mu'_{20} &= \bar{\mu}'_{20}-\tfrac{1}{12}h_1^2, \quad \mu'_{11} = \bar{\mu}'_{11}, \quad \mu'_{02} = \bar{\mu}'_{02}-\tfrac{1}{12}h_2^2;\\ \mu'_{30} &= \bar{\mu}'_{30}-\mu'_{10}h_1^2/4, \quad \mu'_{21} = \bar{\mu}'_{21}-\bar{\mu}'_{01}h_1^2/12;\\ \mu'_{40} &= \bar{\mu}'_{40}-\bar{\mu}'_{20}h_1^2/2+7h_1^4/240, \quad \mu'_{31} = \bar{\mu}'_{31}-\bar{\mu}'_{11}h_1^2/4;\\ \mu'_{22} &= \bar{\mu}'_{22}-\bar{\mu}'_{20}h_2^2/12-\bar{\mu}'_{02}h_1^2/12+h_1^2h_2^2/144. \end{aligned}\right\} \tag{3.83}$$

For cumulants we have

$$\left.\begin{aligned} \kappa_{rs} &= \bar{\kappa}_{rs}, & r,s>0\\ \kappa_{r0} &= \bar{\kappa}_{r0}-B_r h_1^r/r, & r\geqslant 2\\ \kappa_{0s} &= \bar{\kappa}_{0s}-B_s h_2^s/s. & s\geqslant 2 \end{aligned}\right\} \tag{3.84}$$

Measures of skewness

3.31 We have considered measures of location and dispersion in Chapter 2. With the aid of the moments we can now proceed to consider measures of other qualities of the population, and in particular its departure from symmetry.

In a symmetrical population, mean, median and mode coincide. It is thus natural to take the deviation mean to mode or mean to median as measuring the skewness of the distribution. K. Pearson proposed the measure

$$\frac{\text{mean}-\text{mode}}{\sigma}$$

which is subject to the inconvenience of determining the mode. For a wide class of frequency-distributions known as Pearson's (cf. Chapter 6), this measure may, however, be expressed exactly in terms of the first four moments. We define

$$\beta_1 = \frac{\mu_3^2}{\mu_2^3} \tag{3.85}$$

$$\beta_2 = \frac{\mu_4}{\mu_2^2}. \tag{3.86}$$

It will be seen in **6.2** that for Pearson distributions

$$\frac{\text{mean}-\text{mode}}{\sigma} = \frac{\sqrt{\beta_1}(\beta_2+3)}{2(5\beta_2-6\beta_1-9)} \tag{3.87}$$

and this equation may be taken as defining a measure of skewness applicable to all distributions whose moments up to and including the fourth exist.

Johnson and Rogers (1951) show that for unimodal distributions, $(\text{mean}-\text{mode})/\sigma \leqslant \sqrt{3}$. Exercise 3.22 concerns the measure of skewness $(\text{mean}-\text{median})/\sigma$ for which general bounds of variation are given.

The coefficient β_1 itself is also a measure of skewness. Clearly, if the distribution is symmetrical, β_1 vanishes with μ_3, and the ratio of μ_3 to $\mu_2^{3/2}$ (i.e., $\sqrt{\beta_1}$) will give some indication of the extent of departure from symmetry. However, Ord (1968b) gives some asymmetrical distributions with as many zero odd-order moments as desired, and Exercise

3.26 gives one with odd-order moments all zero, so the value of $\sqrt{\beta_1}$ must be interpreted with some caution.

Generally we may define

$$\beta_{2n+1} = \frac{\mu_3 \mu_{2n+3}}{\mu_2^{n+3}}, \qquad \beta_{2n} = \frac{\mu_{2n+2}}{\mu_2^{n+1}}, \tag{3.88}$$

quantities which are not in general use but will be found to occur occasionally in statistical literature.

More convenient quantities than β_1 and β_2 for certain purposes are

$$\gamma_1 = \sqrt{\beta_1} = \frac{\mu_3}{\mu_2^{3/2}} = \frac{\kappa_3}{\kappa_2^{3/2}}, \tag{3.89}$$

$$\gamma_2 = \beta_2 - 3 = \frac{\mu_4}{\mu_2^2} - 3 = \frac{\kappa_4}{\kappa_2^2}. \tag{3.90}$$

If the distribution is expressed in standard measure, γ_1 and γ_2 are its third and fourth cumulants.

Kurtosis

3.32 In the so-called " normal " distribution

$$dF = \frac{1}{\sigma\sqrt{(2\pi)}} e^{-\frac{1}{2}x^2/\sigma^2} dx, \qquad -\infty \leqslant x \leqslant \infty$$

β_2 attains the value 3 and γ_2 is zero. Distributions for which $\gamma_2 = 0$ are called mesokurtic. Those for which $\gamma_2 > 0$ are called leptokurtic; those for which $\gamma_2 < 0$ are called platykurtic.

These names were originally conferred because among certain regular symmetric distributions, the more sharply peaked are leptokurtic, and the more flat-topped platykurtic. This is not necessarily so for other symmetrical or for asymmetrical distributions, and although the terms are useful they are better regarded as describing the sign of γ_2 rather than the shape of the distribution. Cf. Exercises 3.20–21 and 3.25.

By putting $a_i = (x_i - \mu_1')^2, b_i \equiv 1$ in the Cauchy–Schwarz inequality in **2.7**, we see that $\mu_4 \geqslant \mu_2^2$, with equality only when the squared deviations a_i do not vary at all, i.e. when the distribution is wholly concentrated at two values of x. Thus $\beta_2 \geqslant 1$ ($\gamma_2 \geqslant -2$) always.

Exercise 3.18 shows that for regular unimodal symmetric distributions, $\beta_2 \geqslant 1{\cdot}8$, while Exercise 3.19 gives a general relation between β_1 and β_2.

Example 3.18

For the distribution of Australian marriages considered in Example 3.1 we found, for the raw moments about the mean in units of three years,

$$\bar{\mu}_2 = 7{\cdot}056{,}977, \quad \bar{\mu}_3 = 36{\cdot}151{,}595, \quad \bar{\mu}_4 = 408{\cdot}738{,}210.$$

With Sheppard's corrections these become

$$\mu_2 = 6{\cdot}973{,}644, \quad \mu_3 = 36{\cdot}151{,}595, \quad \mu_4 = 405{\cdot}238{,}888.$$

From these values we find

$$\beta_1 = 3{\cdot}854, \quad \gamma_1 = 1{\cdot}963$$
$$\beta_2 = 8{\cdot}333, \quad \gamma_2 = 5{\cdot}333$$

indicating considerable skewness and leptokurtosis.

Example 3.19

From the formulae for the moments of the binomial distribution considered in Example 3.2 we find

$$\gamma_1 = \frac{q-p}{\sqrt{(npq)}}$$
$$\gamma_2 = \frac{1-6pq}{npq}$$

so that, as $n \to \infty$, γ_1 and $\gamma_2 \to 0$. This is in accordance with a result we shall prove later, that the binomial distribution in standard measure tends to the normal form as n tends to infinity.

Moments as characteristics of a distribution

3.33 The use of moments and cumulants in determining the nature of a frequency-distribution will be abundantly illustrated in later chapters, but some general remarks may be made at this stage.

It has been noted that the characteristic function determines the moments when they exist, and it will be proved in Chapter 4 that the characteristic function also determines the distribution function. It does not, however, follow that the moments completely determine the distribution, even when moments of all orders exist. Only under certain conditions will a set of moments determine a distribution uniquely, but, fortunately for statisticians, those conditions are obeyed by the distributions commonly arising in statistical practice. For all ordinary purposes, therefore, a knowledge of the moments, when they all exist, is equivalent to a knowledge of the distribution function: equivalent, that is, in the sense that it should be possible *theoretically* to exhibit all the properties of the distribution in terms of the moments.

3.34 In particular we expect that if two distributions have a certain number of moments in common they will bear some resemblance to each other. If, say, moments up to those of order n are identical we expect that as n tends to infinity the distributions approach identity, and consequently we expect that by identifying the lower moments of two distributions we bring them to approximate equality. Some mathematical support for this so-called Principle of Moments may be derived from the following approach.

It is known that a function which is continuous in a finite range a to b can be represented in that range by a uniformly convergent series of polynomials in x, say $\sum_{n=0}^{\infty} P_n(x)$ where $P_n(x)$ is of degree n. Suppose we wish to represent such a function approximately by the *finite* series of powers $\sum_{n=0}^{s} a_n x^n$. The coefficients a_n may be determined by the principle of least squares, i.e. so as to make

$$\int_a^b (f - \Sigma a_n x^n)^2 \, dx \tag{3.91}$$

a minimum. Differentiating by a we have

$$2\int_a^b (f-\Sigma a_n x^n)\, x^j\, dx = 0$$

or

$$\int_a^b f x^j\, dx = \mu_j' = \int_a^b \Sigma a_n x^{n+j}\, dx. \tag{3.92}$$

If now two distributions have moments up to order s equal they must have the same least-squares approximation, for the coefficients a_n are determined by the moments in virtue of (3.92). Furthermore, if in the range the distribution f_1 differs from $\Sigma a_n x^n$ by ε_1 and f_2 by ε_2, then f_1 differs from f_2 by not more than $\varepsilon_1+\varepsilon_2$.

A similar line of approach may be adopted when the range is infinite, the distributions in such cases being, under certain general conditions, capable of representation by a series of terms such as $e^{-x^2}P_n(x)$.(*) The same conclusion is reached.

Thus distributions which have a finite number of the lower moments in common will, in a sense, be approximations one to another. We shall encounter many cases where, although we cannot determine a distribution function explicitly, we may ascertain its moments at least up to some order ; and hence we shall be able to approximate to the distribution by finding another distribution of known form which has the same lower moments. In practice, approximations of this kind often turn out to be remarkably good, even when only the first three or four moments are equated.

Inequalities of the Chebyshev type

3.35 It is also possible, given certain moments or expected values of a distribution, to make precise statements about the frequency or distribution functions in terms of inequalities. Let F represent a distribution with an existent second moment. For any real t we have

$$\int_{-\infty}^{\infty}\left(1-\frac{x^2}{t^2}\right)dF = \left\{\int_{-\infty}^{-t}+\int_{t}^{\infty}\right\}\left(1-\frac{x^2}{t^2}\right)dF+\int_{-t}^{t}\left(1-\frac{x^2}{t^2}\right)dF.$$

In the first two integrals on the right $1-x^2/t^2$ is $\leqslant 0$ and hence

$$1-\frac{\mu_2'}{t^2} = \int_{-\infty}^{\infty}\left(1-\frac{x^2}{t^2}\right)dF \leqslant \int_{-t}^{t}\left(1-\frac{x^2}{t^2}\right)dF \leqslant \int_{-t}^{t} dF.$$

Thus the frequency from $-t$ to t is no less than $1-\mu_2'/t^2$. In particular, if we take an origin at the mean of the distribution,

$$F(t)-F(-t)\geqslant 1-\mu_2/t^2. \tag{3.93}$$

In the language of probability this is more usually written in the form, taking $\mu_2 = \sigma^2$ and $t = \lambda\sigma$,

$$\mathrm{P}\{|x-\mathrm{E}(x)|\leqslant \lambda\sigma\}\geqslant 1-1/\lambda^2 \tag{3.94}$$

or

$$\mathrm{P}\{|x-\mathrm{E}(x)|\geqslant \lambda\sigma\}\leqslant 1/\lambda^2. \tag{3.95}$$

This is the inequality given by I. J. Bienaymé in 1853. It is usually named after P. L. Chebyshev, who published it in 1867, but it is better called the Bienaymé–Chebyshev inequality. It gives us a lower limit to the proportion of frequency lying within a range of $\lambda\sigma$ on either side of the mean. Being expressed with great generality, the upper

(*) A theorem of Vera Myller-Lebedeff (1907, *Math. Ann.*, **64**, 388) states that a function can be expanded in a series of derivatives of $\exp(-x^2)$. This is not to be confused with the Gram–Charlier series of Chapter 6, where a function is expanded in terms of derivatives of $\exp(-\frac{1}{2}x^2)$.

limits are usually too wide to be of much use in determining frequencies in particular cases, but such inequalities are invaluable in proving limiting properties and are not without their uses for determining probabilities in a very rough way.

3.36 There are by now a large number of inequalities of the Chebyshev type. We shall meet several of them in the sequel. Here we will record a few of the commoner ones for reference.

If ν_r is the absolute moment of order r it is easy to show by a direct extension of the foregoing method that

$$\mathrm{P}\{|x-\mathrm{E}(x)|/\nu_r^{1/r}\leqslant\lambda\}\geqslant 1-1/\lambda^r. \tag{3.96}$$

This inequality, given by K. Pearson (1919), gives (3.94) when $r = 2$. When the variate does not take negative values we may likewise show (cf. Exercise 2.15) that its d.f. satisfies, for all $t > 0$,

$$F(t)>1-\mu_1'/t \tag{3.97}$$

which is sometimes called Markov's inequality. This gives (3.94) if we take $\{x-\mathrm{E}(x)\}^2$ as the variate and $t = \lambda^2\sigma^2$.

For general distributions these inequalities cannot be improved upon, but it is rather remarkable that by imposing quite broad restrictions on the form of the frequency function we can improve them considerably. For example, if a continuous frequency function has a single mode, say m_0, and the origin is taken there,

$$\mathrm{P}\{|x-m_0|\leqslant\lambda\tau\}\geqslant 1-4/(9\lambda^2) \tag{3.98}$$

where τ is the square root of the second moment about the mode, as compared with (3.94). This expression, in a different form, is traceable to Gauss.

3.37 By making further assumptions (e.g., that several moments are known), we may obtain more refined inequalities. For an account of these inequalities see a comprehensive review of the subject by Godwin (1955) and the monograph by the same author *Inequalities on Distribution Functions* (1964, Griffin).

EXERCISES

3.1 Show that the discontinuous (Poisson) distribution whose frequencies corresponding to the values 0, 1, . . . , j, . . are

$$e^{-\lambda}\left(1, \frac{\lambda}{1!}, \ldots, \frac{\lambda^j}{j!}, \ldots\right)$$

has, for the moments about the mean,

$$\mu_2 = \lambda,\ \mu_3 = \lambda,\ \mu_4 = \lambda(1+3\lambda),\ \mu_5 = \lambda(1+10\lambda),\ \mu_6 = \lambda(1+25\lambda+15\lambda^2).$$

3.2 For the distribution

$$dF = kx^{-p}e^{-\gamma/x}\,dx, \qquad 0 \leqslant x \leqslant \infty\ ;\ \gamma > 0,$$

show that, about the origin,

$$\mu'_r = \frac{\gamma^r\,\Gamma(p-r-1)}{\Gamma(p-1)}$$

if $r < p-1$, and does not exist in the contrary case.

3.3 In the distribution

$$dF = k\left(1+\frac{x^2}{a^2}\right)^{-m}\exp\{-\nu\arctan(x/a)\}\,dx, \qquad -\infty \leqslant x \leqslant \infty,$$

show that, about the origin,

$$\mu'_r = ka^{r+1}\int_{-\frac{1}{2}\pi}^{\frac{1}{2}\pi}\cos^{2m-r-2}\theta\,\sin^r\theta\,e^{-\nu\theta}\,d\theta$$

and hence that

$$\mu'_r = \frac{a}{2m-r-1}\{(r-1)a\mu'_{r-2}-\nu\mu'_{r-1}\}.$$

3.4 $f(x)$ is a continuous frequency function, symmetrical about $x = 0$ and ranging from $-a$ to $+a$. It has a single mode at $x = 0$. Show that

$$\mu_{2r} < a^{2r}/(2r+1), \qquad r \geqslant 1.$$

3.5 Show that for any symmetrical distribution the cumulants of odd order (except κ_1) are zero if they exist.

3.6 Show that for the exponential distribution

$$dF = e^{-x/\sigma}\,dx/\sigma, \qquad 0 \leqslant x \leqslant \infty\ ;\ \sigma > 0,$$

$$\kappa_r = \sigma^r(r-1)!$$

3.7 Show that e^{itx} may be expanded in an infinite series, valid in $-\infty \leqslant x \leqslant \infty$,

$$1+(e^{it}-1)x^{[1]}+(e^{it}-1)^2\frac{x^{[2]}}{2!}+\ldots+(e^{it}-1)^r\frac{x^{[r]}}{r!}+\ldots$$

the factorials being taken with unit interval; and hence that

$$\mu'_{[r]} = [D^r\phi(t)]_{t=0}$$

where

$$D = \frac{d}{d(e^{it})}.$$

Hence show that, for the binomial $(q+p)^n$ about the origin,

$$\mu'_{[r]} = n^{[r]}p^r.$$

3.8 Show that for the distribution

$$dF = \frac{k\,dx}{1+x^{2r}}, \quad -\infty \leqslant x \leqslant \infty,\ r \text{ a positive integer,}$$

the moments μ_p exist for $p \leqslant 2(r-1)$. Show that for finite moments,

$$\mu_{2s-1} = 0, \quad s = 1, 2, \dots$$

$$\mu_{2s} = \sin\left(\frac{\pi}{2r}\right) \Big/ \sin\left\{\frac{(2s+1)\pi}{2r}\right\}.$$

3.9 Show that

$$\mu'_r = \sum_{j=1}^{r} \binom{r-1}{j-1} \mu'_{r-j} \kappa_j$$

and hence that

$$\kappa_r = (-1)^{r-1} \begin{vmatrix} \mu'_1 & 1 & 0 & 0 & \cdot & 0 \\ \mu'_2 & \binom{1}{0}\mu'_1 & 1 & 0 & \cdot & \cdot & 0 \\ \mu'_3 & \binom{2}{0}\mu'_2 & \binom{2}{1}\mu'_1 & 1 & \cdot & \cdot & 0 \\ \cdot & \cdot & \cdot & \cdot & \cdot & \cdot & 1 \\ \mu'_r & \binom{r-1}{0}\mu'_{r-1} & \binom{r-1}{1}\mu'_{r-2} & \cdot & \cdot & \cdot & \binom{r-1}{r-2}\mu'_1 \end{vmatrix}$$

3.10 Show that for the distribution $dF = dx$, $0 \leqslant x \leqslant 1$, grouped into an integral number of intervals of equal width h, the corrections to the second and fourth moments about the mean are

$$\mu_2 = \bar{\mu}_2 + \frac{h^2}{12}, \quad \mu_4 = \bar{\mu}_4 + \bar{\mu}_2\frac{h^2}{2} + \frac{h^4}{80}.$$

(Cf. Elderton, 1938. Note that the first is exactly, and the second approximately, the Sheppard correction *with sign reversed.*)

3.11 If ∂_p stands for the operator such that

$$\partial_p \mu'_r = r^{[p]} \mu'_{r-p} \qquad r \geqslant p$$
$$= 0, \qquad r < p,$$

and ∂_p is distributive when applied to products, e.g.

$$\partial_p(AB) = B(\partial_p A) + A(\partial_p B),$$

show that ∂_p annihilates every cumulant (considered as a function of the moments) except κ_p, and that $\partial_p \kappa_p = p!$

3.12 If $f(x)$ is an odd function of x of period $\frac{1}{2}$, show that

$$\int_0^\infty x^r x^{-\log x} f(\log x)\,dx = 0$$

for all integral values of r. Hence show that the distributions

$$dF = x^{-\log x}\{1 - \lambda \sin(4\pi \log x)\}\,dx, \qquad 0 \leqslant x \leqslant \infty;\ 0 \leqslant \lambda \leqslant 1,$$

have the same moments whatever the value of λ. (Stieltjes, 1918)

3.13 Show that if the frequencies of a discontinuous distribution are distributed at equal intervals h/m, with m in each grouping-interval h, the average grouping corrections to the cumulants are given by

$$\kappa_r = \bar{\kappa}_r - \frac{B_r h^r}{r}\left(1 - \frac{1}{m^r}\right).$$

(C. C. Craig, 1936b)

3.14 Show that for the bivariate normal distribution

$$dF = \frac{1}{2\pi\sigma_1\sigma_2(1-\rho^2)^{\frac{1}{2}}}\exp\left[-\frac{1}{2(1-\rho^2)}\left\{\frac{x_1^2}{\sigma_1^2}-\frac{2\rho x_1 x_2}{\sigma_1\sigma_2}+\frac{x_2^2}{\sigma_2^2}\right\}\right]dx_1\,dx_2, \qquad -\infty \leqslant x_1,\ x_2 \leqslant \infty,$$

all cumulants κ_{rs}, r, $s>2$, vanish; and further, if

$$\lambda_{rs} = \frac{\mu_{rs}}{\sigma_1^r\sigma_2^s}$$

$$\lambda_{rs} = (r+s-1)\rho\lambda_{r-1,s-1}+(r-1)(s-1)(1-\rho^2)\lambda_{r-2,s-2}$$

$$\lambda_{2r,2s} = \frac{(2r)!\,(2s)!}{2^{r+s}}\sum_{j=0}^{t}\frac{(2\rho)^{2j}}{(r-j)!\,(s-j)!\,(2j)!}$$

$$\lambda_{2r+1,2s+1} = \frac{(2r+1)!\,(2s+1)!}{2^{r+s}}\rho\sum_{j=0}^{t}\frac{(2\rho)^{2j}}{(r-j)!\,(s-j)!\,(2j+1)!}$$

$$\lambda_{2r,2s+1} = \lambda_{2r+1,2s} = 0,$$

where t is the smaller of r and s. In particular,

$$\lambda_{11} = \rho,\quad \lambda_{31} = 3\rho,\quad \lambda_{51} = 15\rho,\quad \lambda_{71} = 105\rho,\quad \lambda_{91} = 945\rho$$
$$\lambda_{22} = (1+2\rho^2),\quad \lambda_{24} = 3(1+4\rho^2),\quad \lambda_{26} = 15(1+6\rho^2),$$
$$\lambda_{28} = 105(1+8\rho^2),\quad \lambda_{2,10} = 945(1+10\rho^2);$$
$$\lambda_{33} = 3\rho(3+2\rho^2),\quad \lambda_{35} = 15\rho(3+4\rho^2),$$
$$\lambda_{44} = 3(3+24\rho^2+8\rho^4).$$

3.15 Regarding the absolute moment ν_r' at (3.21) as a function of r, show that $\dfrac{\partial^2 \log \nu_r'}{\partial r^2} > 0$ in non-degenerate cases, and hence that for any $a \geqslant b \geqslant c \geqslant 0$

$$\log \nu_b' \leqslant \{(a-b)\log \nu_c' + (b-c)\log \nu_a'\}/(a-c),$$

whence Liapounoff's inequality (3.22) follows. (Belz, 1947)

3.16 Verify the expressions for κ_{42} and κ_{33} in equation (3.81).

3.17 A distribution is grouped in intervals of width h. Show that as the grouping grid moves along the variate axis the raw moments have period h. Hence, writing

$$\bar{\mu}_r' = \sum_{j=-\infty}^{\infty}\zeta^r\int_{\zeta-\frac{1}{2}h}^{\zeta+\frac{1}{2}h} f(x)\,dx$$

$$= A_0 + \sum_{j=1}^{\infty} A_j \sin j\theta + \sum_{j=1}^{\infty} B_j \cos j\theta$$

where
$$\zeta = (j+\theta/2\pi)h,$$
show that

$$A_0 = \frac{1}{h}\int_{-\infty}^{\infty} x^r \int_{-\frac{1}{2}h}^{\frac{1}{2}h} f(x+\zeta)\,d\zeta\,dx$$

and hence that this leads to Sheppard's corrections.

Show also that

$$A_s = \frac{2}{h}\int_{-\infty}^{\infty} f(x)\,dx\int_{x-\frac{1}{2}h}^{x+\frac{1}{2}h}\zeta^r \sin(2\pi s\zeta/h)\,d\zeta$$

with a similar expression for B_s.

For the distribution $\dfrac{1}{\sqrt{(2\pi)}}e^{-\frac{1}{2}x^2}dx$, $-\infty \leqslant x \leqslant \infty$, show that, for corrections to the mean, the B's vanish and $A_s = (-1)^{s+1}\dfrac{h}{\pi s}\exp\left(-\dfrac{2s^2\pi^2}{h^2}\right)$ and hence that, even for a coarse grouping $h = 1$, the correction to the mean is not greater than $e^{-2\pi^2}/\pi$ approximately.

(Fisher, 1921a)

3.18 *The Gauss–Winckler inequality.* By considering $1-f(x)/f(\tilde{x})$ as a distribution function and using Liapounoff's inequality, show that for a frequency function which is differentiable and has a single mode $\tilde{x}$, the absolute moments about $\tilde{x}$ obey the relation

$$\{(r+1)\nu_r'\}^{1/r} \leqslant \{(n+1)\nu_n'\}^{1/n}, \quad r<n.$$

In particular show that, about the mode, $\mu_4'/\mu_2'^2 \geqslant 1{\cdot}8$.

3.19 Show that

$$\begin{vmatrix} \mu_0' & \mu_1' & \mu_2' \\ \mu_1' & \mu_2' & \mu_3' \\ \mu_2' & \mu_3' & \mu_4' \end{vmatrix} > 0$$

and hence that $\beta_2 > 1+\beta_1$.

3.20 Show that if two distributions $f(x)$, $g(x)$ are symmetrical with zero means and unit variances and

$$f(x) < g(x) \quad \text{for } a < |x| < b,$$
$$f(x) > g(x) \quad \text{otherwise,}$$

then μ_4 is greater for $f(x)$ than for $g(x)$. (Finucan, 1964)

3.21 The following four unimodal symmetrical f.f.'s have zero mean and unit variance. Show that they have the values of μ_4 and the maximum ordinates indicated.

		Value of μ_4	Maximum ordinate
(1)	$\frac{1}{3\sqrt{\pi}}\left(\frac{9}{4}+x^4\right)e^{-x^2}$	2·75	0·423
(2)	$\frac{3}{2\sqrt{(2\pi)}}e^{-\frac{1}{2}x^2}-\frac{1}{6\sqrt{\pi}}\left(\frac{9}{4}+x^4\right)e^{-x^2}$	3·125	0·387
(3)	$\frac{1}{6\sqrt{\pi}}(e^{-\frac{1}{4}x^2}+4e^{-x^2})$	4·5	0·470
(4)	$\frac{3\sqrt{3}}{16\sqrt{\pi}}(2+x^2)e^{-\frac{3}{4}x^2}$	2·667	0·366

Comparing these with the normal values $\mu_4 = 3$, max. ordinate $= 0{\cdot}399$, show that any combination of higher or lower ordinate at the centre and lepto- or platykurtosis is possible. (Kaplansky, 1945)

3.22 x is a variate with mean μ, median M, variance σ^2, $\int_{\mu+}^{\infty} dF = p$, $\int_{-\infty}^{\mu-} dF = q$, $p+q \leqslant 1$.

Putting $a = \int_{\mu+}^{\infty}(x-\mu)dF = -\int_{-\infty}^{\mu-}(x-\mu)dF$, show that $a^2\left(\frac{1}{p}+\frac{1}{q}\right) \leqslant \sigma^2$, that $a \geqslant \frac{1}{2}(M-\mu)$ if $M>\mu$, $a \geqslant -\frac{1}{2}(M-\mu)$ if $M<\mu$, and hence that

$$\left|\frac{\mu-M}{\sigma}\right| \leqslant 2\left(\frac{1}{p}+\frac{1}{q}\right)^{-\frac{1}{2}} \leqslant 1.$$

(Majindar (1962); he also shows that the left-hand equality is not attainable. Hotelling and Solomons (1932) gave $\left|\frac{\mu-M}{\sigma}\right| \leqslant 1$.)

3.23 Using (3.61) show that the c.f. of the rectangular distribution $dF = dx/h$, $-\frac{1}{2}h \leqslant x \leqslant \frac{1}{2}h$ is given by

$$\phi_x(t) = \sinh(\tfrac{1}{2}ith)/(\tfrac{1}{2}ith)$$

with cumulants

$$\kappa_{2r-1} = 0, \qquad \kappa_{2r} = B_{2r} h^{2r}/(2r), \qquad r = 1, 2, \ldots.$$

Similarly for the discrete rectangular distribution

$$P(y = r) = 1/n, \qquad r = 1, 2, \ldots, n,$$

show that the c.f. about the mean is

$$\phi_y(t) = \sinh(\tfrac{1}{2}itn)/\{n \sinh(\tfrac{1}{2}it)\}$$

with cumulants about zero

$$\kappa_1 = \tfrac{1}{2}(n+1)$$
$$\kappa_{2r+1} = 0,$$
$$\kappa_{2r} = B_{2r}(n^{2r}-1)/(2r), \qquad r = 1, 2, \ldots.$$

Deduce $\phi_y(t)$ from $\phi_x(t)$ by considering the sum of independent variates x (with $h = 1$) and y.

3.24 Show that if the variate takes integer values 0, 1, 2, . . ., n only, (3.29) may be inverted to give

$$f_j = \frac{1}{j!} \sum_{r=0}^{n-j} (-1)^r \mu'_{[j+r]}/r!$$

Invert the factorial moments obtained in Example 3.8 by this means.

3.25 A sequence of variates x_K is defined by the d.f.'s

$$F_K(x) = \left(1 - \frac{1}{K^2-1}\right)\Phi(x) + \frac{1}{K^2-1}\Phi\left(\frac{x}{K}\right), \qquad K = 2, 3, \ldots,$$

where $\Phi(x) = \int_{-\infty}^{x} (2\pi)^{-\frac{1}{2}} e^{-\frac{1}{2}u^2} du$ is the d.f. of the standardized normal distribution. Show that each x_K is symmetric about the origin, with variance 2, and that its kurtosis coefficient $\gamma_2 = \frac{3}{4}(K^2-2)$, increasing monotonically to infinity as $K \to \infty$, when $F_K(x) \to \Phi(x)$.

(Ali, 1974)

3.26 Using the result for $p > 0$

$$\int_0^\infty x^{p-1} e^{-x} \sin x \, dx = 2^{-\frac{1}{2}p} \Gamma(p) \sin(p\pi/4),$$

show that the distribution

$$dF = \tfrac{1}{48}\{1 - \operatorname{sgn} x \sin(|x|^{1/4})\} \exp\{-|x|^{1/4}\}, \qquad -\infty \leqslant x \leqslant \infty,$$

has all odd-order moments zero, but that it is not symmetrical.

(Cf. Churchill (1946)—the result is due to Stieltjes.)

CHAPTER 4

CHARACTERISTIC FUNCTIONS

4.1 In 3.5, we introduced the characteristic function

$$\phi(t) = \int_{-\infty}^{\infty} e^{itx}\, dF \tag{4.1}$$

as a moment-generating function. It has many other useful and important properties which give it a central rôle in statistical theory and in this chapter we shall give an account of them. To establish our general theorems we shall, however, have to use somewhat advanced mathematics; and the reader who is interested principally in the statistical applications may prefer to omit the proofs, especially at first reading, and merely note the results. Broadly speaking, this part of our theory may be regarded as a special case of the general theory of Fourier transforms—special because our functions are frequency and distribution functions. Several of the proofs in this chapter are based on Lévy (1925). Lukacs (1970) has published an elegant monograph on the subject. See also Lukacs and Laha (1964).

4.2 We recall, in the first instance, that $\phi(t)$ always exists, since

$$|\phi(t)| = \left|\int_{-\infty}^{\infty} e^{itx}\, dF\right| \leqslant \int_{-\infty}^{\infty} |e^{itx}|\, dF = \int_{-\infty}^{\infty} dF = 1, \tag{4.2}$$

so that the defining integral converges absolutely. Further, $\phi(t)$ is uniformly continuous in t and differentiable j times under the integral sign if the resulting expressions exist and are uniformly convergent, for which it is sufficient that ν_j exists. For then

$$\begin{aligned} |\phi^{(j)}(t)| &= \left|\int_{-\infty}^{\infty} x^j e^{itx}\, dF\right| \\ &\leqslant \int_{-\infty}^{\infty} |x^j|\, dF = \nu_j. \end{aligned} \tag{4.3}$$

$\phi(t)$ is real if and only if the f.f. is symmetrical—cf. Exercise 4.1.

The Inversion Theorem

4.3 We now prove the fundamental theorem of the theory of characteristic functions, which will be called the Inversion Theorem, namely that the characteristic function (c.f.) uniquely determines the distribution function; more precisely, if $\phi(t)$ is given by (4.1) then

$$F(x) - F(0) = \frac{1}{2\pi}\int_{-\infty}^{\infty} \frac{1-e^{-itx}}{it}\phi(t)\, dt, \tag{4.4}$$

the integral being understood as a principal value, i.e. as

$$\lim_{c\to\infty} \frac{1}{2\pi}\int_{-c}^{c} \frac{1-e^{-itx}}{it}\phi(t)\, dt.$$

Further, if $F(x)$ is continuous everywhere and $dF = f(x)\,dx$

$$f(x) = \frac{1}{2\pi}\int_{-\infty}^{\infty} e^{-itx}\,\phi(t)\,dt, \tag{4.5}$$

the integral, as before, being a principal value if there is not separate convergence at the limits. Equation (4.5) may be compared with the form

$$\phi(t) = \int_{-\infty}^{\infty} e^{itx} f(x)\,dx, \tag{4.6}$$

the comparison exhibiting the kind of reciprocal relationship which exists between $f(x)$ and $\phi(t)$.

Equation (4.4) contains $F(0)$ and we may avoid introducing this quantity by an alternative form

$$F(x) = \frac{1}{2}+\frac{1}{2\pi}\int_{0}^{\infty} \frac{e^{itx}\phi(-t)-e^{-itx}\phi(t)}{it}\,dt. \tag{4.7}$$

We require as a preliminary result the theorem that for a real number ξ,

$$\frac{1}{\pi}\int_{-\infty}^{\infty} \frac{\sin t\xi}{t}\,dt = \left.\begin{array}{rl} -1, & \xi < 0 \\ 0, & \xi = 0 \\ 1, & \xi > 0 \end{array}\right\} \tag{4.8}$$

The quantity on the right of (4.8) is written sgn ξ, read as " signum ξ ". The result is, perhaps, most easily established by evaluating the complex integral $\int \frac{e^{iz\xi}}{iz}\,dz$ round a contour consisting of a large semicircle, the real axis and a small semicircle round the origin. We note that for a variate u and some fixed x

$$\begin{aligned}\int_{-\infty}^{\infty} \text{sgn}\,(u-x)\,dF(u) &= -\int_{-\infty}^{x} dF(u)+\int_{x}^{\infty} dF(u) \\ &= 1-2F(x).\end{aligned} \tag{4.9}$$

Now we have, for a positive number c, the uniformly convergent integral

$$\begin{aligned} I_c \equiv \int_{-c}^{c} \frac{1-e^{-ixt}}{it}\phi(t)\,dt &= \int_{-c}^{c} \frac{1-e^{-ixt}}{it}\int_{-\infty}^{\infty} e^{itu}\,dF(u)\,dt \\ &= \int_{-c}^{c}\int_{-\infty}^{\infty} \frac{e^{itu}-e^{it(u-x)}}{it}\,dF(u)\,dt \\ &= \int_{-c}^{c}\int_{-\infty}^{\infty} \frac{\sin tu-\sin\{t(u-x)\}}{t}\,dF(u)\,dt. \end{aligned}$$

Because the integral is uniformly convergent, we may change the order of integration to obtain

$$I_c = \int_{-\infty}^{\infty}\int_{-c}^{c}\left\{\frac{\sin tu}{t}-\frac{\sin\{t(u-x)\}}{t}\right\}dt\,dF(u). \tag{4.10}$$

The integral with respect to t is continuous in c and bounded. We may therefore let c tend to infinity to obtain, from (4.8) and (4.9),

$$\begin{aligned}\lim_{c\to\infty} I_c &= \pi\int_{-\infty}^{\infty} \{\text{sgn}\,u-\text{sgn}\,(u-x)\}\,dF(u) \\ &= \pi[1-2F(0)-\{1-2F(x)\}] = 2\pi\{F(x)-F(0)\}.\end{aligned}$$

Equation (4.4) follows. Equation (4.7) may be obtained in a similar way by use of the result

$$\frac{2}{\pi}\int_0^\infty \frac{\sin t\xi}{t}\,dt = \operatorname{sgn}\xi. \tag{4.11}$$

Further, if $f(x)$ exists we may differentiate (4.4) under the integral sign, since the integral converges uniformly, to obtain (4.5). We also have, on differentiating (4.7),

$$f(x) = \frac{1}{2\pi}\int_0^\infty \{e^{itx}\phi(-t)+e^{-itx}\phi(t)\}\,dt. \tag{4.12}$$

4.4 From the definition of the c.f. we note that $\phi(t)$ and $\phi(-t)$ are conjugate quantities and hence, if $\mathscr{R}(t)$ and $\mathscr{I}(t)$ are the real and imaginary parts of $\phi(t)$ we have, from (4.4),

$$F(x)-F(0) = \frac{1}{2\pi}\int_{-\infty}^{\infty} \frac{\mathscr{R}(t)\sin xt+\mathscr{I}(t)(1-\cos xt)}{t}\,dt, \tag{4.13}$$

and, from (4.7),

$$F(x) = \tfrac{1}{2}+\frac{1}{\pi}\int_0^\infty \frac{\mathscr{R}(t)\sin xt-\mathscr{I}(t)\cos xt}{t}\,dt. \tag{4.14}$$

Similarly we have

$$f(x) = \frac{1}{2\pi}\int_{-\infty}^{\infty} \{\mathscr{R}(t)\cos xt+\mathscr{I}(t)\sin xt\}\,dt. \tag{4.15}$$

4.5 Consider now the expression J_c defined by

$$J_c = \frac{1}{2c}\int_{-c}^{c} e^{-itx}\phi(t)\,dt. \tag{4.16}$$

If the distribution function $F(x)$ has a derivative $f(x)$ we have, from (4.5),

$$\lim_{c\to\infty} J_c = 2\pi \lim f(x)/2c = 0$$

and hence J_c tends to zero at all points where $F(x)$ is continuous and differentiable, i.e. if the frequency function is continuous.

If the distribution function is discrete, consider one point of discontinuity, say where the frequency is f_j. The contribution of this part of the frequency to $\phi(t)$ is $f_j \exp(itx_j)$ and thus the contribution to J_c is

$$\frac{1}{2c}\int_{-c}^{c} f_j \exp(itx_j - itx)\,dt$$
$$= \frac{1}{2c} f_j \left[\frac{\exp\{it(x_j-x)\}}{i(x_j-x)}\right]_{-c}^{c}.$$

If $x_j \neq x$ this clearly tends to zero as $c\to\infty$. But if $x_j = x$ the integral is simply

$$\frac{1}{2c}\int_{-c}^{c} f_j\,dt = f_j.$$

Thus J_c tends to f_j at $x = x_j$. Hence, if J_c tends to zero at a point, there is no discontinuity in the distribution function at that point; but if it tends to a positive number

f_j the d.f. is discontinuous at that point and the frequency is f_j. This gives us a criterion whether a given c.f. represents a continuous distribution or not.

4.6 Looking similarly to (4.6) we see that if the distribution has discrete frequencies at x_j ($j = 1, 2, \ldots$) the c.f. is given by

$$\phi(t) = \Sigma f_j \exp(itx_j)$$

and

$$\begin{aligned} F(x) &= \frac{1}{2}+\frac{1}{2\pi}\int_0^\infty \frac{\Sigma f_j \exp\{it(x-x_j)\} - \Sigma f_j \exp\{-it(x-x_j)\}}{it}\,dt \\ &= \frac{1}{2}+\frac{1}{\pi}\Sigma f_j \int_0^\infty \frac{\sin t(x-x_j)}{t}\,dt \\ &= \tfrac{1}{2}+\tfrac{1}{2}\Sigma f_j \operatorname{sgn}(x-x_j). \end{aligned} \tag{4.17}$$

This has a saltus at $x = x_j$ equal to f_j. We note, however, that for $x = x_k$

$$\begin{aligned} F(x_k) &= \tfrac{1}{2}+\tfrac{1}{2}\sum_{j=1}^{k-1} f_j - \tfrac{1}{2}\sum_{j=k+1}^{\infty} f_j \\ &= \tfrac{1}{2}\left(1-\sum_{k+1}^{\infty} f_j\right)+\tfrac{1}{2}\sum_{1}^{k-1} f_j \\ &= \sum_{1}^{k-1} f_j + \tfrac{1}{2}f_k. \end{aligned} \tag{4.18}$$

The value assigned to $F(x)$ at the saltus is thus the frequency up to that point plus one-half of the increase in frequency at that point. On our ordinary definition, with $F(x)$ continuous on the right, the value of $F(x_k)$ would be $\sum_1^k f_j$. This is a point to remember when the distribution function of a discrete distribution is obtained from the characteristic function.

Example 4.1

We found in Example 3.4 that the characteristic function of the normal distribution

$$dF = \frac{1}{\sigma\sqrt{(2\pi)}} e^{-x^2/2\sigma^2}\,dx, \qquad -\infty \leqslant x \leqslant \infty,$$

is

$$\phi(t) = \exp(-\tfrac{1}{2}t^2\sigma^2).$$

Suppose we are given this function and require to find the distribution, if any, of which it is the characteristic function.

In the first place we note that the distribution, if any, is continuous. For

$$J_c = \frac{1}{2c}\int_{-c}^{c} \exp\{-\tfrac{1}{2}t^2\sigma^2 - itx\}\,dt$$

and the integral is less in modulus than $\int_{-c}^{c} \exp(-\tfrac{1}{2}t^2\sigma^2)\,dt$, which is less than

$\int_{-\infty}^{\infty} \exp(-\frac{1}{2}t^2\sigma^2)\,dt = \sqrt{(2\pi)}/\sigma$. Thus $J_c \to 0$ everywhere. We have for the frequency function, if any,

$$f(x) = \frac{1}{2\pi}\int_{-\infty}^{\infty} \exp\{-\tfrac{1}{2}t^2\sigma^2 - itx\}\,dt$$

$$= \frac{1}{2\pi}e^{-x^2/2\sigma^2}\int_{-\infty}^{\infty} \exp\{-\tfrac{1}{2}(t\sigma + ix/\sigma)^2\}\,dt.$$

This may be regarded as an integral in the complex plane along a line parallel to the real axis. Taking $t\sigma + ix/\sigma = u$ as a new variable we find that the integral is $\sqrt{(2\pi)}/\sigma$. Thus

$$f(x) = \frac{1}{\sigma\sqrt{(2\pi)}}e^{-\frac{1}{2}x^2/\sigma^2}.$$

This is everywhere positive and $\int_{-\infty}^{\infty} dF$ converges to unity. Hence it is, in fact, a frequency function with the given expression for its c.f.

Example 4.2

To find the frequency function, if any, for which

$$\phi(t) = e^{-|t|}.$$

We note that J_c tends to zero and that the distribution, if any, is continuous. We then have for $f(x)$, if it exists,

$$f(x) = \frac{1}{2\pi}\int_{-\infty}^{\infty} e^{-|t|}e^{-itx}\,dt$$

and using symmetries, we therefore have

$$\pi f(x) = \int_0^{\infty} e^{-t}\cos tx\,dt$$

$$= \Big[-e^{-t}\cos tx\Big]_0^{\infty} - x\int_0^{\infty} e^{-t}\sin tx\,dt$$

$$= 1 - x\left\{\Big[-e^{-t}\sin tx\Big]_0^{\infty} + x\int_0^{\infty} e^{-t}\cos tx\,dt\right\}$$

$$= 1 - x^2\pi f(x).$$

Thus

$$f(x) = \frac{1}{\pi(1+x^2)}, \qquad -\infty \leqslant x \leqslant \infty.$$

As before, this function can represent a frequency function, and Example 3.13 showed that $f(x)$ has, in fact, the required characteristic function. This f.f. is known as the Cauchy distribution, although it was discussed by Poisson in 1824, nearly 30 years earlier than by Cauchy.

Example 4.3

Does there exist a frequency function for which $\phi(t) = e^{it}$?

We have

$$J_c = \frac{1}{2c}\int_{-c}^{c} e^{it-itx}\,dt = \frac{1}{2c}\int_{-c}^{c} e^{it(1-x)}\,dt.$$

If $1-x$ is not zero the integral is

$$\int_{-c}^{c} [\cos\{(1-x)t\} + i\sin\{(1-x)t\}]\,dt.$$

Since $\sin t$ is an odd function this is equal to

$$\int_{-c}^{c} \cos\{(1-x)t\}\,dt = \left[-\frac{\sin(1-x)t}{1-x}\right]_{-c}^{c}.$$

This does not converge, but it is bounded and hence $J_c \to 0$.

If, however, $x = 1$, the integral is simply

$$\int_{-c}^{c} dt$$

and thus $J_c = 1$.

Thus there is unit frequency at $x = 1$ and it is seen at once that this accounts for the whole of the frequency, so that there is no frequency elsewhere. The distribution thus consists of a unit at $x = 1$. This is otherwise evident from the consideration that $\log \phi(t) = it$, so that the second cumulant is zero and there is no dispersion.

Example 4.4

For what distribution, if any, are the cumulants given by $\kappa_r = (r-1)!$?

The series

$$\sum_{j=1}^{\infty} \kappa_j \frac{(it)^j}{j!} = \Sigma \frac{(it)^j}{j}$$

converges absolutely for $|t| < 1$ and is thus equal to the c.g.f., say $\psi(t)$, if such a function exists. We have

$$\psi(t) = \Sigma \frac{(it)^j}{j} = -\log(1-it),$$

and thus

$$\phi(t) = \frac{1}{1-it}.$$

If the frequency function exists we have

$$f(x) = \frac{1}{2\pi}\int_{-\infty}^{\infty} \frac{e^{-ixz}}{1-it}\,dt.$$

This integral may be evaluated by integrating the complex function $\frac{e^{-ixz}}{1-iz}$ round a contour consisting of the real axis and the infinite semicircle below that axis. The first part reduces to the integral we are seeking with its sign reversed. On the semicircle of radius R we have $z = R(\cos\theta + i\sin\theta)$ and the integrand becomes

$$\frac{\exp(-ixR\cos\theta + xR\sin\theta)}{1 - iR\cos\theta + R\sin\theta}.$$

θ here lies between π and 2π and hence $\sin\theta$ is negative. Hence *if x is positive* the expression is less in modulus than

$$\frac{e^{-xR|\sin\theta|}}{R},$$

i.e. tends to zero as $R \to \infty$.

Now the function $\dfrac{e^{-ixz}}{1-iz}$ has a pole within the domain of integration at $z = -i$ and the residue there is ie^{-x}. Hence

$$\begin{aligned} f(x) &= -\frac{1}{2\pi}. -2\pi e^{-x} \\ &= e^{-x}, \qquad 0 \leqslant x \leqslant \infty. \end{aligned}$$

More generally, if $\kappa_r = p(r-1)!, p>0$, it will be found that the residue of $\dfrac{e^{-ixz}}{(1-iz)^p}$ is $\dfrac{i\,x^{p-1}e^{-x}}{\Gamma(p)}$, so that the distribution is

$$f(x) = \frac{e^{-x}x^{p-1}}{\Gamma(p)}, \qquad 0 \leqslant x \leqslant \infty;\ p>0.$$

Example 4.5

For what distribution, if any, are all cumulants of odd order zero and those of even order a constant, say $2a$?

We have for the c.g.f.

$$\psi(t) = 2a\left\{\frac{(it)^2}{2!}+\frac{(it)^4}{4!}+\dots\right\}.$$

This series converges and

$$\psi(t) = 2a(\cos t - 1).$$

Hence

$$\phi(t) = e^{2a(\cos t-1)}$$

If we try to integrate $\int_{-\infty}^{\infty} e^{2a(\cos t-1)}e^{-itx}\,dt$ in the ordinary way, we fail. Let us then look into the question of continuity of the distribution function.

We have

$$\begin{aligned} J_c &= \frac{e^{-2a}}{2c}\int_{-c}^{c} e^{2a\cos t}e^{-itx}\,dt \\ &= \frac{e^{-2a}}{2c}\int_{-c}^{c}\sum_{j=0}^{\infty}\frac{(2a)^j}{j!}\cos^j t\, e^{-itx}\,dt. \end{aligned}$$

The series is uniformly convergent and hence

$$J_c = \frac{e^{-2a}}{2c}\sum_{j=0}^{\infty}\int_{-c}^{c}\frac{(2a)^j}{j!}\cos^j t\, e^{-itx}\,dt$$

$$= \frac{e^{-2a}}{2c} \sum_{j=0}^{\infty} \int_{-c}^{c} \frac{(2a)^j}{j!} \cos^j t \cos xt \, dt,$$

since $\sin xt$ is an odd function.

Consider now the integral $\int_{-\infty}^{\infty} 2^j \cos^j t \cos xt \, dt$. By a well-known expansion

$$2^j \cos^j t \cos xt = \tfrac{1}{2}(e^{it}+e^{-it})^j (e^{ixt}+e^{-ixt})$$

$$= \tfrac{1}{2}(e^{ixt}+e^{-ixt})\left\{e^{ijt}+\binom{j}{1}e^{(j-2)it}+\ldots+e^{-ijt}\right\}.$$

The only part of this expression of present interest is the constant term, the others not contributing more than a finite amount. The coefficient of e^0 is zero unless x is integral in absolute value, and in that case is

$$\frac{1}{2}\left\{\binom{j}{\frac{1}{2}(j-x)}+\binom{j}{\frac{1}{2}(j+x)}\right\} = \binom{j}{\frac{1}{2}(j-x)}.$$

Thus J_c tends to zero unless x is integral in absolute value, and in the latter case

$$J_c \to e^{-2a} \sum_{j=x}^{\infty} \frac{a^j}{j!} \binom{j}{\frac{1}{2}(j-x)}.$$

Thus, if x is even, say $2r$, the frequency at $x = \pm 2r$ is

$$e^{-2a}\left\{\frac{a^{2r}}{(2r)!}+\frac{a^{2r+2}}{(2r+2)!}\binom{2r+2}{1}+\frac{a^{2r+4}}{(2r+4)!}\binom{2r+4}{2}+\ldots\right\}$$

$$= e^{-2a}\left\{\frac{a^{2r}}{(2r)!}+\frac{a^{2r+2}}{(2r+1)!\,1!}+\frac{a^{2r+4}}{(2r+2)!\,2!}+\ldots\right\},$$

and if x is odd the frequency at $x = 2r+1$ is

$$e^{-2a}\left\{\frac{a^{2r+1}}{(2r+1)!}+\frac{a^{2r+3}}{(2r+2)!\,1!}+\frac{a^{2r+5}}{(2r+3)!\,2!}+\ldots\right\}.$$

We may now verify that these frequencies account for the whole of the characteristic function and hence that all frequencies have been found. (Cf. Irwin, 1937.)

Conditions for a function to be a characteristic function

4.7 Any function which is not negative in its range of definition and which is integrable in the Stieltjes sense can be a frequency function ; and any non-decreasing function which increases from 0 to 1 in its range of definition can be a distribution function. There are much more restrictive conditions to be obeyed before a given function can be a characteristic function.

In the first place, let us note that it is a necessary and sufficient condition for a function $\phi(t)$ to be a characteristic function that the integral at (4.4)

$$\frac{1}{2\pi}\int_{-\infty}^{\infty} \frac{1-e^{-itx}}{it}\phi(t)\,dt$$

shall (except for an additive constant $F(0)$) be a distribution function. This, however, is not a very helpful criterion in practice.

Looking to the definition of $\phi(t)$ as $\int_{-\infty}^{\infty} e^{itx}\,dF$, we see that necessary conditions for $\phi(t)$ to be a characteristic function are

(a) that $\phi(t)$ must be continuous in t,
(b) that $\phi(t)$ is defined in every finite t-interval,
(c) that $\phi(0) = 1$,
(d) that $\phi(t)$ and $\phi(-t)$ shall be conjugate quantities,
(e) that $|\phi(t)| \leqslant \int_{-\infty}^{\infty} |e^{itx}|\,dF = 1 = \phi(0)$.

4.8 A general theorem of Cramér (1937) states that for a bounded and continuous function $\phi(t)$ to be a c.f. it is necessary and sufficient that $\phi(0) = 1$ and that

$$\int_0^A \int_0^A \phi(t-u) \exp\{ix(t-u)\}\,dt\,du$$

is real and non-negative for all real x and all $A > 0$.

There are also some notable specialized theorems on the subject. For example Marcinkiewicz (1938) proved that $\exp\{P(t)\}$, where $P(t)$ is a polynomial in t, cannot be a c.f. unless $P(t)$ is of the first or second degree—cf. Exercise 4.5 for a simplified version, and Exercise 4.11. Lukacs and Szász (1952) have given conditions under which a *rational* function can be a c.f. See also Exercise 4.10.

Limiting properties of distribution and characteristic functions

4.9 We now proceed to discuss a set of results involving the tendency of d.f.'s and c.f.'s to limiting forms. Suppose there is given a sequence of distribution functions $F_n(x)$ depending on a parameter n which can increase indefinitely. To each F_n there will correspond a characteristic function ϕ_n. The type of question to be discussed is this: if F_n tends to a limit G, will ϕ_n tend to a limit ϕ and is ϕ the characteristic function of G? Conversely, if ϕ_n tends to a limit ϕ, does F_n tend to a limit G and is G a distribution function having ϕ for its characteristic function? The answers to these questions, as will be seen below, are affirmative under certain general conditions.

4.10 Let us first of all consider what is meant by a distribution function tending to another. If both are continuous, $F_n(x)$ is said to tend to $G(x)$ if, given any ε, there is an n_0 such that $|F_n(x) - G(x)| < \varepsilon$ for all $n > n_0$ and for all x. If there are discontinuities present, F_n will be said to tend to G if it does so in every point of continuity of G. Since by definition our functions are taken to be continuous on the right at saltuses, this evidently conforms to the definition for the continuous case and to the common-sense requirements of the situation.

4.11 We require a number of results concerning the convergence of d.f.'s. The first is that if a sequence of d.f.'s $\{F_n(x)\}$ converges to the d.f. $G(x)$ at all its points of continuity, then the convergence is uniform in every closed interval of continuity of $G(x)$.

Let $G(x)$ be continuous in the range a to b. Divide the range into a finite number of parts, say at $a = \xi_1, \xi_2, \ldots, \xi_h = b$, such that the increase of $G(x)$ in each interval is at most ε. Choose n_0 so large that $|F_n(\xi_j) - G(\xi_j)| < \varepsilon$, $j = 1, \ldots, h$, $n > n_0$. Then for such an n and any x in the interval (a, b) there is an r such that x lies in (ξ_r, ξ_{r+1}) and

$$G(x)-2\varepsilon \leqslant G(\xi_r)-\varepsilon < F_n(\xi_r) \leqslant F_n(x) \leqslant F_n(\xi_{r+1}) < G(\xi_{r+1})+\varepsilon \leqslant G(x)+2\varepsilon \quad (4.19)$$

and hence

$$|F_n(x)-G(x)| < 2\varepsilon.$$

This is true for all x in (a, b) and hence the theorem is established.

In particular, if $G(x)$ is continuous the convergence is uniform throughout the whole interval $-\infty \leqslant x \leqslant \infty$.

4.12 The second theorem we require (the Montel–Helly theorem) is that if the sequence $\{F_n(x)\}$ is monotonic and bounded for all x (which is so for distribution functions) then we can pick out a subsequence $\{F_{n'}(x)\}$ which converges to some monotonic increasing function G (not necessarily a distribution function itself, for it may not vary from 0 to 1).

Consider first of all a series of values $x_1, x_2, \ldots$ It is known that every bounded set of numbers contains a convergent sequence. Hence we can pick out from the sequence $\{F_n(x_1)\}$ a convergent sequence, say $\{F_{n_1}(x_1)\}$. Then from the subsequence $\{F_{n_1}(x_2)\}$ we can pick out a subsequence $\{F_{n_2}(x_2)\}$ and $\{F_{n_2}(x)\}$ is thus convergent at both x_1 and x_2. Continuing in this way we may, by picking out the first function in $\{F_{n_1}(x)\}$, the second in $\{F_{n_2}(x)\}$, and so on, arrive at a sequence of functions $G_1(x)$, $G_2(x) \ldots$ which converges at each of the values $x_1, x_2, \ldots$, etc. This is the so-called Weierstrassian diagonal process.

It follows that the sequence G_n is convergent at every *rational* point x. Since $G_n(a) \leqslant G_n(x) \leqslant G_n(b)$ for every x between a and b, we see that if $G_n(a)$ and $G_n(b)$ converge, the limiting values of $G_n(x)$ lie between those limits, say $G(a)$ and $G(b)$.

Then the function $u(x)$ = upper bound of $G_n(x)$ (x not necessarily rational) is well defined and non-decreasing and so has no more than an enumerable number of points of discontinuity. If u is continuous at x, we take y and z such that $y<x<z$ and $u(z)-u(y)<\varepsilon$. Then if a and b are rational points such that $y<a<x<b<z$ it follows that $u(y) \leqslant G(a) \leqslant G(b) \leqslant u(z)$. Moreover, as all the limiting values of $G_n(x)$ are between $G(a)$ and $G(b)$, they are between $u(y)$ and $u(z)$. Hence, as ε can be arbitrarily small, we see that $G(x)$ tends to $u(x)$ at every point of continuity of u. Finally, by the diagonal process, we can select a sequence which will also be convergent at the points of discontinuity of $u(x)$. The theorem is established.

4.13 It is not obvious that if a sequence of functions $\{F_n(x)\}$ all lie between 0 and 1 then their limit $G(x)$ can attain those values. A counter-example is furnished by the distribution

$$\left.\begin{aligned} F_n(x) &= 0, & x &< -n, \\ &= \tfrac{1}{2}, & -n \leqslant x &\leqslant n, \\ &= 1, & x &> n. \end{aligned}\right\} \quad (4.20)$$

The limit of $F_n(x)$ here is $\frac{1}{2}$, $-\infty \leqslant x \leqslant \infty$. A condition which is sufficient (but not necessary) has been given by D. G. Kendall and Rao (1950). If $F_n(x)$ has a second moment $\mu'_{2(n)} \leqslant$ some finite A, all $n > n_0$, then $G(x)$ is a d.f.

For, from the Bienaymé–Chebyshev inequality (**3.35**),

$$F_n(-X)+1-F_n(X)\leqslant A/X^2, \qquad n>n_0,$$

and hence, for sufficiently large X,

$$F_n(-X)+1-F_n(X)<\varepsilon$$

and since this is so uniformly in X

$$G(-X)+1-G(X)\leqslant\varepsilon \tag{4.21}$$

provided that (as we can ensure without loss of generality) $G(x)$ is continuous at $\pm X$. Thus $G(-\infty) = 0$ and $G(\infty) = 1$.

The First Limit Theorem

4.14 We now prove the theorem: if a sequence of distribution functions $\{F_n\}$ tends to a continuous distribution function G, then the corresponding sequence of characteristic functions ϕ_n tends to ϕ uniformly in any finite t-interval, where ϕ is the characteristic function of G.

It is required to prove that, given ε, there is an n_0 independent of t such that

$$|\phi(t)-\phi_n(t)| = \left|\int_{-\infty}^{\infty} e^{itx}(dG-dF_n)\right| < \varepsilon, \qquad n>n_0.$$

Select two points of continuity of G, X and $-X$. We can make X as large as we please. We then split the integral

$$\int_{-\infty}^{\infty} e^{itx}(dG-dF_n) \tag{4.22}$$

into two parts, that in the range $-X$ to $+X$ and that in the remaining portion of the range. Now

$$\left|\int_{x<-X}^{x>X} e^{itx}\,dG\right| \leqslant \int_{x<-X}^{x>X} dG \leqslant 1-G(X)+G(-X),$$

and by taking X large enough we can make this quantity less than $\varepsilon/6$.

Similarly

$$\left|\int_{x<-X}^{x>X} e^{itx}\,dF_n\right| \leqslant 1-F_n(X)+F_n(-X),$$

and since F_n tends to G (and that uniformly) this, for some large X, will be less than $\varepsilon/3$. Hence for some n_0 the portion of (4.22) outside the range $-X$ to $+X$ will be less in modulus than $\varepsilon/6+\varepsilon/3 = \frac{1}{2}\varepsilon$. Consider now the other part

$$\int_{-X}^{X} e^{itx}(dG-dF_n). \tag{4.23}$$

This expression is the limit of the sum

$$\Sigma\, e^{itx_j}[\{G(\xi_{j+1})-G(\xi_j)\}-\{F_n(\xi_{j+1})-F_n(\xi_j)\}], \tag{4.24}$$

ξ_j, ξ_{j+1} being the boundaries of the interval into which the range is subdivided and x_j a value in that interval. The difference between this sum and the limiting value can be made less than $\frac{1}{4}\varepsilon$ if the intervals are small enough; for if they are less than η in width the difference of e^{itx_j} and $e^{it\xi_j}$ is less in modulus than $\eta|t|$, by the mean value theorem, and thus in any t-range $\pm T$ the difference of (4.23) and (4.24) is less in modulus than

$$\eta T \,|\, \Sigma [\{G(\xi_{j+1}) - G(\xi_j)\} - \{F_n(\xi_{j+1}) - F_n(\xi_j)\}] \,| < 2\eta\, T,$$

which is less than $\frac{1}{4}\varepsilon$ if $\eta < \varepsilon/8T$.

Now the sum (4.24) will itself be less than $\frac{1}{4}\varepsilon$ for some $n > n_0$, for it is the sum of a finite number of terms each of which tends to zero. Consequently (4.23) is less than $\frac{1}{2}\varepsilon$ and hence

$$|\phi(t) - \phi_n(t)| < \varepsilon, \qquad n > n_0.$$

Converse of the First Limit Theorem

4.15 The converse result is even more important.

Let $\{\phi_n\}$ be a sequence of characteristic functions corresponding to the sequence of distribution functions $\{F_n\}$. Then if $\phi_n(t)$ tends to $\phi(t)$ for all real t, and uniformly in a finite t-interval $|\,t\,| < a$,(*) $\{F_n\}$ tends to a distribution function G and ϕ is the characteristic function of G.

As a preliminary lemma, let us prove that if G is a distribution function with characteristic function ϕ, then for all real ξ and all $h > 0$

$$\frac{1}{h}\int_{\xi}^{\xi+h} G(u)\,du - \frac{1}{h}\int_{\xi-h}^{\xi} G(u)\,du = \frac{1}{\pi}\int_{-\infty}^{\infty} \left(\frac{\sin t}{t}\right)^2 e^{-\frac{2it\xi}{h}}\,\phi\left(\frac{2t}{h}\right) dt. \qquad (4.25)$$

In fact, put

$$H(x) = \frac{1}{h}\int_{x}^{x+h} G(u)\,du.$$

This is a continuous distribution function and its characteristic function is

$$\int_{-\infty}^{\infty} e^{itx}\,dH = \frac{1}{h}\int_{-\infty}^{\infty} e^{itx}\,\{G(x+h) - G(x)\}\,dx,$$

which by a partial integration becomes

$$\frac{1}{h}\left[\frac{\{G(x+h) - G(x)\}\,e^{itx}}{it}\right]_{-\infty}^{\infty} - \frac{1}{ith}\int_{-\infty}^{\infty} e^{itx}\,\{dG(x+h) - dG(x)\}$$

$$= -\frac{1}{ith}\int_{-\infty}^{\infty} \{e^{it(x-h)}\,dG(x) - e^{itx}\,dG(x)\}$$

$$= \phi(t)\,\frac{1 - e^{-ith}}{ith}.$$

Substituting for $H(x)$ in (4.4) we get

$$\frac{1}{h}\int_{\xi+h}^{\xi+2h} G(u)\,du - \frac{1}{h}\int_{\xi}^{\xi+h} G(u)\,du$$

$$= \frac{1}{2\pi h}\int_{-\infty}^{\infty} \left(\frac{1 - e^{-ith}}{it}\right)^2 e^{-it\xi}\,\phi(t)\,dt,$$

whence, writing ξ for $\xi + h$, we find, after a little re-arrangement, equation (4.25).

Reverting now to the theorem required to be proved, note that it is sufficient to

---

(*) Or equivalently, if $\phi(t)$ is continuous at $t = 0$ or if $\phi(t)$ is a characteristic function.

establish that if $\phi_n \to \phi$ uniformly in *some* interval $|t| < a$, then $\{F_n\}$ tends to some distribution function G in every point of continuity of G. When this is established it follows from the First Limit Theorem that ϕ is the characteristic function of G and that ϕ_n converges to ϕ uniformly in *every* finite t-interval.

As shown in **4.12**, given a sequence $\{F_n\}$ we may always choose from it a subsequence $\{F_{n'}\}$ such that $\{F_{n'}\}$ converges to a non-decreasing function G in every continuity point of G.

Let us then choose such a sequence. We have of necessity $0 \leqslant G \leqslant 1$, and G may be supposed everywhere continuous on the right. It is then a distribution function if $G(+\infty) - G(-\infty) = 1$, and this we proceed to prove. From (4.25) with $\xi = 0$ we have

$$\frac{1}{h}\int_0^h F_{n'}(u)\,du - \frac{1}{h}\int_{-h}^0 F_{n'}(u)\,du = \frac{1}{\pi}\int_{-\infty}^{\infty}\left(\frac{\sin t}{t}\right)^2 \phi_{n'}\left(\frac{2t}{h}\right)dt.$$

By hypothesis ϕ_n tends uniformly to ϕ for $|t| < a$ and hence $\phi_{n'}$ does so, and it is easily seen that the integral on the right is uniformly convergent. Thus, given ε, we can find h_0 such that for $h > h_0$

$$\left(\frac{1}{h}\int_0^h - \frac{1}{h}\int_{-h}^0\right) G(u)\,du = \frac{1}{\pi}\int_{-\frac{1}{2}ah}^{\frac{1}{2}ah}\left(\frac{\sin t}{t}\right)^2 \phi\left(\frac{2t}{h}\right)dt + \eta,$$

where $|\eta| < \varepsilon$. Now let h tend to infinity. As G is a non-decreasing function the left-hand side tends to $G(+\infty) - G(-\infty)$. The right-hand side tends, in virtue of the uniformity of $\phi_{n'}$ and the consequent continuity of ϕ near $t = 0$, to

$$\frac{1}{\pi}\int_{-\infty}^{\infty}\left(\frac{\sin t}{t}\right)^2 dt,$$

which is equal to unity.

Hence G, the limit of the subsequence $\{F_{n'}\}$, is a distribution function whose characteristic function is ϕ.

But any sequence of ϕ_n tends to ϕ, in virtue of the uniformity of the convergence, and hence any convergent subsequence of $\{F_n\}$ tends to G. Consequently $\{F_n\}$ tends to G in every point of continuity of G and the theorem follows.

Example 4.6

In Example 3.5, it was shown in effect that the binomial distribution has the characteristic function

$$\phi(t) = (q + pe^{it})^n.$$

Now the frequency at $x = j$ is $\binom{n}{j} q^{n-j} p^j$. This is greater than the ordinate at $x = j+1$ if

$$\binom{n}{j} q^{n-j} p^j > \binom{n}{j+1} q^{n-j-1} p^{j+1}$$

or

$$j > pn - q.$$

For large n the maximum frequency will then be in the neighbourhood of $j = pn$, and is there

$$\binom{n}{pn} q^{qn} p^{pn}.$$

In virtue of Stirling's approximation to the factorial this approximates to

$$\frac{n^n e^{-n} \sqrt{(2\pi n)}\, q^{qn} p^{pn}}{(pn)^{pn} e^{-pn} \sqrt{(2\pi pn)}\,(qn)^{qn} e^{-qn} \sqrt{(2\pi qn)}} \sim \frac{1}{\sqrt{(2\pi pqn)}}$$

and therefore tends to zero.

Thus every frequency in the binomial distribution tends to zero and the distribution does not tend to any limiting distribution.

Suppose, however, that we express the distribution in standard measure. Putting $q = (x-\mu_1')/\sigma$ we have

$$\phi_x(t) = \int_{-\infty}^{\infty} e^{itx}\, dF(x) = \int_{-\infty}^{\infty} \exp\{it(\sigma y + \mu_1')\}\, dF(y)$$
$$= e^{it\mu_1'} \phi_y(\sigma t).$$

Hence
$$\phi_y(t) = \exp(-it\mu_1'/\sigma)\, \phi_x(t/\sigma).$$

The effect on $\phi(t)$ of transferring to standard measure is then to replace t by t/σ and to multiply by $\exp(-it\mu_1'/\sigma)$.

For the binomial, $\mu_1' = np$, $\mu_2 = npq$, and thus the characteristic function of the binomial expressed in standard measure is

$$\exp\left\{-\frac{itnp}{(npq)^{\frac{1}{2}}}\right\}\left\{q + p\exp\frac{it}{(npq)^{\frac{1}{2}}}\right\}^n.$$

Thus

$$\begin{aligned}
\log\phi &= \frac{-itnp}{(npq)^{\frac{1}{2}}} + n\log\left\{1 + p\left(\exp\frac{it}{(npq)^{\frac{1}{2}}} - 1\right)\right\} \\
&= \frac{-itnp}{(npq)^{\frac{1}{2}}} + n\log\left\{1 + p\frac{it}{(npq)^{\frac{1}{2}}} - \frac{pt^2}{2npq} + \frac{p\theta t^3}{6(npq)^{\frac{3}{2}}}\right\}, \qquad 0 \leqslant |\theta| \leqslant 1, \\
&= n\left\{\frac{-pt^2}{2npq} - \frac{1}{2}\frac{(pit)^2}{npq}\right\} + O(t^3 n^{-\frac{1}{2}}) \\
&= -\tfrac{1}{2}t^2 + O(t^3 n^{-\frac{1}{2}}).
\end{aligned}$$

Thus for any finite t, $\log\phi$ tends uniformly to $-\frac{1}{2}t^2$ and hence

$$\phi(t) \to e^{-\frac{1}{2}t^2}.$$

Thus the binomial distribution expressed in standard measure tends to the distribution whose characteristic function is $e^{-\frac{1}{2}t^2}$, i.e. to the form

$$dF = \frac{1}{\sqrt{(2\pi)}} e^{-\frac{1}{2}x^2}\, dx, \qquad -\infty \leqslant x \leqslant \infty.$$

Multivariate characteristic functions

4.16 The characteristic function of a bivariate distribution $F(x_1, x_2)$ is defined as

$$\phi(t_1, t_2) = \int_{-\infty}^{\infty}\int_{-\infty}^{\infty} e^{it_1x_1 + it_2x_2}\, dF(x_1, x_2) \tag{4.26}$$

and generally, that of a multivariate distribution $F(x_1, x_2, \ldots, x_n)$ as

$$\phi(t_1, t_2, \ldots, t_n) = \int_{-\infty}^{\infty}\int_{-\infty}^{\infty}\cdots\int_{-\infty}^{\infty} e^{it_1x_1+it_2x_2\ldots+it_nx_n}\, dF(x_1, x_2, \ldots, x_n). \quad (4.27)$$

If any subset of the t_j are put equal to zero, we obtain the c.f. of the (marginal) distribution of the other variables; e.g., $\phi(t_1, 0)$ is the c.f. of x_1 alone, and so on.

If $x_1, x_2, \ldots, x_n$ are independent we have

$$\begin{aligned}\phi(t_1, t_2, \ldots, t_n) &= \int_{-\infty}^{\infty} e^{it_1x_1}\, dF_1(x_1)\int_{-\infty}^{\infty} e^{it_2x_2}\, dF_2(x_2)\ldots\int_{-\infty}^{\infty} e^{it_nx_n}\, dF_n(x_n)\\ &= \phi_1(t_1)\phi_2(t_2)\ldots\phi_n(t_n) \quad (4.28)\end{aligned}$$

where ϕ_r is the c.f. of F_r. Similarly

$$\psi(t_1, t_2, \ldots, t_n) = \sum_{j=1}^{n} \log \phi_j(t_j). \quad (4.29)$$

Thus the characteristic function of the joint distribution of a number of independent variables is the product of their characteristic functions ; and the cumulant-generating function is the sum of their cumulant-generating functions. This is a fundamentally important result in the theory of sampling.

4.17 In generalization of (4.4) we have

$$F(x_1, x_2, \ldots, x_n) - F(0, x_2, \ldots, x_n)\ldots - F(x_1, x_2, \ldots, 0) + F(0, 0, x_3, \ldots, x_n) + \ldots$$
$$+F(x_1, x_2, \ldots, 0, 0) - \ldots = \frac{1}{(2\pi)^n}\int_{-\infty}^{\infty}\cdots\int_{-\infty}^{\infty}\frac{1-e^{-ix_1t_1}}{it_1}\cdots\frac{1-e^{-ix_nt_n}}{it_n}$$
$$\phi(t_1, t_2, \ldots, t_n)\, dt_1\ldots dt_n. \quad (4.30)$$

There are 2^n terms on the left of (4.30), $\binom{n}{r}$ of which have r zeros among the arguments of F. The integrals are to be understood as principal values $\lim\limits_{c\to\infty}\int_{-c}^{c}\cdots\int_{-c}^{c}$.

The proof is similar to that for the univariate case. For example, with two variates we have, in an obvious generalization of (4.10),

$$I_c = \int_{-\infty}^{\infty}\int_{-\infty}^{\infty}\int_{-c}^{c}\int_{-c}^{c}\left\{\frac{\sin t_1u_1}{t_1} - \frac{\sin t_1(u_1-x_1)}{t_1}\right\}\left\{\frac{\sin t_2u_2}{t_2} - \frac{\sin t_2(u_2-x_2)}{t_2}\right\}$$
$$dt_1\, dt_2\, dF(u_1, u_2)$$

and
$$\begin{aligned}\lim I_c &= \lim\int_{-\infty}^{\infty}\int_{-c}^{c}\left\{\frac{\sin t_1u_1}{t_1} - \frac{\sin t_1(u_1-x_1)}{t_1}\right\} dt_1(2\pi)\{dF(u_1, x_2) - dF(u_1, 0)\}\\ &= (2\pi)^2\{F(x_1, x_2) - F(0, x_2) - F(x_1, 0) + F(0, 0)\}.\end{aligned}$$

Similarly, (4.5) generalizes to

$$f(x_1, x_2, \ldots, x_n) = \frac{1}{(2\pi)^n}\int_{-\infty}^{\infty}\cdots\int_{-\infty}^{\infty}\exp\left(-i\sum_{j=1}^{n} t_jx_j\right)\phi(t_1, t_2, \ldots, t_n)\, dt_1\ldots dt_n. \quad (4.31)$$

As in the univariate case, the c.f. defines the distribution uniquely and the corresponding analogues of the limit theorems also remain true. Thus the converse of the result of **4.16** is ensured by the unique correspondence of the c.f. and the d.f. : if a joint c.f. factorizes, the variables are independent.

4.18 If we have a distribution $F(x)$ and some function of the variate such as $y\ (= y(x))$, we define the c.f. of y as

$$\phi_y(t) = \int_{-\infty}^{\infty} \exp(ity)\, dF(x). \tag{4.32}$$

The distribution of y is then given by inversion ; for instance, the frequency function of y, say $g(y)$, is given by

$$g(y) = \frac{1}{2\pi}\int_{-\infty}^{\infty} \exp(-ity)\, \phi_y(t)\, dt, \tag{4.33}$$

and a similar result holds for a function g of several variates.

Conditional characteristic functions

4.19 We may express the c.f. of a conditional distribution in terms of the underlying multivariate c.f. In the bivariate case, e.g., (4.26) may be written, using **1.32–3**, as

$$\begin{aligned}\phi(t_1, t_2) &= \int_{-\infty}^{\infty}\int_{-\infty}^{\infty} e^{it_1x_1+it_2x_2} p(x_1 \mid x_2) f_2(x_2)\, dx_1\, dx_2 \\ &= \int_{-\infty}^{\infty} e^{it_2x_2} \phi_{12}(t_1) f_2(x_2)\, dx_2,\end{aligned}$$

where ϕ_{12} is the c.f. of the conditional distribution $p(x_1 \mid x_2)$. We see that here $\phi_{12}(t_1) f_2(x_2)$ plays the role usually taken by $f_2(x_2)$ in obtaining the c.f. of the marginal distribution of x_2, $\phi_2(t_2)$. Thus, applying (4.5),

$$2\pi\phi_{12}(t_1) f_2(x_2) = \int_{-\infty}^{\infty} \phi(t_1, t_2)\, e^{-ix_2t_2}\, dt_2.$$

Dividing this equation by its value at $t_1 = 0$, when $\phi_{12}(0) = 1$, we find the required conditional c.f.

$$\phi_{12}(t_1) = \int_{-\infty}^{\infty} \phi(t_1, t_2)\, e^{-ix_2t_2}\, dt_2 \Big/ \int_{-\infty}^{\infty} \phi(0, t_2)\, e^{-ix_2t_2}\, dt_2. \tag{4.34}$$

Essentially the same result holds if x_1, x_2 are sets of variables, now using (4.31)—the general result may be found in **27.6**, Vol. 2.

The problem of moments

4.20 It is of some interest to consider how far a set of moments (assuming that they all exist) determine a distribution uniquely. To give some point to the discussion let us note that in some circumstances it is possible for two different distributions to have the same set of moments.

Consider the integral

$$\int_0^{\infty} t^{p-1} e^{-qt}\, dt = \Gamma(p)/q^p, \qquad p > 0, \qquad \mathscr{R}(q) > 0.$$

Put $p = (n+1)/\lambda$, n a non-negative integer; $0<\lambda<\frac{1}{2}$; $q = \alpha+i\beta$; $\beta/\alpha = \tan\lambda\pi$; $x^\lambda = t$. We find on substitution, since $(1+i\tan\lambda\pi)^{(n+1)/\lambda}$ is real, that the imaginary part of the integral is zero and thus that

$$\int_0^\infty x^n \exp(-\alpha x^\lambda)\sin(\beta x^\lambda)\,dx = 0.$$

Hence the distributions

$$f(x) = k\exp(-\alpha x^\lambda)\{1+\varepsilon\sin(\beta x^\lambda)\},\quad 0\leqslant x\leqslant\infty;\quad \alpha>0;\quad 0<\lambda<\tfrac{1}{2};\quad |\varepsilon|<1; \tag{4.35}$$

have the same moments for all ε in the range $|\varepsilon| < 1$. Exercise 6.21 below deals with an important distribution which is not determined by its moments.

We may derive a similar family with range infinite in both directions by putting $p = (2n+1)/\rho$, $q = \alpha+i\beta$, $\beta/\alpha = \tan\frac{1}{2}\rho\pi$, $x^\rho = t$, $\rho = 2s/(s+1)$, s a positive integer. The family then becomes

$$f(x) = k\exp\{-\alpha|x|^\rho\}\{1+\varepsilon\cos(\alpha|x|^\rho)\},\qquad -\infty\leqslant x\leqslant\infty. \tag{4.36}$$

The problem of moments in its full generality considers a set of constants $c_0, c_1, c_2, \ldots$ and inquires whether they can be the moments of a distribution. For statistical purposes this is not of particular interest.(*) We are more concerned with the problem: given that the set of constants are, in fact, the moments of a distribution, can any other distribution have the same set?

Note in the first instance that this problem need only be considered when absolute moments of all orders exist. It is not difficult to see that more than one distribution can exist having a limited number of given moments finite and the remainder infinite. Thus we consider only the case when all absolute moments exist.

4.21 We will prove in the first place the theorem that a set of moments determines a distribution uniquely if the series $\sum_{j=0}^{\infty} \nu_j t^j/j!$ converges for some real non-zero t. We write ν without the prime in this and the following sections to save printing, but the result is true for moments about any origin.

The characteristic function is continuous in t and by **3.5** its derivatives exist at $t = 0$ if the moments exist. We have then in the neighbourhood of $t = 0$

$$\phi(t) = \sum_{j=0}^{r} (it)^j \mu'_j/j! + R_r, \tag{4.37}$$

where R_r is less in absolute value than $\dfrac{\nu_{r+1}|t|^{r+1}}{(r+1)!}$

Thus if $\Sigma\nu_j t^j/j!$ converges, $\nu_j t^j/j!$ tends to zero and hence $\phi(t)$ is equal to the sum of the infinite series $\sum_{j=0}^{\infty} (it)^j\mu'_j/j!$ if it exists. Moreover, this series is majorated by $\Sigma\nu_j t^j/j!$ and hence is absolutely convergent if the latter is convergent. Hence we have

$$\phi(t) = \sum_{j=0}^{\infty} (it)^j \mu'_j/j! \tag{4.38}$$

(*) For some results on the general problem, see Stieltjes (1918) and Hamburger (1920). For a general review, see Shohat and Tamarkin (1943) and Chapter III of Widder (1941).

and thus $\phi(t)$ is uniquely determined in the neighbourhood of $t = 0$. In the neighbourhood of $t = t_0$ we have

$$\phi(t) = \sum_{j=0}^{r} \left\{ \frac{i^j (t-t_0)^j}{j!} \int_{-\infty}^{\infty} x^j e^{it_0 x} dF \right\} + R_r'$$

and the modulus of the coefficient of $(t-t_0)^j/j!$ is not greater than ν_j. Consequently $\phi(t)$ can be expanded everywhere as a convergent Taylor series and is equal to the sum of that series. Hence $\phi(t)$ may be extended from the neighbourhood $t = t_0$ by analytic continuation through any finite t-interval. Hence $\phi(t)$ is everywhere uniquely defined. But $\phi(t)$ determines the distribution function and hence the latter is uniquely determined.

4.22 A few simple but effective results follow as corollaries.

(a) The moments uniquely determine the distribution if the upper limit of $(\nu_n^{1/n})/n$ is finite. For the series whose general term is $\nu_n t^n/n!$ is convergent if

$$\limsup (\nu_n t^n/n!)^{1/n} < 1.$$

Replacing the factorial by its Stirling approximation we see that this will be true if

$$\limsup (\nu_n^{1/n})/n < k/t$$

where k is some constant. If the upper limit is finite the inequality can be satisfied for some non-zero t.

(b) It is also sufficient for the upper limit of $(\mu_{2n}'^{1/2n})/2n$ to be finite, a form which enables us to disregard the absolute moments. It is, in fact, easy to see that this upper limit and the upper limit of $(\nu_n^{1/n})/n$ are finite or infinite together.

(c) The moments uniquely determine the distribution if the range is finite. For, taking an origin at the start of the distribution (of range h, say), we see that $\nu_r \leqslant h^r$ and hence $(\nu_n^{1/n})/n \leqslant h/n$, which tends to zero.

4.23 Two further criteria may be mentioned. The first is due to Carleman (1925). A set of moments determines a distribution uniquely if (in the case of limits $-\infty$ to $+\infty$)

$$\sum_{j=0}^{\infty} \frac{1}{(\mu_{2j})^{1/2j}} \tag{4.39}$$

diverges. For the limits 0 to ∞ the corresponding series is

$$\sum_{j=0}^{\infty} \frac{1}{(\mu_j)^{1/2j}}. \tag{4.40}$$

Secondly, if there exists a frequency function, the moments determine it uniquely if, for limits $-\infty$ to $+\infty$,

$$f(x) < M|x|^{\beta-1} \exp(-\alpha|x|^{\lambda}) \text{ for } |x| > x_0, \quad M, \beta, \alpha, > 0, \ \lambda \geqslant 1, \tag{4.41}$$

and for limits 0 to ∞,

$$f(x) < M|x|^{\beta-1} \exp(-\alpha|x|^{\lambda}) \text{ for } |x| > x_0, \quad M, \beta, \alpha, > 0, \ \lambda \geqslant \tfrac{1}{2}. \tag{4.42}$$

This result is due ultimately to Stieltjes. It follows without difficulty from the Carleman criterion.

It is interesting to note that if for some x_0

$$f(x) > \exp(-\alpha|x|^{\lambda}), \qquad \alpha > 0, \ x > x_0, \tag{4.43}$$

then the problem of moments is necessarily indeterminate (as usual, $\lambda < \frac{1}{2}$ for the range 0 to ∞ and $\lambda < 1$ for the range $-\infty$ to $+\infty$). This follows from the examples in equations (4.35) and (4.36), for we can add to (4.43), *without rendering any frequency negative*, a function all of whose moments are zero.

Example 4.7

The moments of the distribution

$$dF = \frac{1}{\sigma\sqrt{(2\pi)}} \exp(-x^2/2\sigma^2), \qquad -\infty \leqslant x \leqslant \infty,$$

are given (cf. Example 3.4) by

$$\mu_{2r+1} = 0, \qquad \mu_{2r} = \frac{(2r)!}{2^r r!}\sigma^{2r}.$$

Thus the upper limit of $(\mu_{2n}{}^{1/2n})/2n$ is, from the Stirling approximation to the factorial, asymptotically equivalent to

$$\frac{\sigma}{2n\sqrt{2}}\left[\frac{e^{-2n}(2n)^{2n}\sqrt{(4\pi n)}}{\{e^{-n}n^n\sqrt{(2\pi n)}\}}\right]^{1/2n} \sim \frac{\sigma}{(2en)^{\frac{1}{2}}}.$$

The upper limit is then zero and the distribution is uniquely determined by its moments.

4.24 Cramér and Wold (1936) have extended Carleman's criterion to multivariate distributions. We need only consider the moments of the marginal distributions, $\mu'_{r00\ldots}$, $\mu'_{0r0\ldots}$, $\mu'_{00r\ldots}$, etc. If the sum of these for all variates is λ_r then the criteria of (4.39) and (4.40) remain with λ_r instead of μ_r; for example, a distribution ranging from $-\infty$ to ∞ is completely determined by its moments if

$$\sum_{j=0}^{\infty} \frac{1}{(\lambda_{2j})^{1/2j}} \tag{4.44}$$

diverges.

4.25 If a moment of order r exists it must be given by the rth derivative of the c.f. at $t = 0$. Thus, if $\phi(t)$ can be expanded as an infinite Taylor series, that series can only be $\Sigma(it)^j \mu'_j/j!$. And if this series does not converge, $\phi(t)$ cannot be so expanded. But it can always be expanded asymptotically in the finite form with remainder

$$\phi(t) = \sum_{j=0}^{r} (it)^j \mu'_j/j! + R.$$

This illustrates the source of a difficulty in discussing limiting properties when the infinite series does not converge, for it is known that there exist an infinite number of functions which have a given set of coefficients in an asymptotic expansion. For instance, if $\alpha(t)$ has an asymptotic expansion in t the functions $\alpha(t) + kt^{-\log t}$ all have the same expansion. It is therefore hardly surprising that when $\Sigma(it)^j \mu'_j/j!$ fails to converge, there may be more than one frequency or distribution function with the same set of moments.

But it does not follow from what has been said that there *must* be more than one frequency distribution. There must be more than one function, but those functions may not qualify as frequency distributions, e.g. they may be negative in part of the range. In the example just given, $t^{-\log t}$ cannot be a characteristic function, for it does not obey the well-known condition that $\phi(t)$ and $\phi(-t)$ should be conjugate.

4.26 We now proceed to what is known as the Second Limit Theorem, which is concerned with the way in which a sequence of distribution functions $\{F_n(x)\}$ tends to a limit if the corresponding sequence of moments of order j, say $\mu_j(n)$, tends to a limit μ_j. (To save printing, we shall throughout write the moments without primes.) Our method of proof is due to D. G. Kendall and Rao (1950).

We require in the first place a result concerning the expansion of a c.f. which is rather more precise than those which have served our purpose so far. If μ_{2m} exists then the c.f. can be expressed in the form

$$\phi(t) = \sum_{j=0}^{2m-1} (it)^j \mu_j/j! + \rho (it)^{2m} \mu_{2m}/(2m)! \tag{4.45}$$

for all real t, ρ being such that $|\rho|<1$. Also $\lim_{t\to 0} \rho = 1$.

For $m = 0$ this is true trivially. For $m>0$ we have for real x, t,

$$e^{ixt} = \sum_{j=1}^{n} \frac{(ixt)^j}{j!} + \rho' \frac{(ixt)^{n+1}}{(n+1)!}, \qquad |\rho'| \leqslant 1, \tag{4.46}$$

and on substitution in the integral defining $\phi(t)$ we get, for $n = 2m-1$, equation (4.45). For t, μ_{2m} zero we may define $\rho = 1$; for other values, considering the difference of (4.46) taken to n and $n+1$ terms we find

$$|\rho'-1| \leqslant |xt|/(n+2). \tag{4.47}$$

Since

$$|\rho-1|\,\mu_{2m} = \int_{-\infty}^{\infty} (\rho'-1)\, x^{2m}\, dF$$

we have

$$|\rho-1|\,\mu_{2m} < \varepsilon + \int_{-X}^{X} |\rho'-1|\, x^{2m}\, dF,$$

for any given ε, if X is large enough (for otherwise μ_{2m} would not converge). But for fixed X the integral can be made as small as we please by taking t sufficiently small, in virtue of (4.47). Thus we have

$$|\rho-1|\,\mu_{2m} < 2\varepsilon$$

and thus $\rho \to 1$ when $t \to 0$.

4.27 A second result we require (a converse of the previous one) is as follows: If for some positive integer m

$$\phi(t) = \sum_{j=0}^{2m} (it)^j \lambda_j/j! + o(t^{2m}) \tag{4.48}$$

as t tends to zero through real values, then the first $2m$ moments exist and $\mu_r = \lambda_r$, $r = 0, 1, \ldots, 2m$.

Let δ^2 denote the operation of taking central differences with interval $2h$, i.e.

$$\delta^2 A(t) = A(t+2h) - 2A(t) + A(t-2h).$$

Then if the remainder term in (4.48) is $t^{2m} B(t)$, where $B(t)$ tends to zero as $t \to \infty$, a little calculation shows that for $t = 0$

$$\left(\frac{\delta^2}{4h^2}\right)^m t^{2m} B(t) = \sum_{j=0}^{2m} \binom{2m}{j} (-1)^j (m-j)^{2m} B(2mh-2jh).$$

Thus for $|B(t)|<\varepsilon$

$$\left(\frac{\delta^2}{4h^2}\right)^m t^{2m} B(t) \leqslant \sum_{j=0}^{2m} \binom{2m}{j} |m-j|^{2m} \varepsilon \tag{4.49}$$

when $|h|<\eta/2m$ and $|t|\leqslant\eta$. Hence if (4.48) is true

$$\left[\lim_{h\to 0}\left(\frac{\delta^2}{4h^2}\right)^m t^{2m} B(t)\right]_{t=0} = 0. \tag{4.50}$$

But $\lim_{h\to 0}\left(\frac{\delta^2}{4h^2}\right)^m$, $= D^{2m}$ say, acting on a polynomial is easily seen to be the $(2m)$th derivative. Thus applying the operation to (4.48) we find, remembering (4.50), that when $t = 0$

$$D^{2m}\phi(t) = (-1)^m \lambda_{2m}. \tag{4.51}$$

Now

$$\delta^2 e^{ixt} = -(2\sin xh)^2 e^{ixt}$$

and thus

$$\lim_{h\to 0}\int_{-\infty}^{\infty}\left(\frac{\sin xh}{xh}\right)^{2m} x^{2m}\, dF = \lambda_{2m}. \tag{4.52}$$

From the uniform convergence to unity of $(\sin xh/xh)^{2m}$ we see that for any finite interval (a, b)

$$\int_a^b x^{2m}\, dF \leqslant \lambda_{2m},$$

and thus the moment of order $2m$ must be finite. Hence from (4.45), as t tends to zero,

$$\phi(t) = \sum_{j=0}^{2m-1}(it)^j \mu_j/j! + \frac{(it)^{2m}\mu_{2m}}{(2m)!}[1+o(1)]. \tag{4.53}$$

Comparing this with (4.48) we see that

$$\mu_r = \lambda_r, \qquad r = 0, 1, \ldots, 2m.$$

The result of this section implies that if the $(2m)$th derivative of the c.f. exists at $t = 0$, the $(2m)$th moment also exists. We have seen in **3.5** that if the rth moment exists the rth derivative of the c.f. will exist at $t = 0$, a statement of which the above is the converse for even moments. For odd moments the existence of the derivative at $t = 0$ is not sufficient for the existence of the corresponding moment. Pitman (1956) has shown that necessary and sufficient conditions for the existence of the derivative are (1) that the principal value of μ_r' exists and (2) that

$$\lim_{x\to\infty} x^r\{F(-x)+1-F(x)\} = 0.$$

4.28 The results of the two preceding sections are necessary to establish the uniqueness of asymptotic expansions of the c.f. under certain conditions. We now prove some theorems on sequences of distributions and moments.

Let $F_n(x)$, the nth member of a sequence $\{F_n(x)\}$, possess a finite moment of the jth order, say $\mu_j(n)$, for all $n>n_0$. Let

$$\lim_{n\to\infty}\mu_j(n) = \lambda_j$$

for every value of j; and let $\{F_n(x)\}$ converge to a limit function $G(x)$ at all its points of continuity. ($G(x)$ is necessarily bounded, monotonically increasing and may be taken to be continuous on the right.) Then $G(x)$ is a distribution function possessing moments of all orders ; $\{\lambda_j\}$ is a sequence of moments ; and λ_j is the jth moment of $G(x)$.

That $G(x)$ is a d.f. follows from **4.13**, for the second moments $\mu_2(n)$ form a convergent and therefore bounded sequence. Hence, by the First Limit Theorem of **4.14**, the sequence of c.f.'s $\{\phi_n(t)\}$ converges to the c.f. of $G(x)$, say $\phi(t)$.

Now when $n > n_0$

$$\phi_n(t) = \sum_{j=0}^{2m-1} (it)^j \mu_j(n)/j! + \rho\,(it)^{2m} \mu_{2m}(n)/(2m)! \tag{4.54}$$

and the last term on the right must approach a limit since all the other terms do so. Thus

$$\phi(t) = \sum_{j=0}^{2m-1} (it)^j \lambda_j/j! + R\,(it)^{2m} \lambda_{2m}/(2m)! \quad |R| \leqslant 1. \tag{4.55}$$

This is true for all m and by comparing each of these formulae with the next following one we have, as in **4.26**, $R \to 1$ as $t \to 0$. From the results of **4.27** the theorem now follows.

Example 4.8

Note that the result of **4.28** follows even if $F_n(x)$ does not possess moments of all orders. For example, the distribution

$$dF = \frac{k\,dt}{(1+t^2/\nu)^{\frac{1}{2}(\nu+1)}}, \qquad -\infty \leqslant t \leqslant \infty\,;\ \nu > 1,$$

possesses moments only up to and including $\mu_{\nu-1}$. In fact (cf. Example 3.3)

$$\mu_{2r} = \frac{\Gamma(\frac{1}{2}\nu - r)\,\Gamma(r+\frac{1}{2})}{\Gamma(\frac{1}{2}\nu)\,\Gamma(\frac{1}{2})}\nu^r. \tag{4.56}$$

By use of the Stirling approximation to the Gamma-function we find that as $\nu \to \infty$

$$\mu_{2r} \to \frac{(2r)!}{2^r r!}, \tag{4.57}$$

namely to the moments of the standardized normal distribution (Example 3.4), all of which exist. If, then, our limit function converges it must do so to the normal form, i.e. to a form for which all moments exist.

It is, in fact, easy to see directly that the frequency function tends to $k\exp(-\frac{1}{2}t^2)$.

The Second Limit Theorem

4.29 Let $\{F_n(x)\}$ converge to $G(x)$; let $\mu_j(n)$ exist for $n > n_0$ and for all $j \geqslant 0$; and let $\mu_j(n)$ be bounded above by some constant A_j. Then all the moments λ_j of $G(x)$ exist and $\mu_j(n) \to \lambda_j$ as $n \to \infty$.

From the First Limit Theorem we know that the sequence of c.f.'s $\{\phi_n(t)\}$ converges to $\phi(t)$, the c.f. of $G(x)$. Also

$$\phi_n(t) = 1 + it\mu_1(n) + O(t^2), \qquad n > n_0,$$

where the constant implied by $O(t^2)$ can be taken as independent of n and t. Thus if m is some other value $> n_0$,

$$|\mu_1(n)-\mu_1(m)| \leqslant A_1|t|+\delta_{m,n}$$

where $\delta_{m,n} \to 0$ as m, n tend independently to infinity. This, being true for every non-zero t, implies that

$$\lim_{m,n\to\infty} |\mu_1(n)-\mu_1(m)| = 0$$

and hence there exists a constant λ_1 such that

$$\lim \mu_1(n) = \lambda_1. \tag{4.58}$$

We now proceed inductively by the same kind of argument to establish the limit of $\mu_2(n)$. We require

$$\phi_n(t) = 1+it\mu_1(n)+(it)^2\mu_2(n)/2!+O(t^3),$$

an expression which follows from (4.54) because the sixth moment is bounded. We then find

$$|\mu_2(n)-\mu_2(m)| \leqslant \tfrac{2}{3}A_6^{\frac{1}{2}}|t|+\delta'_{m,n}$$

and, as before, this implies the limit:

$$\lim \mu_2(n) = \lambda_2. \tag{4.59}$$

Thus we establish that the moment-sequence $\{\lambda_j\}$ exists. It then follows from the theorem of **4.28** that these moments are, in fact, the moments of $G(x)$.

4.30 For most purposes we require a converse form of the Second Limit Theorem: let the moments $\mu_j(n)$ exist and the limits $\lim \mu_j(n) = \lambda_j$ be the moments of a distribution $G(x)$ *which is uniquely determined by its moments*. Then $\{F_n(x)\}$ converges to $G(x)$ in all points of continuity.

We can always find a sequence in $\{F_n(x)\}$ which converges to a distribution function. It then follows from the theorem of **4.28** that this function must be the unique function having the moments λ_j and thus it is $G(x)$.

Suppose now that at some point of continuity of x, say a,

$$\alpha \equiv \limsup F_n(a) \neq G(a).$$

We can choose a subsequence converging to α at $x = a$; and a further sub-subsequence which converges both to α and also to $G(a)$, because $G(x)$ is continuous at $x = a$. This contradiction can only be avoided if

$$\limsup F_n(a) = G(a);$$

and by a similar argument

$$\liminf F_n(a) = G(a).$$

The converse theorem is thus established.

Example 4.9

The discontinuous (Poisson) distribution whose frequency at $x = j$ $(j = 0, 1, \ldots)$ is $e^{-\lambda}\lambda^j/j!$ has a characteristic function (cf. Example 3.10)

$$\phi(t) = \exp\{\lambda(e^{it}-1)\},$$

and hence all cumulants equal to λ.

The distribution is evidently the only one with such cumulants, for $\Sigma \kappa_j(it)^j/j!$ is convergent and equals $\lambda(e^{it}-1)$, so that the cumulative function and the characteristic function are uniquely determined.

Now as λ tends to infinity the frequency at x_j, $e^{-\lambda}\lambda^j/j!$, tends to zero and thus the distribution does not tend to a limit. This is consistent with the behaviour of the cumulants, which increase without limit.

Suppose, however, we express the distribution in standard measure. Then

$$\kappa_1 = 0, \qquad \kappa_r = \lambda/\kappa_2^{\frac{1}{2}r} = \lambda^{-(\frac{1}{2}r-1)}, \qquad r \geqslant 2.$$

Hence as $\lambda \to \infty$ all cumulants higher than the second tend to zero, and the cumulants of the distribution tend to those of the "normal distribution"

$$dF = \frac{1}{\sqrt{(2\pi)}}\exp(-\tfrac{1}{2}x^2)\,dx, \qquad -\infty \leqslant x \leqslant \infty.$$

Now we know that this distribution is completely determined by its moments (Example 4.7). We also know that the cumulants determine the moments and vice versa, so that if the cumulants of the discontinuous distribution tend to those of the normal distribution, the moments will tend to the moments of that distribution. Hence the converse form of the Second Limit Theorem is applicable, and the discontinuous distribution does in fact tend to the normal form *when expressed in standard measure.*

4.31 In connection with the dual role of the characteristic function and the distribution function, it is worth remarking that the behaviour of one in the neighbourhood of the origin is related to the behaviour of the other at infinity. In fact, the mth derivative of $\phi(t)$ at $t = 0$ is the mth moment, the existence of which depends on the behaviour of $x^m f(x)$ at infinity. Conversely, from (4.5) we see that the mth derivative of $f(x)$, if it exists, is given by $\int_{-\infty}^{\infty}(-it)^m\phi(t)e^{-itx}\,dt/2\pi$, which in modulus is not greater than $\int_{-\infty}^{\infty}t^m\phi(t)\,dt/2\pi$, which depends for its convergence on the behaviour of $t^m\phi(t)$ at infinity.

4.32 It is natural to inquire whether there is some other transform of the distribution function which enjoys properties similar to those of the c.f. The answer appears to be in the negative. Lukacs (1952) has proved a theorem to the following effect: Let $\kappa(x, t)$ be a complex function defined for all real x, t, bounded and measurable in t. Define

$$\phi(t) = \int_{-\infty}^{\infty}\kappa(x,t)\,dF(x);$$

and let the two following conditions hold:

(1) $\phi_1(t) \equiv \phi_2(t)$ if and only if $F_1(x) = F_2(x)$.

(2) If $F(x) = \int_{-\infty}^{\infty}F_1(x-y)\,dF_2(y)$, then $\phi(t) = \phi_1(t)\phi_2(t)$.

Then $\kappa(x, t)$ must have the form $\exp\{itA(x)\}$ where $A(x)$ is a real-valued function which assumes all values of a set which is dense on a real line. It is very remarkable that we do not require among the conditions either an inversion theorem or limit theorems. The Fourier transform seems, as it were, to present us gratuitously with these useful and desirable properties.

Infinitely divisible and stable characteristic functions

4.33 We shall see in **7.18** that a c.f. is sometimes the product of other c.f.'s, and in particular may be a power of another c.f., so that

$$\phi(t) = \{\phi_j(t)\}^n. \tag{4.60}$$

If a given c.f. $\phi(t)$ can thus be represented as the nth power of some other c.f., $\phi_j(t)$, for every positive integer n, $\phi(t)$ is called an *infinitely divisible* c.f., and the corresponding d.f. is also called infinitely divisible. Since $\phi(t)$ is fixed, $\phi_j(t)$ is a function of n through the relation (4.60), and we re-label it $\phi_n(t)$, so that (4.60) becomes

$$\phi(t) = \{\phi_n(t)\}^n. \tag{4.61}$$

Example 4.10

The Cauchy c.f. of Example 4.2,

$$\phi(t) = \exp(-|t|) = \left\{\exp\left(-\left|\frac{t}{n}\right|\right)\right\}^n$$

for any positive integer n, and

$$\phi_n(t) = \exp\left(-\left|\frac{t}{n}\right|\right) = \phi\left(\frac{t}{n}\right)$$

is a c.f., so $\phi(t)$ is infinitely divisible.

Example 4.11

The normal c.f. of Example 4.1,

$$\phi(t) = \exp(-\tfrac{1}{2}t^2\sigma^2) = \{\exp(-\tfrac{1}{2}t\ \ \sigma^2/n)\}^n$$

for any positive integer n, and

$$\phi_n(t) = \exp(-\tfrac{1}{2}t^2\sigma^2/n) = \phi\left(\frac{t}{n^{1/2}}\right)$$

is a c.f., so $\phi(t)$ is infinitely divisible.

4.34 Now consider two linear functions of independent variates x_j $(j = 1, 2, \ldots)$ with the same c.f. $\phi(t)$. Let $y_j = a_j x_j + b_j$, where the $a_j > 0$. The c.f. of y_j is

$$\phi_j(t) = \phi(a_j t)\mathrm{e}^{itb_j} \tag{4.62}$$

and thus

$$\phi_j(t)\phi_k(t) = \phi(a_j t)\phi(a_k t)\exp\{it(b_j+b_k)\}. \tag{4.63}$$

If the product (4.63) is of the same form as (4.62), i.e.

$$\phi(a_j t)\phi(a_k t)\exp\{it(b_j+b_k)\} = \phi(a_0 t)e^{itb_0}, \quad a_0 > 0, \tag{4.64}$$

$\phi(t)$ is called a *stable* c.f., and the corresponding d.f. a stable d.f. (4.64) simplifies to

$$\phi(a_j t)\phi(a_k t) = \phi(a_0 t)e^{itb}, \tag{4.65}$$

where $b = b_0 - b_j - b_k$. Clearly, if (4.65) holds, we may extend it to

$$\prod_{j=1}^{n} \phi(a_j t) = \phi(at)e^{itb}, \quad a > 0, \tag{4.66}$$

by defining a and b appropriately. Thus, with $a_j \equiv 1$, we have

$$\phi(at) = \{\phi(t)\}^n e^{-itb} = \{\phi(t)\, e^{-itb/n}\}^n$$

or

$$\phi(t) = \left\{\phi\left(\frac{t}{a}\right)\exp[-(itb)/(na)]\right\}^n. \tag{4.67}$$

The expression in braces on the right of (4.67) is a c.f., the exponential factor merely representing a shift in location. Since (4.67) holds for any positive integer n, it satisfies (4.61). Thus any stable distribution is infinitely divisible.

4.35 Infinitely divisible and stable c.f.'s are of great importance in probability theory, but we shall here mention only a few of their properties that are useful in Statistics, referring the reader to Lukacs (1970) for a general exposition and a bibliography.

First, we observe that an infinitely divisible c.f. $\phi(t)$ can have no real zeros, for as $n \to \infty$ through the positive integers, we see from (4.61) that $\phi_n(t) = \{\phi(t)\}^{1/n}$ can only tend to limiting values of 1 (if $\phi(t) \neq 0$) and 0 (if $\phi(t) = 0$) because $|\phi_n(t)| \leqslant 1$. Moreover the limit of $\phi_n(t)$ is itself a c.f., and (cf. **4.7**) equals 1 at $t = 0$ and is continuous in t. It cannot therefore jump to the value 0. Thus $\phi(t) \neq 0$ for all t.

4.36 Whereas infinitely divisible distributions may be discrete (cf. Exercise 4.25), all stable distributions are continuous and unimodal.

There is an explicit general form for a stable c.f., which we write

$$\log\phi(t) = ait - c|t|^\alpha\left\{1 + i\beta\frac{t}{|t|}w(|t|, \alpha)\right\}, \tag{4.68}$$

where

$$w(|t|, \alpha) = \begin{cases} \tan(\tfrac{1}{2}\pi\alpha), & \alpha \neq 1, \\ \dfrac{2}{\pi}\log|t|, & \alpha = 1, \end{cases}$$

a, $c \geqslant 0$, $|\beta| \leqslant 1$ and $0 < \alpha \leqslant 2$ are real.

In (4.68), α is called the characteristic exponent and β is a skewness parameter. When $\beta = 0$, or $\alpha = 2$, (4.68) reduces to

$$\log\phi(t) = ait - c|t|^\alpha, \tag{4.69}$$

so that if we take a as origin, as we shall, $\phi(t)$ is real and, by Exercise 4.1, the stable distribution is symmetric then and only then, with $c^{1/\alpha}$ as a scale parameter.

In (4.68) with $a = 0$, changing the sign of the variate changes only sgn t. Thus if x has a stable distribution with skewness parameter β, $(-x)$ has the same distribution with skewness parameter $(-\beta)$, c and α remaining unchanged.

Example 4.12

In (4.69), putting $\alpha = 1$, $a = 0$, $c = 1$, gives the Cauchy c.f. of Example 4.10, while $\alpha = 2$, $a = 0$, $c = \frac{1}{2}\sigma^2$ gives the normal c.f. of Example 4.11. The location and scale parameters a and $c^{1/\alpha}$ may, of course, be varied.

4.37 Apart from the two distributions in Example 4.12, no other symmetric stable distribution has a known elementary form for its f.f. Even in the much wider class (4.68), only one such case is known.

Example 4.13

In (4.68), put $\alpha = \frac{1}{2}$, $\beta = -1$, $a = 0$, $c = 1$. We find

$$\log \phi(t) = -|t|^{\frac{1}{2}} (1 - i \operatorname{sgn} t).$$

Since $(1 - i \operatorname{sgn} t)^2 = -2i \operatorname{sgn} t$, we may write this

$$\log \phi(t) = -|t|^{\frac{1}{2}} (-2i \operatorname{sgn} t)^{\frac{1}{2}} = -(-2it)^{\frac{1}{2}}.$$

This stable c.f. is discussed in Exercise 11.25 below, where it may be seen that its f.f. is

$$f(y) = (2\pi)^{-1/2} \exp\left(-\frac{1}{2y}\right) y^{-3/2}, \qquad 0 \leqslant y \leqslant \infty, \tag{4.70}$$

a distribution none of whose moments exist; it is sometimes called the Lévy distribution. It is evident from (4.68) that the mean of a stable distribution only exists if $\alpha > 1$, and the variance only in the extreme case $\alpha = 2$, which is the normal distribution.

4.38 Because the stable distributions' f.f.'s cannot, except in the three cases discussed above, be expressed simply, they are usually studied through the c.f. However, by numerical evaluation of integrated convergent series expansions for the f.f., Fama and Roll (1968) constructed 4 d.p. tables of the symmetric stable d.f. corresponding to (4.69) for $\alpha = 1{\cdot}0\ (0{\cdot}1)\ 1{\cdot}9,\ 1{\cdot}95,\ 2{\cdot}0$; $a = 0$, $c = 1$ and the variate (their $|u|$) $= 0{\cdot}05\ (0{\cdot}05)\ 1{\cdot}0\ (0{\cdot}1)\ 2{\cdot}0\ (0{\cdot}2)\ 4{\cdot}0\ (0{\cdot}4)\ 6, 7, 8, 10, 15, 20$, with an additional table of the $100p$ and $100(1-p)$ percentiles, $p = 0{\cdot}52\ (0{\cdot}02)\ 0{\cdot}94\ (0{\cdot}01)\ 0{\cdot}97\ (0{\cdot}005)\ 0{\cdot}995,\ 0{\cdot}9995$. Holt and Crow (1973) use various approximations to give 4 d.p tables and charts of the general stable f.f. corresponding to (4.68) for $\alpha = 0{\cdot}25\,(0{\cdot}25)\ 1{\cdot}75$; $\beta = -1\,(0{\cdot}25)\ 1$; $a = 0$, $c = 1$ and the variate $x \geqslant 0$ in steps varying from 0·001 and 0·01 near the origin to 10 and 100 in the extreme tails. (For negative x, the sign of β must be changed—cf. **4.36**.) A curiosity appears because of the discontinuity of the function $w(|t|, \alpha)$ at $\alpha = 1$: for $\alpha = 1$ and $\beta > 0$, the stable distributions have larger positive tails, but for $\alpha \neq 1$, $\beta > 0$ larger negative tails.

EXERCISES

4.1 Show that if a frequency function $f(x)$ is symmetrical the characteristic function, apart from a term involving the mean, is an even function, i.e. $\phi(t) = \phi(-t)$, and that therefore $\phi(t)$ is real; and conversely, if $\phi(t)$ is real the frequency function, if any, is symmetrical.

4.2 Show that the characteristic function

$$\phi(t) = \left(\frac{e^{it}-1}{it}\right)^n, \quad n \text{ a positive integer}$$

is that of the frequency function

$$f(x) = \frac{1}{(n-1)!}\sum_{j=0}^{[x]}(-1)^j\binom{n}{j}(x-j)^{n-1}$$

4.3 Show that the distribution whose c.f. is $1/(1+t^2)$ has frequency function $\frac{1}{2}e^{-|x|}$, $-\infty \leqslant x \leqslant \infty$, and that the distribution whose c.f. is $\cos\frac{1}{2}\pi t/(1-t^2)$ has f.f. $\frac{1}{2}\cos x$, $-\frac{1}{2}\pi \leqslant x \leqslant \frac{1}{2}\pi$.

4.4 If for a certain distribution

$$\kappa_r = \lambda a^r,$$

a and λ being positive constants, show that the distribution is discontinuous (the Poisson) with variate-values $0, a, 2a, \ldots, ra, \ldots$ and the frequency at ra equal to $e^{-\lambda}\lambda^r/r!$.

4.5 Show that the function $\exp(-t^\alpha)$ cannot be a characteristic function unless $\alpha = 2$.

4.6 Show that there is only one distribution with moments given by

$$\mu'_r = \Gamma(\nu+r)/\Gamma(\nu)$$

and that it is

$$dF = \frac{1}{\Gamma(\nu)}e^{-x}x^{\nu-1}\,dx, \quad 0 \leqslant x \leqslant \infty.$$

4.7 Show that the distribution

$$dF = \frac{\sqrt{2}}{\pi}\frac{dx}{1+x^4}, \quad -\infty \leqslant x \leqslant \infty,$$

has c.f.

$$\phi(t) = \sqrt{2}\exp(-|t|/\sqrt{2})\sin(|t|/\sqrt{2}+\tfrac{1}{4}\pi).$$

4.8 Show that the distribution

$$dF = \frac{rb^{2r-1}\sin(\pi/2r)}{\pi(b^{2r}+x^{2r})}dx, \quad -\infty \leqslant x \leqslant \infty;\ b>0;\ r \text{ a positive integer},$$

has c.f.

$$\phi(t) = \sum_{s=0}^{r-1}\exp\left(-b|t|\sin\frac{2s+1}{2r}\pi\right)\sin\left(\frac{2s+1}{2r}\pi+b|t|\cos\frac{2s+1}{2r}\pi\right)\sin(\pi/2r).$$

4.9 Show that the distribution

$$dF = \frac{dx}{\cosh \pi x}, \quad -\infty \leqslant x \leqslant \infty,$$

has c.f.

$$\phi(t) = \operatorname{sech}\tfrac{1}{2}t.$$

4.10 Show that $(1+t^{2k})^{-1}$ cannot be a characteristic function for $k>1$; and also that $(1-t^2)^{-1}$ cannot be a c.f.

4.11 Using the theorem of Marcinkiewicz referred to in **4.8** show that, if all cumulants κ_r vanish for r greater than some $r_0>2$, then all cumulants vanish for $r>2$; and hence that the distribution is normal.

4.12 Show that the distribution

$$F_n(x) = \left(1-\frac{1}{n}\right)\frac{1}{\sqrt{(2\pi)}}\int_{-\infty}^{x} e^{-\frac{1}{2}u^2}du+\frac{1}{2n}\{1+\operatorname{sgn}(x-n)\}$$

tends to normality as $n\to\infty$ but that the higher moments tend to infinity.

(D. G. Kendall and Rao, 1950)

4.13 A theorem due to Weierstrass states that any function continuous in the range (a, b) can be represented by a uniformly convergent series of polynomials $\sum\limits_{n=0}^{\infty} P_n(x)$, $P_n(x)$ being of degree n in x. Deduce that if two continuous frequency functions, f_1 and f_2, have the same moments of all orders,

$$\int_a^b (f_1-f_2)^2\,dx = 0,$$

and hence that the moments determine a distribution uniquely if it is continuous and of finite range.

4.14 If θ is a non-negative function of the variate x and

$$\alpha(t) = \int_{-\infty}^{\infty} \theta^t\,dF(x),$$

show that the frequency function of θ, if any, is given by

$$f(\theta) = \frac{1}{2\pi i}\int_{-i\infty}^{i\infty} \theta^{-t-1}\alpha(t)\,dt.$$

4.15 Show that if a characteristic function $\phi(t)$ possesses derivatives up to and including the second order, then

$$\left|\left(\frac{d\phi}{dt}\right)^2_{t=0}\right| \leqslant \left|\left(\frac{d^2\phi}{dt^2}\right)_{t=0}\right|$$

and generalize this result.

4.16 *Mixtures of Poisson distributions.* λ is a positive random variable with c.f. equal to $\phi(t)$. Another variable y is conditionally distributed in the (Poisson) form of Exercise 3.1 with parameter λ. Show that the unconditional c.f. of y is $\phi\{(e^{it}-1)/i\}$ and that its mean equals that of λ, while its variance equals $\operatorname{var}(\lambda)+E(\lambda)$.

4.17 Show that the distribution

$$dF = \frac{dx_1\,dx_2}{2\pi(1-\rho^2)^{\frac{1}{2}}}\exp\left\{\frac{-1}{2(1-\rho^2)}(x_1^2-2\rho x_1 x_2+x_2^2)\right\}, \qquad -\infty\leqslant x_1, x_2\leqslant\infty,$$

is uniquely determined by its moments.

4.18 If a variate is distributed in the normal form

$$dF = \frac{1}{\sigma\sqrt{(2\pi)}}e^{-\frac{1}{2}x^2/\sigma^2}\,dx, \qquad -\infty\leqslant x\leqslant\infty,$$

show that the c.f. of its square is given by $\phi(t) = (1-2it\sigma^2)^{-\frac{1}{2}}$.

4.19 If two variates are distributed as in Exercise 4.17 show that the joint c.f. of x_1^2 and x_2^2 (with variables t_1 and t_2 respectively) is $\{(1-2it_1)(1-2it_2)+4\rho^2 t_1 t_2\}^{-\frac{1}{2}}$.

4.20 A discontinuous distribution defined at integral values in the range $-\frac{1}{2}n(n-1)$ to $\frac{1}{2}n(n-1)$ has a frequency-generating function

$$P(t) = \prod_{j=1}^{n} \frac{t^j - t^{-j}}{j(t-t^{-1})}.$$

By considering the cumulant-generating function, and using the relation for Bernoulli numbers

$$|B_{2k}| = \frac{2(2k)!}{(2\pi)^{2k}} \sum_{j=1}^{\infty} \left(\frac{1}{j^{2k}}\right),$$

show that in standard measure the distribution tends to the normal form.

4.21 Show that the c.f. of the *logistic distribution*

$$dF = \frac{dx}{4\cosh^2(\frac{1}{2}x)} = \frac{\exp(-x)\,dx}{\{1+\exp(-x)\}^2}, \qquad -\infty \leqslant x \leqslant \infty,$$

is $\phi(t) = \Gamma(1-it)\Gamma(1+it)$ and that its variance is $\pi^2/3$.

4.22 If x is an integer-valued random variable, show that the limits of integration in the inversion formula (4.5) may be replaced by $-\pi$ and π. Use this fact to invert the c.f. of a Poisson variate, given in Example 3.10.

4.23 *Mixtures by location parameter.* x is distributed with f.f. $f(x-\mu)$ conditional upon the value of μ, which is a random variable with f.f. $g(\mu)$. Show using c.f.'s that the unconditional distribution of x is that of the sum of independent variables y, z, with f.f.'s $f(y)$, $g(z)$.

4.24 *Mixtures by scale parameter.* x is distributed with f.f. $\sigma f(x\sigma)$ $(\sigma > 0)$ conditional upon the value of σ, which is a random variable with f.f. $g(\sigma)$. Show using c.f.'s that the unconditional distribution of x is that of the ratio of independent variables y, z, with f.f.'s $f(y)$, $g(z)$.

4.25 Show that the Poisson c.f. in Example 4.9 is infinitely divisible.

CHAPTER 5

STANDARD DISTRIBUTIONS

5.1 There are certain distribution and frequency functions which, for both theoretical and practical reasons, occupy a central position in statistical theory. In this and the next chapter we shall consider their properties, leaving their statistical uses to be developed and illustrated later in the book. We shall, however, indicate briefly some of the ways in which they arise, even at the expense of anticipating ideas introduced at a subsequent stage. This will not impair the logical continuity of our development and will give concreteness to a treatment which might otherwise appear somewhat abstract.

Comprehensive accounts of all the major theoretical distributions are given by Johnson and Kotz (1969, 1970).

The binomial distribution

5.2 Suppose we have a large population of members each of which exhibits either some quality P or a complementary quality Q (= not-P), for example, a population of men who are either blue-eyed or not-blue-eyed. Suppose that the proportion of individuals with quality P is p and that with quality Q is q, where of course $p+q = 1$. If we take a random sample of N members from the population we expect that on the average Nq members will exhibit Q and Np will exhibit P. We may thus array the members according to the quality as

$$N(q+p).$$

Now suppose we choose N *pairs* of individuals. There will be pairs PP, pairs PQ, pairs QP and pairs QQ. Of the Nq pairs for which the first member is Q there will, on the average, be a proportion q for which the second member is Q and p for which it is P. Similarly for the Np exhibiting P in the first member. Thus the pairs may be arrayed as

$$Nq(q+p)+Np(q+p) = N(q+p)^2.$$

Generally if we choose N sets of n the array will be $N(q+p)^n$. That is to say, the proportion of cases containing j P's and $(n-j)$ Q's will be $\binom{n}{j}p^j q^{n-j}$, the term in $p^j q^{n-j}$ in $(q+p)^n$. We are then led to consider the distribution given by expanding the binomial expression

$$1 \equiv (q+p)^n \tag{5.1}$$

as a discontinuous frequency-distribution, the variate being the number of P's in the set of n, which may vary from 0 to n. The relative frequency of j P's is then

$$f_j = \binom{n}{j} p^j q^{n-j}, \quad j = 0, 1, 2, \ldots, n, \tag{5.2}$$

the so-called *binomial distribution.*

5.3 Distributions very close to the binomial form occur in practice, particularly in artificial experiments with coin-tossing or dice-throwing. Table 5.1 shows some data due to Weldon, who threw 12 dice 26,306 times and noted the values at each throw. This is equivalent to the drawing of samples of 12 from a large population. The occurrence of a 5 or a 6 on any die was regarded as the exhibition of the quality P, a "success" as we may call it.

Table 5.1—Frequency-distribution of 26,306 throws of 12 dice, the occurrence of a 5 or 6 being counted a success

No. of successes	Observed frequency	Theoretical frequency from the binomial 26,306 $(0\cdot6623+0\cdot3377)^{12}$	No. of successes	Observed frequency	Theoretical frequency from the binomial 26,306 $(0\cdot6623+0\cdot3377)^{12}$
0	185	187	6	3,067	3,043
1	1,149	1,146	7	1,331	1,330
2	3,265	3,215	8	403	424
3	5,475	5,465	9	105	96
4	6,114	6,269	10 and over	18	16
5	5,194	5,115			
			TOTALS	26,306	26,306

If the dice were perfect (a condition rarely realized in practice) the proportion p of successes would be $\frac{1}{3}$; and the appropriate binomial would be, in the form (5.2), $(\frac{2}{3}+\frac{1}{3})^{12}$. In this particular case the dice were not quite perfect, the proportion of cases exhibiting a 5 or a 6 being $0\cdot3377$. Taking this as the value of p, we get the frequency function $(0\cdot6623+0\cdot3377)^{12}$, which when multiplied by the total frequency 26,306 gives the theoretical frequencies shown in the third column of Table 5.1. The agreement with observation is evidently fairly good.

5.4 We have already found the moments and factorial moments of the distribution in Examples 3.2 and 3.8. The characteristic function of the distribution is given by

$$\phi(t) = \sum_{j=0}^{n} \binom{n}{j} q^{n-j} p^j e^{ijt}$$
$$= (q+pe^{it})^n. \tag{5.3}$$

Taking logarithms and expanding we have for the cumulant-generating function (writing $\theta = it$)

$$\psi(t) = n\log(1+p\theta+\tfrac{1}{2}p\theta^2+\tfrac{1}{6}p\theta^3+\tfrac{1}{24}p\theta^4+\ldots).$$

Expanding the logarithm and identifying powers in θ we then find

$$\kappa_1 = \mu_1' = np\,;\quad \kappa_2 = \mu_2 = np(1-p) = npq\,;\quad \kappa_3 = \mu_3 = np(1-p)(1-2p)$$
$$= npq(q-p)\,;\quad \kappa_4 = np(1-p)(1-6p+6p^2) = npq(1-6pq)\,;\ \text{etc.} \tag{5.4}$$

so that

$$\mu_4 = 3n^2p^2q^2+npq(1-6pq)\,; \tag{5.5}$$

$$\gamma_1 = \mu_3/\mu_2^{3/2} = \frac{q-p}{(npq)^{\frac{1}{2}}} = \sqrt{\beta_1}\,; \tag{5.6}$$

$$\gamma_2 = \kappa_4/\kappa_2^2 = \frac{1-6pq}{npq} = \beta_2 - 3. \tag{5.7}$$

The factorial moments about the origin are particularly simple. We have already found (Example 3.8) that

$$\left.\begin{aligned} \mu'_{[r]} &= p^r n^{[r]},\ r \leqslant n \\ &= 0,\ r > n. \end{aligned}\right\} \tag{5.8}$$

5.5 There are some interesting recurrence relations connecting the moments of the binomial, due to Romanovsky (1923).

Writing $\theta = it$ we have, for the characteristic function referred to the mean as origin,

$$\phi(t) = e^{-np\theta}(q+pe^{\theta})^n. \tag{5.9}$$

Differentiating with respect to θ we find

$$\begin{aligned} \sum_{j=1}^{\infty} \frac{\mu_j \theta^{j-1}}{(j-1)!} &= -npe^{-np\theta}(q+pe^{\theta})^n + ne^{-np\theta}(q+pe^{\theta})^{n-1}pe \\ &= -np\sum_{j=0}^{\infty}\frac{\mu_j\theta^j}{j!} + \frac{npe^{\theta}}{q+pe^{\theta}}\sum_{j=0}^{\infty}\frac{\mu_j\theta^j}{j!}, \end{aligned}$$

and hence, after a little rearrangement,

$$(q+pe^{\theta})\left\{\sum_{j=1}^{\infty}\frac{\mu_j\theta^{j-1}}{(j-1)!}\right\} - npq(e^{\theta}-1)\sum_{j=0}^{\infty}\frac{\mu_j\theta^j}{j!} = 0.$$

Identifying coefficients in θ^{r-1} we get

$$\mu_r = npq\sum_{j=0}^{r-2}\binom{r-1}{j}\mu_j - p\sum_{j=0}^{r-2}\binom{r-1}{j}\mu_{j+1}, \tag{5.10}$$

giving the moment of order r about the mean in terms of those of lower orders.

Furthermore, writing the moment about the mean as

$$\mu_r = \sum_{j=0}^{n}(j-np)^r\binom{n}{j}(1-p)^{n-j}p^j$$

we have, differentiating with respect to p,

$$\begin{aligned} \frac{d\mu_r}{dp} = -rn\Sigma(j-np)^{r-1}\binom{n}{j}q^{n-j}p^j - \Sigma(j-np)^r\binom{n}{j}q^{n-j-1}(n-j)p^j \\ + \Sigma(j-np)^r\binom{n}{j}q^{n-j}jp^{j-1}. \end{aligned}$$

The first term on the right is $-rn\mu_{r-1}$. The sum of the other two will be found to be $\frac{1}{pq}\Sigma(j-np)^{r+1}\binom{n}{j}q^{n-j}p^j = \frac{1}{pq}\mu_{r+1}$. Hence we find

$$\mu_{r+1} = pq\left(nr\mu_{r-1} + \frac{d\mu_r}{dp}\right). \tag{5.11}$$

For example, $\mu_1 = 0$, $\mu_2 = npq = np(1-p)$ and hence, as stated in (5.4),

$$\mu_3 = pq(n-2np) = npq(q-p).$$

Exercise 5.1 gives an even simpler recurrence relation for the cumulants.

Exercise 5.2 shows that (5.11) holds for the incomplete moments, and Exercise 5.3 generalizes (5.10) similarly.

5.6 If $p = q = \frac{1}{2}$, the binomial distribution is obviously symmetrical, but if $p \neq q$ the distribution is skew. It will be unimodal unless $p(n+1) \leqslant 1$ (which holds for $p = \frac{1}{2}$ only at $n = 1$) since the frequency of r "successes" exceeds that of $(r-1)$ if and only if

$$\binom{n}{r} p^r q^{n-r} > \binom{n}{r-1} p^{r-1} q^{n-r+1} \quad \text{or} \quad \frac{r}{n+1} < p.$$

This is equivalent to

$$r - p < np = \mu_1'. \tag{5.12}$$

Thus the frequency of r "successes" increases steadily so long as $r < \mu_1' + p$, and decreases steadily for $r > \mu_1' + p$ with maximum frequency at $r = [p(n+1)]$. There are equal maximum frequencies at $(r-1)$ and r if $r = p(n+1)$, when $r-1 < \mu_1' < r$. Some typical distributions are shown in Table 5.2.

Table 5.2—Terms of the binomial distribution for $n = 20$ and values of p from 0·1 to 0·5 (each term has been multiplied by 10,000)

Number of successes	$p = 0{\cdot}1$ $q = 0{\cdot}9$	$p = 0{\cdot}2$ $q = 0{\cdot}8$	$p = 0{\cdot}3$ $q = 0{\cdot}7$	$p = 0{\cdot}4$ $q = 0{\cdot}6$	$p = 0{\cdot}5$ $q = 0{\cdot}5$
0	1216	115	8	—	—
1	2702	576	68	5	—
2	2852	1369	278	31	2
3	1901	2054	716	123	11
4	898	2182	1304	350	46
5	319	1746	1789	746	148
6	89	1091	1916	1244	370
7	20	545	1643	1659	739
8	4	222	1144	1797	1201
9	1	74	654	1597	1602
10	—	20	308	1171	1762
11	—	5	120	710	1602
12	—	1	39	355	1201
13	—	—	10	146	739
14	—	—	2	49	370
15	—	—	—	13	148
16	—	—	—	3	46
17	—	—	—	—	11
18	—	—	—	—	2
19	—	—	—	—	—
20	—	—	—	—	—

The ordinates of the binomial are most directly calculated from the formula $\binom{n}{j} q^{n-j} p^j$; for low values of n the calculation is straightforward and for high values tables listed in **5.7** may be used.

5.7 The calculation of the d.f. $F(x) = \sum_{j=0}^{x} f_j$, i.e. the summation of terms of the binomial, is tedious to perform directly, but use may be made of E. S. Pearson and N. L. Johnson's *Tables of the Incomplete Beta Function* (2nd edn, Cambridge U.P., 1968).

Consider the integral

$$B_{1-p}(n-r+1, r) = \int_0^{1-p} u^{n-r}(1-u)^{r-1}\,du. \tag{5.13}$$

When r and n are positive integers and $n \geqslant r$, integration by parts gives for (5.13)

$$\frac{(1-p)^{n-r+1}}{n-r+1}.p^{r-1}+\frac{r-1}{n-r+1}B_{1-p}(n-r+2,\ r-1) \tag{5.14}$$

and $(r-1)$ further such integrations yield

$$B_{1-p}(n-r+1,\ r) = \sum_{j=0}^{r-1} \frac{(r-1)^{[r-1-j]}}{(n-j)^{[r-j]}}\, p^j(1-p)^{n-j}. \tag{5.15}$$

Division of (5.15) by $B(n-r+1,\ r)$, i.e. multiplication by $\dfrac{n!}{(r-1)!(n-r)!}$, gives

$$I_q(n-r+1,\ r) = \frac{B_{1-p}(n-r+1,\ r)}{B(n-r+1,\ r)} = \sum_{j=0}^{r-1}\binom{n}{j}p^j(1-p)^{n-j} = F(r-1), \tag{5.16}$$

the sum of the first r terms of the binomial distribution. Since $I_x(a, b) = 1 - I_{1-x}(b, a)$, (5.16) may also be written as $1 - I_p(r, n-r+1)$, so that $I_p(r, n-r+1)$ is the remainder after the first r terms.

The sum of the first $r+1$ terms is, similarly, $I_q(n-r,\ r+1)$, and hence the $(r+1)$th term is

$$\binom{n}{r}q^{n-r}p^r = I_q(n-r, r+1) - I_q(n-r+1, r). \tag{5.17}$$

Example 5.1

When $n = 20$, $r = 11$, $p = 0{\cdot}4$ we have for the remainder after 11 terms $I_{0\cdot4}(11, 10)$ which from the tables is found to be 0·127,521,2. The value given by summing the last six terms in the appropriate column of Table 5.2 is 0·1276, the error in the last place being due to rounding up. The remainder after 12 terms is $I_{0\cdot4}(12, 9)$ = 0·056,526,4. The 12th term (11 "successes") is then the difference of these two remainders = 0·0710, as shown in Table 5.2 for the frequency per 10,000 of 11 successes.

Tables of the binomial distribution

(a) The *Biometrika Tables* give the individual terms of the distribution for p = 0·01, 0·02 (0·02) 0·1 (0·01) 0·5 and n = 5 (5) 30, to 5 decimal places.

(b) *Tables of the Binomial Probability Distribution* (National Bureau of Standards, Applied Mathematics Series, 6, Washington, 1950) gives the individual terms and the distribution function for p = 0·01 (0·01) 0·50 and n = 2 (1) 49, to 7 d.p.

(c) Romig (1953) extends (b) above to $n = 50\,(5)\,100$, to 6 d.p.

(d) *Tables of the Cumulative Binomial Probabilities* (U.S. Army Ordnance Corps Pamphlet ORD P 20–1, Washington, 1952) gives the distribution function for $p = 0{\cdot}01\,(0{\cdot}01)\,0{\cdot}50$ and $n = 1\,(1)\,150$, to 7 d.p.

(e) *Tables of the Cumulative Binomial Probability Distribution* (Harvard University Press, 1955) gives the distribution function for $p = 0{\cdot}01\,(0{\cdot}01)\,0{\cdot}50$, $p = \frac{1}{16}\,(\frac{1}{16})\,\frac{1}{2}$, $p = \frac{1}{12}\,(\frac{1}{12})\,\frac{1}{2}$ and $n = 1\,(1)\,50\,(2)\,100\,(10)\,200\,(20)\,500\,(50)\,1000$.

(f) Weintraub (1963) gives the d.f. to 9 d.p. for $p = 0{\cdot}00001, 0{\cdot}0001\,(0{\cdot}0001)\,0{\cdot}001\,(0{\cdot}001)\,0{\cdot}1$ and $n = 1\,(1)\,100$.

(g) Robertson (1960) gives the d.f for $p = 0{\cdot}001\,(0{\cdot}001)\,0{\cdot}020$, $n = 2\,(1)\,100\,(2)\,200\,(10)\,400\,(200)\,1{,}000$, and for $p = 0{\cdot}021\,(0{\cdot}001)\,0{\cdot}050$, $n = 2\,(1)\,50\,(2)\,100\,(5)\,200\,(10)\,300\,(20)\,500$ or $600\,(50)\,1{,}000$.

(h) Miller (1954) gives the coefficients $\binom{n}{r}$ for $2 \leqslant r \leqslant \frac{1}{2}n \leqslant 100$; $r = 2\,(1)\,12$, $n \leqslant 500$; $r = 2\,(1)\,11$, $n = 500\,(1)\,1000$; $r = 2\,(1)\,5$, $n = 1000\,(1)\,2000$; $r = 2, 3$, $n = 2000\,(1)\,5000$.

As we have seen in Example 4.6, the binomial distribution tends to the "normal" distribution as n increases. Raff (1956) showed that if $np^{3/2} > 1{\cdot}07$, the error in using the normal distribution function instead of the binomial never exceeds $0{\cdot}05$ for any r, and he compared other approximations to the binomial, as did Gebhardt (1969). Peizer and Pratt (1968) and Pratt (1968) gave an approximation accurate to order $n^{-3/2}$.

The Poisson distribution

5.8 Cases sometimes occur in which the proportion p of "successes" in the population is very small. We may suppose our number n large enough to render np itself appreciable though p is small; and we are thus led to consider the limiting form of the binomial distribution as $n \to \infty$, $p \to 0$ subject to the condition that np remains finite, and tends to λ, say.

Under these conditions the term

$$\binom{n}{r} p^r q^{n-r} = \frac{n!}{(n-r)!\,r!}\frac{\lambda^r}{n^r}\left(1-\frac{\lambda}{n}\right)^{n-r}$$
$$\sim \frac{\sqrt{(2\pi)}\,e^{-n}\,n^{n+\frac{1}{2}}}{\sqrt{(2\pi)}\,(n-r)^{n-r+\frac{1}{2}}\,e^{-n+r}\,n^r}\cdot\frac{\lambda^r}{r!}e^{-\lambda}$$
$$\sim \frac{1}{\left(1-\frac{r}{n}\right)^n e^r}\cdot\frac{\lambda^r}{r!}e^{-\lambda} \sim \frac{\lambda^r}{r!}e^{-\lambda}.$$

Thus the terms of the binomial become

$$f_r = e^{-\lambda}\frac{\lambda^r}{r!}, \quad r = 0, 1, 2, \ldots\, ad\ inf.\,; \ \lambda > 0. \tag{5.18}$$

The distribution (5.18) is called the Poisson distribution, having been first given by S. D. Poisson in 1837.

For the characteristic function we have (writing $\theta = it$)

$$\phi(t) = \lim_{n\to\infty} (q+pe^{\theta})^n = \lim_{n\to\infty} \left\{1+\frac{\lambda}{n}(e^{\theta}-1)\right\}^n$$
$$= \exp\{\lambda(e^{\theta}-1)\} \tag{5.19}$$

which has already been shown in Example 3.10 to be the c.f. of the distribution (5.18).

Thus the cumulant-generating function is given by

$$\psi(t) = \lambda(e^{\theta}-1) = \lambda \Sigma\, \theta^j/j!$$

and hence all cumulants of the Poisson distribution are equal to λ. We thus find

$$\mu_1' = \lambda, \quad \mu_2 = \lambda, \quad \mu_3 = \lambda, \quad \mu_4 = \lambda+3\lambda^2. \tag{5.20}$$

If we let $n \to \infty$, $p \to 0$, $np \to \lambda$ in (5.10) and (5.11) we find

$$\mu_r = \lambda \sum_{j=0}^{r-2} \binom{r-1}{j} \mu_j \tag{5.21}$$

and

$$\mu_{r+1} = r\lambda\mu_{r-1} + \frac{\lambda\, d\mu_r}{d\lambda}. \tag{5.22}$$

The factorial moments are given in Exercise 5.26.

5.9 Since $f_r/f_{r-1} = \lambda/r$, the frequency at r increases steadily so long as $r<\lambda$ (the mean) and declines steadily to zero for $r>\lambda$. The maximum frequency is at $r = [\lambda]$, with an equal adjacent maximum at $\lambda-1$ if λ is an integer. For low values of λ the frequency-polygons of the distribution are very skew, being J-shaped for $\lambda<1$, but become nearer to unimodal symmetry as λ increases.

The d.f. $F(x) = \sum_{j=0}^{x} f_j$ may be evaluated in a manner similar to that of **5.7.**

If r is a positive integer, the integral $\int_{\lambda}^{\infty} e^{-u}u^{r-1}\,du$ may be repeatedly integrated by parts to give

$$\int_{\lambda}^{\infty} e^{-u}u^{r-1}\,du = e^{-\lambda} \sum_{=0}^{r-1} \lambda^s (r-1)^{[r-1-s]}.$$

Division by $\Gamma(r) = (r-1)!$ yields

$$\frac{1}{\Gamma(r)}\int_{\lambda}^{\infty} e^{-u}u^{r-1}\,du = e^{-\lambda}\sum_{s=0}^{r-1} \lambda^s/s! = F(r-1), \tag{5.23}$$

the sum of the first r terms of the Poisson distribution. In the notation of K. Pearson's *Tables of the Incomplete Gamma Function*, (5.23) is $1-I\left(\frac{\lambda}{\sqrt{r}}, r-1\right)$, so that the argument used in these tables is a difficult one to work with in the present case. Special tables of the d.f. are listed below. It is easier to calculate $e^{-\lambda}\lambda^r/r!$ directly rather than to use an expression analogous to (5.17) for the individual terms.

Tables of the Poisson distribution

(a) *Biometrika Tables for Statisticians* gives the terms $e^{-\lambda}\lambda^r/r!$ for $\lambda = 0{\cdot}1\,(0{\cdot}1)\,15$, $r = 0\,(1)\,k$ to 6 d.p. where k is the value for which the function is not zero to six places of decimals.

(b) The d.f. can be obtained using (5.23) from the tables of the χ^2 distribution described in **16.4** below—cf. Exercise 16.7. Thus *Biometrika Tables for Statisticians* gives $\sum_{j=0}^{r-1} e^{-\lambda}\lambda^j/j!$ to 5 d.p. for r up to 35 and $\lambda = 0{\cdot}0005$ (0·0005) 0·005 (0·005) 0·05 (0·05) 1·0 (0·1) 5·0 (0·25) 10·0 (0·5) 20 (1) 60. These can be used to supplement the tables of the individual terms, especially in the range $0 \leqslant \lambda \leqslant 1$.

(c) Molina (1942) gives the terms and the sums to 6 or 7 decimal places for $\lambda = 0{\cdot}001$ (0·001) 0·010 (0·010) 0·30 (0·10) 15 (1) 100.

(d) *Tables of the Individual and Cumulative Terms of Poisson Distribution* (Van Nostrand, Princeton, 1962) gives 8 d.p. tables of the f_r, the d.f. $F(r)$ and $1-F(r-1)$ for λ-values similar to those in (c) and also $\lambda = 100$ (5) 200.

As $\lambda \rightarrow \infty$, we see in **5.8** that this effectively removes the limitation to small p, and we should thus expect the "normal" approximation to the binomial (cf. **5.7**) to apply to the Poisson also; it does, as we saw in Example 4.9. Peizer and Pratt (1968) and Pratt (1968) give a much closer normal approximation.

5.10 We now consider a generalization of the binomial and the Poisson distributions. In **5.2** our approach was based on the drawing of sets of n from the same population. Suppose, however, that each set of n consists of an observation drawn from each of n different populations with proportions $p_1, p_2, \ldots, p_n$. Then our proportional frequencies will be arrayed by the form

$$(p_1+q_1)(p_2+q_2)\ldots(p_n+q_n) = \prod_{j=1} (p_j+q_j) \tag{5.24}$$

which of course reduces to the binomial distribution if all the p's are equal.

The characteristic function of this distribution is

$$\phi(t) = \Pi(q_j+p_j e^{\theta})$$

from which we have

$$\begin{aligned}\psi(t) &= \Sigma \log(q_j+p_j e^{\theta}) \\ &= \Sigma\log(1+p_j\theta+\tfrac{1}{2}p_j\theta^2+\ldots) \\ &= \theta\Sigma p_j+\tfrac{1}{2}\theta^2\Sigma(p_j-p_j^2)+\ldots, \text{ etc.,}\end{aligned}$$

giving

$$\mu_1' = \Sigma p_j, \qquad \kappa_2 = \mu_2 = \Sigma p_j q. \tag{5.25}$$

Writing now $\bar{p}$ for the mean of the p's in the different populations, we have

$$\begin{aligned}\mu_1' &= n\bar{p} \\ \mu_2 &= \Sigma pq = \Sigma p - \Sigma p^2 \\ &= \Sigma p - \frac{1}{n}(\Sigma p)^2 - \left\{\Sigma p^2 - \frac{1}{n}(\Sigma p)^2\right\} \\ &= n\bar{p} - n\bar{p}^2 - n\sigma_p^2\end{aligned}$$

(where σ_p^2 is written for the variance of p)

$$= n\bar{p}\bar{q} - n\sigma_p^2. \tag{5.26}$$

A comparison of these results with those for the binomial distribution shows that the variance of the distribution arrayed by (5.24) is *less* than that of the binomial with the same average p by an amount equal to n times the variance of p.

Taking limits, we see that, for the Poisson distribution in such a case,

$$\left.\begin{aligned}\mu_1' &= \bar{\lambda}, \\ \text{and} \qquad \mu_2 &= \bar{\lambda} - \sigma_\lambda^2/n \\ &= \bar{\lambda} + O(n^{-1}).\end{aligned}\right\} \tag{5.27}$$

The Poisson value for the variance thus holds for (5.24) notwithstanding the inequality of the p's, provided that the variance of λ is small compared with n, which will be so if all the p's are small.

The generalization which we have developed is a special form of *stratified sampling*, to be discussed in Chapter 39, Vol. 3.

5.11 Consider now a second generalization, when each complete set of n is drawn from one of k different populations, with proportions $p_1, p_2, \ldots, p_k$. (In the previous case we supposed any set of n obtained by taking one observation from each of n populations. We now suppose that any set is drawn from one population only, but that different sets come from different populations.) Our array of frequencies will now be

$$\frac{1}{k}\sum_{j=1}^{k}(q_j+p_j)^n \tag{5.28}$$

and evidently the moments of this array are the means of the moments of the $(q+p)^n$, that is to say, from (5.4),

$$\mu_1' = \frac{1}{k}\Sigma np,$$

$$\mu_2' = \frac{1}{k}\{\Sigma np+\Sigma n(n-1)p^2\}.$$

Writing $\bar{p}$ for the mean of the p's as before, we have

$$\begin{aligned}\mu_1' &= n\bar{p},\\ \mu_2 &= n\bar{p}+\frac{1}{k}\Sigma n(n-1)p^2-n^2\bar{p}^2\\ &= n\bar{p}\bar{q}+\frac{1}{k}\Sigma n(n-1)p^2-n(n-1)\bar{p}^2\\ &= n\bar{p}\bar{q}+n(n-1)\left(\frac{1}{k}\Sigma p^2-\bar{p}^2\right)\\ &= n\bar{p}\bar{q}+n(n-1)\sigma_p^2. \end{aligned} \tag{5.29}$$

In this case the variance is *greater* than what it would be if the distribution were of the ordinary binomial type by an amount $n(n-1)$ var p.

For the Poisson distribution we have, on taking limits,

$$\mu_1' = \lambda, \qquad \mu_2 = \lambda+\sigma_\lambda^2, \tag{5.30}$$

and here also the variance of the distribution is affected.

This second generalization is a special case of *cluster sampling*, also to be discussed in Chapter 39, Vol. 3.

5.12 It is often found in practice that data derived from the sampling of attributes over an extended area or an extended period of time do not conform to the simple binomial type. For example, suppose we regard the possession of blue eyes as a success, and take a number of samples of n from the population of the United Kingdom in different localities. We should probably find that the proportions in these samples did not conform to the simple binomial form. The variance calculated from the known

n and average observed p would probably turn out to be smaller than the variance observed among the different samples. If so, we should conclude from (5.29) that the proportion p varied from place to place in the population, the excess in the variance of the proportions observed being due to the variance of p itself in the sections of the population from which the samples were chosen. We are assuming for the time being that these differences are not explicable on the basis of sampling fluctuation alone; but a full discussion will have to wait until later chapters.

Mixtures of distributions

5.13 The same effect is found in distributions which at first sight might be expected to be of the Poisson type. For example, suicide is a rare event and it might be supposed that if we took a series of large samples, say the population of the United Kingdom in successive years, the frequencies of suicides would follow the Poisson distribution. This, however, is not necessarily so, for all members of the population are not equally exposed to risk, and the temptation to suicide may vary from year to year, e.g. being greater in years of trade depression. This inequality of risk is typical of one field in which the Poisson distribution has been freely applied, namely, industrial accidents. Table 5.3 shows, in the second column, the frequency of accidents occurring to women working on the manufacture of shells. The Poisson frequencies shown in the third column provide a very poor fit. One possible reason is that the liability of individuals to accident varies.

Table 5.3—Accidents to 647 women working on H.E. shells in 5 weeks
(Greenwood and Yule (1920))

Number of accidents	Observed frequency	Poisson distribution with same mean	Distribution given by (5.33)
0	447	406	442
1	132	189	140
2	42	45	45
3	21	7	14
4	3	1	5
5 and over	2	0·1	2
TOTALS	647	648	648

As a working hypothesis (cf. Greenwood and Yule, 1920) suppose that the population is a mixture of individuals with different degrees of accident proneness, represented by different values of λ in a Poisson distribution; and suppose that in the population the distribution of λ is of the Gamma form of Example 3.6,

$$dF = \frac{c^r}{\Gamma(r)} e^{-c\lambda} \lambda^{r-1} d\lambda, \qquad 0 \leqslant \lambda \leqslant \infty; r>0, c>0. \tag{5.31}$$

The frequency of j successes is then

$$\int_0^\infty f(\lambda) e^{-\lambda} \frac{\lambda^j}{j!} d\lambda = \int_0^\infty \frac{c^r}{\Gamma(r)} e^{-c\lambda} \lambda^{r-1} e^{-\lambda} \frac{\lambda^j}{j!} d\lambda,$$

or the coefficient of t^j in the frequency-generating function

$$P(t) = \frac{c^r}{\Gamma(r)} \int_0^\infty e^{-c\lambda} \lambda^{r-1} e^{-\lambda+\lambda t} d\lambda,$$

which, on the substitution of $(c+1-t)\lambda = u$, becomes

$$P(t) = \frac{c^r}{(c+1-t)^r} = \left(\frac{c}{c+1}\right)^r\left(1-\frac{t}{c+1}\right)^{-r}. \tag{5.32}$$

The frequency of 0, 1, 2, . . . successes is therefore

$$\left(\frac{c}{c+1}\right)^r\left\{1, \frac{r}{c+1}, \frac{r(r+1)}{2!\,(c+1)^2}, \ldots\right\}. \tag{5.33}$$

The mean of the distribution (5.33) is

$$\mu_1' = \left(\frac{c}{c+1}\right)^r\left\{\frac{r}{c+1}+\frac{r(r+1)}{(c+1)^2}+\ldots\right\} = \left(\frac{c}{c+1}\right)^r\cdot\frac{r}{c+1}\cdot\left(\frac{c+1}{c}\right)^{r+1} = \frac{r}{c}. \tag{5.34}$$

Similarly $$\mu_2' = \frac{r}{c}\left(\frac{r+c+1}{c}\right),$$

so that $$\mu_2 = \frac{r(c+1)}{c^2} = \frac{r}{c}+\frac{r}{c^2}. \tag{5.35}$$

If we now put the observed mean and variance of Table 5.3 equal to the values of (5.34) and (5.35) we have two equations which can be solved for r and c. The distribution (5.33) can then be found. The frequencies are given in the fourth column of Table 5.3 and evidently give a much better agreement with the facts.

(5.33) is called a *mixture* of the Poisson distributions, using (5.31) as the weighting (or mixing) distribution of λ. Such mixtures were the subject of Exercise 4.16, which may be used to obtain (5.33)—this is left to the reader in Exercise 5.23.

Exercise 5.9 shows that any unimodal *continuous* mixture of Poissons is itself unimodal, as here.

Generalizing, we may form a mixture of any f.f. containing a parameter θ, say $f(x\,|\,\theta)$, by weighting it by a distribution for θ, say $p(\theta)$, and then integrating with respect to θ. Thus we obtain

$$g(x) = \int_{-\infty}^{\infty} f(x\,|\,\theta)p(\theta)\,d\theta.$$

Essentially, we now have a bivariate distribution of x and θ, and making use of the basic ideas in **1.32–3**, $f(x\,|\,\theta)$ is the conditional distribution of x given θ, while $p(\theta)$ is the marginal distribution of θ. Their product, the integrand above, is the bivariate distribution, as at (1.29). The marginal distribution $g(x)$ is obtained as at (1.30).

Exercises 4.23–4, 5.17 and 5.24 are concerned with mixtures. See also Exercise 5.8.

The negative binomial distribution

5.14 An interesting feature of the distribution arrayed in (5.33) is that it is generated by a binomial $\left(\frac{c+1}{c}-\frac{1}{c}\right)^{-r}$ *with negative index*. In the approach adopted in **5.2** the index is necessarily positive. Cases not infrequently arise, however, in which data are represented by such “ negative binomials ”. A very simple generative process which we shall use when we come to consider sequential sampling was first given by Yule (1910); it does not depend on any arbitrary assumptions about distributions such as that embodied in (5.31).

In the sampling procedure of **5.2**, suppose that n is not fixed, but is determined as the sample size required to achieve r "successes". Evidently, the final (nth) individual will then always be a "success" (since otherwise sampling would have stopped earlier), and the previous $(n-1)$ individuals must contain a total of $(r-1)$ "successes". Thus the relative frequency with which n individuals will be required to obtain r "successes" is, using (5.2),

$$f_n = \binom{n-1}{r-1} q^{n-r} p^{r-1} \times p = p^r \binom{n-1}{r-1} q^{n-r}, \quad n = r, r+1, r+2, \ldots \tag{5.36}$$

and these are the successive terms in $p^r(1-q)^{-r}$, a binomial with negative index. It will be seen that in **5.13** we had $p = c/(c+1)$ and origin at the value r, and that whereas in our present discussion r is necessarily a positive integer, r could take any positive value in **5.13**. In fact, since $q<1$, $(1-q)^{-r}$ may validly be expanded in series to give (5.36) for non-integral r. Henceforth, whether or not r is an integer, we shall call (5.36) the *negative binomial* (or the Pascal) distribution. When $r = 1$, it is called the *geometric* distribution, then being $f_n = pq^{n-1}$.

Since $f_n/f_{n-1} = (n-1)q/(n-r)$, we see that $f_n > f_{n-1}$ if and only if $p<(r-1)/(n-1)$. Thus the distribution is J-shaped if $r = 1$, and unimodal or J-shaped otherwise, with maximum frequency at $n = \left[1+\dfrac{r-1}{p}\right]$.

The relative frequency with which *more* than n individuals will be required for r successes is evidently exactly the relative frequency of *fewer* than r successes in n ordinary binomial observations, i.e. is given by (5.16). Thus the negative binomial distribution function is the complement of this, namely

$$F(n) = \sum_{s=r}^{n} f_s = 1 - I_{1-p}(n-r+1, r) = I_p(r, n-r+1).$$

Williamson and Bretherton (1963) have given tables of the negative binomial distribution. Peizer and Pratt (1968) and Pratt (1968) give a very good "normal" approximation.

5.15 With the origin at the value r, the c.f. of the distribution (5.36) is (writing θ for it)

$$\phi(t) = p^r(1-qe^{\theta})^{-r}.$$

Thus for the cumulant-generating function

$$\psi(t) = r\log p - r\log(1-qe^{\theta}) = -r\log\left\{1-\frac{q}{p}(e^{\theta}-1)\right\}. \tag{5.37}$$

Expanding and identifying coefficients, we have

$$\left.\begin{aligned} \kappa_1 &= \frac{rq}{p}; \quad \kappa_2 = \frac{rq}{p^2}; \quad \kappa_3 = \frac{rq(1+q)}{p^3}; \\ \kappa_4 &= \frac{rq(1+4q+q^2)}{p^4}. \end{aligned}\right\} \tag{5.38}$$

These relations may be derived from equations (5.4) by putting therein $-q/p$ for p, $1/p$ for q and $-r$ for n. (5.34–5) are obtained with $p = c/(c+1)$.

The f.m.g.f. is given by

$$P(1+t) = p^r\{1-q(1+t)\}^{-r} = (1-qt/p)^{-r}. \tag{5.39}$$

Hence we find, for the factorial moments about the value r,

$$\mu'_{[s]} = \left(\frac{q}{p}\right)^s (r+s-1)^{[s]}. \tag{5.40}$$

From (5.38), $\mu_2 > \mu'_1$ (in accordance with the general result of Exercise 4.16 for mixtures of Poissons), whereas for the ordinary binomial $\mu_2 < \mu'_1$, and for the Poisson $\mu_2 = \mu'_1$. A preliminary calculation of μ_2/μ'_1 is often a good guide to which of the three is likely to fit the data best. Exercise 5.10 gives another useful distinguishing feature.

If we measure from the origin zero, the mean alone is affected, being increased by r to r/p, and $\mu_2 > \mu'_1$ if and only if $q > p$, i.e. $p < \frac{1}{2}$.

The logarithmic distribution

5.16 There is an interesting limiting form of the negative binomial. Suppose that q represents the probability of the presence of an attribute and p of its absence. In some circumstances we may not observe—we may not even be capable of observing—the situation in which there is complete absence of the attribute. We are then led to consider a *truncated* negative binomial with the zero class missing, i.e. the distribution with frequencies proportional to

$$p^r\left\{rq, \quad \frac{r(r+1)q^2}{2!}, \ldots\right\}. \tag{5.41}$$

The total frequency is $1-p^r$ and we may write the distribution as

$$\frac{rp^r}{1-p^r}\left\{q, \quad \frac{(r+1)}{2!}q^2, \quad \frac{(r+1)(r+2)}{3!}q^3, \ldots\right\}. \tag{5.42}$$

Now let r tend to zero. For the initial factor we have

$$\lim \frac{r}{p^{-r}-1} = \lim \frac{r}{\exp(-r\log p)-1} = \frac{1}{-\log p},$$

by L'Hôpital's rule. The distribution then tends to the form

$$\frac{-1}{\log(1-q)}\left\{q, \quad \frac{q^2}{2}, \quad \frac{q^3}{3}, \ldots\right\}, \tag{5.43}$$

that is to say, the frequency at integral values $s \geqslant 1$ is the coefficient of t^s in the frequency-generating function

$$P(t) = \frac{1}{\log(1-q)}\log(1-qt). \tag{5.44}$$

This distribution is due to Fisher *et al.* (1943). As we have introduced it, the process of derivation looks rather artificial; for we obtained the negative binomial with index r and we are now letting r tend to zero. But as we observed below (5.36), the latter remains valid for any positive r, however close to zero, and the limiting process is therefore not invalid.

The logarithmic distribution has been found to be useful by entomologists, in particular.

An example of a distribution of the logarithmic type is given in Table 5.4; but this is to be regarded as illustrative of goodness of fit, not as evidence that the data are generated by the particular limiting process by which we have derived the distribution.

Table 5.4—Distribution of butterflies in Malaya, with theoretical frequencies given by the logarithmic distribution

(From Fisher *et al.* (1943))

No. of species	Theoretical frequency	Observed frequency
1	135·05	118
2	67·33	74
3	44·75	44
4	33·46	24
5	26·69	29
6	22·17	22
7	18·95	20
8	16·53	19
9	14·65	20
10	13·14	15
11	11·91	12
12	10·89	14
13	10·02	6
14	9·28	12
15	8·63	6
16	8·07	9
17	7·57	9
18	7·13	6
19	6·74	10
20	6·38	10
21	6·06	11
22	5·77	5
23	5·50	3
24	5·25	3

5.17 Replacing t by e^{θ} in (5.44) we find the characteristic function of the distribution to be

$$\phi(t) = 1+\alpha \log\left\{1-\frac{q}{p}(e^{\theta}-1)\right\}$$

where $\alpha = 1/\log(1-q)$. Noting the similarity of this to (5.37) we have for the moments about the origin (not the cumulants—cf. Exercises 5.22 and 5.21)

$$\mu_1' = -\frac{\alpha q}{p};\ \mu_2' = -\frac{\alpha q}{p^2};\ \mu_3' = -\frac{\alpha q(1+q)}{p^3};\ \mu_4' = -\frac{\alpha q(1+4q+q^2)}{p^4};\qquad (5.45)$$

whence

$$\left.\begin{aligned}\mu_2 &= -\frac{\alpha q(1+\alpha q)}{p^2},\\ \mu_3 &= -\frac{\alpha q(1+q+3\alpha q+2\alpha^2 q^2)}{p^3},\\ \mu_4 &= -\frac{\alpha q\{1+4q+q^2+4\alpha q(1+q)+6\alpha^2 q^2+3\alpha^3 q^3\}}{p^4}\end{aligned}\right\}\qquad (5.46)$$

Tables of the f.f. and d.f. are given by Williamson and Bretherton (1964) for $\mu_1 = 1{\cdot}1\,(0{\cdot}1)\,2{\cdot}0\,(0{\cdot}5)\,5{\cdot}0\,(1{\cdot}0)\,10{\cdot}0$. G. P. Patil *et al.* (1964) give extensive tables of the f.f. and d.f.

The hypergeometric distribution

5.18 Consider now the generalization of the approach of **5.2** when samples of n are drawn from a population of N individuals, where N is not necessarily large. If we take a sample which contains r P's and $n-r$ Q's, it can arise in

$$\binom{n}{r}\frac{Np(Np-1)\ldots(Np-r+1)Nq(Nq-1)\ldots(Nq-n+r+1)}{N(N-1)\ldots(N-n+1)} = \binom{n}{r}\frac{(Np)^{[r]}(Nq)^{[n-r]}}{N^{[n]}} \tag{5.47}$$

ways. For there are $\binom{N}{n}$ ways of selecting the sample, and the r P's can be chosen in $\binom{Np}{r}$ ways, and the $n-r$ Q's in $\binom{Nq}{n-r}$ ways, if $r \leqslant Np$ and $n-r \leqslant Nq$, the expression given in (5.47) being equal to $\binom{Np}{r}\binom{Nq}{n-r}\Big/\binom{N}{n}$.

Hence we are led to consider the discontinuous distribution whose general term is

$$f_j = \frac{1}{N^{[n]}}\binom{n}{j}(Np)^{[j]}(Nq)^{[n-j]}, \quad j=0,1,\ldots, \min(n, Np), \tag{5.48}$$

a form in which the analogy with the binomial distribution is evident. As $N \to \infty$ the form (5.48) approaches the binomial form.

The frequency-generating function

$$P(t) = \frac{1}{N^{[n]}}\sum_{j=0}^{n}\left\{\binom{n}{j}(Np)^{[j]}(Nq)^{[n-j]}t^j\right\} \tag{5.49}$$

is equal to

$$\frac{(Nq)^{[n]}}{N^{[n]}}\Sigma\frac{(Np)^{[j]}n^{[j]}}{(Nq-n+j)^{[j]}}\frac{t^j}{j!},$$

that is to say, to the hypergeometric function

$$\frac{(Nq)^{[n]}}{N^{[n]}}F(\alpha,\beta,\gamma,t)$$

if

$$\alpha = -n, \quad \beta = -Np, \quad \gamma = Nq-n+1. \tag{5.50}$$

The distribution (5.48) is therefore called hypergeometric. We have (cf. Romanovsky, 1925)

$$F(\alpha,\beta,\gamma,t) = 1+\frac{\alpha\beta}{\gamma}\frac{t}{1!}+\frac{\alpha(\alpha+1)\beta(\beta+1)}{\gamma(\gamma+1)}\frac{t^2}{2!}+\ldots$$

and it is well known that this function satisfies the differential equation

$$t(1-t)\frac{d^2F}{dt^2}+\{\gamma-(\alpha+\beta+1)t\}\frac{dF}{dt}-\alpha\beta F = 0, \tag{5.51}$$

a fact which may be readily verified from the equation itself.

If in (5.49) we put $t = e^\theta$ (where $\theta = it$) we have the characteristic function ϕ of the distribution. On making this substitution in (5.51) we find, after some reduction and replacement of the values of α, β, γ by those of (5.50),

$$(1-e^\theta)\left\{\frac{d^2\phi}{d\theta^2}-(n+Np)\frac{d\phi}{d\theta}+nNp\phi\right\}-Nnp\phi+N\frac{d\phi}{d\theta} = 0. \tag{5.52}$$

Since $\phi = \Sigma\mu_j'\theta^j/j!$ we find, from the coefficient of θ^0 in this expression,

$$-Nnp+N\mu_1' = 0$$

$$\mu_1' = np, \tag{5.53}$$

the same result as for the binomial. The mean of the hypergeometric distribution is independent of N.

Measuring the distribution about its mean, and hence substituting $e^{np\theta}\phi$ for ϕ in (5.52), we find

$$(1-e^{\theta})\left[\frac{d^2\phi}{d\theta^2}+\frac{d\phi}{d\theta}\{n(p-q)-Np\}+(N-n)pqn\phi\right]+N\frac{d\phi}{d\theta} = 0 \tag{5.54}$$

whence, identifying coefficients in θ, θ^2, θ^3, we find

$$\left.\begin{aligned}\mu_2 &= \frac{npq(N-n)}{N-1}, \quad \mu_3 = \frac{npq(q-p)(N-n)(N-2n)}{(N-1)(N-2)}, \\ \mu_4 &= \frac{npq(N-n)}{(N-1)(N-2)(N-3)} \\ &\quad \times [N(N+1)-6n(N-n)+3pq\{N^2(n-2)-Nn^2+6n(N-n)\}],\end{aligned}\right\} \tag{5.55}$$

and generally, if E denotes the operation of raising the order of a moment by unity, i.e. $E\mu_r = \mu_{r+1}$, we have

$$N\mu_{r+1} = \{(1+E)^r-E^r\}[\mu_2-\{Np+n(q-p)\}\mu_1+\{npq(N-n)\mu_0\}]. \tag{5.56}$$

As we expect, when $N\to\infty$ these values tend to those of the binomial distribution

The factorial moments are given in Exercise 5.26.

5.19 An example of the occurrence of the hypergeometric distribution in practice is given in Table 5.5, giving the frequency of occurrence of cards of a certain suit in hands of whist. Here N is the number of cards in the pack, 52, and $n = 13$, $p = \frac{1}{4}$. The appropriate frequencies are thus

$$\frac{1}{52^{[13]}}\binom{13}{j}13^{[j]}39^{[n-j]}$$

giving the figures shown in the third column. The agreement is reasonably good.

Table 5.5—Distribution of 3400 first hands at whist according to number of trumps in the hand (K. Pearson (1924a))

Number of cards in the hand	Observed frequency	Frequency of hypergeometric distribution	Number of cards in the hand	Observed frequency	Frequency of hypergeometric distribution
0	35	43·5	5	444	424·0
1	290	272·2	6	115	141·3
2	696	700·0	7	21	30·0
3	937	973·5	8	11	4·0
4	851	811·3	9 and over	0	0·2
			TOTALS	3400	3400

Lieberman and Owen (1961) give the d.f. and f.f. for $N = 1\,(1)\,50\,(10)\,100$ and all n; and for $N = 200\,(100)\,2000$ and selected n. Wise (1954) developed a quickly convergent expansion for the d.f. in terms of the Incomplete Beta-function. Sandiford (1960) shows that a close approximation is obtained from a binomial distribution with the same mean and variance. Ord (1968a) compares these and gives other approximations.

Other types of hypergeometric distributions are discussed by O. L. Davies (1933–4) and by Ord (1967b)—these include the Beta-binomial of Exercise 5.17.

The normal distribution

5.20 We have already noted in Examples 4.6 and 4.9 that the binomial and the Poisson distributions tend, when expressed in standard measure, to the form

$$dF = \frac{1}{\sqrt{(2\pi)}} e^{-\frac{1}{2}x^2}\, dx, \qquad -\infty \leqslant x \leqslant \infty. \tag{5.57}$$

The more general form

$$dF = \frac{1}{\sigma\sqrt{(2\pi)}} \exp\left\{-\frac{1}{2\sigma^2}(x-\mu_1')^2\right\} dx, \qquad -\infty \leqslant x \leqslant \infty\,;\ \sigma>0, \tag{5.58}$$

is known as the normal distribution.(*) It is the most important theoretical distribution in statistics. The expression (5.57) is the normal distribution in standard measure. We recall some of its properties which have been given in examples.

The c.f. of (5.58) is easily found to be—cf. Example 3.4—

$$\phi(t) = \exp\left(it\mu_1' - \tfrac{1}{2}t^2\sigma^2\right) \tag{5.59}$$

giving

$$\mu_{2r} = \frac{(2r)!}{2^r r!}\sigma^{2r}, \quad \mu_{2r+1} = 0, \quad r \geqslant 1. \tag{5.60}$$

For the cumulant-generating function we have

$$\psi(t) = it\mu_1' - \tfrac{1}{2}t^2\sigma^2$$

so that

$$\kappa_2 = \sigma^2, \qquad \kappa_r = 0, \qquad r>2. \tag{5.61}$$

We also have $\beta_2 = 3$, $\gamma_2 = 0$, which accounts for the standard adopted for mesokurtosis in **3.32**.

5.21 The frequency function of the standardized normal distribution will occur so often that we shall introduce a special symbol for it. We write

$$\alpha(x) = \frac{1}{\sqrt{(2\pi)}} e^{-\frac{1}{2}x^2}. \tag{5.62}$$

For the distribution function we have

$$F(x) = \int_{-\infty}^{x} \alpha(y)\, dy = \tfrac{1}{2} + \int_{0}^{x} \alpha(y)\, dy. \tag{5.63}$$

We may expand $\alpha(x)$ as a power series in x and integrate term by term to obtain

$$F(x) = \tfrac{1}{2} + \frac{1}{\sqrt{(2\pi)}}\left(x - \frac{x^3}{2.3} + \frac{x^5}{2!\,2^2.5} - \dots\right). \tag{5.64}$$

This, however, is not a very useful expression for the calculation of $F(x)$. It converges too slowly except when x is small. A better expansion with similar properties is given in Exercise 5.28.

(*) The description of the distribution as the " normal ", used by Galton, is almost universal in English, although " Gaussian " is becoming a frequent synonym in special contexts. Continental writers refer to it variously by the names of Laplace and/or Gauss. As an approximation to the binomial it was published by Demoivre in 1733 but he did not discuss its properties.

5.22 For large x an asymptotic series may be employed. Let us note that

$$\frac{\exp\{-\frac{1}{2}x^2(1+u^2)\}}{1+u^2} = \int_{\frac{1}{2}x^2}^{\infty} \exp\{-v(1+u^2)\}\,dv.$$

We may integrate both sides for u from $-\infty$ to ∞, and in virtue of the uniform convergence on the right-hand side, may there invert the order of integration. We then have

$$\begin{aligned}\int_{-\infty}^{\infty} \frac{\exp\{-\frac{1}{2}x^2(1+u^2)\}}{1+u^2}\,du &= \int_{\frac{1}{2}x^2}^{\infty}\int_{-\infty}^{\infty} \exp\{-v(1+u^2)\}\,du\,dv \\ &= \sqrt{(2\pi)}\int_{\frac{1}{2}x^2}^{\infty} \frac{1}{\sqrt{(2v)}}e^{-v}\,dv \\ &= \sqrt{(2\pi)}\int_{x}^{\infty} e^{-\frac{1}{2}u^2}\,du.\end{aligned}$$

Hence

$$\begin{aligned}1-F(x) &= \frac{1}{\sqrt{(2\pi)}}\int_{x}^{\infty} e^{-\frac{1}{2}u^2}\,du \\ &= \frac{1}{2\pi}\int_{-\infty}^{\infty} \frac{\exp\{-\frac{1}{2}x^2(1+u^2)\}}{1+u^2}\,du \\ &= \frac{1}{\sqrt{(2\pi)}}e^{-\frac{1}{2}x^2}.\frac{1}{\sqrt{(2\pi)}}\int_{-\infty}^{\infty} \frac{\exp(-\frac{1}{2}x^2u^2)}{1+u^2}\,du. \qquad (5.65)\end{aligned}$$

The ratio of the "tail area" of the distribution, $1-F(x)$, to its bounding ordinate $\alpha(x)$, is known as Mills' ratio (*) and denoted by $R(x)$. Thus

$$\begin{aligned}R(x) &= \{1-F(x)\}\Big/\frac{1}{\sqrt{(2\pi)}}e^{-\frac{1}{2}x^2} \\ &= e^{\frac{1}{2}x^2}\int_{x}^{\infty} e^{-\frac{1}{2}u^2}\,du. \qquad (5.66)\end{aligned}$$

From (5.65) we have

$$R(x) = \frac{1}{\sqrt{(2\pi)}}\int_{-\infty}^{\infty} \frac{\exp(-\frac{1}{2}x^2u^2)}{1+u^2}\,du, \qquad (5.67)$$

and on replacing $\frac{1}{2}x^2u^2$ by t,

$$R(x) = \frac{x}{2\sqrt{\pi}}\int_{0}^{\infty} \frac{e^{-t}t^{-\frac{1}{2}}}{t+\frac{1}{2}x^2}\,dt.$$

Expanding the denominator and expressing the resultant integrals in terms of Gamma-functions we have

$$\begin{aligned}R(x) &= \frac{x}{2\sqrt{\pi}}\int_{0}^{\infty} \frac{2}{x^2}\left(1-\frac{2t}{x^2}+\frac{4t^2}{x^4}-\ldots\right)e^{-t}t^{-\frac{1}{2}}\,dt \\ &= \frac{1}{x}-\frac{1}{x^3}+\frac{1.3}{x^5}-\ldots(-1)^j\frac{1.3.5\ldots(2j-1)}{x^{2j+1}}+R\,(x), \qquad (5.68)\end{aligned}$$

(*) It was tabulated by Mills in 1926, but had, in effect, been considered by earlier writers. Its reciprocal $F'(x)/\{1-F(x)\}$ is called the *failure rate* of a distribution $F(x)$, the name deriving from life-testing applications.

where $$|R_j(x)| = \frac{2x^{-(2j+1)}}{2\sqrt{\pi}}\int_0^\infty \frac{t^{2j+2}e^{-t}t^{-\frac{1}{2}}}{t+x^2}dt.$$

This series does not converge but

$$|R_j(x)| < \frac{2x^{-(2j+1)}}{2\sqrt{\pi}}\int_0^\infty t^{2j+\frac{1}{2}}e^{-t}\,dt$$
$$< x^{-(2j+1)}1.3.5\ldots(2j-1). \quad (5.69)$$

Thus the remainder at any point in the summation is less in absolute value than the last term taken into account, and the series is asymptotic. For large x the expansion is reasonably effective.

Exercise 5.27 indicates how the leading term in (5.68) may be used to obtain an expansion of the inverse of the d.f.

5.23 The most useful forms for the calculation of $F(x)$, or equivalently of $R(x)$, however, are continued fractions. One such, given by Laplace in 1805, is as follows. Define $z(t)$ by

$$\frac{1}{x}z(t) = R\{x(1-t)\} = \frac{1}{\sqrt{(2\pi)}}\int_{-\infty}^{\infty}\frac{\exp\{-\frac{1}{2}x^2u^2(1-t)^2\}\,du}{1+u^2}, \quad (5.70)$$

using (5.67). Then $$\frac{1}{x}\frac{dz}{dt} = -x\{(1-t)z-1\}. \quad (5.71)$$

Hence if z has the expansion in powers of t

$$z = \sum_{j=0}^{\infty} y_j t^j$$

we find, on substitution in (5.70) and identification of coefficients,

$$\frac{1}{x^2}(r+1)y_{r+1}+y_r-y_{r-1} = 0. \quad (5.72)$$

Hence $$\frac{y_r}{y_{r-1}} = \frac{1}{1+\frac{r+1}{x^2}\frac{y_{r+1}}{y_r}}.$$

Repeated applications of this formula give

$$\frac{y_1}{y_0} = \frac{x}{x+}\,\frac{2}{x+}\,\frac{3}{x+}\cdots\frac{n}{x+}\cdots$$

Also, from (5.70) we find $y_0 = xR(x)$, $y_1/x^2 = 1-y_0$. Hence, substituting for y_1 we have, after a little rearrangement,

$$R(x) = y_0/x = \frac{1}{x+}\,\frac{1}{x+}\,\frac{2}{x+}\,\frac{3}{x+}\cdots\frac{n}{x+}\cdots \quad (5.73)$$

This expression was used by Sheppard (1939, posthumous) in calculating his tables of the normal distribution, described in **5.25** below. At the end of this volume we give some tables which will suffice to illustrate the theory and examples given in this book.

5.24 Another continued fraction which converges much more rapidly than Laplace's has been given by Shenton (1954). Defining

$$\overline{R}(x) = \int_0^x \alpha(y)\,dy/\alpha(x) = \frac{1}{2\alpha(x)} - R(x),$$

Shenton finds
$$\overline{R}(x) = \frac{x}{1-}\,\frac{x^2}{3+}\,\frac{2x^2}{5-}\,\frac{3x^2}{7+}\cdots \tag{5.74}$$

Ruben (1962, 1963, 1964), Ray and Pitman (1963) and Rabinowitz (1969) give other asymptotic expressions for $R(x)$. Further related results are given in Exercises 5.12–16 and 5.18.

Hastings (1955) gives Chebyshev polynomial approximations to the normal f.f. and d.f. The former is approximated within 0·00025 by $f(x) = \{\sum_{r=0}^{5} b_{2r}x^{2r}\}^{-1}$, and the latter within 0·000,000,15 for $x \geqslant 0$ by $F(x) = 1-0{\cdot}5\{1+\sum_{r=1}^{6} a_r x^r\}^{-16}$, where the a_r, b_r are constants. He also gives a rational approximation to the *inverse* d.f., i.e. the solution for x of $F(x) = c$. See also Exercise 5.27.

Tables of the normal distribution

5.25 (a) Sheppard's (1939) tables give, among other things, Mills' ratio to 12 decimal places by intervals of 0·01 of x; the same to 24 decimal places by intervals of 0·1; and the negative natural logarithm of the tail area to 16 places by intervals of 0·1.

(b) The *Biometrika Tables*, in a table based on Sheppard's work, give the ordinate and the distribution function of the standardized distribution, to 7 decimal places for $x = 0\ (0{\cdot}01)\ 4{\cdot}50$ and to 10 d.p. for $x = 4{\cdot}50\ (0{\cdot}01)\ 6{\cdot}00$, as well as an auxiliary table which extends to $x = 500$ and inverse tables giving the deviate and the ordinate from the value of the distribution function.

(c) *Tables of Normal Probability Functions* (National Bureau of Standards, Applied Mathematics Series, 23, Washington, 1953) gives the ordinate and $\{2F(x)-1\}$, where F is the distribution function, for $x = 0\ (0{\cdot}0001)\ 1\ (0{\cdot}001)\ 7{\cdot}800$ to 15 d.p., and the ordinate and $2\{1-F(x)\}$ to 7 *significant figures* for $x = 6\ (0{\cdot}01)\ 10$, as well as auxiliary tables for large values of x.

(d) *The Kelley Statistical Tables* (Harvard U.P., 1948) gives an inverse table of the standardized deviate and the ordinate, each to 8 d.p., for values of the distribution function 0·0001 (0·0001) 0·9999, with a finer tabulation for 10 lower and 10 higher values extending to 0·000,000,001 and 0·999,999,999.

(e) Smirnov (1965) gives 7 d.p. tables of the d.f. and f.f. and also a 5 d.p. table of $-\log\{1-F(x)\}$ for $x = 5\,(1)\,50\,(10)\,100\,(50)\,500$.

(f) J. S. White (1970) gives inverse tables to 20 d.p. of x for $F(x) = 0{\cdot}5\ (0{\cdot}005)\ 0{\cdot}995$ and for $1-F(x) = 10^{-k}$, $2{\cdot}5\times10^{-k}$ and 5×10^{-k} with $k = 1\,(1)\,20$.

5.26 The shape of the standardized normal distribution

$$\alpha(x) = \frac{1}{\sqrt{(2\pi)}}e^{-\frac{1}{2}x^2}$$

is illustrated in Fig. 5.1. It is symmetrical and unlimited in range, falling off to zero very rapidly as the variate increases. There are points of inflection at unit distance on either side of the mean.

For the mean deviation we have

$$\frac{1}{\sqrt{(2\pi)}}\int_{-\infty}^{\infty} |x|\,e^{-\frac{1}{2}x^2}\,dx = \sqrt{\frac{2}{\pi}}\int_0^{\infty} x\,e^{-\frac{1}{2}x^2}\,dx = \sqrt{\frac{2}{\pi}} = 0{\cdot}79788. \tag{5.75}$$

The variance is of course unity, because the distribution is expressed in standard measure. The quartiles are distant 0·674,489,75 from the mean, as may be found from the Tables.

5.27 As an illustration of the occurrence in practice of a distribution which is very close to the normal, the height data of Table 1.7 may be taken. Table 5.6 shows the actual frequencies and those given by the normal curve with the same mean and standard deviation (67·46 and 2·56 inches respectively).

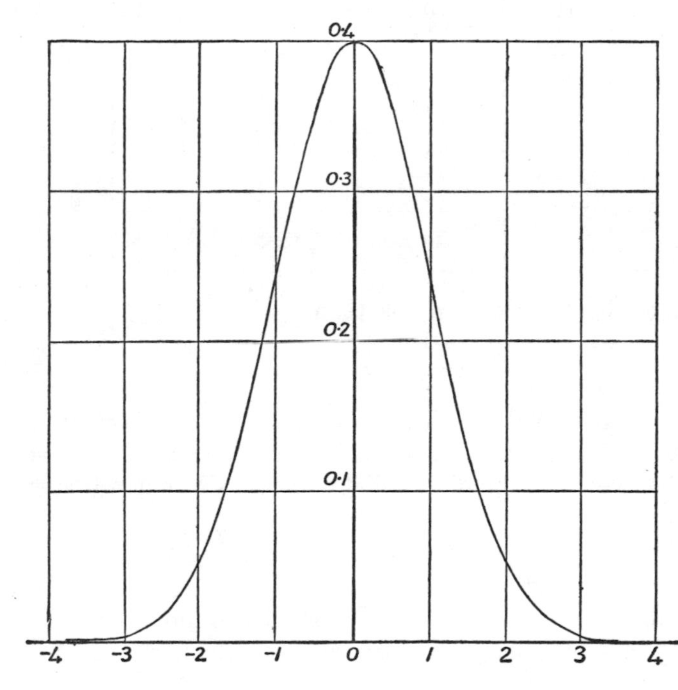

Fig. 5.1—The normal frequency function
$$y = \frac{1}{\sqrt{(2\pi)}} e^{-\frac{1}{2}x^2}$$

The correspondence is evidently fairly good. It must, however, be noted that whereas the theoretical distribution has infinite range, the practical distribution has not, since it is impossible to have a negative height. In this particular case the relative frequency of the normal distribution outside the range 57–77 inches is so small that the point is unimportant ; but when distributions of finite range are represented by those of infinite range it is as well to remember that the fit near the tails may not be very close.

5.28 Since the normal distribution may be considered as the limit of the binomial, it is natural to inquire into the limiting forms, if any, of the hypergeometric distribution. From (5.48) we see that the difference between two successive terms in the distribution is

$$\frac{1}{N^{[n]}} \frac{n!}{r!(n-r-1)!} (Np)^{[r]} (Nq)^{[n-r-1]} \left\{ \frac{Np-r}{r+1} - \frac{Nq-n+r+1}{n-r} \right\}$$
$$= \frac{1}{N^{[n]}} \frac{n!}{r!(n-r)!} (Np)^{[r]} (Nq)^{[n-r]} \left\{ \frac{Nnp-Nq+n-1-r(N+2)}{(r+1)(Nq-n+r+1)} \right\}$$

Table 5.6—Frequency-distribution of 8585 men according to height (Table 1.7) compared with theoretical frequencies of a normal distribution with the same mean and variance

Height (inches)	Observed frequency	Theoretical frequency	Height (inches)	Observed frequency	Theoretical frequency
57–	2	1	68–	1230	1234
58–	4	3	69–	1063	989
59–	14	11	70–	646	682
60–	41	33	71–	392	405
61–	83	88	72–	202	207
62–	169	200	73–	79	91
63–	394	395	74–	32	34
64–	669	669	75–	16	11
65–	990	976	76–	5	3
66–	1223	1227	77–	2	1
67–	1329	1326			
			TOTALS	8585	8586

The ratio of this difference to the $(r+1)$th term is then

$$\frac{\Delta y_r}{y_r} = \frac{A+Br}{C+Dr+Er^2}, \tag{5.76}$$

where the quantities $A \ldots E$ are constants. In the limit when the distribution is expressed in standard measure, Δy_r is the increment when r increases by a small quantity, and we are thus led to consider the differential equation defining a frequency function

$$\frac{df}{f} = \frac{A+Bx}{C+Dx+Ex^2}dx.$$

This is the equation of a family of functions—the Pearson distributions—which will be considered from a slightly different standpoint in the next chapter.

Some extensions

5.29 The distributions we have studied in this chapter may be generalized in various ways. The extension of the normal distribution to the multivariate case is so important that we shall later devote a chapter to it. We conclude our present account with a brief reference to two extensions of the binomial distribution.

The multinomial distribution

5.30 Suppose that the members of a population, instead of being classified into P's and not-P's as in **5.2**, can be classified into $k+1$ classes, $P_1, P_2, \ldots, P_k$ and neither -P_1 nor -$P_2 \ldots$ nor -P_k, which we may denote by P_0. Thus, for example, with $k = 3$, P_1 might denote the possession of blue eyes, P_2 that of grey eyes, P_3 that of brown eyes and P_0 that of eyes of some other colour, or perhaps no eyes at all. If we denote the corresponding proportions by $p_1, \ldots, p_k, p_0$, it is easy to see by an extension of the argument of **5.2** that the frequencies in sets of n individuals are given by the expansion of the multinomial

$$(p_0+p_1+p_2+\ldots+p_k)^n, \tag{5.77}$$

that is to say, the frequency of a set having r_0 P_0's, r_1 P_1's, etc., is the term

$$\frac{n!}{r_0!r_1!\dots r_k!} p_0^{r_0} p_1^{r_1} \dots p_k^{r_k}. \tag{5.78}$$

The characteristic function, attaching variables t_i to p_i $(i = 1, \dots, k)$, is given by

$$(p_0+p_1 e^{it_1}+p_2 e^{it_2}+\dots+p_k e^{it_k})^n \tag{5.79}$$

and we may obtain the moments and product-moments in the usual way. It will be sufficient if we write down certain of the cumulants up to the fourth order for $k = 3$. The others follow by symmetry.

$$\left.\begin{aligned}
&\kappa_{1000} = np_0; \quad \kappa_{2000} = np_0 q_0; \quad \kappa_{1100} = -np_0 p_1; \\
&\kappa_{3000} = np_0 q_0(q_0-p_0); \quad \kappa_{2100} = -np_0 p_1(q_0-p_0); \quad \kappa_{1110} = 2np_0 p_1 p_2; \\
&\kappa_{4000} = np_0 q_0(1-6p_0 q_0); \quad \kappa_{3100} = -np_0 p_1(1-6p_0 q_0); \\
&\kappa_{2200} = -np_0 p_1 \{(q_0-p_0)(q_1-p_1)+2p_0 p_1\}; \\
&\kappa_{2110} = 2np_0 p_1 p_2(q_0-2p_0); \quad \kappa_{1111} = -6np_0 p_1 p_2 p_3.
\end{aligned}\right\} \tag{5.80}$$

Here, as usual, we write $q_i = 1-p_i$ for brevity. For a method of deriving (5.80), see Exercise 5.19.

The bivariate binomial distribution

5.31 We may also have a population classified by two qualities, e.g. blue-eyed or not-blue-eyed and male or female. (More generally, we could have several qualities and several categories in each, but this would give us a cumbrous complexity without raising any essentially new points.)

Suppose the presence or absence of one quality is denoted by $P, Q\ (= 1-P)$ and those of the other quality by $P', Q'\ (= 1-P')$. The proportions of the four possible combinations may be represented thus:

	P'	Q'	TOTAL
P	p_{11}	p_{10}	p
Q	p_{01}	p_{00}	q
TOTAL	p'	q'	1

(5.81)

From some points of view we may regard this distribution as a multinomial arrayed by

$$(p_{00}+p_{01}+p_{10}+p_{11})^n. \tag{5.82}$$

We are, however, usually more interested in the distribution of p and p' than those of p_{00}, etc. The joint c.f. of the numbers of P's and P''s is given by

$$\phi(t) = (p_{00}+p_{01}e^{\theta_1}+p_{10}e^{\theta_2}+p_{11}e^{\theta_1+\theta_2})^n, \tag{5.83}$$

where, as usual, $\theta_1 = it_1$, $\theta_2 = it_2$. If we now define

$$p_{(11)} = p_{11}-pp' \tag{5.84}$$

we find, on expansion of $\log\phi(t)$, the bivariate cumulants

$$\left.\begin{aligned} \kappa_{11} = np_{(11)}; \quad \kappa_{21} = np_{(11)}(q-p); \quad \kappa_{31} = np_{(11)}(1-6pq); \\ \kappa_{22} = np_{(11)}\{(q-p)(q'-p')-2p_{(11)}\}. \end{aligned}\right\} \tag{5.85}$$

Other bivariate cumulants such as κ_{12} are obtainable by symmetry; and, of course, such cumulants as κ_{20} are the cumulants of the univariate (marginal) binomials $(q+p)^n$ and $(q'+p')^n$.

5.32 A little calculation shows that if $p_{(11)}$ of equation (5.84) is zero the c.f. becomes

$$(q+p\,e^{\theta_1})(q'+p'\,e^{\theta_2}). \tag{5.86}$$

The qualities are then independent. If we transfer the origin to the means of the two variates (5.83) becomes

$$\begin{aligned} \log\phi(t) &= n\log(p_{00}+p_{01}e^{\theta_1}+p_{10}e^{\theta_2}+p_{11}e^{\theta_1+\theta_2})-np\,\theta_1-np'\,\theta_2 \\ &= \tfrac{1}{2}n\,(pq\,\theta_1^2+2p_{(11)}\theta_1\theta_2+p'\,q'\,\theta_2^2)+o\,(n). \end{aligned} \tag{5.87}$$

If we then transform to standard measure, the variances of the two variates being npq and $np'q'$, and then let n tend to infinity, we find

$$\log\phi(t) = \tfrac{1}{2}(\theta_1^2+2\rho\,\theta_1\theta_2+\theta_2^2) \tag{5.88}$$

where

$$\rho = \frac{p_{(11)}}{(pqp'\,q')^{\frac{1}{2}}}. \tag{5.89}$$

This, as was seen, in effect, in Example 3.17, is the c.f. of the distribution

$$dF = \frac{1}{2\pi(1-\rho^2)^{\frac{1}{2}}}\exp\left\{\frac{-1}{2(1-\rho^2)}(x_1^2-2\rho\,x_1x_2+x_2^2)\right\}dx_1\,dx_2, \qquad -\infty \leqslant x_1, x_2 \leqslant \infty. \tag{5.90}$$

Thus the bivariate binomial distribution tends to the form (5.90) which is the bivariate analogue of the normal distribution.

The variates x_1 and x_2 become independent if and only if $\rho = 0$, which implies that $p_{(11)} = 0$, or $p_{11} = pp'$. This in turn means that the proportion of P's among the P' is the same as the proportion among the Q', for then $p_{10} = p-p_{11} = qp'$. The table (5.81) then assumes the form

	P'	Q'	TOTAL
P	pp'	pq'	p
Q	qp'	qq'	q
TOTAL	p'	q'	1

(5.91)

We shall have more to say about this type of situation in discussing the theory of categorized data in Chapter 33 (Vol. 2). It is enough at this stage to remark that "independence" as we have defined it agrees with the ordinary meaning which we should assign to it in speaking of qualities.

Wishart (1949) has given general formulae for the cumulants of multivariate multinomial distributions. Exercise 5.11 discusses the bivariate Poisson distribution.

Distribution on the circle and on the sphere

5.33 We have so far discussed only distributions of variates whose possible values form some interval of the real line. Statistical frequency distributions are almost always of this type, but there is a class of situations for which representation on the line is not natural, namely distributions of directions measured from a fixed line as origin. For example, the direction of movement of an animal and the direction of magnetization of a rock are common observations in zoological and geological research. Such observations are naturally recorded as angles over the range $(0, 2\pi)$ or $(0, 360°)$, so that they may be regarded as points on the circumference of the unit circle or of the unit sphere.

One is immediately tempted to ask whether any special treatment of such data on the unit circle is necessary. After all, one can record the observations as angles, and then draw an ordinary histogram on the line from 0 to 360°, and analyse it by familiar linear methods. It is easy to see that this leads to artificiality that may cause trouble in the analysis, if the data are really spread out over the whole circumference of the circle, and not merely a short arc of the circle. First, where should we " cut " the circle? If we cut it at the fixed line origin, and analyse the results linearly, any two observations at $\alpha°$ and $(360°-\alpha°)$ would have an arithmetic mean of 180°, whereas if $\alpha<180$ they are both closer to 0° (i.e., 360°) than to 180°. The same essential phenomenon occurs if we choose any other cutting-point for the circle. In fact, directional observations have a natural invariance (modulo 2π) on the circle, and no linear representation can in general do them justice.

5.34 For observations on the circle, even when represented circularly, the familiar measures of location and dispersion will not do. For example, a measure of location on the circle should possess the same invariance property (mod 2π) as the observations themselves : and, e.g., the range of the observations must be re-defined as the shortest arc covering the observations. Similarly, the d.f. of the variate x on the circle must satisfy

$$F(x+2\pi)-F(x) = 1 \tag{5.92}$$

identically in x, since a complete tour of the circle covers all the observations, wherever we start from ; it follows that the c.f. of x,

$$\phi(t) = E(e^{itx}) = E(e^{it(x+2\pi)}),$$

whence if $|\phi(t)|>0$,

$$1 = e^{it2\pi} = \cos 2\pi t+i\sin 2\pi t. \tag{5.93}$$

(5.93) holds, and the c.f. need only be defined, for integer values of t.

5.35 If the direction is uniformly distributed on the circle,

$$f(x) = \frac{1}{2\pi}, \quad 0\leqslant x\leqslant 2\pi, \tag{5.94}$$

the simplest form possible. A more general distribution is the *Von Mises distribution*

$$f(x) = \{2\pi I_0(\kappa)\}^{-1}\exp\{\kappa\cos(x-\mu)\}, \tag{5.95}$$
$$0\leqslant x, \mu\leqslant 2\pi\,;\quad \kappa>0,$$

where

$$I_0(\kappa) = \sum_{s=0}^{\infty} \frac{\kappa^{2s}}{2^{2s}(s!)^2} \tag{5.96}$$

is the modified Bessel function of the first kind and of order zero. In (5.95), μ is the fixed direction from which x is measured and is thus a location parameter, and κ is a dispersion parameter. (5.95) is a unimodal distribution, symmetrical about μ. When $\kappa = 0$, (5.95) reduces to (5.94), and when $\kappa \to \infty$, (5.95) concentrates wholly in the point $x = \mu$. In fact, if we write $z = \kappa^{1/2}(x-\mu)$ we see from (5.95) that the distribution of z is

$$g(z) \propto \kappa^{-1/2} \exp\{\kappa \cos(\kappa^{-1/2} z)\}$$

and as $\kappa \to \infty$, $\cos(\kappa^{-1/2} z) \sim 1 - z^2/(2\kappa)$, so

$$g(z) \doteqdot \text{const.} \exp\{-z^2/2\}, \tag{5.97}$$

so that x is approximately normal with mean μ and variance κ^{-1}.

Mardia (1972) reproduces 5 d.p. tables, due to E. Batschelet, of the d.f. of (5.95), with $\mu = 180°$, $\kappa = 0\,(0{\cdot}2)\,10$ and $x = 0°\,(5°)\,180°$; and also gives the 100δ per cent points, $\delta = 0{\cdot}0005, 0{\cdot}005, 0{\cdot}025, 0{\cdot}05$, for $\kappa = 0\,(0{\cdot}1)\,5\,(0{\cdot}2)\,10{\cdot}0\,(0{\cdot}5)\,13, 14, 15, 20\,(10)\,50, 100$.

5.36 In some circumstances, three-dimensional angles are measured, so that instead of distributions on a circle we have distributions on the surface of a unit sphere. Two angles (x, y) are needed to specify a point on the sphere relative to a fixed direction, and they are customarily assumed to be independently distributed.

To extend the uniform distribution on the circle at (5.94) to that on the sphere, let y be distributed as at (5.94); we need only observe that the second angle being introduced, x, has its cosine uniformly distributed on the interval $(-1, 1)$. Transforming from $\cos x$ to x, the Jacobian of the transformation is $\sin x$, and we thus have

$$f(x, y) = \frac{1}{2\pi} \cdot \tfrac{1}{2} \sin x = \frac{1}{4\pi} \sin x, \qquad 0 \leqslant x \leqslant \pi\,;\; 0 \leqslant y \leqslant 2\pi, \tag{5.98}$$

the uniform distribution on the sphere. Putting $\mu = 0$ in (5.95), it similarly generalizes on the sphere to

$$f(x, y) \propto \exp\{\kappa \cos x\} \sin x$$

which, on evaluating the constant, becomes

$$f(x, y) = \frac{\kappa}{4\pi \sinh \kappa} \exp\{\kappa \cos x\} \sin x, \qquad 0 \leqslant x \leqslant \pi\,;\; 0 \leqslant y \leqslant 2\pi, \tag{5.99}$$

which is *Fisher's (spherical) distribution.* It is symmetrical if rotated about its directional axis. When $\kappa = 0$, (5.99) reduces to the uniform (5.98), and as $\kappa \to \infty$, concentrates about its axis.

(5.99) is the product of the uniform marginal distribution of y on $(0, 2\pi)$ and the independent marginal distribution of x,

$$f(x) \propto \frac{\kappa}{\sinh \kappa} \exp(\kappa \cos x) \sin x, \qquad 0 \leqslant x \leqslant \pi. \tag{5.100}$$

We approximate as before for large κ, first putting $z = \kappa^{1/2}x$. We find

$$g(z) \propto \frac{\kappa^{1/2}}{\sinh \kappa} \exp\{\kappa \cos(\kappa^{-1/2} z)\} \sin(\kappa^{-1/2} z)$$

which since $\sinh \kappa \sim \frac{1}{2}e^{\kappa}$, $\sin(\kappa^{-1/2}z) \sim \kappa^{-1/2}z$, $\cos(\kappa^{-1/2}z) \sim 1 - z^2/(2\kappa)$, becomes

$$g(z) \propto z \exp(-\tfrac{1}{2}z^2). \tag{5.101}$$

Thus $u = \frac{1}{2}z^2$ has distribution $h(u) = e^{-u}$, the distribution of Example 4.4 with $p = 1$. We express this by saying that $z^2 = \kappa x^2$ is asymptotically distributed " as χ^2 with 2 degrees of freedom "—we shall explain this terminology in **16.2** below. Essentially, as we shall see, it implies that κx^2 is distributed like the sum of squares of *two* independent normal variates, whereas for the Von Mises distribution (5.95), with $\mu = 0$ as here, κx^2 was asymptotically the square of a single normal variable.

Mardia (1972) gives the 100δ per cent. points, with parameters as for the Von Mises percentiles in **5.35**.

5.37 Mardia's (1972) monograph is a comprehensive discussion of the statistical analysis of directions.

EXERCISES

5.1 Show that the c.f. of the binomial distribution satisfies the relation

$$\frac{\partial^2 \log \phi(t)}{\partial (it)^2} = pq \frac{\partial^2 \log \phi(t)}{\partial p \, \partial (it)}$$

and hence that

$$\kappa_{r+1} = pq \frac{d\kappa_r}{dp}, \qquad r \geqslant 1.$$

(Cf. Frisch (1926), and Haldane (1939) who gives formulae up to κ_{12}.)

5.2 For the *incomplete* moments about the mean of the binomial, starting from some variate-value ρ,

$$\mu_r = \sum_{j=\rho}^{n} (j-np)^r \binom{n}{j} p^j q^{n-j},$$

show by differentiating with respect to p that equation (5.11) holds. (Frisch, 1925)

5.3 Writing $T_j = \binom{n}{j} p^j q^{n-j}$, show that the incomplete moments of the binomial are given by

$$\mu_0 = \sum_{j=\rho}^{n} T_j, \qquad \mu_1 = \rho q T_\rho, \qquad \mu_2 = \rho q T_\rho \{\rho - (n+1)p\} + npq\mu_0,$$

$$\mu_3 = \rho q T_\rho [\{\rho - (n+1)p\}^2 + pq(2n-1)] + npq(q-p)\mu_0,$$

and generally

$$\mu_r = \rho q T_\rho (\rho - np)^{r-1} + npq \sum_{j=0}^{r-2} \binom{r-1}{j} \mu_j - p \sum_{j=0}^{r-2} \binom{r-1}{j} \mu_{j+1}.$$

(Frisch (1926). This is the generalization of equation (5.10) to incomplete moments.)

5.4 Writing $f_{r,n}$ for the binomial frequency $\binom{n}{r} p^r q^{n-r}$, show that

$$(np - r) f_{r,n} = npq(f_{r,n-1} - f_{r-1,n-1}), \quad r \geqslant 1$$

$$np f_{0,n} = npq f_{0,n-1},$$

and hence that the mean deviation of the binomial is (cf. **2.18**)

$$\delta_1 = 2 \sum_{r=0}^{[np]} (np - r) f_{r,n} = 2npq \, f_{[np],\, n-1}.$$

Show that when np is an integer, $f_{np,n-1} = f_{np,n}$ (the maximum frequency of the $f_{r,n}$ (cf. **5.6**)), so that the relation (A) of Exercise 2.25 holds exactly in this case.

(cf. Frame (1945); also Johnson (1957). Ramasubban (1958) obtains the mean deviation and mean difference for most of the discrete distributions in this chapter.)

5.5 From equation (5.56) derive the recurrence formula for the moments of the binomial distribution $\{(1+E)^r - E^r\}(npq\mu_0 - p\mu_1) = \mu_{r+1}$; and that for the Poisson distribution $\{(1+E)^r - E^r\}\lambda\mu_0 = \mu_{r+1}$. (K. Pearson, 1924*a*)

5.6 Show that for the Poisson distribution $f(x) = e^{-\lambda}\lambda^x/x!$, the mean deviation is $\delta_1 = 2\lambda f([\lambda]) = 2\mu_2 f([\mu_1'])$. (Cf. (A) of Exercise 2.25.)

5.7 *Neyman's Type A contagious distribution.* The frequency function of a variate r defined at $r = 0, 1, \ldots$ is given by

$$f_r = \frac{\lambda_2^r}{r!} e^{-\lambda_1} \sum_{j=0}^{\infty} \frac{j^r}{j!} (\lambda_1 e^{-\lambda_2})^j, \qquad \lambda_1, \lambda_2 > 0.$$

Show that the factorial cumulant g.f. is $\lambda_1(e^{\lambda_2 t}-1)$ and that $\mu_1' = \lambda_1\lambda_2$; $\mu_2 = \lambda_1\lambda_2(1+\lambda_2)$. Hence write down the c.f. and invert it, using Exercise 4.22.

(Neyman, 1939 ; Barton (1957) shows that the distribution may be highly multimodal.)

5.8 *The Pólya–Aeppli distribution*—cf. Anscombe (1950). The frequency function of a variate defined at $r = 0, 1, \ldots$ is given by

$$f_0 = e^{-\lambda_1}$$

$$f_r = e^{-\lambda_1}\tau^r \sum_{j=1}^{r} \binom{r-1}{j-1} \frac{1}{j!}\left\{\frac{\lambda_1(1-\tau)}{\tau}\right\}^j, \qquad r \geqslant 1;\ 0<\tau<1,\ \lambda_1>0.$$

Show that this may be written

$$f_r = \exp(-\lambda_1/\tau)\tau^r \sum_{j=0}^{\infty} \binom{j+r-1}{r}\left\{\frac{\lambda_1(1-\tau)}{\tau}\right\}^j \Big/ j!, \quad r\geqslant 0,$$

and hence or otherwise that the factorial c.g.f. is $\lambda_1 t/(1-\tau-\tau t)$, with mean $\lambda_1(1-\tau)^{-1}$ and variance $\lambda_1(1+\tau)(1-\tau)^{-2}$. Show that f_0 is a peak frequency if and only if $\lambda_1(1-\tau)<1$, and that, irrespective of this, there is a mode elsewhere if $\lambda_1>2$.

(This and Exercise 5.7 are instances of *compound* distributions of the total number of individuals contained in a number n of clusters which itself has a distribution. In both Exercises, n has a Poisson distribution with parameter λ_1. In Exercise 5.7, the number of individuals in a cluster is also a Poisson variate with parameter λ_2, but in Exercise 5.8 it has the geometric form $(1-\tau)\tau^{x-1}$, $x\geqslant 1$. Cf. Exercise 5.22 for a general result on compounds. Thus, in our terminology, compounds are the special case of mixtures (discussed in **5.13**) in which n is the parameter used for mixing. Other terminologies are often used in the literature—in particular " compound " is often used for our " mixture " also.)

5.9 Show that a mixture of Poisson distributions as in **5.13**, using any unimodal continuous mixing distribution for λ, is a unimodal discrete distribution.

(Holgate, 1970 ; this result does not hold for unimodal discrete mixing, as can be seen by observing that Neyman's distribution in Exercise 5.7 can also be derived as a mixture of Poisson distributions with parameters $\lambda = \lambda_2 x$, where x itself is Poisson with parameter λ_1.)

5.10 Show that the f.f. ratio $f(x)/f(x-1)$ is a linear function of $1/x$ if $f(x)$ is (a) the binomial, (b) the Poisson, (c) the negative binomial with origin where the first non-zero probability occurs, and (d) the logarithmic distribution. Using the slope and intercept of the linear function in each case, show how these four distributions may be distinguished graphically.

(Ord, 1967a)

5.11 *The bivariate Poisson distribution.* Show that when p_{01}, p_{10} and p_{11} in equation (5.82) are small, but $np_{11} = \lambda_3$, $np_{10} = \lambda_2-\lambda_3$ and $np_{01} = \lambda_1-\lambda_3$ are finite, the distribution tends to the form whose general term is

$$\frac{\lambda_3^i(\lambda_1-\lambda_3)^j(\lambda_2-\lambda_3)^k}{i!j!k!}e^{-\lambda_1-\lambda_2+\lambda_3}.$$

5.12 Use the Laplace continued fraction (5.73) to show that

$$\frac{x}{x^2+1} < R(x) < \frac{x^2+2x}{x^3+3x}.$$

5.13 Show that Mills' ratio

$$R(x) = x\int_{-\infty}^{\infty} \frac{1}{u^2+x^2}\frac{1}{\sqrt{(2\pi)}}e^{-\frac{1}{2}u^2}\,dv.$$

Replacing the integral by the sum $h\sum_{j=-\infty}^{\infty}\frac{1}{\sqrt{(2\pi)}}\frac{e^{-\frac{1}{2}j^2h^2}}{x^2+j^2h^2}$, show that for $h=\frac{1}{4}$, $x=1$, the sum to 13 terms gives $R=0{\cdot}158{,}655{,}24$. (Das (1956). The true value is 0·158,655,25.)

5.14 For the Laplace continued fraction (5.73) for Mills' ratio, show that if the sth convergent c_s is denoted by a_s/b_s,

$$a_0=0,\ a_1=1,\ b_0=1,\ b_1=x \text{ and}$$
$$c_s-c_{s-1}=\frac{(-1)^{s-1}(s-1)!}{b_{s-1}b_s}$$
$$c_s-c_{s-2}=\frac{(-1)^{s-1}x(s-2)!}{b_{s-2}b_s}.$$

Hence show that $c_0<c_2<c_4<\ldots<R(x)<\ldots<c_5<c_3<c_1$. (Shenton 1954)

5.15 In the previous exercise show that

$$R(x)=\int_0^\infty \exp(-\tfrac{1}{2}t^2-tx)\,dt$$

and hence that

$$R(x)\,b_s-a_s=\frac{(-1)^s\int_x^\infty (t-x)^s e^{-\frac{1}{2}t^2}\,dt/\sqrt{(2\pi)}}{\frac{1}{\sqrt{(2\pi)}}e^{-\frac{1}{2}x^2}}.$$

In the Schwarzian inequality

$$\begin{vmatrix}\int f_1^2\,dx & \int f_1f_2\,dx\\ \int f_1f_2\,dx & \int f_2^2\,dx\end{vmatrix}>0$$

put
$$f_1=(t-x)^{\frac{1}{2}s}e^{-\frac{1}{4}t^2},\qquad f_2=(t-x)^{\frac{1}{2}s+1}e^{-\frac{1}{4}t^2},$$

and hence show that
$$R>\frac{\alpha_{2s}+(2s)!\sqrt{(x^2+8s+4)}}{2\beta_{2s}},$$
$$R<\frac{2\gamma_{2s+1}}{\alpha_{2s+1}+(2s+1)!\sqrt{(x^2+8s+8)}},$$

where
$$\alpha_s=b_sa_{s+2}+b_{s+2}a_s-2b_{s+1}a_{s+1},$$
$$\beta_s=b_sb_{s+2}-b_{s+1}^2,\quad \gamma_s=a_sa_{s+2}-a_{s+1}^2.$$

Hence derive Birnbaum's inequality

$$\tfrac{1}{2}\{-x+\sqrt{(x^2+4)}\}<R(x)<4/\{3x+\sqrt{(x^2+8)}\},\quad x>0,$$

the positive sign of the radical being taken. (Shenton, 1954)

5.16 If $\nu(x)=1/R(x)$ and $\lambda(x)=d\nu/dx=\nu(\nu-x)$ show that $0<\lambda<1$. If $\pi(x)$ is defined by

$$\pi(x)=e^{-x^2}\Big/\int_{-\infty}^{x}e^{-\frac{1}{2}u^2}\,du\int_x^\infty e^{-\frac{1}{2}u^2}\,du$$

show that
$$d\pi/dx=2x\pi(x)\{\lambda(x')-1\},$$

where $-|x|\leqslant x'\leqslant|x|$. Hence show that $\pi(x)$ is a decreasing function of x^2.

(Sampford, 1953)

5.17 *Mixtures of binomials.* If in a binomial distribution (5.2), the parameter p is itself distributed in the form

$$g(p) = \frac{p^{a-1}(1-p)^{b-1}}{B(a,b)}, \quad 0 \leqslant p \leqslant 1; \; a, b > 0, \tag{A}$$

show that the unconditional distribution of the number of "successes" is

$$h(x) = \frac{1}{n+1}\,\frac{B(a+x,\, b+n-x)}{B(x+1, n-x+1)\,B(a,b)}, \quad x = 0, 1, 2, \ldots, n,$$

with mean $na/(a+b)$ and variance

$$n\,a\,b\,(a+b+n)/\{(a+b)^2(a+b+1)\}.$$

(This f.f. is sometimes called the *Beta-binomial*; if (A) is applied to the negative binomial, we similarly obtain the *Beta-Pascal* distribution.)

5.18 Show that the function

$$\overline{R}(x) = y(x) = e^{\frac{1}{2}x^2}\int_0^x e^{-\frac{1}{2}u^2}\,du$$

satisfies the differential equation $\quad \dfrac{dy}{dx} = xy+1$

with initial condition $y(0) = 0$. Hence show that for $r = 0, 1, 2, \ldots$

$$y^{(2r)}(0) = 0, \quad y^{(2r+1)}(0) = 2^r r! \quad \text{and}$$

$$y(x) = \sum_{r=0}^{\infty} \frac{x^{2r+1}}{(2r+1)!}\,2^r r! = \sum_{r=0}^{\infty} \frac{x^{2r+1}}{(2r+1)(2r-1)\ldots 3.1}.$$

(Pólya, 1945)

5.19 For the multinomial (5.77) show that if $a_i = p_i/p_0$ then

$$\kappa(r_1, r_2, \ldots, r_i+1, \ldots, r_k) = a_i \frac{\partial}{\partial a_i}\kappa(r_1, r_2, \ldots, r_i, \ldots, r_k)$$

where the cumulant on the right is of order r_1 in p_1 etc. (Guldberg, 1935). Hence derive equations (5.80).

5.20 x is the number of "successes" in n binomial trials with parameter p, where n itself is a random variable. Show that if n is the number of successes in another series of N binomial trials with parameter P, x will also be distributed as a binomial based on N trials with parameter Pp. Hence or otherwise show that if, instead, n is a Poisson variate with parameter λ, x is also a Poisson variate with parameter λp, and that in this case x is distributed independently of the number of "failures" $n-x$.

5.21 In a certain experiment clusters are observed, each cluster consisting of individuals. Show that if the number of clusters follows a Poisson distribution and the number of individuals in a cluster follows a logarithmic distribution, the total number of individuals observed follows a negative binomial distribution.

(Quenouille, 1949; thus a negative binomial can be produced as a mixture (cf. **5.13**) or as a compound from the Poisson. By Exercises 5.7, 5.9, the same holds for Neyman's Type A.)

5.22 *Compound distributions.* z is the sum of n independent variates x_i, each with c.f. $\phi_1(t)$, where n is a random variable, taking the values 0, 1, 2, . . . , with c.f. $\phi_2(t)$. Show that the unconditional c.f. of z is $\phi_2\left(\dfrac{\log\phi_1(t)}{i}\right)$, and that $E(z) = E(n)E(x)$, $\operatorname{var} z = E(n)\operatorname{var} x + \operatorname{var} n\,\{E(x)\}^2$. Show that the d.f. of z has a saltus at zero even if the x_i are continuous.

If n is a Poisson variable with parameter λ, show that the c.f. of z is $\exp\{\lambda[\phi_1(t)-1]\}$ Derive the results of Exercises 5.7, 5.8, 5.20 and 5.21 from this. Show that the cumulants of z are λ times the moments of x, and hence that z has kurtosis $\beta_2-3 = c\beta_1$, where $c \geqslant 1$ (cf. the general inequality in Exercise 3.19).

5.23 Use the result of Exercise 4.16 to establish (cf. **5.13**) that a Poisson distribution with parameter λ itself distributed in a Gamma distribution (5.31) has the negative binomial distribution (5.33).

5.24 *Mixtures of normal distributions.* x is conditionally normally distributed with mean μ and variance λ. μ and λ are random variables with joint c.f. $\phi(t,u) = E\{\exp(it\mu+iu\lambda)\}$. Show that the unconditional c.f. of x is $\phi(t, \frac{1}{2}it^2)$.

5.25 In Exercise 5.24, show that if μ and λ are distributed independently of each other with respective c.f.'s

$$\phi_1(t) = (1+t^2\sigma^2)^{-1}, \qquad \phi_2(t) = (1-2iu\sigma^2)^{-m}, \qquad m \text{ a positive integer,}$$

(cf. Exercises 4.3, 4.18 for these c.f.'s) then x is distributed as the sum of $(m+1)$ independent variates, each distributed exactly like μ, and that if instead μ is constant at zero and $m = 1$, x has the c.f. $\phi_1(t)$.

If μ is constant at zero and $\phi_2(t) = \exp\{-(-2it)^{1/2}\}$, show that x has the Cauchy distribution of Example 4.2, with mean infinite by Example 3.3, even though each component of the mixture has zero mean.

5.26 Show that for the Poisson f.f. (5.18), $\mu'_{[r]} = \lambda^r$; and that for the hypergeometric f.f. (5.48), $\mu'_{[r]} = n^{[r]}(Np)^{[r]}/N^{[r]}$.

5.27 Using $R(x) \sim x^{-1}$ from (5.68), show that for the normal distribution the solution for x of the equation $1-F(x) = z^{-1}$ has the expansion, as $x(z) \to \infty$,

$$x \sim (2\log z)^{1/2}\{1-\tfrac{1}{2}(\log 4\pi+\log\log z)/(2\log z)\}.$$

Verify numerically that this gives $x = 2{\cdot}36$ for $F(x) = 0{\cdot}99$, against the true value $2{\cdot}33$ obtained from Appendix Table 2.

5.28 By expanding $f(z) = \exp(-\frac{1}{2}z^2)$ in infinite Taylor series about $z = \frac{1}{2}x$, and integrating with respect to z from 0 to x, show that

$$\int_0^x f(z)\,dz = x\exp(-x^2/8)\sum_{r=0}^{0}\frac{\theta_{2r}(x/2)}{2r+1},$$

where $\theta_r(x) = H_r(x)x^r/r!$ and $H_r(x)$ is the polynomial defined at (6.21) below. Show from (6.26) that

$$\theta_{r+1}(x) = x^2\{\theta_r(x)-\theta_{r-1}(x)\}/(r+1),$$

facilitating computation of the series, which converges most rapidly for small x.

(Kerridge and Cook, 1976)

CHAPTER 6

SYSTEMS OF DISTRIBUTIONS

6.1 In this chapter we continue the account, begun in the last, of the standard distributions of statistical theory. From the variety of forms assumed by the frequency distributions of experience, as exemplified in Chapter 1, it is evident that an elastic system would be required to describe them all in mathematical terms. Three main approaches will be considered : the first, due to Karl Pearson, seeks to ascertain a *family* of distributions that will satisfactorily represent observed data ; the second, due to Bruns, Gram, Charlier and Edgeworth, seeks to represent a given frequency function as a series in the derivatives of the normal frequency function ; the third, due to Edgeworth and later writers, seeks a transformation of the variate that will throw the distribution at least approximately into a known form.

General accounts of the various approaches are given by Elderton and Johnson (1969) and by Ord (1972), who give extensive bibliographies.

Pearson distributions

6.2 It was noted in **5.28** that in the limiting case the hypergeometric distribution can be expressed in the form

$$\frac{df}{dx} = \frac{(x-a)f}{b_0+b_1x+b_2x^2}. \tag{6.1}$$

This equation may be considered from a slightly different standpoint. The unimodal distributions of Chapter 1 suggest that it might be worth while examining the class of frequency functions which (a) have a single mode, so that df/dx vanishes at some point $x = a$; (b) have smooth contact with the x-axis at the extremities, so that df/dx vanishes when $f = 0$. Evidently these conditions are in general obeyed by any distribution of the family (6.1). In actual fact, as will be seen below, there are also solutions of (6.1) in particular cases which are J- or U-shaped.

The family of frequency functions defined by (6.1) are known as Pearson distributions. Before obtaining explicit solutions of the equation, we consider certain general results which are true of all members of the system. We have immediately

$$(b_0+b_1x+b_2x^2)\,df = (x-a)f\,dx$$

or

$$x^n(b_0+b_1x+b_2x^2)\frac{df}{dx}dx = x^n(x-a)f\,dx.$$

Integrating the left-hand side by parts over the range of the distribution, we find, assuming that the integrals exist,

$$\Big[x^n(b_0+b_1x+b_2x^2)f\Big]_{-\infty}^{\infty} - \int_{-\infty}^{\infty}\{nb_0x^{n-1}+(n+1)b_1x^n+(n+2)b_2x^{n+1}\}f\,dx$$
$$= \int_{-\infty}^{\infty}x^{n+1}f\,dx - a\int_{-\infty}^{\infty}x^nf\,dx. \tag{6.2}$$

Let us assume that the expression in square brackets vanishes at the extremities of the distribution, i.e. that $\lim_{x\to\pm\infty} x^{n+2}f\to 0$ if the range is infinite. We then have, substituting moments for integrals in (6.2) :—

$$n-b_0\mu'_{n-1}-(n+1)b_1\mu'_n-(n+2)b_2\mu'_{n+1} = \mu'_{n+1}-a\mu_n$$

or
$$nb_0\mu'_{n-1}+\{(n+1)b_1-a\}\mu'_n+\{(n+2)b_2+1\}\mu'_{n+1} = 0. \tag{6.3}$$

The m.g.f., c.g.f. and cumulants obey similar recurrence relations—cf. Exercise 6.7.

(6.3) permits of the determination of any moment from those of lower orders. In fact, all moments can be expressed in terms of a, b_0, b_1, b_2, μ_0 $(\equiv 1)$ and μ'_1. Conversely we can express the four constants in terms of the moments μ'_1 to μ'_4, or, if we take the mean μ'_1 as origin, the three central moments μ_2 to μ_4. Putting $\mu'_1 = 0$ and $n = 0, 1, 2, 3$ successively in (6.3), we find equations for a, b_0, b_1, b_2 which result in

$$\left.\begin{aligned} b_1 = a &= -\frac{\mu_3(\mu_4+3\mu_2^2)}{A} = -\frac{\sqrt{\mu_2}\sqrt{\beta_1}(\beta_2+3)}{A'} \\ b_0 &= -\frac{\mu_2(4\mu_2\mu_4-3\mu_3^2)}{A} = -\frac{\mu_2(4\beta_2-3\beta_1)}{A'} \\ b_2 &= -\frac{(2\mu_2\mu_4-3\mu_3^2-6\mu_2^3)}{A} = -\frac{(2\beta_2-3\beta_1-6)}{A'} \end{aligned}\right\} \tag{6.4}$$

where
$$\left.\begin{aligned} A &= 10\mu_4\mu_2-18\mu_2^3-12\mu_3^2 \\ A' &= 10\beta_2-18-12\beta_1. \end{aligned}\right\} \tag{6.5}$$

b_0 in (6.4) can be positive or negative but not zero, since by Exercise 3.19 its numerator is positive. Further, $b_1 = a = 0$ if $\beta_1 = 0$, e.g. if the distribution is symmetrical.

In equation (6.1) the mode is evidently at the point $x = a$. From (6.4), we have for the Pearson measure of skewness **(3.31)**

$$\frac{\text{mean}-\text{mode}}{\sigma} = \frac{-a}{\sqrt{\mu_2}} = \frac{\sqrt{\beta_1}(\beta_2+3)}{10\beta_2-12\beta_1-18}, \tag{6.6}$$

the form given at (3.87).

6.3 If, instead of using μ'_1 as origin, we use the mode a thus, by putting $X = x-a$, the constants b_0, b_1 and b_2 will take values different from those in (6.4) : we shall denote these values by B_0, B_1 and B_2. (6.1) now becomes, putting $a = 0$,

$$\frac{\partial \log f}{\partial X} = \frac{X}{B_0+B_1X+B_2X^2}. \tag{6.7}$$

By equating coefficients in $b_0+b_1x+b_2x^2 \equiv B_0+B_1(x-a)+B_2(x-a)^2$ we obtain, using $b_1 = a$ from (6.4),

$$\left.\begin{aligned} B_0 &= b_0+a^2(1+b_2), \\ B_1 &= a(1+2b_2), \\ B_2 &= b_2. \end{aligned}\right\}. \tag{6.8}$$

We find from (6.7) that

$$\frac{d^2f}{dX^2} = \frac{d}{dX}\frac{Xf}{B_0+B_1X+B_2X^2} = \frac{f}{(B_0+B_1X+B_2X^2)^2}(B_0-(B_2-1)X^2).$$

Thus any points of inflection in the frequency curve are given by

$$X^2 = B_0/(B_2-1).$$

Hence there cannot be more than two of them, and if two exist, they are equidistant from the mode. It is not to be inferred that a curve of the family cannot have a single point of inflection, for one solution may be outside the permissible range of x.

6.4 The explicit expression of the frequency function f requires the integration of the right-hand side of (6.7). We first assume $B_2 > 0$, if necessary changing the sign of X to achieve this—the case $B_2 = 0$ is discussed in **6.9** below and in Exercise 6.1.

We may distinguish three main types of distribution according as the quadratic in the denominator on the right of (6.7) has real roots of opposite sign, real roots of the same sign, or complex roots. If the roots are real, $B_1^2 \geqslant 4B_0B_2$ and B_0B_2, the product of the roots, is negative if the roots have opposite sign (i.e. lie on opposite sides of the mode, our origin) and is positive if they have the same sign. Thus

(i) if the real roots have opposite sign, $K = B_1^2/(4B_0B_2) \leqslant 0$;
(ii) if the real roots have the same sign, $K \geqslant 1$;
(iii) if the roots are complex, $B_1^2 < 4B_0B_2$ and $0 \leqslant K < 1$.

K is thus a criterion for assigning a distribution to a type of the Pearson system. Using (6.8), we see that $B_1^2-4B_0B_2 = b_1^2-4b_0b_2$, reflecting the fact that the reality of the roots of the quadratic cannot be affected by a change of origin, here from the mode to the mean. However, a new origin can affect the signs of the roots and K must then be replaced by the criterion appropriate to the new origin. Using the mean, we see that

$$K = \frac{B_1^2}{4B_0B_2} = 1+\frac{\kappa-1}{1+4b_2(1+b_2)\kappa}, \tag{6.9}$$

where $\kappa = b_1^2/(4b_0b_2)$ is the same function of the b's as K is of the B's.

From (6.9), we see that κ and K are 0 or 1 together, while since $b_2 > 0$,

(i) $\kappa < 0$ does not imply $K < 0$ unless

$$-\{4b_2(1+b_2)\}^{-1} < \kappa < 0; \text{ otherwise } K > 1.$$

(ii) $\kappa > 1$ implies $K > 1$.
(iii) $0 < \kappa < 1$ implies $0 < K < 1$.

(iii) reflects the fact that the complexity of the roots is independent of the origin; (ii) shows that if both roots are on the same side of the mean, they must be on the same side of the mode, but (i) implies that the converse is not true.

Thus the value of κ gives useful sufficient conditions for the value of K. From (6.4), we see that

$$\kappa = \frac{b_1^2}{4b_0b_2} = \frac{\beta_1(\beta_2+3)^2}{4(4\beta_2-3\beta_1)(2\beta_2-3\beta_1-6)}. \tag{6.10}$$

TYPE I (*Beta distribution*)

6.5 If the roots are real and of opposite sign, we have, using the mode as origin, $K \leqslant 0$, with $K = 0$ if and only if $B_1 = 0$, i.e. from (6.8) and (6.4) if $\beta_1 = 0$, as it is for any symmetric distribution. Generally, for $K \leqslant 0$, we have

$$B_0 + B_1 X + B_2 X^2 = B_2(X+\alpha_1)(X-\alpha_2), \qquad \alpha_1, \alpha_2 > 0.$$

Then
$$\frac{d}{dX}(\log f) = \frac{X}{B_2(X+\alpha_1)(X-\alpha_2)}$$
$$= \frac{\alpha_1}{B_2(\alpha_1+\alpha_2)} \cdot \frac{1}{(X+\alpha_1)} + \frac{\alpha_2}{B_2(\alpha_1+\alpha_2)} \cdot \frac{1}{(X-\alpha_2)},$$

giving
$$f = k(X+\alpha_1)^{\frac{\alpha_1}{B_2(\alpha_1+\alpha_2)}}(X-\alpha_2)^{\frac{\alpha_2}{B_2(\alpha_1+\alpha_2)}}.$$

This may be written in the form

$$f = k\left(1+\frac{X}{a_1}\right)^{m_1}\left(1-\frac{X}{a_2}\right)^{m_2} \tag{6.11}$$

where
$$\frac{m_1}{a_1} = \frac{m_2}{a_2} > 0.$$

The range for which $f \geqslant 0$ is from $-a_1$ to a_2, and by integrating between these values we find

$$f = \frac{a_1^{m_1} a_2^{m_2}}{(a_1+a_2)^{m_1+m_2+1} B(m_1+1, m_2+1)}\left(1+\frac{X}{a_1}\right)^{m_1}\left(1-\frac{X}{a_2}\right)^{m_2}. \tag{6.12}$$

The origin here is the mode. Taking an origin at the start of the distribution and measuring in units (a_1+a_2) times the original, (6.12) reduces to

$$f = \frac{1}{B(m_1+1, m_2+1)} x^{m_1}(1-x)^{m_2}, \quad 0 \leqslant x \leqslant 1. \tag{6.13}$$

This is Pearson's Type I. Type II is the symmetrical case with $K = 0$—cf. Exercise 6.1.

6.6 (6.13) is usually written in the standard form

$$f = \frac{1}{B(p, q)} x^{p-1}(1-x)^{q-1}, \quad 0 \leqslant x \leqslant 1; \; p, q > 0. \tag{6.14}$$

Since its d.f. is the Incomplete Beta Function, (6.14) is called the *Beta distribution* with parameters p and q. Its percentage points are tabulated in the *Biometrika Tables*, extended by Amos (1963). If $p, q > 1$, f has a unique mode at $(p-1)/(p+q-2)$ and is zero at $x = 0, 1$. If p or $q = 1$, f has a corresponding terminal value q or p; if $p = q = 1$, (6.14) becomes the uniform distribution

$$dF = dx, \qquad 0 \leqslant x \leqslant 1. \tag{6.15}$$

If p or q is between 0 and 1, f is infinite at the corresponding terminal. Thus a variety of U-shaped distributions are obtainable with $0 < p, q \leqslant 1$, and J-shaped distributions with $0 < p \leqslant 1 < q$ or $0 < q \leqslant 1 < p$.

Since the range of (6.14) is finite, all its moments exist; we obtained the first two in Examples 2.2 and 2.8, and the general expression is similarly seen to be

$$\mu_r' = B(p+r, q)/B(p, q) = \Gamma(p+r)\Gamma(p+q)/\{\Gamma(p)\Gamma(p+q+r)\}.$$

TYPE VI

6.7 If the real roots of the quadratic have the same sign and are not equal, $K>1$ and simple changes in the argument of **6.5** show that we may write the f.f. in the form

$$f = kx^{-q_1}(x-c)^{q_2}, \quad 0<c\leqslant x\leqslant \infty;\ q_1>q_2+1>0. \tag{6.16}$$

This was called Type VI by Pearson; the transformation $y = c/x$ reduces it to the Beta form (6.14) with $p = q_1-q_2-1$, $q = q_2+1$. In the standard form

$$f = \frac{1}{B(p,q)}\frac{x^{p-1}}{(1+x)^{p+q}}, \quad 0\leqslant x\leqslant \infty;\ p, q>0, \tag{6.17}$$

it is called a Beta distribution of the second kind ((6.14) is of the first kind.) It is discussed in slightly different form as the F-distribution in **16.15–21** below. (6.17) is reducible to (6.14) by putting $x = y/(1-y)$.

The rth moment of (6.17) exists only if $r<q$, and is then found to be

$$\mu_r' = B(p+r, q-r)/B(p,q) = \Gamma(p+r)\Gamma(q-r)/\{\Gamma(p)\Gamma(q)\}.$$

Exercise 6.6 treats the case when the roots are equal.

TYPE IV

6.8 If the roots of $B_0+B_1X+B_2X^2$ are complex we have $0\leqslant K<1$ and

$$\frac{d}{dX}(\log f) = \frac{X}{B_2\left\{\left(X+\frac{B_1}{2B_2}\right)^2+\frac{B_0}{B_2}-\frac{B_1^2}{4B_2^2}\right\}}$$

$$= \frac{X}{B_2\{(X+\gamma)^2+\delta^2\}},$$

since $B_0B_2>B_1^2$. Integrating, we have

$$\log f = \log k+\frac{1}{2B_2}\log\{(X+\gamma)^2+\delta^2\}-\frac{\gamma}{B_2\delta}\arctan\frac{X+\gamma}{\delta},$$

$$f = k\{(X+\gamma)^2+\delta^2\}^{\frac{1}{2B_2}}\exp\left\{-\frac{\gamma}{B_2\delta}\arctan\frac{X+\gamma}{\delta}\right\}.$$

This is Pearson's Type IV and is usually written in the form

$$f = k\left(1+\frac{x^2}{a^2}\right)^{-m}\exp\{-\nu\arctan(x/a)\}, \qquad m>\tfrac{1}{2}. \tag{6.18}$$

The distribution has unlimited range in both directions and is unimodal. Its first four moments exist if $m>\frac{5}{2}$. It is very difficult to handle comfortably in practice, notwithstanding the existence of certain special tables, owing to the impossibility of expressing the distribution function in terms of ordinary functions. When the distribution function is required it is most easily derived by quadrature from the frequency function.

Type VII (Student's distribution) is the symmetrical case $K = 0$, when B_1 and ν are also zero, and the normal distribution is its extreme case $B_1 = B_2 = 0$—cf. Exercise 6.1.

TYPE III (*Gamma distribution*)

6.9 Pearson distinguished nine other types, some entirely trivial, some no longer of interest. We need mention only one, which has extensive theoretical applications. Some particulars of the others will be found in Exercises 6.1, 6.4 and 6.6.

If, in (6.7), $B_2 = 0$ the distribution becomes

$$f = k\left(1+\frac{x}{a}\right)^p e^{-px/a}, \qquad -a \leqslant x \leqslant \infty. \tag{6.19}$$

Here the origin is at the mode. If we transfer the origin to the start of the distribution we arrive at a form in which we have already encountered it (Examples 3.6, 3.12). A scale-change then gives the standard form

$$f = \frac{1}{\Gamma(\lambda)} x^{\lambda-1} e^{-x}, \qquad \lambda > 0,\ 0 \leqslant x \leqslant \infty. \tag{6.20}$$

In this case the criterion K of (6.9) is infinite—cf. Exercise 6.5. The curve is unimodal except for values of λ less than or equal to unity, when it is J-shaped. It is known as Type III or the Gamma distribution, the latter name being due to the fact that its distribution function is an Incomplete Γ-function. Tables of the d.f. and of percentage points of the distribution of $2x$ in (6.20) are described in **16.4** below.

6.10 It has been found that Pearson distributions sometimes give a good representation of observational data. Apart from this, they (and in particular Types I and III) are sometimes useful in approximating to theoretical distributions from their known moments, as we shall see in later chapters.

A systematic account of the technique of fitting is given by Elderton and Johnson (1969), and an extensive table of percentage points for the whole family of distributions is given by Johnson *et al.* (1963) and reproduced with many enlargements in Vol. II of the *Biometrika Tables*. We will here merely indicate the general principles and give one example of fitting in what is, perhaps, the most troublesome case.

6.11 From (6.4), we see that all the Pearson distributions are determined by the first four moments; some of the degenerate types are determined by fewer moments. The family is sometimes represented by plotting β_1 against β_2, as in the *Biometrika Tables*, or against δ of Exercise 6.22. To fit a Pearson distribution, we

(1) calculate the values of the first four moments, β_1 and β_2 of the observed distribution;
(2) determine the type to which the observed distribution belongs, either using the criterion κ of (6.10) or plotting (β_1, β_2) in the graph of the family in the *Biometrika Tables*;
(3) equate the observed moments to the moments of this type of distribution expressed in terms of its parameters; and
(4) solve the resulting equations for those parameters, whereupon the fitted distribution is determined.

The following example will illustrate the process :—

Example 6.1

In Table 1.15 there are shown, in the column totals, a distribution of 9440 beans according to length. The figures are repeated in Table 6.1 on page 160. It is required to fit a Pearson distribution to these data.

For the moments it is found that, with Sheppard's corrections (in units of 0·5 mm.),

$$\mu_1' \text{ (origin at the mode 14·5)} = -0·190,783,898 \qquad \mu_3 = -5·306,566,352$$
$$\mu_2 = 3·238,424,951 \qquad \mu_4 = 50·999,624,044$$

$$\beta_1 = 0·829,135,838, \qquad \gamma_1 = \sqrt{\beta_1} = -0·910,569$$
$$\beta_2 = 4·862,944,362.$$

To determine the Pearson type, we first calculate (6.10),

$$\kappa = \frac{51·262}{84·040}.$$

Since $0<\kappa<1$, it follows from **6.4** that $0<K<1$, and Type IV is appropriate. Putting $\tan\theta = x/a$ and $2m-2 = r$ in equation (6.18) we find

$$\mu_n' = k\int_{-\frac{1}{2}\pi}^{\frac{1}{2}\pi} a^{n+1}\cos^{r-n}\theta\sin^n\theta\, e^{-\nu\theta}\,d\theta,$$

whence, integrating by parts with $\cos^{r-n}\theta\sin\theta$ as one part,

$$\mu_n' = \frac{a}{r-n+1}\{(n-1)\,a\mu_{n-2}'-\nu\mu_{n-1}'\},$$

a particular case of (6.3). Hence, in terms of moments about the mean,

$$\mu_1' = -\frac{a\nu}{r}, \qquad \mu_2 = \frac{a^2}{r^2(r-1)}(r^2+\nu^2),$$

$$\mu_3 = -\frac{4a^3\nu(r^2+\nu^2)}{r^3(r-1)(r-2)}, \qquad \mu_4 = \frac{3a^4(r^2+\nu^2)\{(r+6)(r^2+\nu^2)-8\nu^2\}}{r^4(r-1)(r-2)(r-3)},$$

whence it is found that

$$r = \frac{6(\beta_2-\beta_1-1)}{2\beta_2-3\beta_1-6},$$

$$\nu = \frac{r(r-2)\sqrt{\beta_1}}{\sqrt{\{16(r-1)-\beta_1(r-2)^2\}}},$$

$$a = \sqrt{\left[\frac{\mu_2}{16}\{16(r-1)-\beta_1(r-2)^2\}\right]}.$$

Substituting for β_2, β_1 and μ_2 we find

$$r = 14·697,72, \qquad m = 8·348,86$$
$$\nu = 18·380,43, \qquad a = 4·159,49.$$

The signs here want a little watching. r and m present no difficulty; but a is to be taken positive and ν positive since μ_3 is negative.

So far the process is straightforward, but to evaluate the constant k we require tables of the distribution function.(*) On evaluation of k we find

$$f = 0{\cdot}395{,}121\left(1+\frac{x^2}{17{\cdot}301{,}34}\right)^{-8{\cdot}348{,}86}\exp\left(-18{\cdot}380{,}43\arctan\frac{x}{4{\cdot}159{,}49}\right).$$

With sufficient arithmetical patience the ordinates of the frequency-function and the areas in given ranges can now be calculated. Table 6.1, column (3), gives the results for comparison with the observed frequencies.

6.12 Ord (1967b) has derived and studied in detail the discrete analogue of the Pearson system. Instead of the limiting differential equation (6.1), he solves the original difference equation (5.76). The resulting types of discrete distributions include the binomial, Poisson and negative binomial (all of which, interestingly, are members of the analogue of Pearson's Type III) as well as a variety of hypergeometric distributions. This discrete system is complicated by the presence of end-effects which vanish in the continuous case.

6.13 The fitting of Pearson distributions by the method of moments as above may be regarded from two rather different standpoints. If our object (for instance in actuarial work) is to obtain a mathematical expression which will satisfactorily represent observation and allow of accurate graduation and interpolation, fitting by moments is generally satisfactory. The method has, however, been criticized when the observed data are samples from a population, and it is desired to find a mathematical representation *of that population.* In such cases the moments calculated from observation are only *estimates* of population-moments and they do not, in general, lead to the most efficient estimates of the population parameters. We shall have to defer a fuller discussion of this point until Chapter 17, Volume 2.

Other systems of distributions have been studied, with a view to representing frequency functions by expansions in series. It is well known in mathematical and physical work that functions can often be usefully expressed as a series of terms such as powers of the variable (Taylor's series) or trigonometrical functions (Fourier's series). Neither of these forms is very suitable for frequency functions, but we proceed to consider another set of functions with more promising possibilities.

Chebyshev–Hermite polynomials

6.14 Writing, as in **5.21**,

$$\alpha(x) = \frac{1}{\sqrt{(2\pi)}}e^{-\frac{1}{2}x^2}$$

and

$$D = \frac{d}{dx},$$

(*) The tables were given in the first two editions of K. Pearson's *Tables for Statisticians*, Part I, but omitted from later editions. See also Woodward (1976).

consider successive derivatives of $\alpha(x)$ with respect to x. We have

$$\begin{aligned} D\alpha(x) &= -x\alpha(x) \\ D^2\alpha(x) &= (x^2-1)\alpha(x) \\ D^3\alpha(x) &= (3x-x^3)\alpha(x), \end{aligned}$$

and so on. The result will obviously be, in general, a polynomial in x multiplied by $\alpha(x)$. We then define the Chebyshev–Hermite polynomial $H_r(x)$ by the identity

$$(-D)^r\alpha(x) = H_r(x)\alpha(x). \tag{6.21}$$

Evidently $H_r(x)$ is of degree r in x and the coefficient of x^r is unity. By convention $H_0 = 1$. We have

$$\alpha(x-t) = \frac{1}{\sqrt{(2\pi)}}\exp\left(-\tfrac{1}{2}x^2+tx-\tfrac{1}{2}t^2\right) = \alpha(x)\exp\left(tx-\tfrac{1}{2}t^2\right)$$

and also, by Taylor's theorem,

$$\alpha(x-t) = \sum_{j=0}^{\infty}\frac{(-1)^j}{j!}t^jD^j\alpha(x) = \sum_{j=0}^{\infty}\frac{t^j}{j!}H_j(x)\alpha(x).$$

Consequently $H_r(x)$ is the coefficient of $\dfrac{t^r}{r!}$ in $\exp(tx-\tfrac{1}{2}t^2)$. It follows that

$$H_r(x) = x^r-\frac{r^{[2]}}{2.1!}x^{r-2}+\frac{r^{[4]}}{2^2.2!}x^{r-4}-\frac{r^{[6]}}{2^3.3!}x^{r-6}+\dots \tag{6.22}$$

The first ten polynomials are

$$\left.\begin{aligned} H_0 &= 1 \\ H_1 &= x \\ H_2 &= x^2-1 \\ H_3 &= x^3-3x \\ H_4 &= x^4-6x^2+3 \\ H_5 &= x^5-10x^3+15x \\ H_6 &= x^6-15x^4+45x^2-15 \\ H_7 &= x^7-21x^5+105x^3-105x \\ H_8 &= x^8-28x^6+210x^4-420x^2+105 \\ H_9 &= x^9-36x^7+378x^5-1260x^3+945x \\ H_{10} &= x^{10}-45x^8+630x^6-3150x^4+4725x^2-945 \end{aligned}\right\} \tag{6.23}$$

6.15 The polynomials have a number of interesting properties. Differentiating the identity

$$\exp\left(tx-\tfrac{1}{2}t^2\right) = \sum_{j=0}^{\infty}\frac{t^jH_j(x)}{j!}$$

with respect to x and identifying coefficients in t^r we have for $r \geqslant 1$

$$\frac{d}{dx}H_r(x) = rH_{r-1}(x) \tag{6.24}$$

and generally for $r \geqslant j$

$$D^jH_r(x) = r^{[j]}H_{r-j}(x). \tag{6.25}$$

Differentiating the identity with respect to t and identifying coefficients in t^{r-1} we have for $r \geqslant 2$

$$H_r(x) - xH_{r-1}(x) + (r-1)H_{r-2}(x) = 0. \tag{6.26}$$

From (6.24) and (6.26) together we find

$$\frac{d^2 H_r(x)}{dx^2} - x\frac{dH_r(x)}{dx} + rH_r(x) = 0, \qquad r \geqslant 2. \tag{6.27}$$

It is also known that the equation in x, $H_r(x) = 0$, has r real roots, each not greater in absolute value than $\sqrt{\{\frac{1}{2}r(r-1)\}}$. (Cf. Charlier, 1931.)

6.16 The polynomials have an important orthogonal property, namely, that

$$\left.\begin{aligned}\int_{-\infty}^{\infty} H_m(x) H_n(x) \alpha(x)\, dx &= 0, \qquad m \neq n \\ &= n!, \qquad m = n.\end{aligned}\right\} \tag{6.28}$$

In fact, integrating by parts, we have, if $m \leqslant n$,

$$\begin{aligned}\int_{-\infty}^{\infty} H_m H_n \alpha\, dx &= (-1)^n \int_{-\infty}^{\infty} H_m D^n \alpha\, dx \\ &= (-1)^n \Big[H_m D^{n-1}\alpha\Big]_{-\infty}^{\infty} + (-1)^{n-1}\int_{-\infty}^{\infty} \frac{dH_m}{dx} D^{n-1}\alpha\, dx.\end{aligned}$$

The term in square brackets vanishes and, in virtue of (6.24), the integral becomes

$$m(-1)^{n-1}\int_{-\infty}^{\infty} H_{m-1} D^{n-1}\alpha\, dx.$$

Continuing the process, we find either zero, if m is not equal to n, or $m!$ if $m = n$.

The Gram–Charlier series of Type A

6.17 Suppose now that a frequency function can be expanded formally in a series of derivatives of $\alpha(x)$. (We shall discuss in **6.22** the conditions under which such an expansion is valid.) We have then

$$f(x) = \sum_{j=0}^{\infty} c_j H_j(x) \alpha(x). \tag{6.29}$$

Multiplying by $H_r(x)$ and integrating from $-\infty$ to ∞ we have, in virtue of the orthogonality relationship (6.28),

$$c_r = \frac{1}{r!}\int_{-\infty}^{\infty} f(x) H_r(x)\, dx.$$

The reader familiar with harmonic analysis will recognize the resemblance between this procedure and the evaluation of constants in a Fourier series.

Substituting the explicit value of $H_r(x)$ from (6.22) we find

$$c_r = \frac{1}{r!}\left\{\mu_r' - \frac{r^{[2]}}{2.1!}\mu_{r-2}' + \frac{r^{[4]}}{2^2.2!}\mu_{r-4}' - \dots\right\}. \tag{6.30}$$

In particular, for moments about the mean,

$$\left.\begin{aligned}
c_0 &= 1\\
c_1 &= 0\\
c_2 &= \tfrac{1}{2}(\mu_2-1)\\
c_3 &= \tfrac{1}{6}\mu_3\\
c_4 &= \tfrac{1}{24}(\mu_4-6\mu_2+3)\\
c_5 &= \tfrac{1}{120}(\mu_5-10\mu_3)\\
c_6 &= \tfrac{1}{720}(\mu_6-15\mu_4+45\mu_2-15)\\
c_7 &= \tfrac{1}{5040}(\mu_7-21\mu_5+105\mu_3)\\
c_8 &= \tfrac{1}{40320}(\mu_8-28\mu_6+210\mu_4-420\mu_2+105)
\end{aligned}\right\} \tag{6.31}$$

Thus we find the formal expansion

$$f(x) = \alpha(x)\{1+\tfrac{1}{2}(\mu_2-1)H_2+\tfrac{1}{6}\mu_3H_3+\tfrac{1}{24}(\mu_4-6\mu_2+3)H_4+\dots\}. \tag{6.32}$$

If $f(x)$ is in standard measure the series becomes

$$f(x) = \alpha(x)\{1+\tfrac{1}{6}\mu_3H_3+\tfrac{1}{24}(\mu_4-3)H_4+\dots\}. \tag{6.33}$$

This is the so-called Gram–Charlier series of Type A. If only the three terms shown in (6.33) are used, it makes allowance for skewness and kurtosis.

Edgeworth's form of the Type A series

6.18 Consider the Fourier transformation of a term $H_r(x)\alpha(x)$.

Since
$$\sqrt{(2\pi)}\,\alpha(t) = e^{-\frac{1}{2}t^2} = \int_{-\infty}^{\infty} e^{itx}\frac{1}{\sqrt{(2\pi)}}e^{-\frac{1}{2}x^2}\,dx$$

we have
$$\sqrt{(2\pi)}\frac{d^r}{dt^r}\alpha(t) = (-1)^r\sqrt{(2\pi)}\,H_r(t)\,\alpha(t) = \int_{-\infty}^{\infty} i^r x^r\frac{e^{itx}}{\sqrt{(2\pi)}}e^{-\frac{1}{2}x^2}\,dx$$

and thus the transform of $x^r\alpha(x)$ is $i^r\sqrt{(2\pi)}\,H_r(t)\,\alpha(t)$. Conversely

$$x^r\alpha(x) = \frac{1}{2\pi}\int_{-\infty}^{\infty} e^{-itx}i^r\sqrt{(2\pi)}\,H_r(t)\,\alpha(t)\,dt.$$

Interchanging x and t, we find

$$\sqrt{(2\pi)}(-i)^r t^r\alpha(t) = \int_{-\infty}^{\infty} e^{-ixt}H_r(x)\,\alpha(x)\,dx$$

and hence, changing the sign of t, that the transform of $H_r(x)\alpha(x)$ is $\sqrt{(2\pi)}\,i^r t^r\alpha(t)$.

Consider now the expression

$$\{\exp(\kappa_r D^r)\}\,\alpha(x). \tag{6.34}$$

Its characteristic function is

$$\int_{-\infty}^{\infty} e^{itx}\exp(\kappa_r D^r)\,\alpha(x)\,dx = \int_{-\infty}^{\infty} e^{itx}\Sigma\left(\frac{\kappa_r^j D^{rj}}{j!}\right)\alpha(x)\,dx$$
$$= \Sigma\frac{\kappa_r^j}{j!}\int_{-\infty}^{\infty} e^{itx}D^{rj}\,\alpha(x)\,dx$$

$$= \Sigma\frac{\kappa_r^j}{j!}\int_{-\infty}^{\infty} e^{itx}(-1)^{rj}H_{rj}(x)\alpha(x)\,dx$$

$$= \Sigma\frac{\kappa_r^j}{j!}\sqrt{(2\pi)}(-i)^{rj}t^{rj}\alpha(t)$$

$$= \sqrt{(2\pi)}\alpha(t)\exp\{\kappa_r(-it)^r\}. \tag{6.35}$$

In a similar way it will be seen that the characteristic function of

$$\exp\left\{-\frac{\kappa_1-a}{1!}D+\frac{\kappa_2-b}{2!}D^2-\frac{\kappa_3}{3!}D^3+\frac{\kappa_4}{4!}D^4\ldots\right\}\alpha(x) \tag{6.36}$$

is equal to

$$\sqrt{(2\pi)}\alpha(t)\exp\left\{\frac{\kappa_1-a}{1!}it+\frac{\kappa_2-b}{2!}(it)^2+\frac{\kappa_3}{3!}(it)^3+\frac{\kappa_4}{4!}(it)^4+\ldots\right\} \tag{6.37}$$

More generally, if we replace $\alpha(x)$ by the unstandardized

$$\beta(x) = \frac{1}{\sigma\sqrt{(2\pi)}}e^{-\frac{1}{2}\frac{(x-m)^2}{\sigma^2}} = \alpha\{(x-m)/\sigma\}/\sigma,$$

the characteristic function of

$$\exp\left\{-\frac{\kappa_1-a}{1!}D+\frac{\kappa_2-b}{2!}D^2-\frac{\kappa_3}{3!}D^3+\frac{\kappa_4}{4!}D^4\ldots\right\}\beta(x) \tag{6.38}$$

is equal to

$$\sqrt{(2\pi)}\alpha(t\sigma)e^{imt}\exp\left\{\frac{\kappa_1-a}{1!}it+\frac{\kappa_2-b}{2!}(it)^2+\frac{\kappa_3}{3!}(it)^3+\frac{\kappa_4}{4!}(it)^4\ldots\right\} \tag{6.39}$$

as may be seen by the same line of argument.

Now suppose that (6.38) represents a frequency function. Its cumulant-generating function is then the logarithm of (6.39), i.e. is equal to

$$\frac{(\kappa_1-a+m)it}{1!}+\frac{\kappa_2-b+\sigma^2}{2!}(it)^2+\frac{\kappa_3}{3!}(it)^3+\frac{\kappa_4}{4!}(it)^4+\ldots$$

and hence its cumulants are κ_1-a+m, $\kappa_2-b+\sigma^2$, κ_3, $\kappa_4\ldots$, $\kappa_r\ldots$, etc. We may take $a = m$ and $b = \sigma^2$ and thus we obtain a distribution whose cumulants are κ_1, κ_2, . . . etc. Now if these are in fact the cumulants of a distribution the series (6.38) must be equal to that distribution, provided that (1) the series converges to a frequency function, and (2) it is uniquely determined by its moments.

If we take the frequency function to be expressed in standard measure, then $\kappa_1 = 0$, $\kappa_2 = 1$ and (6.38) becomes

$$f(x) = \exp\left\{-\kappa_3\frac{D^3}{3!}+\kappa_4\frac{D^4}{4!}-\ldots\right\}\alpha(x), \tag{6.40}$$

where we have written $\alpha(x)$ for $\beta(x)$ because now m vanishes and $\sigma^2 = 1$.

6.19 A series of this kind was derived by Edgeworth (1904), though from an entirely different approach through the theory of elementary errors. Equation (6.40) is formally identical with (6.33), and the reader who consults the original memoirs on this subject may be puzzled by the fact that Edgeworth claimed his series to be different from the Type A series and better as a representation of frequency functions.

The explanation is that for practical purposes it is necessary to take only a finite number of terms in the series and to neglect the remainder. If we take the first k terms in (6.33) the result is in general different from that obtained by taking the first $(k-1)$ terms of the operator in the exponential of (6.40). The argument centred on the fact (cf. Example 6.3 below) that the terms in (6.33) do not tend regularly to zero from the point of view of elementary errors, so that in general no term is negligible compared with a preceding term.

6.20 In (6.29), the coefficients c_j evaluated at (6.31) become, when expressed in standard measure and converted into cumulants using (3.38),

$$\left.\begin{aligned} c_0 &= 1, \quad c_1 = c_2 = 0 \\ c_3 &= \frac{\kappa_3}{6} \\ c_4 &= \frac{\kappa_4}{24} \\ c_5 &= \frac{\kappa_5}{120} \\ c_6 &= \frac{1}{720}(\kappa_6+10\kappa_3^2) \\ c_7 &= \frac{1}{5040}(\kappa_7+35\kappa_4\kappa_3) \\ c_8 &= \frac{1}{40320}(\kappa_8+56\kappa_5\kappa_3+35\kappa_4^2) \end{aligned}\right\} \tag{6.41}$$

We see that c_j depends on κ_j, and we shall find in Chapter 10 that for $j>4$ this is unreliable owing to sampling fluctuations. When sampling effects are not in question the series may be taken to more terms, usually not higher than the term in H_6. We should then have to investigate how far the observed distribution can be represented by the series

$$f(x) = \alpha(x)\left(1+\frac{\kappa_3}{6}H_3+\frac{\kappa_4}{24}H_4+\frac{\kappa_5}{120}H_5+\frac{\kappa_6+10\kappa_3^2}{720}H_6\right) \tag{6.42}$$

in the hope that the remainder after these terms could be neglected in comparison. If cumulants above the fourth are neglected, (6.40) differs from (6.33) by the term $\kappa_3^2 H_6/72$ in parentheses. (6.42) is often called the Edgeworth form of the Type A series.

It may be noted in passing that, in contrast to the Pearson system, the distribution function of such a series is easy to obtain. If

$$f(x) = \sum_{r=0}^{\infty} a_r H_r(x)\alpha(x),$$

then using (6.21),

$$\begin{aligned} \int_{-\infty}^{x} f(x)dx &= \sum_{r=0}^{\infty} a_r \int_{-\infty}^{x} H_r(x)\alpha(x)\,dx \\ &= \int_{-\infty}^{x} \alpha(x)\,dx - \sum_{r=1}^{\infty} a_r H_{r-1}(x)\alpha(x). \end{aligned} \tag{6.43}$$

Table 6.1—Fitting of certain frequency functions to the distribution of length of beans (Table 1.15) (from Pretorius, 1930, Johnson, 1949a)

Length of beans (mm) (1)	Observed frequency (2)	Pearson Type IV (3)	Gram-Charlier Type A (three terms) (4)	Gram-Charlier Type A (four terms) (5)	Gram-Charlier Type A (five terms) (6)	Lognormal (7)	Johnson Type S_u (8)
>17·25	—	{	16·3	−15·2	2·0	—	0·1
17·0	6	1·4	12·8	13·7	−35·3	{	2·0
16·5	55	28·5	25·6	116·6	22·3	10·1	32·2
16·0	275	299·3	241·7	370·4	438·1	280·5	290·1
15·5	1129	1181·6	1012·7	926·2	1214·0	1255·2	1151·5
15·0	2082	2132·6	2155·4	1833·0	1866·9	2163·8	2130·3
14·5	2294	2229·8	2593·0	2506·4	2112·8	2179·6	2240·6
14·0	1787	1638·9	1788·4	2082·6	1916·7	1598·5	1642·5
13·5	929	968·9	713·4	921·3	1183·4	965·0	970·6
13·0	437	503·6	280·7	199·0	371·2	515·3	508·7
12·5	199	243·7	258·7	132·1	66·9	254·5	249·3
12·0	115	113·8	206·2	178·1	101·2	119·5	118·0
11·5	70	52·5	98·7	117·0	107·1	54·4	55·2
11·0	36	24·2	29·6	43·5	54·0	24·3	25·7
10·5	18	11·3	5·9	10·0	15·4	10·7	12·1
10·0	7	5·4	{	{	{	4·8	5·8
9·5	1	2·6	·9	1·7	3·3	2·1	2·7
<9·25	—	1·9	—	—	—	1·7	2·6
TOTALS	9440	9440	9440	9440	9440	9440	9440

(The brackets mean that the frequencies shown are rounded up and include some small frequency in blank rows covered by the brackets.)

Example 6.2

Consider the fitting of a Type A series to the bean data of Example 6.1.

We have already found the first four moments. In standard measure we have

$$\mu_3 = -\ 0{\cdot}910{,}569$$
$$\mu_4 = 4{\cdot}862{,}944$$

and we also find

$$\mu_5 = -12{\cdot}574{,}125$$
$$\mu_6 = 53{\cdot}221{,}083.$$

Hence the series

$$9440\alpha(x)\{1-0{\cdot}151{,}762\,H_3+0{\cdot}077{,}622{,}7\,H_4-0{\cdot}028{,}903{,}6\,H_5+0{\cdot}014{,}273{,}5\,H_6\}.$$

Columns (4) to (6) of Table 6.1 above show the frequencies given by taking the first three, the first four and the first five terms of this series. A glance at the figures will show that the four- and five-term series are, if anything, rather worse than the three-term. Furthermore, the five-term series gives negative frequencies at one terminal and a mode at 12 mm., which is contrary to the data. The representation is clearly not very satisfactory and no better than that given by the Pearson Type IV curve.

Tetrachoric functions

6.21 The numerical evaluation of $H_r(x)\alpha(x) = (-D)^r\alpha(x)$ may be carried out (a) directly by evaluating the polynomial in (6.23) and multiplying by $\alpha(x)$; (b) from tables of $H_r(x)$ and $\alpha(x)$; (c) from tables of $D^r\alpha(x)$. The *Biometrika Tables*, Vol. II, give $D^r\alpha(x)$ for $x = 0(0{\cdot}02)4{\cdot}00(0{\cdot}05)6{\cdot}20$ for $r = 1,2$ (6 d.p.); $r = 3,4$ (5 d.p.); $r = 5,6$ (4 d.p.); $r = 7$ (3 d.p.); $r = 8$ (2 d.p.) and $r = 9$ (1 d.p.). A similar function was tabulated by K. Pearson in the form

$$\tau_r(x) = \frac{H_{r-1}(x)\alpha(x)}{(r!)^{\frac{1}{2}}}. \tag{6.44}$$

This was known as a tetrachoric function, for reasons which will become apparent in the second volume when we discuss the estimation of correlation coefficients from a 2×2 table. Smirnov (1965) gives 7 d.p. values of (6.44) for $r = 2(1)21$, $x = 0(0{\cdot}002)4$, together with a 10 d.p. table of the coefficients in the recurrence relations derived from (6.27).

6.22 Up to this point it has been assumed that a frequency function possesses a convergent Type A series. We shall not here enter into a discussion of the conditions under which this is so, except to warn the reader that mistakes have been made on the subject and to quote some theorems without proof.

(1) Cramér (1926). If $f(x)$ is a function which has a continuous derivative such that

$$\int_{-\infty}^{\infty} \left(\frac{df}{dx}\right)^2 e^{\frac{1}{2}x^2}\,dx$$

converges and if $f(x)$ tends to zero as $|x|$ tends to infinity, then $f(x)$ may be developed in the series

$$f(x) = \sum_{j=0}^{\infty} \frac{c_j}{j!} D^j\alpha(x), \tag{6.45}$$

where $$c_j = \int_{-\infty}^{\infty} f(x)H_j(x)\,dx.$$

This series is absolutely and uniformly convergent for $-\infty \leqslant x \leqslant \infty$.

(2) A theorem by Cramér (1926) based on one by Galbrun. If $f(x)$ is of bounded variation in every finite interval and if

$$\int_{-\infty}^{\infty} |f(x)|\, e^{\frac{1}{2}x^2}\,dx$$

exists, then the expansion of $f(x)$ in the series (6.45) converges everywhere to the sum $\frac{1}{2}\{f(x+0)+f(x-0)\}$. The convergence is uniform in every finite interval of continuity.

Cramér has also shown that this last theorem cannot be substantially improved upon as regards the behaviour of $f(x)$ at infinity. Consider in fact the function $f(x) = e^{-\lambda x^2}$. We have, in virtue of (6.26) and (6.24),

$$\begin{aligned}\int_{-\infty}^{\infty} e^{-\lambda x^2} H_r(x)\,dx &= \int_{-\infty}^{\infty} e^{-\lambda x^2}.xH_{r-1}\,dx - (r-1)\int_{-\infty}^{\infty} e^{-\lambda x^2}.H_{r-2}\,dx \\ &= \left[\frac{e^{-\lambda x^2}}{-2\lambda}H_{r-1}\right]_{-\infty}^{\infty} + \frac{r-1}{2\lambda}\int_{-\infty}^{\infty} e^{-\lambda x^2}H_{r-2}\,dx - (r-1)\int_{-\infty}^{\infty} e^{-\lambda x^2}H_{r-2}\,dx \\ &= (r-1)\left(\frac{1}{2\lambda}-1\right)\int_{-\infty}^{\infty} e^{-\lambda x^2}H_{r-2}(x)\,dx.\end{aligned}$$

If r is odd the integral vanishes because H_r is an odd function of x. If r is even, say $2r$, the integral becomes

$$(2r-1)(2r-3)\ldots 1\left(\frac{1}{2\lambda}-1\right)^r \cdot \sqrt{\frac{\pi}{\lambda}} = \sqrt{\frac{\pi}{\lambda}}\frac{(2r)!}{2^r r!}\left(\frac{1}{2\lambda}-1\right)^r.$$

The appropriate coefficient of H_{2r} in the Type A series is then $\left(\frac{1}{2\lambda}-1\right)^r/(2^r r!)$. Now when $x = 0$, $H_{2r} = (-1)^r(2r)!/(2^r r!)$. The series then becomes

$$\sum_{r=0}^{\infty}\sqrt{\frac{\pi}{\lambda}}\frac{(2r)!}{2^{2r}(r!)^2}\left(1-\frac{1}{2\lambda}\right)^r.$$

In virtue of the Stirling approximation to the factorial, the rth term of this, say u_r, becomes in the limit

$$u_r \sim \left(1-\frac{1}{2\lambda}\right)^r \frac{1}{\sqrt{(\lambda r)}}$$

so that

$$u_r^{1/r} \sim 1-(1/2\lambda).$$

Hence, for $\lambda < \frac{1}{4}$ the series is divergent.

Exercise 6.9 is to show that a convergent Type A expansion requires that the distribution be determined by its moments.

6.23 From the statistical viewpoint, however, the important question is not whether an *infinite* series can represent a frequency function, but whether a finite number of terms can do so to a satisfactory approximation. Two things seem clear :—

(a) The sum of a finite number of terms of the series may give negative frequencies, particularly near the tails (as, for instance, in Example 6.2).

(b) The series in the Charlier form (6.33) may behave irregularly in the sense that the sum of k terms may give a worse fit than the sum of $(k-1)$ terms.

Barton and Dennis (1952)—see also Draper and Tierney (1972)—show that if cumulants above the fourth are neglected, the three-term expansion (6.33) and the four-term (6.42) do not give a unimodal f.f., or even non-negative frequencies, unless the skewness and kurtosis coefficients β_1 and β_2 are close enough to zero. For example, (6.42) is non-unimodal if $|\beta_1| \geqslant 0{\cdot}25$, and has negative frequencies if $|\beta_1| \geqslant 0{\cdot}5$; (6.33) is non-unimodal if $|\beta_1| \geqslant 0{\cdot}7$ and has negative frequencies if $|\beta_1| > 1{\cdot}2$; neither is unimodal if $\beta_2 > 2{\cdot}4$ and both have negative frequencies for $\beta_2 > 4$. Berndt (1957) makes a similar investigation when only κ_3 is used.

Thus the finite series is useful only in cases of moderate skewness, and in such cases a Pearson distribution, or a transformed distribution of the type we consider later, may be just as good. For many statistical purposes we are most interested in the tails of a distribution, and it is here that the inadequacies of this approach are most serious.

Example 6.3

As an illustration of the irregular behaviour of terms in the Type A series, consider the Gamma distribution

$$dF = \frac{1}{\Gamma(\lambda)}e^{-x}x^{\lambda-1}\,dx, \qquad 0 \leqslant x \leqslant \infty;\ \lambda > 0.$$

Its characteristic function is (cf. Example 3.6)

$$\phi(t) = 1/(1-it)^{\lambda}$$

and thus $\kappa_r = \lambda(r-1)!$ or, in standard measure,

$$\kappa_r = \frac{(r-1)!}{\lambda^{\frac{1}{2}r-1}}.$$

From the manner of the formation of terms in (6.41) it is evident that the coefficient c_r is the sum of terms $\kappa_r, \kappa_{r-3}\kappa_3, \ldots (\kappa_{q_1}\kappa_{q_2}\ldots\kappa_{q_m})$, where $(q_1 \ldots q_m)$ is a partition of r such that no q is less than 3. It will then be clear that, since κ_q is of order $\lambda^{1-\frac{1}{2}q}$, the term of greatest order in λ is that with the greatest number of parts in $(\kappa_{q_1}\ldots\kappa_{q_m})$. For example, if $r = 9$ it is (3^3), if $r = 8$ it is (4^2), and so on.

From these considerations we can find the order in λ of the terms in the Type A series. They are

Term	c_0	c_3	c_4	c_5	c_6	c_7	c_8	c_9	c_{10}	c_{11}	c_{12}
Order in λ	0	$-\frac{1}{2}$	-1	$-1\frac{1}{2}$	-1	$-1\frac{1}{2}$	-2	$-1\frac{1}{2}$	-2	$-2\frac{1}{2}$	-2

The terms decrease in order of λ, but not at all regularly, and it is clear that in general no coefficient will be negligible compared with a preceding one if λ is large. The asymptotic qualities of such series obviously require careful investigation in particular cases.

Other expansions

6.24 Charlier also proposed a Type B series based on derivatives of the Poisson frequency $e^{-\lambda}\lambda^x/x!$ with respect to λ, or equivalently, on first differences of that frequency with respect to x. (Cf. Exercise 6.8.) This had certain mathematical attractions and has been used to graduate a few examples of discontinuous variation resembling the Poisson distribution. It has not come into general use. A further form (Type C) proposed by Charlier to avoid negative frequencies has also failed to stand the test of experience.

Expansions can also be obtained in terms of derivatives of the Gamma and Beta distributions. The appropriate polynomials are those of Laguerre and Jacobi respectively. See G. Szegö, *Orthogonal Polynomials*, New York, 1939, and Durbin and Watson (1951), especially p. 172. Wallace (1958) discusses the validity of asymptotic expansions generally, including non-normal f.f.'s as generating distributions, and also the problem of normalization discussed in **6.25–6** below. Gray *et al.* (1975) give an expansion of any theoretical $f(x)$ in terms of any $\alpha(x)$ (not necessarily normal) using derivatives but not cumulants, which therefore need not exist. In the Gamma case of Example 6.3, this expansion improves on Edgeworth's in the tails of the distribution.

Polynomial transformations to normality

6.25 Many of the important theoretical distributions occurring in statistics depend on some variable n in such a way that as n tends to infinity the distribution tends to normality. For large n it is often a sufficient approximation to assume the distribution normal, but for small or moderate n this may be hardly exact enough. In such a case we are nevertheless able to use the normal integral by seeking a polynomial variate-transformation

$$\xi = b_0 + b_1 x + b_2 x^2 + b_3 x^3 + \ldots \tag{6.46}$$

where the b's are of order $n^{-\frac{1}{2}}$ or smaller. By choosing the b's appropriately we can bring the distribution of ξ much nearer to normality than that of x and hence from (6.46) find the distribution function of x from that of ξ, assumed normal.

Consider in fact the Edgeworth form of the Type A expansion (6.38)

$$\exp\left\{-\frac{\kappa_1-m}{1!}D+\frac{\kappa_2-\sigma^2}{2!}D^2-\frac{\kappa_3}{3!}D^3+\dots\right\}\beta(x). \tag{6.47}$$

We have retained the terms in D and D^2 because the approximation may perhaps be slightly improved by taking m and σ^2 in the ξ-distribution not quite equal to the mean and variance of x.

We now assume that the cumulant κ_r is of order n^{1-r}, a case of fairly common occurrence; by choice of m and σ^2, we may write

$$\kappa_1-m = l_1\sigma, \quad l_1 = O(n^{-\frac{1}{2}}),$$
$$\kappa_2-\sigma^2 = l_2\sigma^2, \quad l_2 = O(n^{-1}).$$

Then σ^2 is of order κ_2, i.e. n^{-1}, and thus

$$\frac{\kappa_r}{\sigma^r} = O(n^{1-\frac{1}{2}r}).$$

Thus (6.47) may be written

$$\exp\{-l_1\sigma D+\tfrac{1}{2}l_2\sigma^2D^2-\tfrac{1}{6}l_3\sigma^3D^3+\tfrac{1}{24}l_4\sigma^4D^4-\tfrac{1}{120}l_5\sigma^5D^5$$
$$+\tfrac{1}{720}l_6\sigma^6D^6-\dots\}\beta(x), \tag{6.48}$$

where $\quad l_1$ and l_3 are $O(n^{-\frac{1}{2}})$, $\quad l_5$ is $O(n^{-\frac{3}{2}})$,

$\qquad l_2$ and l_4 are $O(n^{-1})$, $\quad l_6$ is $O(n^{-2})$, etc.

Expanding the operator and retaining only terms up to and including $O(n^{-2})$ in the l's we find for the operator

$$1-l_1\sigma D+\tfrac{1}{2}l_2\sigma^2D^2-\tfrac{1}{6}l_3\sigma^3D^3+\tfrac{1}{24}l_4\sigma^4D^4-\tfrac{1}{120}l_5\sigma^5D^5+\tfrac{1}{720}l_6\sigma^6D^6+\tfrac{1}{2}(l_1^2\sigma^2D^2$$
$$+\tfrac{1}{4}l_2^2\sigma^4D^4+\tfrac{1}{36}l_3^2\sigma^6D^6+\tfrac{1}{576}l_4^2\sigma^8D^8-l_1l_2\sigma^3D^3+\tfrac{1}{3}l_1l_3\sigma^4D^4-\tfrac{1}{12}l_1l_4\sigma^5D^5$$
$$+\tfrac{1}{60}l_1l_5\sigma^6D^6-\tfrac{1}{6}l_2l_3\sigma^5D^5+\tfrac{1}{24}l_2l_4\sigma^6D^6-\tfrac{1}{72}l_3l_4\sigma^7D^7+\tfrac{1}{360}l_3l_5\sigma^8D^8)+\tfrac{1}{6}(-l_1^3\sigma^3D^3$$
$$-\tfrac{1}{216}l_3^3\sigma^9D^9+\tfrac{3}{2}l_1^2l_2\sigma^4D^4-\tfrac{1}{2}l_1^2l_3\sigma^5D^5-\tfrac{1}{8}l_1^2l_4\sigma^6D^6+\tfrac{1}{288}l_3^2l_4\sigma^{10}D^{10}-\tfrac{1}{12}l_1l_3^2\sigma^7D^7$$
$$+\tfrac{1}{24}l_2l_3^2\sigma^8D^8+\tfrac{1}{2}l_1l_2l_3\sigma^6D^6+\tfrac{1}{24}l_1l_3l_4\sigma^8D^8)+\tfrac{1}{24}(l_1^4\sigma^4D^4+\tfrac{1}{1296}l_3^4\sigma^{12}D^{12}$$
$$+\tfrac{2}{3}l_1^3l_3\sigma^6D^6+\tfrac{1}{6}l_1^2l_3^2\sigma^8D^8+\tfrac{1}{54}l_1l_3^3\sigma^{10}D^{10}). \tag{6.49}$$

The result of this operation is a similar expression, which we will not bother to write out at length, with the operator σ^rD^r replaced by $(-1)^r\mathrm{H}\left(\frac{x-m}{\sigma}\right)$ and multiplied by $\beta(x)$.

The distribution function is given by integrating this expression, and we then have for the frequency less than or equal to $m+\sigma x$ (arranging the terms in order of magnitude in n)

$$\int_{-\infty}^{x}\alpha(x)\,dx+\alpha(x)[-(l_1+\tfrac{1}{6}l_3H_2)-(\tfrac{1}{2}l_1^2H_1+\tfrac{1}{2}l_2H_1+\tfrac{1}{6}l_1l_3H_3+\tfrac{1}{24}l_4H_3$$
$$+\tfrac{1}{72}l_3^2H_5)-(\tfrac{1}{6}l_1^3H_2+\tfrac{1}{2}l_1l_2H_2+\tfrac{1}{12}l_1^2l_3H_4+\tfrac{1}{12}l_2l_3H_4+\tfrac{1}{24}l_1l_4H_4$$
$$+\tfrac{1}{120}l_5H_4+\tfrac{1}{72}l_1l_3^2H_6+\tfrac{1}{44}l_3l_4H_6+\tfrac{1}{1296}l_3^3H_8)-(\tfrac{1}{24}l_1^4H_3+\tfrac{1}{8}l_2^2H_3+\tfrac{1}{4}l_1^2l_2H_3$$
$$+\tfrac{1}{36}l_1^3l_3H_5+\tfrac{1}{12}l_1l_2l_3H_5+\tfrac{1}{48}l_1^2l_4H_5+\tfrac{1}{48}l_2l_4H_5+\tfrac{1}{120}l_1l_5H_5+\tfrac{1}{720}l_6H_5$$
$$+\tfrac{1}{144}l_1^2l_3^2H_7+\tfrac{1}{144}l_2l_3^2H_7+\tfrac{1}{1152}l_4^2H_7+\tfrac{1}{144}l_1l_3l_4H_7+\tfrac{1}{720}l_3l_5H_7+\tfrac{1}{129}\,l_1l_3^3H_9$$
$$+\tfrac{1}{1728}l_3^2l_4H_9+\tfrac{1}{31104}l_3^4H_{11})]. \tag{6.50}$$

6.26 Now let ξ be a normal variate. We will determine ξ in terms of x so that

$$G(\xi) = \int_{-\infty}^{\xi} \alpha(y)\,dy = F(x), \tag{6.51}$$

$F(x)$ being the distribution function given by the Type A expansion (6.50).

We have, by Taylor's theorem,

$$G(\xi) = G\{x+(\xi-x)\} = G(x)+\sum_{r=1}^{\infty} \frac{(\xi-x)^r}{r!}\frac{d^r}{dx^r}G(x)$$

and on using (6.21), this becomes

$$G(\xi) = G(x)-\sum_{r=1}^{\infty} \frac{(x-\xi)^r}{r!}H_{r-1}(x)\alpha(x) \tag{6.52}$$

and this is equal to (6.50).

The next step is to invert this series to give $(x-\xi)$ in powers of x. Assuming

$$x-\xi = a_0+a_1x+a_2x^2+\dots, \tag{6.53}$$

which is (6.46) slightly re-arranged for the sake of convenience, we see that when $x=0$, $\xi = -a_0$, $x-\xi$ is of order $n^{-\frac{1}{2}}$; and hence, to order n^{-2}, we have from (6.52) with $x = 0$, on using (6.23),

$$\xi-\frac{\xi^2}{2}(+0)+\frac{\xi^3}{6}(-1) = -\left(a_0-\frac{a_0^3}{6}\right)$$

and this is equal to the expression in square brackets in (6.50) with $x = 0$.

We then find

$$a_0 = l_1-\tfrac{1}{6}l_3-\tfrac{1}{2}l_1l_2+\tfrac{1}{6}l_1^2l_3+\tfrac{1}{4}l_2l_3+\tfrac{1}{8}l_1l_4+\tfrac{1}{40}l_5-\tfrac{7}{36}l_1l_3^2-\tfrac{15}{144}l_3l_4+\tfrac{52}{648}l_3^3.$$

We can now find a_1 in (6.50) by identifying coefficients in x, and so on. After some algebraic reduction we find, writing the terms in descending order in n,

$$\begin{aligned}
x-\xi = {} & l_1+\tfrac{1}{6}l_3(x^2-1)+\tfrac{1}{2}l_2x-\tfrac{1}{3}l_1l_3x+\tfrac{1}{24}l_4(x^3-3x)-\tfrac{1}{36}l_3^2(4x^3-7x)-\tfrac{1}{2}l_1l_2 \\
& +\tfrac{1}{6}l_1^2l_3-\tfrac{1}{12}l_2l_3(5x^2-3)-\tfrac{1}{8}l_1l_4(x^2-1)+\tfrac{1}{120}l_5(x^4-6x^2+3)+\tfrac{1}{36}l_1l_3^2(12x^2-7) \\
& -\tfrac{1}{144}l_3l_4(11x^4-42x^2+15)+\tfrac{1}{648}l_3^3(69x^4-187x^2+52)-\tfrac{3}{8}l_2^2x+\tfrac{5}{6}l_1l_2l_3x \\
& +\tfrac{1}{8}l_1^2l_4x-\tfrac{1}{48}l_2l_4(7x^3-15x)-\tfrac{1}{30}l_1l_5(x^3-x)+\tfrac{1}{720}l_6(x^5-10x^3+15x) \\
& -\tfrac{1}{3}l_1^2l_3^2x+\tfrac{1}{72}l_2l_3^2(36x^3-49x)-\tfrac{1}{384}l_4^2(5x^5-32x^3+35x)+\tfrac{1}{36}l_1l_3l_4(11x^3-21x) \\
& -\tfrac{1}{360}l_3l_5(7x^5-48x^3+51x)-\tfrac{1}{324}l_1l_3^3(138x^3-187x) \\
& +\tfrac{1}{864}l_3^2l_4(111x^5-547x^3+456x)-\tfrac{1}{7776}l_3^4(948x^5-3628x^3+2473x).
\end{aligned} \tag{6.54}$$

This is our required expression of the variate ξ in terms of the variate x. To order n^{-2} at least, ξ will be normally distributed.

It is often more convenient to express x in terms of ξ. This may be done by noting that

$$\begin{aligned}
x-\xi &= g(x) = g(\xi+x-\xi) \\
&= g(\xi)+(x-\xi)g'(\xi)+\dots \\
&= g(\xi)+g'(\xi)\{g(\xi)+(x-\xi)g'(\xi)\}+\dots
\end{aligned}$$

and by continuing the process

$$x-\xi = g(\xi)+g(\xi)g'(\xi)+g(\xi)g'^2(\xi)+\tfrac{1}{2}g^2(\xi)g''(\xi)+g(\xi)g'^3(\xi)+\tfrac{3}{2}g^2(\xi)g'(\xi)g''(\xi) + \tfrac{1}{6}g^3(\xi)g'''(\xi)+\dots \tag{6.55}$$

Hence, using the value of ξ given by (6.54), we find, after some reduction,

$$\begin{aligned} x-\xi = {} & l_1+\tfrac{1}{6}l_3(\xi^2-1)+\tfrac{1}{2}l_2\xi+\tfrac{1}{24}l_4(\xi^3-3\xi)-\tfrac{1}{36}l_3^2(2\xi^3-5\xi)-\tfrac{1}{6}l_2 l_3(\xi^2-1) \\ & +\tfrac{1}{120}l_5(\xi^4-6\xi^2+3)-\tfrac{1}{24}l_3 l_4(\xi^4-5\xi^2+2)+\tfrac{1}{324}l_3^3(12\xi^4-53\xi^2+17) \\ & -\tfrac{1}{8}l_2^2\xi-\tfrac{1}{16}l_2 l_4(\xi^3-3\xi)+\tfrac{1}{720}l_6(\xi^5-10\xi^3+15\xi)+\tfrac{1}{72}l_2 l_3^2(10\xi^3-25\xi) \\ & -\tfrac{1}{384}l_4^2(3\xi^5-24\xi^3+29\xi)-\tfrac{1}{180}l_3 l_5(2\xi^5-17\xi^3+21\xi) \\ & +\tfrac{1}{288}l_3^2 l_4(14\xi^5-103\xi^3+107\xi)-\tfrac{1}{7776}l_3^4(252\xi^5-1688\xi^3+1511\xi). \end{aligned} \tag{6.56}$$

l_1 appears only in the first term on the right of (6.56), since it represents the location of x, and cannot affect the cumulants other than the mean.

The approach of these sections is due to Cornish and Fisher (1937). Fisher and Cornish (1960) give higher-order expansions and tables to facilitate their use.

The reader may have noticed that in (6.54) and (6.56) the coefficient of l_r is $H_{r-1}/r!$ This is one of several formulae for the coefficients proved by Hill and Davis (1968), who have computerized the expansions. They also generalize the polynomial transformation to any desired regular continuous distribution (e.g. the χ^2 distribution of Chapter 16—cf. **42.11**, Vol. 3, 3rd edn).

Example 6.4

Consider again the Gamma distribution of Example 6.3 :—

$$dF = \frac{1}{\Gamma(\lambda)}e^{-x}x^{\lambda-1}\,dx, \qquad 0\leqslant x\leqslant\infty;\ \lambda>0.$$

We saw that, in standard measure, κ_r is of order $\lambda^{1-\frac{1}{2}r}$. Thus, as $\lambda\to\infty$, $\kappa_r\to 0$ for $r>1$ and the distribution tends to normality by (5.61) and the Inversion Theorem.

We will take l_1 and l_2 of (6.48) to be zero, which implies that our normal variate ξ is to have the same mean and variance as that of x. We have

$$l_3 = 2\lambda^{-\frac{1}{2}},\quad l_4 = 6\lambda^{-1},\quad l_5 = 24\lambda^{-3/2},\quad l_6 = 120\lambda^{-2}.$$

(6.50) then becomes

$$G(x)-\alpha(x)\left\{\frac{1}{3\lambda^{\frac{1}{2}}}H_2+\frac{1}{4\lambda}H_3+\frac{1}{18\lambda}H_5+\frac{1}{5\lambda^{\frac{3}{2}}}H_4+\frac{1}{12\lambda^{\frac{3}{2}}}H_6+\frac{1}{162\lambda^{\frac{3}{2}}}H_8 + \frac{1}{6\lambda^2}H_5+\frac{1}{32\lambda^2}H_7+\frac{1}{15\lambda^2}H_7+\frac{1}{72\lambda^2}H_9+\frac{1}{1944\lambda^2}H_{11}\right\}.$$

Let us, as a simple illustration, find the distribution function of x for $\lambda = 9$, $x = 12$. The mean of the distribution is then 9 and its variance 9, so that this corresponds to a deviation $(12-9)/\sqrt{9}$ in standard measure, equal to unity. It is found from (6.23) and an additional equation for H_{11} that

$$H_2 = 0,\ H_3 = -2,\ H_4 = -2,\ H_5 = 6,\ H_6 = 16,\ H_7 = -20,\ H_8 = -132,\ H_9 = 28,\ H_{10} = 1216,\ H_{11} = 936.$$

We then find for the distribution function

$$\int_{-\infty}^{1}\alpha(x)\,dx+\alpha(x)(0{\cdot}015{,}163{,}5).$$

The values for the normal function are obtained from the tables and we get the value

$$0{\cdot}841{,}345+(0{\cdot}241{,}970{,}7)(0{\cdot}015{,}163{,}5) = 0{\cdot}8450,$$

which is exact to four places. The approximation is evidently fairly good, even for values of λ as low as 9.

We could have found the same result by using (6.54). Substituting $x = 1$ in that equation we find $\xi = 1{\cdot}015{,}386$, and the distribution function for the normal integral with deviate equal to this value of ξ is $0{\cdot}8450$ as before.

Suppose now that we wish to find the deviate x whose distribution function is $F(x) = 0{\cdot}99$ when $\lambda = 15$.

The normal deviate ξ corresponding to such a value is found from tables to be $2{\cdot}326{,}348$. We then have from (6.56)

$$x-\xi = \frac{1}{3\lambda^{\frac{1}{2}}}(\xi^2-1)+\frac{1}{4\lambda}(\xi^3-3\xi)+\dots$$

which will be found to give $x = 2{\cdot}697{,}22$. This is the value in standard measure. The deviate in ordinary measure is $15+x\sqrt{15} = 25{\cdot}45$. This is exact to two places of decimals.

The example shows that, notwithstanding the non-convergence of the infinite Type A series, a satisfactory approximation may be obtained from its first few terms, at least in certain cases. We may remark without proof that by an adaptation of a procedure given by Cramér (1928) it may be shown that an asymptotic expansion does in fact exist for the distribution of this example.

Haldane (1938) considers power transformations $y = \{x/E(x)\}^h$, with h chosen to minimize skewness and kurtosis, for variates x with all cumulants of order n, the sample size. His results are given in Exercise 37.24 (Vol. 3, 3rd edn). **16.7** below treats a special case.

Functional transformations to normality

6.27 The idea of transforming from one variate to another with a more convenient frequency function may be developed further. We need not confine ourselves to a polynomial expression of the type of (6.46); nor is that kind of representation necessarily the best. Several writers have studied more general types. Our treatment follows that of Johnson (1949a).

We shall consider a transformation of the type

$$\xi = \gamma+\delta g\left(\frac{x-\mu}{\lambda}\right) \tag{6.57}$$

where μ, λ, γ and δ are parameters at choice and g is some convenient function. For practical purposes it is desirable that g should not itself depend on variable parameters and that it should be a monotonic function; otherwise it might be troublesome to match the ranges of ξ and x. Without loss of generality we may suppose that g is monotone increasing. If we write the transformation in the form

$$\frac{\xi-\gamma}{\delta} = g\left(\frac{x-\mu}{\lambda}\right) \tag{6.58}$$

the parameters μ, γ appear as location parameters and λ, δ as scale parameters. We may suppose the latter pair to be positive. We shall try to choose the parameters and the function g so that ξ is normal or approximately so.

6.28 Without loss of generality we write

$$y = \frac{x-\mu}{\lambda}. \tag{6.59}$$

If ξ is normal with frequency function $\alpha(\xi)$ we have for the frequency function of y

$$dF(y) = \alpha(\xi)\left|\frac{d\xi}{dy}\right| dy$$
$$= \frac{1}{\sqrt{(2\pi)}}\exp[-\tfrac{1}{2}\{\gamma+\delta g(y)\}^2]\,\delta g'(y)\,dy \tag{6.60}$$

where $g'(y) = dg(y)/dy$. We shall examine three types of system :—

(1) The S_L or lognormal system : $g(y) = \log y$.
(2) The S_B system : $g(y) = \log\{y/(1-y)\}$.
(3) The S_U system : $g(y) = \sinh^{-1}y = \log\{y+\sqrt{(y^2+1)}\}$.

Other systems, of course, are possible. The variety of shapes given by these three types is, however, quite as great as that of the Pearson system.

The lognormal distribution

6.29 We have

$$\xi = \gamma+\delta\log y. \tag{6.61}$$

As ξ goes from $-\infty$ to ∞, y goes from 0 to ∞. It has infinitely high-order contact at each terminal of the range; for the distribution, from (6.60), is

$$dF(y) = \frac{\delta}{\sqrt{(2\pi)}\,y}\exp\{-\tfrac{1}{2}(\gamma+\delta\log y)^2\}\,dy, \qquad 0\leqslant y\leqslant\infty$$
$$= p(y)\,dy, \text{ say.}$$

The limit for any positive r

$$\lim_{y\to\infty} y^r p(y) = \lim \frac{\delta e^{-\frac{1}{2}\gamma^2}}{\sqrt{(2\pi)}}\,\frac{y^{r-1-\gamma\delta}}{\exp\{\frac{1}{2}(\delta\log y)^2\}} = 0$$

and

$$\lim_{y\to 0} y^r p(y) = \lim\frac{\delta e^{-\frac{1}{2}\gamma^2}}{\sqrt{(2\pi)}}\,\frac{y^{r-1-\gamma\delta}}{y^{\frac{1}{2}\delta^2\log y}} = 0.$$

Also the distribution is unimodal, for

$$\frac{dp(y)}{dy} = \frac{\delta}{\sqrt{(2\pi)}}\exp\{-\tfrac{1}{2}(\gamma+\delta\log y)^2\}\left\{-\frac{1}{y^2}-\frac{\delta}{y^2}(\gamma+\delta\log y)\right\},$$

which, apart from the values $y = 0$ and $y = \infty$, vanishes where

$$1+\delta(\gamma+\delta\log y) = 0,$$

i.e. at the single modal value $y = \exp\left\{-\frac{\gamma}{\delta}-\frac{1}{\delta^2}\right\}$.

6.30 For the moments of y about zero we have

$$\begin{aligned}\mu_r'(y) &= \int_0^\infty y^r p(y)\,dy \\ &= \frac{1}{\sqrt{(2\pi)}}\int_{-\infty}^{\infty} \exp\left\{\frac{r(\xi-\gamma)}{\delta}\right\}\exp(-\tfrac{1}{2}\xi^2)\,d\xi \\ &= \exp\left(\frac{1}{2}\frac{r^2}{\delta^2}-\frac{r\gamma}{\delta}\right),\end{aligned} \tag{6.62}$$

using (3.15). Thus we find, putting $\omega = \exp(1/2\delta^2)$, $\rho = \exp(-\gamma/\delta)$,

$$\left.\begin{aligned}\mu_1' &= \omega\rho \\ \mu_2 &= \omega^2\rho^2(\omega^2-1) \\ \mu_3 &= \omega^3\rho^3(\omega^2-1)^2(\omega^2+2) \\ \mu_4 &= \omega^4\rho^4(\omega^2-1)^2(\omega^8+2\omega^6+3\omega^4-3),\end{aligned}\right\} \tag{6.63}$$

and

$$\beta_1 = (\omega^2-1)(\omega^2+2)^2 \tag{6.64}$$

$$\beta_2-3 = (\omega^2-1)(\omega^6+3\omega^4+6\omega^2+6). \tag{6.65}$$

The median is the value of y corresponding to $\xi = 0$, since the transformation is monotonic, i.e. from (6.61) the median is $\exp(-\gamma/\delta)$. Thus the lognormal distribution has mode, median and mean in that order, the spacings between them being approximately 2 : 1 in ratio, as at (2.13). The skewness is always positive and the distribution is always leptokurtic. In general the frequency function of y will rise from zero to a maximum and then tail off more slowly as y tends to infinity. The distribution is not uniquely determined by its moments—cf. Exercise 6.21.

6.31 The values of μ_2 and μ_3 determine ω and ρ, and hence δ and γ. Thus, apart from location and scale parameters, we may fit a lognormal distribution from the second and third moments of the x-distribution. The simplest procedure, suggested by Wicksell (1917), is as follows:

Using (6.64) we find $t \equiv (\omega^2-1)^{\frac{1}{2}}$ as the positive root of

$$t^3+3t-\sqrt{\beta_1} = 0. \tag{6.66}$$

ρ is found from μ_2 and t, and hence γ and δ are found from the definitions of ω and ρ. Yuan (1933) has given tables to facilitate the solution of (6.66).

In point of fact, the lognormal distribution does not have the generality of the family (6.57). It depends on three, not four parameters. For

$$\xi = \gamma+\delta\log\{(x-\mu)/\lambda\}$$

may be written

$$\xi = (\gamma-\delta\log\lambda)+\delta\log(x-\mu)$$

and thus $\gamma-\delta\log\lambda$ is a single parameter, the scale parameter λ disappearing from the problem. The moments of x are therefore identical with those of y, apart from their means, which differ by the location parameter μ which is determined, by using (6.63), as

$$\mu = \mu_1'(x)-\frac{\sqrt{\mu_2}}{t}. \tag{6.67}$$

For a more extensive study of the history, theory and applications of the distribution, reference may be made to Aitchison and Brown (1957).

Example 6.5

For the purposes of comparison we will fit a lognormal distribution to the bean data which have been used in previous examples in this chapter. As we have considered the bean distribution, from low to high values of the variate " length ", it is skewed to the left. We will therefore fit a lognormal distribution with the variate-range reversed, i.e. so that negative values of ξ correspond to the greater values of x.

From Example 6.1 we have $\sqrt{\beta_1} = -0{\cdot}910{,}569$. As we are reversing the variate range we change the sign of this, and (6.66) then becomes

$$t^3+3t-0{\cdot}910{,}569 = 0.$$

The positive root is $t = 0{\cdot}294{,}968$. Also we have (as given in Example 6.1) $\mu_2 = 3{\cdot}238{,}425$. Then from the second equation of (6.63) we have $\rho = 5{\cdot}8516$. We notice in passing that from (6.65) we shall then have, for ξ, $\beta_2 = 4{\cdot}510$ as against the β_2 found for x in Example 6.1 $= 4{\cdot}863$. Our transformation happens to bring the values of β_2 roughly into agreement but not completely so.

We then have

$$1/(2\delta^2) = \log_e \omega = \tfrac{1}{2}\log_e(1+t^2) = 0{\cdot}041{,}718$$
$$-\gamma/\delta = \log_e \rho = 1{\cdot}7667$$

giving

$$\delta = 3{\cdot}462, \quad \gamma = -6{\cdot}116.$$

The units (width of class-intervals) are $\frac{1}{2}$ millimetre and hence, from the data of Example 6.1, the variate x has a mean $14{\cdot}5-\frac{1}{2}(0{\cdot}190{,}78) = 14{\cdot}405$. The location parameter μ is by (6.67) equal to

$$14{\cdot}405+\tfrac{1}{2}\sqrt{\mu_2}/t = 17{\cdot}455,$$

the sign here again being changed because the " start " of the distribution is above the mean, and the $\frac{1}{2}$ mm. unit again being taken into account. Thus the fitted distribution " starts " at 17.455 mm.

Column (7) of Table 6.1 on page 160 gives the corresponding frequencies. The fit is not particularly good, being rather worse than that of Type IV but better than those of the Gram–Charlier series.

Exercises 6.13–15 apply the transformation (6.61) to the Gamma and the two kinds of Beta distributions.

6.32 Distributions of the S_B type with $g(y) = \log\{y/(1-y)\}$ have the variate-transformation

$$\xi = \gamma+\delta\log\frac{x-\mu}{\lambda+\mu-x}. \tag{6.68}$$

This may itself be regarded, of course, as a simple transformation of the lognormal $\xi = \gamma+\delta\log(x-\mu)$ with $y = 1-\mu/x$. We may also write

$$\xi = \gamma+\delta\log\{y/(1-y)\} = \gamma+2\delta\tanh^{-1}(2y-1), \tag{6.69}$$

the latter form enabling numerical transformations to be made with the aid of tables of inverse hyperbolic tangents. The frequency function of y, say $p(y)$, is given by

$$p(y) = \frac{\delta}{\sqrt{(2\pi)}}\frac{1}{y(1-y)}\exp\left\{-\tfrac{1}{2}\left(\gamma+\delta\log\frac{y}{1-y}\right)^2\right\}, \qquad 0 \leqslant y \leqslant 1. \qquad (6.70)$$

It may be shown without difficulty that this has contact of infinitely high order at both terminals of its range.

6.33 The frequency functions may, however, be bimodal. For the modal values of y we find, on differentiating (6.70),

$$2y-1 = \delta\left(\gamma+\delta\log\frac{y}{1-y}\right).$$

Putting $y = \frac{1}{2}(y'+1)$ we find

$$y' - \gamma\delta = \delta^2\log\frac{1+y'}{1-y'}.$$

The straight line, in Cartesian co-ordinates (y', u), $u = y'-\gamma\delta$ meets

$$u = \delta^2 \log\{(1+y')/(1-y')\}$$

in one or three points. In the first case the distribution is unimodal, in the second bimodal. It may be shown (we omit the proof) that the necessary and sufficient conditions for bimodality are

$$\delta < 1/\sqrt{2}, \qquad |\gamma| < \delta^{-1}\sqrt{(1-2\delta^2)} - 2\delta\tanh^{-1}\sqrt{(1-2\delta^2)}. \qquad (6.71)$$

6.34 The moments of the distribution are very difficult to determine in a manageable form. Tables to facilitate fitting are given in Vol. II of the *Biometrika Tables* and by Johnson and Kitchen (1971a, b) and, using quantiles, by Bukač (1972). The simplest method of fitting appears to be to identify with observed quantities certain quantiles, as in the following example.

Example 6.6 (Johnson, 1949a)

Consider the data of Table 1.11 (cloudiness at Greenwich). On the assumption that degree of cloudiness stated as 0, 1, 2, . . . can be regarded as groupings $-0{\cdot}5$ to $0{\cdot}5$, $0{\cdot}5$ to $1{\cdot}5$, etc., we may fix the location parameter μ at $-0{\cdot}5$ and the scale parameter λ at 11. We then choose γ and δ so as to provide exact agreement in the two terminal classes.

Table 6.2—Frequency functions of various types fitted to the data of Table 1.11

Degree of cloudiness	Observed frequencies	S_B (1)	S_B (2)	Pearson Type I
0	320	320·0	320·0	321·7
1	129	100·9	120·9	121·5
2	74	73·9	72·0	75·1
3	68	63·8	57·5	61·4
4	45	59·8	52·1	56·0
5	45	59·9	51·6	55·2
6	55	63·4	54·9	57·8
7	65	72·0	63·9	65·5
8	90	90·0	85·5	83·2
9	148	135·4	160·7	139·6
10	676	676·0	676·0	678·0
TOTALS	1715	1715·1	1714·9	1715·0

Thus, y for the first class is 1/11 and $\log_e \{y/(1-y)\}$ becomes $-2{\cdot}302{,}585$. The normal deviate corresponding to a frequency $320/1715 = 0{\cdot}186{,}589$ is $-0{\cdot}8965$ and hence, from (6.69),

$$-0{\cdot}8965 = \gamma - 2{\cdot}3026\delta.$$

Likewise for the frequency at the other end, $y = 10/11$ and the deviate is $0{\cdot}2684$. Hence

$$0{\cdot}2684 = \gamma + 2{\cdot}3026\delta.$$

These two equations yield

$$\gamma = 0{\cdot}3110, \quad \delta = 0{\cdot}25166.$$

Table 6.2 shows the results of fitting this distribution, referred to as S_B (1). The distribution S_B (2) was fitted in the same way except that it was assumed that the groups were 0 to 0·5, 0·5 to 1·5, etc., and μ was taken as 0, λ as 10. The results of fitting a Pearson Type I distribution are also shown. All three distributions fail to give sufficiently small frequencies in the trough in the centre of the range.

6.35 The S_U family of distributions are defined by

$$\xi = \gamma + \delta \sinh^{-1} y \tag{6.72}$$

and for the frequency function we have

$$p(y) = \frac{\delta}{\sqrt{(2\pi)}} \frac{1}{\sqrt{(y^2+1)}} \exp\left[-\tfrac{1}{2}[\gamma + \delta \log \{y + \sqrt{(y^2+1)}\}]^2\right] \quad -\infty \leqslant y \leqslant \infty. \tag{6.73}$$

Again there is high-order contact at the ends of the distribution. It may be verified without much difficulty that the distributions are unimodal and that the mode lies between the median and zero. Thus the distribution is positively or negatively skewed according as γ is negative or positive.

For the moments we have

$$\mu_r' = \frac{1}{\sqrt{(2\pi)}} \int_{-\infty}^{\infty} \exp(-\tfrac{1}{2}\xi^2) . \frac{1}{2^r} \left\{ \exp\left(\frac{\xi-\gamma}{\delta}\right) - \exp\left(-\frac{\xi-\gamma}{\delta}\right) \right\}^r d\xi,$$

leading to

$$\left.\begin{aligned}
\mu_1' &= -\omega \sinh \Omega \\
\mu_2 &= \tfrac{1}{2}(\omega^2-1)(\omega^2 \cosh 2\Omega + 1) \\
\mu_3 &= -\tfrac{1}{4}\omega^2(\omega^2-1)^2 \{\omega^2(\omega^2+2)\sinh 3\Omega + 3 \sinh \Omega\} \\
\mu_4 &= \tfrac{1}{8}(\omega^2-1)^2 \{\omega^4(\omega^8+2\omega^6+3\omega^4-3)\cosh 4\Omega \\
&\qquad + 4\omega^4(\omega^2+2)\cosh 2\Omega + 3(2\omega^2+1)\}
\end{aligned}\right\} \tag{6.74}$$

where $\omega = \exp(1/2\delta^2)$, $\Omega = \gamma/\delta$. The method of fitting is illustrated in the following example. Tables to facilitate fitting are given by Johnson (1965) and reproduced in the *Biometrika Tables*, Vol. II.

Example 6.7

Consider once more the distribution of beans by length which has served in previous examples. For the observed distribution we have, in working units of half a millimetre,

$$\text{mean} = 14{\cdot}405 \text{ mm.}, \quad \beta_1 = 0{\cdot}829,$$
$$\mu_2 = (0{\cdot}9036)^2, \quad \beta_2 = 4{\cdot}863.$$

We could substitute in equations (6.74) and solve for ω and Ω. This is a tedious process and Johnson (1949a) constructed an abac for the purpose. He gives the values $\delta = 2{\cdot}64$, $\gamma = 2{\cdot}38$. With these values we find, using the relations

$$\text{mean } \xi = \text{E}\{(x-\mu)/\lambda\} = 1{\cdot}1029, \quad \text{var } (x/\lambda) = (0{\cdot}5948)^2,$$

the values of μ and λ as

$$\lambda = 1{\cdot}5192, \text{ mean } \xi = 14{\cdot}399 + 1{\cdot}1029\lambda = 16{\cdot}0745.$$

Column (8) of Table 6.1 on page 160 shows the frequencies given by an S_U distribution with these values of the constants. The fit seems about as good as the Pearson Type IV distribution.

6.36 The fitting of distributions to observational data has a certain intrinsic interest which is apt to outrun its statistical usefulness. In order to keep this chapter within bounds we shall refer only briefly to a number of further topics which have been discussed in the literature, leaving the student who wishes to specialize his knowledge to follow the references.

Burr's distributions

6.37 The three systems we have considered in this chapter all attempt to fit curves to the *frequency* function; and we have seen that in some cases, at least, this involves troublesome summation or integration when we require the *distribution* function, as we usually do, if only to calculate theoretical frequencies for comparison with observation. The question naturally arises whether we could not directly fit to the distribution function and obtain the frequency function, if we need it, by the relatively simple process of differentiation.

Such an approach has been considered by Burr (1942) and Hatke (1949). Using a generalization of the Pearson equation (6.1), we consider

$$dF = F(1-F)g(x)\,dx, \tag{6.75}$$

where $g(x)$ is some convenient function, which must be non-negative in $0 \leqslant F \leqslant 1$ and the range of x. (6.75) may be written

$$dF\left(\frac{1}{F}+\frac{1}{1-F}\right) = g(x)\,dx,$$

and integration gives the solution

$$F(x) = [1+\exp\{-G(x)\}]^{-1}, \tag{6.76}$$

where

$$G(x) = \int_{-\infty}^{x} g(t)\,dt.$$

For example, one convenient form is

$$F(x) = 1-\frac{1}{(1+x^{\alpha})^{\beta}}, \quad 0 \leqslant x \leqslant \infty; \quad \alpha, \beta > 0, \tag{6.77}$$

and many others can be chosen. Some are given in Exercise 6.19. Burr and Cislak (1968) study the shape characteristics of (6.77).

6.38 Burr's own method of fitting was to consider *cumulative moments* typified by

$$M_j(a) = \int_a^{\infty} (x-a)^j \{1-F(x)\}\,dx - \int_{-\infty}^{a} (x-a)^j F(x)\,dx \tag{6.78}$$

and to equate these quantities, calculated for the data, to the theoretical values expressed in terms of the parameters of the distribution ; in fact, to do with cumulative moments what Pearson did with ordinary moments. This can be tedious, although tables were provided by Burr and Hatke for the purpose. The simpler method of fitting in the manner of Example 6.6, by equating frequencies in certain ranges, does not seem to have been tried.

Fitting by frequency moments

6.39 Fitting frequency distributions by the method of moments has certain disadvantages, especially when the range is unlimited and observed data are only a sample from the possible observations ; outlying members and " tail " frequencies exert an undue effect on the calculated moments. It is possible to fit by identifying moments taken about the x axis, that is to say, by equating observed and theoretical values of the so-called *frequency moments* or *probability moments* typified by

$$\omega_r = \int_{-\infty}^{\infty} \{f(x)\}^r\,dx. \tag{6.79}$$

This method has been examined at length by Sichel (1949), especially in one of the cases where the Pearson distributions are somewhat inadequate. The point concerning efficiency in estimation, made in **6.13** for Pearson distributions, applies here also. From the standpoint adopted in the present chapter, the appropriate method would be to equate the observed and theoretical values of (6.79). However, grouping corrections are required and the method has not been much tried.

Moriguti (1952) has shown that for any continuous distribution with variance σ^2

$$\sigma^{r-1}\,\omega_r \geqslant \frac{2r}{3r-1}\left[\left(\frac{r-1}{3r-1}\right)^{\frac{1}{2}} \Big/ B\left(\frac{r}{r-1}, \frac{1}{2}\right)\right]^{r-1}. \tag{6.80}$$

Truncated forms

6.40 We have referred briefly to the fact that some observed distributions may be fitted by mathematical expressions which extend over non-observed values of the variate ; or, to put the matter another way, in which the data may be fitted to a truncated theoretical distribution. Fitting in such cases may be very troublesome, because in general the point of truncation has to be estimated and the moments will depend upon it. Estimation for truncated data will be discussed in Chapter 32 (Vol. 2)—cf. also Exercise 17.13 in Vol. 2.

Bivariate families of distributions

6.41 No completely satisfactory method has yet been found of setting up families of bivariate frequency functions in extension of the families of distributions that we have considered in this chapter. Some early efforts are referred to in Pretorius (1930) and Johnson (1949b). A bivariate Gram–Charlier series may also be developed—cf. Exercises 6.16 and 6.17. Mardia (1970a) reviews the various approaches.

6.42 More recently, methods based on the d.f. have been used. Exercise 1.22 above gives a simple construction due to Gumbel (1960). Plackett (1965) proposes what seems to be the most satisfactory solution available. If $F(x)$, $G(y)$ are the desired marginal d.f.'s, $H(x,y)$ the desired bivariate d.f., and $\psi(\geqslant 0)$ a parameter expressing the dependence between the variates, then $H(x, y)$ is the root of the quadratic in $H(x, y)$

$$(F-H)(G-H)\psi = H(1-F-G+H) \tag{6.81}$$

that satisfies the necessary condition (due to Fréchet (1951))

$$\max\{F(x)+G(y)-1,\ 0\} \leqslant H(x, y) \leqslant \min\{F(x),\ G(y)\}. \tag{6.82}$$

Mardia (1967) shows that if $S = 1-(F+G)(1-\psi)$, the required root is

$$H(x, y) = \begin{cases} [\{S^2+4\psi(1-\psi)FG\}^{\frac{1}{2}}-S]/\{2(1-\psi)\}, & \psi \neq 1 \\ FG, \quad \psi = 1. \end{cases} \tag{6.83}$$

Given F and G, ψ is a monotone increasing function of $H(x, y)$, being equal to zero when H attains its lower bound in (6.82), and to ∞ when H attains its upper bound there. (6.83) or (6.81) shows that $\psi = 1$ corresponds to independence.

Unfortunately, (6.83) does not produce a bivariate normal distribution (Example 3.17, or (16.43) below) when F and G are univariate normal, although H is then fairly close to bivariate normality.

Exercises 6.23–4 give further interesting properties of $H(x, y)$. Mardia (1970b) gives methods of fitting $H(x, y)$ to observed distributions.

EXERCISES

6.1 Show that the following are members of the Pearson system of distributions and sketch them for a few values of the constants:

(a) the normal distribution: $B_1 = B_2 = 0$; $\beta_1 = 0, \beta_2 = 3$;

$$f = \frac{1}{\sigma\sqrt{(2\pi)}}\exp\left(-\frac{x^2}{2\sigma^2}\right), \qquad -\infty \leqslant x \leqslant \infty.$$

(b) the Type II distribution: $B_2 > 0, B_1 = 0, B_0 < 0$; $\beta_1 = 0, \beta_2 < 3$;

$$f = \frac{1}{aB(\frac{1}{2}, m+1)}\left(1-\frac{x^2}{a^2}\right)^m \qquad -a \leqslant x \leqslant a.$$

(c) the Type VII distribution (Student's): $B_2 > 0, B_1 = 0, B_0 > 0$; $\beta_1 = 0, \beta_2 > 3$;

$$f = \frac{1}{aB(\frac{1}{2}, m-\frac{1}{2})}\left(1+\frac{x^2}{a^2}\right)^{-m}, \qquad -\infty \leqslant x \leqslant \infty.$$

6.2 With a simple transformation of the variate when necessary, assign the following distributions to one of Pearson's types :—

$$dF = ke^{-\frac{1}{2}\chi^2}\chi^{\nu-1}\,d\chi, \qquad dF = k(1-r^2)^{\frac{1}{2}(n-4)}\,dr,$$

$$dF = \frac{k\,dt}{(1+t^2/\nu)^{\frac{1}{2}(\nu+1)}}, \qquad dF = k\eta^{p-2}(1-\eta^2)^{\frac{1}{2}(N-p-2)}\,d\eta.$$

(All these distributions are important in the theory of sampling.)

6.3 Verify that for the Pearson distributions defined by (6.7), the range is unlimited in both directions if $B_0+B_1X+B_2X^2$ has no real roots; limited in one direction if the roots are real and of the same sign; and limited in both directions if the roots are real and of opposite sign.

6.4 Show that the Pearson Type VI distribution of **6.7** may be written

$$y = y_0\left(1-\frac{x^2}{a^2}\right)^{-m}\exp\{-\nu\tanh^{-1}(x/a)\}$$

and discuss the relationship with the Type IV distribution of **6.8.**

6.5 Show that for a Pearson Type III distribution of **6.9**, $2\beta_2-3\beta_1-6 = 0$ and hence that the criterion κ of (6.10) is infinite.

6.6 In **6.7**, if the real roots of $B_0+B_1X+B_2X^2$ are equal, show that the criterion $K = \kappa = 1$ and that the corresponding Pearson distribution (Type V) may be written

$$f = \frac{\gamma^{p-1}}{\Gamma(p-1)}x^{-p}e^{-\gamma/x}, \qquad 0 \leqslant x \leqslant \infty.$$

6.7 Show that for a Pearson distribution, the characteristic function obeys the relation

$$b_2\theta\frac{d^2\phi}{d\theta^2}+(1+2b_2+b_1\theta)\frac{d\phi}{d\theta}+(a+b_1+b_0\theta)\phi = 0,$$

where $\theta = it$. Deduce the recurrence relation (6.3) between the moments.

Show also that the cumulant-generating function obeys the relation

$$b_2\theta\left\{\frac{d^2\psi}{d\theta^2}+\left(\frac{d\psi}{d\theta}\right)^2\right\}+(1+2b_2+b_1\theta)\frac{d\psi}{d\theta}+(a+b_1+b_0\theta) = 0.$$

Hence show that the cumulants obey the recurrence relations

$$\{1+(r+2)b_2\}\kappa_{r+1}+rb_1\kappa_r+rb_2\left\{\kappa_1\kappa_r+\binom{r-1}{1}\kappa_2\kappa_{r-1}+\binom{r-1}{2}\kappa_3\kappa_{r-2}+\dots\right.$$
$$\left.+\binom{r-1}{j}\kappa_{j+1}\kappa_{r-j}+\dots+\binom{r-1}{1}\kappa_{r-1}\kappa_2+\kappa_r\kappa_1\right\}=0.$$

6.8 *The Charlier Type B series.* Defining

$$\gamma(\lambda,x)=e^{-\lambda}\lambda^x/x!,\qquad \nabla\gamma(\lambda,x-1)=\gamma(\lambda,x)-\gamma(\lambda,x-1),$$

and

$$G_r(\lambda,x)=\left\{\frac{d^r}{d\lambda^r}\gamma(\lambda,x)\right\}\Big/\gamma(\lambda,x),$$

show that

$$G_r(\lambda,x)=\{(-1)^r\nabla^r\gamma(\lambda,x-r)\}/\gamma(\lambda,x)$$

and

$$\sum_{x=0}^{\infty}(G_rG_s\gamma)=0,\qquad r\neq s,$$
$$=r!/\lambda^r,\quad r=s.$$

Hence, if f can be expressed as a sum

$$f=\sum_j(b_jG_j\gamma),$$

show that

$$b_j=\frac{\lambda^j}{j!}\sum_{x=0}^{\infty}(fG_j).$$

6.9 Show that no distribution which is not completely determined by its moments can be expanded in a convergent Type A series.

6.10 Show that if y is a function of x which it is desired to represent approximately by the form

$$y=\sum_{j=0}^{r}c_jH_j(x)\alpha(x),$$

then the values of the c's appropriate to the expansion of y in this form are such as to minimize the sum

$$\int_{-\infty}^{\infty}\left\{\frac{y-\sum_{j=0}^{r}c_jH_j(x)\alpha(x)}{\sqrt{\alpha(x)}}\right\}^2dx.$$

6.11 If (6.42) with $\kappa_r=0$, $r>4$, adequately represents a distribution that is assumed standardized normal, show that the error in the d.f. resulting from the normality assumption is

$$e(x)=\int_{-\infty}^{x}\{f(u)-\alpha(u)\}du=-\alpha(x)\left\{\frac{\kappa_3}{6}H_2(x)+\frac{\kappa_4}{24}H_3(x)+\frac{\kappa_3^2}{72}H_5(x)\right\},$$

and that $e(x)$ has local maxima in its absolute magnitude at those roots in x of

$$\frac{\kappa_3}{6}H_3(x)+\frac{\kappa_4}{24}H_4(x)+\frac{\kappa_3^2}{72}H_6(x)=0$$

where

$$\frac{\kappa_3}{6}H_4(x)+\frac{\kappa_4}{24}H_5(x)+\frac{\kappa_3^2}{72}H_7(x)$$

has the same sign as $e(x)$. The largest such local maximum is a bound for the error in the d.f.

6.12 If $f(x)$ and $g(x)$ are two differentiable frequency functions with cumulants denoted respectively by κ and κ' show that formally

$$f(x) = \exp\left\{\sum_{j=0}^{\infty} \frac{\kappa_j - \kappa_j'}{j!}\left(-\frac{d}{dx}\right)^j\right\} g(x).$$

6.13 If y is distributed in the Gamma form

$$dF = \frac{1}{\Gamma(\lambda)} y^{\lambda-1} e^{-y}\, dy, \qquad 0 \leqslant y \leqslant \infty,$$

and a new variate ξ is defined by

$$\xi = \gamma + \delta \log y,$$

show that for the cumulants of ξ we have

$$\kappa_1(\xi) = \gamma + \delta \frac{d}{d\lambda} \log \Gamma(\lambda),$$

$$\kappa_r(\xi) = \delta^r \frac{d^r}{d\lambda^r} \log \Gamma(\lambda), \qquad r > 1.$$

Hence show that

$$\beta_1(\xi) = \frac{1}{\lambda} + O(\lambda^{-2}),$$

$$\beta_2(\xi) = 3 + \frac{2}{\lambda} + O(\lambda^{-2}).$$

Compare with the corresponding coefficients for y,

$$\beta_1(y) = \frac{4}{\lambda},$$

$$\beta_2(y) = 3 + \frac{6}{\lambda}.$$

(Johnson, 1949a)

6.14 If y is distributed in the second Beta form

$$dF = \frac{\Gamma(\tau)}{\Gamma(\tau-\nu)\Gamma(\nu)} y^{\nu-1}(1+y)^{-\tau}, \qquad 0 \leqslant y \leqslant \infty,$$

show that if $\xi = \gamma + \delta \log y$ we have, for large ν, $\tau - \nu$,

$$\beta_1(\xi) = \frac{1}{\nu} - \frac{4}{\tau} + \frac{1}{\tau - \nu}$$

$$\beta_2(\xi) = 3 + \frac{2}{\nu} - \frac{6}{\tau} + \frac{2}{\tau - \nu}.$$

Compare with

$$\beta_1(y) = \frac{4}{\nu} - \frac{4}{\tau} + \frac{16}{\tau - \nu},$$

$$\beta_2(y) = 3 + \frac{6}{\nu} - \frac{6}{\tau} + \frac{30}{\tau - \nu}.$$

(Johnson, 1949a)

6.15 If R is distributed in the symmetrical form

$$dF = \frac{\{\Gamma(\frac{1}{2}n-1)\}^2}{\Gamma(n-2)} R^{\frac{1}{2}(n-4)}(1-R)^{\frac{1}{2}(n-4)}, \qquad 0 \leqslant R \leqslant 1,$$

show that for

$$\xi = \tfrac{1}{2}\sqrt{(n-3)}\log\{R/(1-R)\}$$

$$\beta_1(\xi) = 0,$$

$$\beta_2(\xi) = 3+\frac{\tfrac{1}{2}\left[\frac{d^2}{d\lambda^2}\log\Gamma(\lambda)\right]_{\lambda=\frac{1}{2}n-1}}{\left[\log\Gamma(\lambda)\right]^2_{\lambda=\frac{1}{2}n-1}}.$$

(Johnson, 1949a)

6.16 If x, y are jointly distributed in the bivariate normal form

$$f(x,y) = \frac{1}{2\pi(1-\rho^2)^{\frac{1}{2}}}\exp\left\{-\frac{1}{2(1-\rho^2)}(x^2-2\rho xy+y^2)\right\}, \qquad -\infty\leqslant x,\ y\leqslant\infty,$$

and $D_1 = d/dx$, $D_2=d/dy$, show that the characteristic function of

$$\exp\left\{\kappa_{11}D_1D_2+\sum_{r+s\geqslant 3}(-1)^{r+s}\kappa_{rs}\frac{D_1^r\,D_2^s}{r!\ s!}\right\}\alpha(x)\,\alpha(y) \qquad \text{(A)}$$

is the exponential of $\quad -\tfrac{1}{2}(t_1^2+2\rho t_1t_2+t_2^2)+\sum\limits_{r+s\geqslant 3}\kappa_{rs}\dfrac{(it_1)^r}{r!}\dfrac{(it_2)^s}{s!}.$

Hence show that the bivariate form of the Edgeworth series may be written either as at (A) or in the form

$$\exp\left\{\sum_{r+s\geqslant 3}(-1)^{r+s}\kappa_{rs}\frac{D_1^rD_2^s}{r!\,s!}\right\}f(x,y). \qquad \text{(B)}$$

(Kendall, 1949b. Multivariate Edgeworth expansions are given by J. M. Chambers (1967).)

6.17 From (A) of the previous exercise show that formally

$$\frac{\partial f}{\partial\kappa_{rs}} = (-1)^{r+s}\frac{D_1^r\,D_2^s}{r!\ s!}f, \qquad r+s\geqslant 3, \text{ or } r = s = 1,$$

and verify this formula directly for the bivariate normal distribution for $r = s = 1$, namely

$$\frac{\partial f}{\partial\rho} = \frac{\partial^2 f}{\partial x\,\partial y}.$$

(Kendall, 1949b)

6.18 Verify that the S_U distributions of **6.35** are unimodal and show that they are positively or negatively skewed according as γ is negative or positive.

6.19 Show that the following distribution functions are admissible solutions of equation (6.76), the parameters in all cases being positive :—

$$F(x) = (1+e^{-x})^{-\alpha}, \qquad -\infty\leqslant x\leqslant\infty;$$

$$F(x) = (1+\alpha\,e^{-\tan x})^{-\alpha}, \qquad -\tfrac{1}{2}\pi\leqslant x\leqslant\tfrac{1}{2}\pi;$$

$$F(x) = \left(\frac{2}{\pi}\arctan e^{-x}\right)^{\alpha}, \qquad -\infty\leqslant x\leqslant\infty;$$

$$F(x) = \left(x-\frac{1}{2\pi}\sin 2\pi x\right)^{\alpha}, \qquad 0\leqslant x\leqslant 1.$$

(Burr, 1942)

6.20 A variate has zero mean, unit variance and the rth cumulant κ_r is of order $n^{1-\frac{1}{2}r}$, $r \geqslant 2$. By considering an Edgeworth expansion show that the mode, to order n^{-1}, is given by $-\frac{1}{2}\kappa_3$ and the median by $-\frac{1}{6}\kappa_3$. Hence derive the approximate relationship (2.13).

(Haldane, 1942)

6.21 In **6.29**, show that for any $r = 0, 1, 2, 3, \ldots$ and $k = 1, 2, 3, \ldots$,

$$\int_0^\infty y^r p(y) \sin\{2\pi k\delta(\gamma+\delta \log y)\}\,dy = 0$$

and hence that the lognormal distribution $p(y)$ is not uniquely determined by its moments.

(Heyde, 1963)

6.22 Defining $\delta = (2\beta_2 - 3\beta_1 - 6)/(\beta_2 + 3)$, show using Exercise 3.19 that $-1 < \delta < 2$.

(Cf. C. C. Craig, 1936a)

6.23 In **6.42**, show that if $F(x)$ is replaced by its complement $1-F(x)$, $H(x,y)$ at (6.83) has ψ replaced by ψ^{-1}.

(Steck, 1968)

6.24 In **6.42**, show that if the marginal distribution of x is dichotomized at any point, and that of y at any point, the frequencies

$$\begin{array}{c|c} a & b \\ \hline c & d \end{array}$$

in the resulting four quadrants satisfy $\psi = bc/(ad)$ whatever the dichotomizing points.

(Plackett, 1965)

CHAPTER 7

THE CALCULUS OF PROBABILITIES

7.1 The previous six chapters have dealt with the theory of statistical distributions from the descriptive point of view. We have seen that the distributions encountered in practice exhibit certain regular features which permit of representation by mathematical forms; that they can be characterized by certain quantities such as moments and cumulants; and that certain general theorems about distribution- and frequency-functions can be deduced. We now begin a study of a different kind, namely the inquiry whether any meaningful and objective statements can be made about a population or a system generating a population when only a sample of the possible observations is available for scrutiny. This, in broad terms, is the problem of statistical inference.

7.2 Except in trivial cases it is not possible to make inferences from sample to population with the certainty of deductive logic. Either our statements must be somewhat vaguely phrased or, if precise, they are surrounded by a band of doubt. In ordinary speech we use various words and phrases to denote the attitude of mind towards propositions of whose truth we are uncertain. We speak of something as being more or less *probable*, or more or less *likely*; we say that we have a certain degree of *confidence* in a proposition or that the *chances* are against it; we refer to it as being surrounded by or open to *doubt* or requiring *confirmation*. Many of these expressions have acquired a technical meaning in statistics which we shall indicate in due course. Fundamentally they all express aspects of the same thing, the uncertainty with which we are compelled by circumstances to regard particular propositions.

7.3 In the next chapter we shall discuss some of these ideas in relation to statistical inference. Before we do so, however, we must give a brief account of a calculus which is required in order to treat them objectively and systematically: the calculus of probabilities.

Some writers use the expression " the theory of probability " to cover all uncertain inference. Others attempt to relate it only to sets of repeatable events. The first approach takes " probability " to be an undefined idea, like the straight line of Euclidean geometry, and builds up the theory from certain axioms. The second usually seeks to define probability in terms of the relative frequency of events and thus to refer the theory to the pure mathematics of abstract sets or to the limiting properties of sequences. Our own opinion (cf. Kendall, 1949a) is that neither approach is sufficient by itself for the statistician, whatever may be its value to the mathematician. It seems, however, that every man must choose for himself and that his psychological make-up, his experience and his fields of interest all determine the kind of axiomatization which he prefers. In statistics it is a mark of immaturity to argue overmuch about the fundamentals of probability theory.

7.4 We are not, in this chapter, concerned with the theory of probability in the broad sense. That is a branch of scientific methodology. We are reviewing the calculus of probabilities, which is a branch of mathematics. The English language is to some extent responsible for a confusion between the two. In medieval times and at the Renaissance there was a distinction between *probabilitas*, which related to the intensity of degrees of belief, and the *doctrine of chances*, which calculated the number of ways in which certain classes of event could occur. In the seventeenth and eighteenth centuries the doctrine of chances became assimilated to " probability," so that nowadays the same word is used to denote both. A useful distinction has been lost. We may well admit " probability " as including chances ; but we shall continue to distinguish between the *theory of probability*, which is concerned with the use of probabilities in making statements about the external world, and the *calculus of probabilities*, which is concerned with the mathematical development of the consequences of certain axioms and postulates. It is the latter which we proceed to describe.

7.5 We shall enunciate the basic rules of the calculus without attempting to derive them from more primitive propositions. Fortunately, whatever the disagreements about fundamentals, there is little or no disagreement about the rules themselves.

In the first place it is assumed that probability is measurable on a continuous scale, so that any probability can be expressed as a real number. We shall, in fact, say that a probability *is* x, a real number. This assumption implies, among other things, that any two probabilities may be compared ; for if they are measured by the numbers x and y we may say that the probability of the first is greater than, equal to, or less than, that of the second according as $x>y$, $x=y$, or $x<y$.

7.6 The probability of a proposition q on data p is written $\mathrm{P}(q\,|\,p)$. We have then *Rule 1* :

$$\text{If } p \text{ entails } q, \quad \mathrm{P}(q\,|\,p) = 1. \tag{7.1}$$

$$\text{If } p \text{ entails not-}q, \quad \mathrm{P}(q\,|\,p) = 0. \tag{7.2}$$

This rule defines the end-points of our scale of probability. Certainty that a proposition is not true is represented by zero, certainty that it is true by unity. Any probability lies in the range 0 to 1.

7.7 *Rule 2.* If the probabilities of n mutually exclusive propositions $q_1, \ldots, q_n$ on data p are $\mathrm{P}_1, \ldots, \mathrm{P}_n$, then the probability on data p that one of them is true is $\mathrm{P}_1+\mathrm{P}_2+\ldots+\mathrm{P}_n$.

This is generally known as the *addition rule* for probabilities. In the language of the textbooks, the probability that one of n mutually exclusive events will happen is the sum of their separate probabilities.

7.8 It follows that if $q_1, \ldots, q_n$ are mutually exclusive and equally probable on data p, and if Q is a subset of m of them, and if one of the q's must hold, the probability that it is one of the subset Q is

$$P(Q\,|\,p) = m/n. \tag{7.3}$$

This proposition is the starting point of the frequency theory of probability. It is usually stated in some such form as: if, of a set of n mutually exclusive and equally probable events, m are distinguished by some characteristic A, the probability of an event bearing A is m/n. As a *definition* of probability it is open to the logical objection of being circular, for it contains the phrase "equally probable."

7.9 *Rule 3.* The probability of two propositions q and r on data p is the product of the probability of q given p and that of r given q and p. Symbolically,

$$P(q, r \mid p) = P(q \mid p) P(r \mid q, p). \tag{7.4}$$

Since q and r appear symmetrically we also have

$$P(q, r \mid p) = P(r \mid p) P(q \mid r, p). \tag{7.5}$$

From the frequency standpoint this rule is almost self-evident. If of a set n, (a) bear the characteristic A, (b) the characteristic B, and (ab) both characteristics, then the rule states that

$$\frac{(ab)}{n} = \frac{(a)}{n}\frac{(ab)}{(a)} = \frac{(b)}{n}\frac{(ab)}{(b)},$$

a simple arithmetical proposition.

More generally we have

$$P(q_1, q_2, \ldots, q_k \mid p) = P(q_1 \mid p) P(q_2 \mid q_1, p) P(q_3 \mid q_2, q_1, p) \ldots P(q_k \mid q_{k-1}, \ldots, q_1, p), \tag{7.6}$$

a result which follows from the repeated application of Rule 3, called the ***multiplication rule*** for probabilities. Provided that $P(q \mid p) \neq 0$, (7.4) gives

$$P(r \mid q, p) = P(q, r \mid p)/P(q \mid p),$$

which is called the conditional probability of r given q, on data p.

If, as a particular case,

$$P(q, r \mid p) = P(q \mid p) P(r \mid p) \tag{7.7}$$

we have, in virtue of (7.4),

$$P(r \mid p) = P(r \mid q, p) \tag{7.8}$$

and q is then said to be independent of r, given p. A knowledge of q does not affect the probability of r on data p.

The terms "conditional" and "independent" here are fundamentally the same as those defined for frequency distributions in **1.32–4.**

7.10 The above three rules and various elaborations of them form the basis of the direct calculus of probability; everything further is pure mathematics. We assume that the reader has some acquaintance with the elements of the subject and give a few examples mainly for the purpose of emphasizing the importance of keeping strictly to the rules.

Example 7.1 Importance of specifying the basic reference-set of probabilities

Three pennies are tossed. What is the probability that they fall either all heads or all tails?

We assume that the probability of a head with any penny is $\frac{1}{2}$ and that the result with one penny is independent of that with the others. Then there are eight possible and equiprobable cases, *HHH*, *HHT*, *HTH*, *HTT*, *THH*, *THT*, *TTH*, *TTT*. Two of these give us all heads or all tails and hence the required probability is $\frac{1}{4}$.

Now consider this argument: there are two possibilities, either the three coins all fall alike or two of them are alike and the other different. Of these two possibilities one is of the type required and therefore the probability is $\frac{1}{2}$.

Consider also this argument: there are four possibilities, three heads, two heads and a tail, two tails and a head, three tails. Two of these four are of the type required and therefore the probability is $\frac{1}{2}$.

Finally, consider this argument: of the three coins, two must fall alike. The other must either be the same as these two or different. Thus there are two possibilities and again the chance is $\frac{1}{2}$.

These three arguments are fallacious. They assume equiprobability among events which are not equiprobable and the application of Rule 2 is not legitimate. For example, in the first case, it is true that there are two possibilities, but they are not equally probable under our assumptions. The reader may care to examine why this is so and how the other two arguments break down on the same point.

Example 7.2

In a hand at bridge, what is the probability that a player has at least two aces in his hand? The question as posed is not unambiguous, and there are in fact four distinct probabilities, in any of which we may be interested:

(a) The probability that a specified player (say, South) has at least two aces, irrespective of the other players' hands.

(b) The probability that some one player of the four has at least two aces, irrespective of the other three players' hands.

(c) The probability that South and no other player has at least two aces.

(d) The probability that exactly one player of the four has at least two aces.

We proceed to evaluate these in turn.

(a) South may have 0, 1, 2, 3 or 4 aces. The pack of 52 cards may be regarded as consisting of a pack of 4 aces and a pack of 48 other cards. The probability that South gets no ace is

$$p_0 = \frac{\binom{4}{0}\binom{48}{13}}{\binom{52}{13}}, \tag{7.9}$$

and the probability of his getting exactly one ace is

$$p_1 = \frac{\binom{4}{1}\binom{48}{12}}{\binom{52}{13}}.$$

The probability we require is

$$p_a = 1-(p_0+p_1)$$

$$= 1-\frac{38.37.11}{49.25.17} = 0{\cdot}257. \tag{7.10}$$

(b) It is tempting to imagine that the second of our probabilities is simply four times the first, but the four players' probabilities are not mutually exclusive. This is otherwise clear from the fact that $4p_a > 1$. In fact, the probability we now seek is simply the complement of the probability that each player has one ace, and is

$$p_b = 1-\frac{\binom{4}{1}\binom{48}{12}\cdot\binom{3}{1}\binom{36}{12}\cdot\binom{2}{1}\binom{24}{12}\cdot\binom{1}{1}\binom{12}{12}}{\binom{52}{13}\binom{39}{13}\binom{26}{13}\binom{13}{13}},$$

whence

$$p_b = 1-\frac{13^3}{49.25.17} = 0{\cdot}895. \tag{7.11}$$

(c) The third probability is less than p_a by the probabilities that (South, North), (South, East) or (South, West) each have two aces. These three events are mutually exclusive, and each has probability equal to

$$p_{2,\,2} = \frac{\binom{4}{2}\binom{48}{11}\cdot\binom{2}{2}\binom{37}{11}}{\binom{52}{13}\binom{39}{13}} = 0{\cdot}02248.$$

The probability we require is

$$p_c = p_a - 3p_{2,\,2} = 0{\cdot}190. \tag{7.12}$$

(d) The fourth probability is the sum of the probabilities of four mutually exclusive events, each with probability p_c. Thus

$$p_d = 4p_c = 0{\cdot}759. \tag{7.13}$$

Example 7.3 Inductive methods

Our first two examples enumerate the basic possibilities directly. A large class of examples for which direct enumeration may be difficult can be solved by inductive or semi-inductive methods. We will illustrate one such method on a very old problem.

n letters, to each of which corresponds an envelope, are placed in the envelopes at random. What is the probability that no letter is placed in the right envelope?

The condition that the letters are put in the envelopes " at random " is to be interpreted as meaning that every possible way of assigning the letters to envelopes is equally probable. The question, under Rule 2, then reduces to the purely algebraic one: in what proportion of the possible cases does no letter get into the right envelope?

Suppose that u_n is the number of ways in which all the letters can go wrong. Consider any particular letter. If this occupies another's envelope and vice versa, which can happen in $n-1$ ways, the number of ways in which the remaining $n-2$ letters can go wrong is u_{n-2}. But if the letter occupies another's place and not vice versa, which can happen in $n-1$ ways, there are u_{n-1} ways in which others can go wrong.

Hence we have the difference equation

$$u_n = (n-1)(u_{n-1}+u_{n-2}).$$

We may re-write this

$$u_n - nu_{n-1} = -\{u_{n-1}-(n-1)u_{n-2}\}$$

and putting

$$v_n = u_n - nu_{n-1},$$

we find

$$v_n = -v_{n-1},$$

$$= (-1)^{n-2}v_2.$$

Thus

$$u_n - nu_{n-1} = (-1)^{n-2}(u_2-2u_1).$$

But $u_1 = 0$ and $u_2 = 1$ and thus

$$u_n - nu_{n-1} = (-1)^n$$

whence

$$u_n = n!\left\{\frac{1}{2!}-\frac{1}{3!}+\ldots+\frac{(-1)^n}{n!}\right\}.$$

The total number of possible ways of arranging the letters is $n!$ and hence the probability required is

$$\frac{1}{2!}-\frac{1}{3!}+\ldots+\frac{(-1)^n}{n!}, \tag{7.14}$$

i.e. (7.14) is the first $n+1$ terms of e^{-1}, the first two cancelling each other and not appearing.

Probability in a continuum

7.11 Up to this point we have considered only probabilities of finite and discrete events; but we may also ask whether any meaning can be attached to probabilities in a continuum. For example, if a square is inscribed in a circle, what is the probability that a point taken at random in the circle is also inside the square? If a line is divided into three segments, what is the probability that they can form a triangle? What is the probability that $x < x_0$ where x is a positive real number less than y_0? And so on.

All probabilities of this kind must be considered as limits. Consider the first example, that of the square inscribed in the circle. Imagine the whole figure divided into small cells of area ε by a rectangular mesh. If we assume that the occurrence of a point in a cell is equally probable for all cells, the probability that a point falls inside both circle and square is the ratio of the number of cells in the latter to those in the former, neglecting the cells at the edges which become of diminishing importance as $\varepsilon \to 0$. In fact, the required probability can be made as near the ratio of the area of the square to that of the circle as we please by taking ε small enough. We may say that the probability *is* that ratio, which is easily seen to be $2/\pi$, an incommensurable number.

We should get the same limiting form of probability if we took other meshes which adequately represented areas; but it is most important to specify the method of proceeding to the limit in speaking of probabilities in a continuum. Otherwise the result has no meaning. The following example will illustrate the point.

Example 7.4

Consider a straight line OA bisected at B. What is the probability that a point chosen at random on the line falls into the segment OB?

Let us suppose in the first place that the line is divided into n equal segments of length OA/n. If we interpret the choosing of a point at random to mean the choice of one of these intervals, the probability is obviously $\frac{1}{2}$ as $n \to \infty$, for there will be half the intervals in the segment OB.

Now let OP be drawn perpendicular to OA and equal to it in length, and imagine a star of $n+1$ lines drawn through P, including OP and PA, so as to divide the angle OPA ($= \pi/4$) into equal angles $\pi/(4n)$. These lines cut off segments on OA, and we may, if we regard equal angles as having equal probability, assign to these segments an equal probability, for they subtend equal angles at P. If we make this convention it is evident that as $n \to \infty$ the probability of a point falling into any segment on OA is proportional to the angle subtended at P. For example, the probability that a point falls in the segment OB is $(\arctan \frac{1}{2})/(\frac{1}{4}\pi)$.

Now this is not the same answer that we got by assuming all small segments of OA equally probable. There is nothing contradictory in this—the two answers are different because the two limiting processes were different. On a little reflection it will be clear that by moving the point P on the perpendicular to OA and taking a star of lines as before we can make the probability of obtaining a point in OB have any value we like. It is thus abundantly clear that the concept of probability in a continuum depends on the limiting process by which that continuum is reached from a finite subdivision of equiprobable intervals.

The concept of the random variable

7.12 Suppose we have a statistical population, finite and discontinuous, distributed according to a variate x. If we take a member at random from this population, the probability that it bears an assigned variate-value x_0 is the frequency function $f(x_0)$, for this is the proportion of members bearing that value. Further, the probability that it bears a value less than or equal to x_0 is the distribution function $F(x_0)$, as follows at once from Rule 2 and the definition of the distribution function.

This is the essential link between probabilities and distributions. The distribution function gives the probability that a member of the population chosen at random will bear a specified value of the variate or less. We must, however, consider whether this statement can still be regarded as true for populations which are infinite or continuous.

Suppose in the first instance that the population is infinite and discontinuous. In such a case we cannot select a member at random, but we may imagine a selection from a finite population which tends to the infinite form under consideration. In this finite population the proportion of members with values less than or equal to some x_0 will be $F(x_0)$ and thus, with due regard to the nature of the limiting process, we may still say that in the infinite population the probability of a value less than or equal to x_0 is $F(x_0)$.

Similarly for a continuous distribution. In Chapter 1 we considered the continuous form as a limiting expression of

$$\Delta F = f(x)\,\Delta x.$$

If a member is chosen at random from this population, the probability that it falls in the range Δx is $f(x)\Delta x$, for the total frequency is unity and $f(x)\Delta x$ is that proportion of it falling in the range Δx. In the limit we may say that the probability of obtaining a value less than or equal to x_0 in taking a member at random from a continuous population is $\int_{-\infty}^{x_0} dF = F(x_0)$. It must, however, be remembered that the nature of the process to the limit should be specified.

Hereafter, in speaking of selecting a member at random from a population $dF = f(x)\,dx$ we shall assume that what is meant is a selection random in the limit for intervals dx.

7.13 We have spoken above of the selection of objects " at random ". In the mathematical theory of probability it is customary to define randomness in terms of probability itself. A member of a population is said to be chosen at random if it is chosen by a random method ; and a random method is one which makes it equally probable that each member of the population will be chosen. Randomness is extremely important in the theory of sampling and we shall consider it at some length in Chapter 9. At this point it is sufficient to note that when we speak of random choice we really mean a method of selection which gives to certain propositions an equal probability and hence allows us to apply the calculus of probability *a priori*. The justification for this is, in the ultimate analysis, empirical. It is found in practice that there exist selective processes which educe members of a population in such a way that the constituent events may be regarded as equiprobable ; and the theory of sampling is largely concerned with samples generated by such processes.

7.14 The idea of a variable which can appear with varying degrees of probability has been elevated into a distinct concept, that of a *random variable*. In ordinary analysis no such idea appears. We usually write "a variable" meaning that we are considering propositions about numbers which may be any of a certain range ; there is no thought that one of these values is to be considered more frequently than others or that it will occur more frequently in practice. The random variable, on the other hand, is to be regarded as defined by a distribution function. It may take any values in a given range, but the values are distinguished by an associated function.

Notwithstanding the nomenclatural dangers, we shall often in the sequel pass over the distinction between " random variable " and " variate ". When we use the expression " variable " alone, an ordinary mathematical variable is to be understood unless the context makes it clear that a random variable is meant. Where there is possible confusion we shall speak of random variable. To avoid drawing distinctions between the random variable and its variate-values, typified respectively by ξ and x, we shall often use roman letters to specify random variables and shall, where the context permits, refer to them as variates.

7.15 Let us consider what is meant by the addition of random variables. In ordinary analysis, given two variables ξ and η, we may define a third variable

$$\zeta = \xi + \eta,$$

which merely means that when $\xi = x$ and $\eta = y$, ζ will be $x+y$. If ξ and η are random variables, can we attach any useful meaning to ζ ?

If the joint distribution function of ξ and η is F_{12}, the frequency of $\xi \leqslant x$ and $\eta \leqslant y$ is $F_{12}(x,y)$. Consider some value z. We may then determine from F_{12} the frequency such that $x+y \leqslant z$ which will, in fact, be the integral

$$\iint dF_{12}(x,y)$$

taken over the region for which $x+y \leqslant z$.

This integral defines a function of z which is in fact a distribution function, for it is zero at $-\infty$, non-decreasing, and unity at $+\infty$. We may then define this as the distribution function of the random variable and say that ζ is the sum of the random variables ξ and η.

7.16 More generally, suppose that we have random variables $\xi_1, \xi_2, \ldots, \xi_n$ with corresponding variates $x_1, x_2, \ldots, x_n$. We may define a random variable ζ, with variate z, by the equation

$$z = z(x_1, x_2, \ldots, x_n). \tag{7.15}$$

The random variable ζ is such that the distribution function of z is the integral of $dF(x_1, x_2, \ldots, x_n)$ over all values of $x_1, x_2, \ldots, x_n$ such that $z(x_1, x_2, \ldots, x_n) \leqslant z$.

In particular we may define the sum of the random variables by

$$z = x_1 + x_2 + \ldots + x_n. \tag{7.16}$$

We may even write

$$\zeta = \xi_1 + \xi_2 + \ldots + \xi_n. \tag{7.17}$$

This notation is justified by the fact that the commutative and associative laws of algebra are valid for such expressions; e.g. it will be evident on examination that $\xi_1 + \xi_2 = \xi_2 + \xi_1$, and $\xi_1 + (\xi_2 + \xi_3) = (\xi_1 + \xi_2) + \xi_3 = \xi_1 + \xi_2 + \xi_3$.

7.17 The notation embodied in (7.17) is not, however, without its dangers. Some writers prefer to avoid the use of the word " sum " and to speak of the *convolution* of random variables. This might be preferable but has not come into general use in this sense among English writers.

We have not yet given any meaning to an expression such as $-\eta$. The most obvious interpretation, which we shall adopt, is that $-\eta$ is a random variable whose variate-values are the negatives of those of a variable η. Our notation has the disadvantage of seeming to imply that if a random variable ζ is the sum of two independent random variables ξ, η, and we subtract a random variable ξ from ζ, we shall necessarily be left with a random variable η. This is not true. As Example 7.5 will make clear, $\xi + \eta_1$ may have the same distribution as $\xi + \eta_2$ (where $\eta_2 \neq \eta_1$) even if the variables are independent.

7.18 We first notice one very simple but powerful result concerning the characteristic function of the sum of independent random variables. If we have n such variables distributed as $dF_1, \ldots, dF_n$, the element of frequency of their sum $\zeta = \xi_1 +$

$\xi_2+\ldots+\xi_n$ is the integral of $dF_1\ldots dF_n$ through the element of volume between z and $z+dz$. Thus the characteristic function of their sum, being the integral of e^{itz} over the range of z, is

$$\begin{aligned}\phi(t) &= \int_{-\infty}^{\infty}\cdots\int_{-\infty}^{\infty} e^{itz}\,dF_1\ldots dF_n \\ &= \int_{-\infty}^{\infty} e^{itx_1}\,dF_1\int_{-\infty}^{\infty} e^{itx_2}\,dF_2\ldots\int_{-\infty}^{\infty} e^{itx_n}\,dF_n \\ &= \phi_1(t)\,\phi_2(t)\ldots\phi_n(t). \qquad (7.18)\end{aligned}$$

Thus for independent random variables the c.f. of a sum is the product of the c.f.'s; and the cumulant-generating function of a sum is the sum of the individual cumulant-generating functions. In particular, if all the ϕ_j are the same, we have simply $\phi(t) = \{\phi_1(t)\}^n$.

Example 7.5

The essential point in this example, which was originally due to A. Khinchin, is that two independent random variables, ξ, η, may have characteristic functions which coincide in some interval of t containing the origin. If the c.f. of ξ is zero outside this interval, (7.18) ensures that the sum of two independent ξ's has the same distribution as $\xi+\eta$.

Starting from the integral

$$\int_0^{\infty}\frac{\sin\alpha x}{x}\,dx = \tfrac{1}{2}\pi, \qquad \alpha>0,$$

we have, on integration with respect to α,

$$\int_0^{\infty}\frac{\cos\beta x-\cos\alpha x}{x^2}\,dx = \tfrac{1}{2}\pi(\alpha-\beta), \qquad \alpha>\beta.$$

It is then easy to show, by the double use of identities such as

$$\sin\theta\cos m\theta = \tfrac{1}{2}\{\sin(m+1)\theta-\sin(m-1)\theta\},$$

that

$$\begin{aligned}\frac{2}{\pi}\int_0^{\infty}\frac{1-\cos x}{x^2}\cos tx\,dx &= 1-|t| \text{ for } |t|\leqslant 1 \\ &= 0 \qquad \text{for } |t|>1.\end{aligned}$$

Thus the distribution

$$dF = \frac{1}{\pi}\frac{1-\cos x}{x^2}\,dx, \qquad -\infty\leqslant x\leqslant\infty,$$

has the c.f.

$$\left.\begin{aligned}\phi_1(t) &= 1-|t|, \qquad &|t|\leqslant 1 \\ &= 0, &|t|>1\end{aligned}\right\} \qquad (7.19)$$

for the integral

$$\frac{1}{\pi}\int_{-\infty}^{\infty}\frac{1-\cos x}{x^2}\sin tx\,dx = 0$$

in virtue of the oddness of the function $\sin tx$. Moreover, since

$$\sum_{n=1}^{\infty} n^{-2} = \pi^2/6 \quad \text{and} \quad \sum_{r=1}^{\infty} \cos\{(2r-1)\theta\}/(2r-1)^2 = (\pi^2 - 2\pi\theta \operatorname{sgn}\theta), \quad -\pi \leqslant \theta \leqslant \pi,$$

we may show that the discrete distribution

$$\left.\begin{aligned} f(0) &= \tfrac{1}{2} \\ f(x) &= \frac{2}{x^2}, \quad x = n\pi, \ n \text{ a positive or negative odd integer,} \end{aligned}\right\} \tag{7.20}$$

has the c.f.

$$\phi_2(t) = \tfrac{1}{2} + \frac{4}{\pi^2} \sum_{r=1}^{\infty} \cos\{(2r-1)\pi t\}/(2r-1)^2,$$

which reduces to

$$\phi_2(t) = 1 - |t|, \qquad |t| \leqslant 1. \tag{7.21}$$

For $|t| > 1$ the c.f. consists of periodic repetitions of the values inside the range $|t| \leqslant 1$.

It follows from (7.19) and (7.21) that *for all* t, $\phi_1(t)\phi_2(t) = \phi_1(t)\phi_1(t)$. Thus the sum of the independent first and second variables is distributed in the same form as the sum of two independent variables like the first. Thus it is not true, in general, that if $\xi + \eta$ and $\xi + \eta_2$ have the same distribution, then $\eta_1 = \eta_2$.

The factorization of a c.f. into two others is unique if all three are infinitely divisible (cf. **4.33**); they cannot then take zero values, by **4.35**.

Example 7.6

The converse of the proposition of **7.18** is untrue. Although we saw in **4.16–17** that, if ξ and η are independent, their joint c.f. is the product of their individual c.f.'s *and conversely*, it is not true that if the c.f. of their sum is the product of their individual c.f.'s they are independent.

Consider the distribution

$$dF = \left\{\frac{1}{2\pi} x^{-\frac{1}{2}} y^{-\frac{1}{2}} e^{-\frac{1}{2}(x+y)} + \varepsilon(x-y)(xy-x-y+2)e^{-(x+y)}\right\} dx\,dy, \quad 0 \leqslant x, y \leqslant \infty. \tag{7.22}$$

Since $x^{\frac{1}{2}} y^{\frac{1}{2}}(x-y)(xy-x-y+2)e^{-\frac{1}{2}(x+y)}$ has a lower bound there will be some value of ε, perhaps small, for which the term in braces on the right in (7.22) is non-negative throughout the range of x and y. That is to say, (7.22) can genuinely represent a frequency function. Moreover, if we interchange x and y in the term involving ε we change its sign and hence the integral of this term over the range $0 \leqslant x, y \leqslant \infty$ is zero. The total frequency is then the integral of the first term, which is clearly $\{2^{\frac{1}{2}}\Gamma(\frac{1}{2})\}^2/(2\pi) = 1$.

The c.f. of x and y is obtained by integrating $\exp(it_1 x + it_2 y)$ over (7.22) and is found to reduce to

$$\phi(t_1, t_2) = \frac{1}{(1-2it_1)^{\frac{1}{2}}(1-2it_2)^{\frac{1}{2}}} + \frac{2\varepsilon t_1 t_2 (t_1 - t_2)}{(1-it_1)^3(1-it_2)^3}. \tag{7.23}$$

To obtain the c.f. of x we put $t_2 = 0$, obtaining

$$\phi_1(t_1) = (1-2it_1)^{-\frac{1}{2}}.$$

Similarly for the c.f. of y we put $t_1 = 0$, obtaining

$$\phi_2(t_2) = (1-2it_2)^{-\frac{1}{2}}.$$

For the c.f. of $x+y$ we put $t_1 = t_2$, obtaining $(1-2it)^{-1}$ which is the product of $\phi_1(t)$ and $\phi_2(t)$. The relationship between the c.f.'s is, in fact, the same as if ε were zero, which is due to the fact that in (7.23) the term containing ε has factors t_1, t_2 and (t_1-t_2).

7.19 The definition of a continuous d.f. $F(x)$ at (1.7) shows that the probability that the continuous variate exactly equals x is zero. Similarly for the probability that two continuous variates take exact values (x, y). Thus the conditional f.f. of one variate given the value of the other, defined at (1.29), will be the ratio of two zeros—more properly put, it will be the limit of the ratio of two quantities each tending to zero. By the argument of **7.11**, the nature of the limiting process can now affect the result obtained, as Exercise 7.22 makes clear.

This paradox is limited to f.f.'s only—the d.f. of any variate remains uniquely defined always.

Sampling distributions

7.20 We have noted that if a member of a population is chosen at random the probability that it will bear a variate-value not greater than x is the distribution function $F(x)$. Similarly, if we choose a member from a multivariate population, the probability that it will bear a value of the first variate not greater than x_1, of the second not greater than $x_2, \ldots,$ of the nth not greater than x_n, is the multivariate distribution function $G(x_1, x_2, \ldots, x_n)$. Further, if the variates are independent, the rth variate having the distribution function $F_r(x_r)$, this probability is equal to

$$F_1(x_1)F_2(x_2)\ldots F_n(x_n).$$

Now suppose that we have a selective process, which we will call sampling, applied to a univariate population in such a way that it abstracts a group of n members. If this process is repeated it will generate a multivariate distribution, each sample exhibiting n values $x_1, x_2, \ldots, x_n$. The nature of this multivariate distribution depends on the sampling process as well as the population. If the distribution is $G(x_1, x_2, \ldots, x_n)$, then this function represents the probability that a random sample will result in n values, the first not greater than x_1, the second not greater than x_2, and so on. We may regard the x's as corresponding to n random variables $\xi_1, \xi_2, \ldots, \xi_n$.

There is one type of sampling process of outstanding importance in statistical theory, namely that in which the distribution $G(x_1, x_2, \ldots, x_n)$ is the product of factors $F_1(x_1)\ F_2(x_2) \ldots F_n(x_n)$. In such a case the sampling is said to be simple. The distributions of the values $x_1, x_2, \ldots, x_n$ are independent one of another, and we may thus say that the selection of any member is independent of that of any other. Moreover, if the sampling is random, every $F_r(x)$ will be equal to $F(x)$, the distribution function of the population from which the sample has been drawn. Thus in this case we have, for the distribution of the variate-values in samples of size n obtained by a simple random method,

$$\begin{aligned} dG(x_1, x_2, \ldots, x_n) &= dF(x_1)\,dF(x_2)\ldots dF(x_n) \\ &= f(x_1)\,f(x_2)\ldots f(x_n)\,dx_1\,dx_2\ldots dx_n, \end{aligned} \qquad (7.24)$$

and $F(x_1)F(x_2)\ldots F(x_n)$ is the probability that in such a sample the first value will not exceed x_1, and so on. Moreover, since the x's appear symmetrically in (7.24) their order is not material.

7.21 If we have a sample of size n, characterized by the variates $x_1, x_2, \ldots, x_n$, we may, as in (7.15), consider the distribution of some function $z = z(x_1, x_2, \ldots, x_n)$ which might, for instance, be their mean or variance. The corresponding random variable ζ will have a distribution which is known as the sampling distribution of ζ (or of z). The probability that a value of ζ less than or equal to z be obtained on random sampling is the distribution function of z. In the case of simple random sampling it is given by integrating the element of (7.24) over the range of x's for which $z(x_1, x_2, \ldots, x_n) \leqslant z$.

Example 7.7

Suppose we draw a simple random sample of size two from the normal population

$$dF = \frac{1}{\sigma\sqrt{(2\pi)}} e^{-x^2/2\sigma^2}, \qquad -\infty \leqslant x \leqslant \infty.$$

In virtue of (7.24) the distribution of the two values, say x_1 and x_2, is given by

$$dF = \frac{1}{2\pi\sigma^2} \exp\{-(x_1^2+x_2^2)/2\sigma^2\}\, dx_1\, dx_2. \tag{7.25}$$

We will integrate this over values $\frac{1}{2}(x_1+x_2) \leqslant z$. This will give us the distribution of the mean of the two values.

Put $\quad u = \frac{1}{2}(x_1+x_2), \qquad v = \frac{1}{2}(x_1-x_2).$

The distribution becomes

$$dF = \frac{1}{\pi\sigma^2} \exp\{-(u^2+v^2)/\sigma^2\}\, du\, dv. \tag{7.26}$$

It so happens that u and v are, in this case, independent. Integrating out for v we find for u the distribution

$$dF = \frac{1}{\sigma\sqrt{\pi}} \exp(-u^2/\sigma^2)\, du. \tag{7.27}$$

Thus the probability that $\frac{1}{2}(x_1+x_2)$ is not greater than some z is simply

$$\frac{1}{\sigma\sqrt{\pi}} \int_{-\infty}^{z} e^{-u^2/\sigma^2}\, du, \tag{7.28}$$

a result which we may express by saying that u is distributed normally with variance $\frac{1}{2}\sigma^2$.

7.22 In the remainder of this chapter we shall be concerned with some statistically important theorems concerning the limit of the sum of n random variables as n tends to infinity. We first of all prove a simple but effective result due to Chebyshev, previously stated at (3.97).

Let x be a random variable and $g(x)$ a non-negative function of it. Then for every $k>0$,

$$\mathrm{P}\{g(x) \geqslant k\} \leqslant \mathrm{E}\{g(x)\}/k. \tag{7.29}$$

In fact, if R is the region for which $g(x) \geqslant k$ we have

$$\mathrm{E}\{g(x)\} = \int_{-\infty}^{\infty} g(x)\, dF \geqslant \int_R g(x)\, dF \geqslant k \int_R dF \geqslant k\mathrm{P}\{g(x) \geqslant k\}.$$

In the particular case when $g(x) = \{x - \mathrm{E}(x)\}^2$ this reduces to the Bienaymé–Chebyshev inequality (3.95).

The Weak Law of Large Numbers

7.23 Let $\xi_1, \xi_2, \ldots$ be a set of independent random variables distributed in the same form with mean μ. Let $\bar{\xi}_n$ be the mean of the first n,

$$\bar{\xi}_n = \frac{1}{n}\sum_{i=1}^{n}\xi_i.$$

Then $\bar{\xi}_n$ also has mean μ, whatever the value of n. The (Weak) Law of Large Numbers states that, in effect, $\bar{\xi}_n$ becomes more and more narrowly dispersed about μ as n increases. More precisely, given any positive ε,

$$\lim_{n\to\infty} P\{|\bar{\xi}_n-\mu|>\varepsilon\} = 0. \tag{7.30}$$

If we may assume that the variance of any ξ exists, and is equal to σ^2, say, the proof is very easy. For $\bar{\xi}_n$ is a random variable with variance σ^2/n and, in virtue of the Bienaymé–Chebyshev inequality (3.95),

$$P\{|\bar{\xi}_n-\mu|>\varepsilon\}\leqslant \sigma^2/n\varepsilon^2.$$

For given ε the probability can therefore be made as small as we please by increasing n.

7.24 The theorem remains true even if we discard the requirement that σ^2 exists. The foregoing proof is evidently easily adaptable to the case when $E(|\xi|^{\alpha})$ exists for $\alpha>1$ but for the case $\alpha = 1$ a different type of proof is required. (Cf. Exercise 7.20.) It is also easily adapted to the case where the ξ's have different means.

The Strong Law of Large Numbers

7.25 The Weak Law of Large Numbers states a limiting property of sums of random variables. The so-called "strong" law states something about the behaviour of the sequence $\Sigma_n \equiv \sum_{i=1}^{n}\xi_i$ for all values of n; something, as it were, about its properties on the way to the limit.

In fact, given any positive numbers ε and δ there is an N such that for every $M>0$

$$P\{|\bar{\xi}_n-\mu|\geqslant\varepsilon\}\leqslant\delta, \quad n = N, N+1,\ldots, N+M. \tag{7.31}$$

The weak law states that $|\bar{\xi}_n-\mu|$ is ultimately small but not that every value is small; it might be that for some n it was large, although such cases could only occur infrequently. The strong law says that the probability of such an event is extremely small. The law is true for independent variables which are identically distributed under the sole condition that μ exists; in other cases further conditions must in general be added.

The Central Limit theorem

7.26 A much more precise theorem is available in cases where the variances exist. In fact, the mean $\bar{\xi}_n$ tends to be distributed normally about μ with variance σ^2/n. More precisely, if $\alpha(x)$ represents the normal frequency function then, for every t_1, t_2,

$$\lim P\left\{t_1\leqslant \frac{\bar{\xi}_n-\mu}{\sigma/\sqrt{n}}\leqslant t_2\right\} = \int_{t_1}^{t_2}\alpha(t)\,dt. \tag{7.32}$$

This theorem is also true if the ξ's have different distributions, μ then being replaced by the mean of individual μ's and $\sigma/\sqrt{n}$ being replaced by the variance of $\bar{\xi}_n$.

Let us suppose that each ξ has a finite absolute third moment. Then as in **3.15** we have, for the c.f. of the rth variate (with first and second moments, say, of μ'_{1r} and μ'_{2r}),

$$\phi_r(t) = 1+\mu'_{1r}(it)+\tfrac{1}{2}\mu'_{2r}(it)^2+R, \tag{7.33}$$

where R is not greater than $\frac{1}{6}\nu'_{3r}|t|^3$. Likewise, for the c.g.f. we have

$$\psi_r(t) = \mu'_{1r}(it) + \tfrac{1}{2}\mu_{2r}(it)^2 + S,$$

where S is some constant times $|t|^3$, say $k|t|^3$. Thus, for the c.f. of all the n variables when summed we have

$$\Psi(t) = \sum_{r=1}^{n} \mu'_{1r}(it) + \tfrac{1}{2}\sum_{r=1}^{n} \mu_{2r}(it)^2 + \Sigma S.$$

Let us now take the mean of the sum as origin and transfer to standard measure by putting $u = \Sigma\,\xi/(\Sigma\,\mu_{2r})^{\frac{1}{2}}$. This gives us for u

$$\psi(t) = -\tfrac{1}{2}t^2 + \Sigma\,S/(\Sigma\,\mu_{2r})^{3/2}.$$

If each ν_{3r} is finite the numerator of the remainder term is bounded by $nK|t|^3$, where K is the largest value of k. The denominator is not less than $n^{3/2}M_{2r}$, where M_{2r} is the smallest variance. Thus the remainder term is of order $n^{-\frac{1}{2}}$ and tends to zero. Thus

$$\psi(t) \to -\tfrac{1}{2}t^2$$

and consequently the distribution to which it relates tends to normality.

7.27 The theorem may be proved under conditions which do not require the existence of third moments. In fact, it is a necessary and sufficient condition that

$$\lim_{n\to\infty} \frac{1}{M_n} \sum_{r=1}^{n} \int_{|x|>\varepsilon\sqrt{M_n}} x^2\,dF_r = 0, \tag{7.34}$$

where $M_n = \Sigma\mu_{2r}$.

This condition, due to Lindeberg and Cramér, implies that the total variance M_n tends to infinity and that every μ_{2r}/M_n tends to zero; in fact that no random variable dominates the others. The theorem may fail to hold for variables which do not possess a second moment; for example, we shall see in Example 11.1 that the mean of n variates, each distributed in the Cauchy form

$$dF = \frac{1}{\pi}\frac{dx}{1+x^2}, \qquad -\infty \leqslant x \leqslant \infty, \tag{7.35}$$

is distributed in precisely the same form. This is easily deduced from the c.f., given in Example 4.2. Exercise 11.25 will give an extreme case where the sample mean has a distribution whose dispersion *increases* with n.

7.28 The Central Limit theorem occupies an important place in statistical theory. We have already met instances of its operation in the tendency of the binomial and the Poisson distributions to normality. We shall meet many more in the sequel. Central Limit theorems are also obtainable for sequences of random variables which are not independent. A general account of these limit theorems will be found in the book by Feller (1950).

7.29 The foregoing methods can also be extended to sets of multivariate random variables. For example, if ξ, η have a joint distribution, a set of n individuals drawn at random from that distribution under simple conditions yields a set of n pairs of variate-values $(x_1, y_1), (x_2, y_2) \ldots, (x_n, y_n)$. We might then enquire after the sampling distribution of some quantity such as their joint product-moment $z = \sum_{i=1}^{n} x_i y_i/n$. This would be obtained by integrating

$$dF(x_1, y_1)\, dF(x_2, y_2) \ldots dF(x_n, y_n)$$

over a region such that $\Sigma x_i y_i / n \leqslant z$. Generalizations to more than two random variables are immediate.

7.30 It is also worth recording that Central Limit theorems can be proved for multivariate random variables. In particular, corresponding to the theorem of **7.26**, the means of n x's and y's tend to joint bivariate normality as n increases under similar very general conditions.

EXERCISES

7.1 Let $E_1, E_2, \ldots E_n$ be n compatible events, i.e. such that any number of them from 0 to n may be simultaneously realized. Let p_i be the probability that E_i is realized (whether the others are realized or not), p_{ij} the probability that E_i and E_j are realized (whether the others are realized or not); and so on. Show that the probability that no event is realized is

$$1-\Sigma p_i+\Sigma p_{ij}-\Sigma p_{ijk}+\ldots+(-1)^n p_{12\ldots n},$$

where summations take place over all possible different values of the suffixes.

7.2 Show that the probability that, of n compatible events, exactly r unspecified events are realized is

$$\Sigma p_{(r)}-\binom{r+1}{1}\Sigma p_{(r+1)}+\binom{r+2}{2}\Sigma p_{(r+2)}+\ldots+(-1)^{n-r}\binom{n}{n-r}p_{(n)},$$

where $p_{(j)}$ denotes a p with j suffixes.

7.3 In the previous exercise show that the probability that at least r unspecified events are realized is

$$\Sigma p_{(r)}-\binom{r}{1}\Sigma p_{(r+1)}+\binom{r+1}{2}\Sigma p_{(r+2)}-\ldots+(-1)^{n-r+1}\binom{n-1}{n-r}p_{(n)}.$$

7.4 Use Exercise 7.2 to solve the envelope problem of Example 7.3. Show also that the probability that exactly r unspecified letters get into the right envelopes is

$$\frac{1}{n!}\left\{\binom{n}{r}(n-r)!-\binom{r+1}{1}\binom{n}{r+1}(n-r-1)!+\text{ etc.}\right\}=\frac{1}{r!}\left\{1-1+\frac{1}{2!}-\ldots+\frac{(-1)^{n-r}}{(n-r)!}\right\}.$$

7.5 Given n independent but compatible events $E_1 \ldots E_n$, and that the probability that E_i alone happens is α_i, show that the probability that E_i happens, whether the others happen or not, p_i, is given by

$$p_i = \alpha_i/(\alpha_i+t),$$

where t is a root of

$$\prod_{i=1}^{n}(\alpha_i+t) = t^{n-1}.$$

Show that in general there are two values of p_i satisfying the conditions.

7.6 From a heap of counters of unknown number N a player takes a handful of n at random. Examine this argument: it is equally probable that N is odd or even. If it is odd, the probability that n is odd is greater than $\frac{1}{2}$, whereas if it is even the probability that n is odd is $\frac{1}{2}$. Thus the probability that n is odd is greater than $\frac{1}{2}$, and the player should bet on getting an odd number.

7.7 An event happens at random on an average once in time t. Regarding occurrences in equal small intervals as equiprobable, show that the probability that it does not happen in a specified interval T is $\exp(-T/t)$.

7.8 If a is equiprobably distributed in the interval $0 \leqslant a \leqslant 6$ and b equiprobably in the interval $0 \leqslant b \leqslant 9$, show that the probability that $x^2-ax+b = 0$ has two real roots is $\frac{1}{3}$.

7.9 A straight line in a plane is represented by the pedal equation

$$x\cos\theta+y\sin\theta = p.$$

A random line is drawn in such a way that elements $d\theta\, dp$ are equally probable. If it intersects a closed convex curve of length l_1, show that the probability that it will also intersect a convex curve of length l_2, lying inside the first, is l_2/l_1.

Deduce that if a random line intersects a circle, the probability that it also intersects a fixed diameter of the circle is $2/\pi$.

7.10 If, in the previous exercise, the co-ordinates are changed by a translation of the origin and a rotation about it, show that $dp'\,d\theta' = dp\,d\theta$ where the primes relate to the new co-ordinates. Hence show that the " randomness " of lines drawn in the foregoing manner is independent of the co-ordinate system.

7.11 Three points are taken at random on a circle. Show that the probability that they lie on the same semicircle is $\frac{3}{4}$. (Assume that in the limit elementary intervals of arc are equally probable.)

Explain the fallacy in the following argument: one pair of points *must* lie on a semicircle terminating at one of them. The probability that the third point lies on this semicircle is $\frac{1}{2}$, which is therefore the required answer.

Consider also this argument: it does not matter where the first point is chosen. Imagine the circle cut at that point and unrolled into a horizontal straight line. The probability that the second and third points fall into the left half of the line is $\frac{1}{2}\times\frac{1}{2} = \frac{1}{4}$; similarly for the right half; and hence the probability required is $\frac{1}{2}$.

7.12 If x and y are independent random variables with characteristic functions ϕ_1 and ϕ_2 respectively, show that the c.f. of $x-y$ is $\phi_1(t)\phi_2(-t)$. Hence show (cf. Exercise 4.1) that if $x+y$ is distributed in the same form as $x-y$, then y is symmetrically distributed about zero.

7.13 x and y are distributed in the bivariate normal form of (5.90). Find the c.f. of $x+y$ and hence show that $x+y$ is normally distributed with variance $2(1+\rho)$. Deduce that $|\rho|\leqslant 1$.

7.14 Show that the sum of a number of independent random variables, each distributed in the Poisson form with possibly different parameters, is also distributed in the Poisson form.

7.15 x is a standardized normal variate and $y = x^2$. Show that the joint c.f. of x and y is

$$\phi_{x,y}(t_1, t_2) = (1-2it_2)^{-\frac{1}{2}}.\exp\left\{-\frac{1}{2}\frac{t_1^2}{1-2it_2}\right\}$$

and deduce the c.f.'s of x and of y alone. Show that x and y are not independent and that the cumulants $\kappa_{rs} = 0$ for any $r \neq 2$ and any s.

7.16 A random sample of n values $x_1, x_2, \ldots, x_n$ is chosen from a population which has mean μ and takes only positive values. Show that the probability that $\sum_{i=1}^{n} x_i$ exceeds some constant λ is not greater than $n\mu/\lambda$.

7.17 x is a variate with mean $\mu > 0$, variance σ^2 and coefficient of variation $c = \sigma/\mu$. Show that

$$P\{x>0\}\,E(x^2)\geqslant\mu^2,$$

so that

$$P\{x>0\}\geqslant\frac{1}{1+c^2}.$$

Hence, taking the median M as origin, show that $\left|\frac{\mu-M}{\sigma}\right|\leqslant 1$ as in Exercise 3.22.

7.18 Find the distribution of the sum of k independent random variables, each distributed in the form

$$dF = e^{-x}\,dx, \qquad 0\leqslant x\leqslant\infty.$$

Graph the distribution for some values of k and observe its tendency to normality.

7.19 The frequency diagram of a distribution is defined by a series of isosceles triangles with bases two units wide and height $A/|x|$ for $x = \pm(p+1)^2, p = 1, 2, \ldots$, together with the remaining part of the x-axis. Show that if A is suitably chosen this gives a properly defined d.f. Show that the expectation of $|x|$ does not exist. Show also that if an expectation be defined as the principal value $\lim_{n\to\infty} \sum_{-n}^{n} xf(x)$, it does exist. By taking a different origin show that the principal value about some other point does not exist. Hence, by taking one random variable to be a constant, show that the expectation of the sum of two independent random variables is not necessarily equal to the sum of their expectations if the principal value be admitted as an expectation.

(Fréchet, 1937)

7.20 By expanding their c.f.'s as

$$\phi_j(t) = 1+\mu it+o(t),$$

show that the mean $\bar{\xi}_n$ of n identical independent variates ξ_j has a c.g.f. for which

$$\lim_{n\to\infty} \psi(t) = \mu it,$$

and hence that $\bar{\xi}_n$ has the unit distribution of Example 4.3 in the limit. This establishes the Weak Law of Large Numbers of **7.23**.

7.21 Using the joint c.f. of Exercise 7.15, or otherwise, show that if $x_1, x_2, \ldots, x_n$ are independent normal variates with means μ_i and variance 1, the c.f. of $\sum_{i=1}^{n} x_i^2$ is

$$\phi(t) = (1-2it)^{-\frac{1}{2}n} \exp\left(\frac{\lambda it}{1-2it}\right)$$

where $\lambda = \sum_{i=1}^{n} \mu_i^2$. (Exercise 4.18 is the special case $n=1$, $\lambda = 0$ if $\sigma^2 = 1$.)

7.22 In the transformation from (x_1, x_2) to (y_1, y_2) in **1.35**, let x_1, x_2 be non-negative, put

$$y_1 = x_1+x_2, \qquad y_2 = x_1/x_2$$

and show that

$$g_1(y_1, y_2) = f\left(\frac{y_1 y_2}{1+y_2}, \frac{y_1}{1+y_2}\right)\frac{y_1}{(1+y_2)^2}.$$

Alternatively, transform from (x_1, x_2) to y_1 and $y_3 = x_1 - kx_2$ to obtain

$$g_2(y_1, y_2) = f\left(\frac{y_3+ky_1}{1+k}, \frac{y_1-y_3}{1+k}\right)\frac{1}{1+k}.$$

Show that the conditional f.f.'s $p_1(y_1 | y_2 = k)$ and $p_2(y_1 | y_3 = 0)$ obtained from (1.29) are not identical, even though $y_2 = k$ if and only if $y_3 = 0$. If $f(x_1, x_2) = 1$, $0 \leqslant x_1, x_2 \leqslant 1$, show that

$$\left.\begin{aligned} p_1(y_1 | y_2 = k) &= 2y_1/(1+k)^2, \\ p_2(y_1 | y_3 = 0) &= 1/(1+k), \end{aligned}\right\} \qquad 0 \leqslant y_1 \leqslant 1+k.$$

(J. S. Williams, 1966)

7.23 $x_1, x_2, \ldots, x_n$ are independently and identically distributed with cumulants κ_s, and γ_1 and γ_2 defined by (3.89–90). If $l = \sum_{j=1}^{n} c_j x_j$, show from **3.13** that the cumulants of l are given by $\lambda_r = \sum_j c_j^r \kappa_r$, and hence that $\gamma_1^* = \lambda_3/\lambda_2^{3/2}$ and $\gamma_2^* = \lambda_4/\lambda_2^2$ satisfy $|\gamma_1^*| \leqslant |\gamma_1|$, $\gamma_2^* \leqslant \gamma_2$. If $c_j \equiv \frac{1}{n}$, so that l is the sample mean, show that γ_1^* and $\gamma_2^* \to 0$, as does every $\lambda_p/\lambda_2^{p/2}$ for $p > 2$. Hence verify the Central Limit Theorem in this case where all cumulants are finite.

CHAPTER 8

PROBABILITY AND STATISTICAL INFERENCE

8.1 The calculus of probabilities, as outlined in the previous chapter, leads us from the given probabilities of primary events to the probabilities of more complex events based upon them. In practice we usually require to make inferences in the reverse direction ; that is to say, given the observations, we require to know something about the parent population from which they emanated or the generating mechanism by which they were produced. In the second volume we shall begin a systematic study of the various methods and inferential processes which are employed in statistics for this purpose. At this stage we shall merely attempt an introductory account, in very broad terms, with the object of giving some point to the topics considered and discussed later in this volume.

Bayes' Theorem

8.2 Let $q_1, q_2, \ldots, q_n$ be a set of exclusive propositions and let H be the information available. Let p represent some further proposition. Then by the multiplicative rule of probabilities (our Rule 3) we have

$$\begin{aligned} \mathrm{P}(q_r, p \mid H) &= \mathrm{P}(p \mid H)\,\mathrm{P}(q_r \mid p, H) \\ &= \mathrm{P}(q_r \mid H)\,\mathrm{P}(p \mid q_r, H). \end{aligned}$$

Thus we have

$$\mathrm{P}(q_r \mid p, H) = \frac{\mathrm{P}(q_r \mid H)\,\mathrm{P}(p \mid q_r, H)}{\mathrm{P}(p \mid H)}. \tag{8.1}$$

If, as we now suppose, the q's are exhaustive so that one of them must hold, we have, on summing (8.1) over the q's,

$$1 = \sum_r \frac{\mathrm{P}(q_r \mid H)\,\mathrm{P}(p \mid q_r, H)}{\mathrm{P}(p \mid H)}. \tag{8.2}$$

On substituting for $\mathrm{P}(p \mid H)$ from (8.2) into (8.1), we then find

$$\mathrm{P}(q_r \mid p, H) = \frac{\mathrm{P}(q_r \mid H)\,\mathrm{P}(p \mid q_r, H)}{\Sigma\,\{\mathrm{P}(q_r \mid H)\,\mathrm{P}(p \mid q_r, H)\}} = \frac{\mathrm{P}(q_r, p \mid H)}{\sum_r \mathrm{P}(q_r, p \mid H)}. \tag{8.3}$$

This is known as Bayes' Theorem. It states that the probability of q_r on data p and H is proportional to the probability of q_r on H multiplied by the probability of p on q_r and H.

8.3 Now suppose that an event can be explained on the mutually exclusive hypotheses represented by $q_1, q_2, \ldots, q_n$. These possibilities have certain probabilities $\mathrm{P}(q_r \mid H)$ of being true. Each of them can give rise to p, but with differing probabilities $\mathrm{P}(p \mid q_r, H)$. The theorem gives us the probabilities of the various q's *when p is known to have happened.* The quantities $\mathrm{P}(q_r \mid H)$ are called *prior* probabilities, those of type $P(q_r | pH)$ are called *posterior* probabilities and $P(p | q_r, H)$ may be called the *likelihood*. Bayes' theorem may then be re-stated in the form: the posterior probability varies as the prior probability multiplied by the likelihood.

Bayes' Postulate

8.4 In this form the theorem is seen to be a simple logical consequence of the multiplication rule of probabilities and is not subject to criticism. What has given rise to criticism in the past has been the use to which the theorem has been put. There is an implied principle that, if we have to choose one of the q's, we take the one with the greatest posterior probability. This is equivalent to choosing the hypothesis which maximizes the *joint probability* of q and p as is seen at once from the extreme right of equation (8.3). The difficulty arises from the fact that to calculate the posterior probabilities we require to know the prior probabilities. These are, in general, unknown, and Bayes suggested that where this is so they should be assumed to be equal; or rather, that they should be assumed equal where nothing was known to the contrary. This assumption, known variously as Bayes' Postulate, the Principle of the Equidistribution of Ignorance and by one or two other names, furnishes one of the most contentious points in the theory of statistical inference. Before we discuss the point it may be useful to give one or two examples.

Example 8.1

An urn contains four balls, which are known to be either (a) all white, or (b) two white and two black. A ball is drawn at random and found to be white. What is the probability that all the balls are white?

We have here two hypotheses, q_1 and q_2. On q_1 the probability of getting a white ball is 1, on q_2 it is $\frac{1}{2}$. From (8.1) we have

$$P(q_1 | p, H) = \frac{P(q_1 | H)}{P(q_1 | H) + \frac{1}{2}P(q_2 | H)}$$

$$P(q_2 | p, H) = \frac{\frac{1}{2}P(q_2 | H)}{P(q_1 | H) + \frac{1}{2}P(q_2 | H)}.$$

Now, in accordance with Bayes' postulate we assume

$$P(q_1 | H) = P(q_2 | H) = \tfrac{1}{2}$$

and find

$$P(q_1 | p, H) = \tfrac{2}{3}$$

$$P(q_2 | p, H) = \tfrac{1}{3}.$$

If we had to choose between the two possibilities (a) and (b) we should select the one with the greater posterior probability, i.e. we proceed on the assumption that the balls are all white.

Now suppose that we replace the ball and draw another at random. If it is found to be black the hypothesis (a) is decisively rejected. But if it turns out to be white we can calculate new posterior probabilities in which our former posterior probabilities become prior. We now have $P(q_1 | H) = \frac{2}{3}$, $P(q_2 | H) = \frac{1}{3}$, where H includes p, and a renewed application of (8.1) gives us for the posterior probabilities of the new event, say p',

$$P(q_1 | p', H) = \tfrac{2}{3}/(\tfrac{2}{3}+\tfrac{1}{2}.\tfrac{1}{3}) = \tfrac{4}{5}$$
$$P(q_2 | p', H) = \tfrac{1}{2}.\tfrac{1}{3}/(\tfrac{2}{3}+\tfrac{1}{2}.\tfrac{1}{3}) = \tfrac{1}{5}.$$

It will be clear that if we repeat the process and again get a white ball the new posterior probability of (a) will be still higher. This is in agreement with the requirements of common sense ; the longer we go on sampling (with replacement) without producing a black ball, the more probable it is that there are no black balls present.

Example 8.2

We may look at the same problem from a slightly different viewpoint. Suppose we draw balls one at a time, replacing them after each drawing, and obtain n white balls in succession. The probability of this event on hypothesis (a) is unity ; that on hypothesis (b) is $1/2^n$. From (8.1) we then have, (p referring to the observation of all n balls as white),

$$P(q_1 | p, H) = \frac{\tfrac{1}{2}}{\tfrac{1}{2}+\tfrac{1}{2}.1/2^n} = 2^n/(2^n+1).$$
$$P(q_2 | p, H) = 1/(2^n+1).$$

As n becomes larger, $P(q_1 | p, H)$ tends to unity and $P(q_2 | p, H)$ to zero.

Moreover, this will be true whatever the original prior probabilities may have been. In fact, if that of hypothesis (a) is t and that of (b) is $1-t$, we find

$$P(q_1 | p, H) = \frac{2^n t}{2^n t+(1-t)}$$

which tends to unity for any non-zero t. This also agrees with common sense. Whatever the original probabilities, the new evidence is so strong as to outweigh them.

Example 8.3

From an urn full of balls of unknown colour a ball is drawn at random and replaced. The process is continued m times and a black ball is drawn each time. What is the probability that if a further ball is drawn it will be black ?

The question as framed does not admit of a definite answer, for, there being an infinite number of possible colours and combinations of colours, we do not know which are the hypotheses to be compared. Let us suppose that the balls are either black or white, and thus consider the hypotheses (1) that all are black, (2) that all but one are black, (3) that all but two are black, and so on. The problem still lacks precision, for the number of balls is not specified. Suppose there are N balls. We shall later let N tend to infinity to get the limiting case.

Consider the hypothesis q_R that there are R black balls and $N-R$ white ones. The probability of choosing a black ball is R/N and that of doing so m times in succession is $(R/N)^m$. If the q's have equal prior probabilities we have, from (8.1),

$$P(q_R | p, H) = \frac{(R/N)^m}{\sum\limits_{R=0}^{N} (R/N)^m}.$$

Now the probability of getting a further black ball on hypothesis q_R is R/N. Since the hypotheses q_R are mutually exclusive, the probability of getting a further black ball is

$$\sum_{R=0}^{N} \frac{R}{N} \mathrm{P}(q_R \mid p, H) = \frac{\Sigma (R/N)^{m+1}}{\Sigma (R/N)^m}. \tag{8.4}$$

This is the answer to the limited form of the question. As $N \to \infty$ this tends to the quotient of definite integrals

$$\int_0^1 x^{m+1}\, dx \Big/ \int_0^1 x^m\, dx = \frac{m+1}{m+2}. \tag{8.5}$$

This is a particular case of the so-called Succession Rule of Laplace. Enthusiasts have applied it indiscriminately in some such unconditioned form as the statement that if an event is observed to happen m times in succession the chances are $m+1$ to 1 that it will happen again. This is clearly unjustified.

8.5 The principal difficulties arising out of Bayes' postulate appear from the standpoint of the frequency theory of probability. If we adopt the approach in which probability is a measure of attitudes of mind, it is reasonable to take prior probabilities to be equal when nothing is known to the contrary, for the mind holds them in equal doubt. The frequency theory, however, would require the states of events corresponding to the various q's to be distributed with equal frequency in some population from which the actual q has emanated, if Bayes' postulate is to be applied. This has appeared to some statisticians, though not to all, to be asking too much of the universe. The postulate is one of the crucial points in the theory of probability. Many adherents of the subjective school accept it. Many of those of the frequency school explicitly reject it. L. J. Savage (1954, 1961) presents a modern plea for the purely subjective use of Bayes' prior distributions. There is some penetrating exposition and discussion of his views in Savage (1962).

The modern versions of the postulate sometimes assume prior distributions which are not equiprobable. This requires some special pleading, and we defer a discussion until Chapter 21, Volume 2.

There is still so much disagreement on this subject that one cannot put forward any set of viewpoints as orthodox. One thing, however, is clear—anyone who rejects Bayes' postulate must put something in its place. The problem which Bayes attempted to solve is supremely important in scientific inference and it scarcely seems possible to have any scientific thought at all without some solution, however intuitive and however empirical, to the problem. We are constantly compelled to assess the degree of credence to be accorded to hypotheses on given data; the struggle for existence, in Thiele's phrase, compels us to consult the oracles.

Maximum likelihood

8.6 Various substitutes for Bayes' postulate have been proposed. Some of these are put forward as solutions to the problem in particular classes of case; such are the principles of least squares and minimum chi-squared which we shall meet later. There is one principle, however, of general application, that of maximum likelihood.

Reverting to (8.3) we may write Bayes' theorem in the form

$$\mathrm{P}(q_r \mid p, H) \propto \mathrm{P}(q_r \mid H)\mathrm{L}(p \mid q_r, H) \tag{8.6}$$

where we now write $\mathrm{L}(p \mid q_r, H)$ for the likelihood. The Principle of Maximum Likelihood states that, when confronted with a choice of hypotheses q_r we choose that one (if any) which maximizes L. In other words, we are to choose that hypothesis which gives the greatest probability to the observed event. Whereas Bayes' theorem enjoins the maximization of the joint probability of q_r and p, maximum likelihood requires the maximization of the conditional probability of p given q_r.

8.7 It is to be particularly noted that this is not the same thing as choosing the hypothesis with the greatest probability. Some advocates of the maximum likelihood principle explicitly deny any meaning to such expressions as " the probability of a hypothesis." We shall see later that in practice the differences between results obtained from maximum likelihood and Bayes' postulate are not so large as might be expected. There is, however, an important conceptual difference involved.

There is, in fact, a shift of emphasis in the way we regard the likelihood function, reflected by our writing it with an L instead of a P. The ordinary probability function gives the probabilities of p on data q_r and H ; p varies, q_r and H are given. From the point of view of likelihood we consider various values of q_r for the observed p and given H ; q_r varies, p and H are given. It is this variation of the function for differing values of the q's that we have in mind in speaking of likelihood.

8.8 Suppose (as is nearly always the case in statistical work) that the hypotheses with which we are concerned assert something about the numerical value of a parameter θ. For instance, the hypotheses might be $q_1 \equiv \theta < 0$, $q_2 \equiv \theta \geqslant 0$, in which case there are two alternatives. Or we might have $q_1 \equiv \theta = 1$, $q_2 \equiv \theta = 2$, and so on, in which case there is a denumerable infinity of hypotheses.

If now θ can have only discontinuous values, we may, confronted with an observed event p, require to estimate θ, or to ask what is the " best " value of θ to take on the evidence p. The method of Bayes would state that the " best " value was the most probable value. In (8.3) we should seek for that q_r which made $\mathrm{P}(q_r \mid p, H)$ a maximum. If we know nothing of the prior probabilities $\mathrm{P}(q_r \mid H)$ we should, in accordance with Bayes' postulate, assume all such probabilities equal. We then merely have to find that q_r which maximizes $\mathrm{L}(p \mid q_r, H)$. In other words, the postulate of Bayes and the principle of maximum likelihood result in the same answer and are equivalent.

8.9 This position apparently does not hold if the permissible values of θ are continuous. We must now replace such expressions as

$$\mathrm{P}(q_r \mid H)$$

by

$$\mathrm{P}(\theta_0 - \tfrac{1}{2}d\theta_0 \leqslant \theta \leqslant \theta_0 + \tfrac{1}{2}d\theta_0 \mid H)$$

and in place of (8.6) we get

$$\mathrm{P}(\theta_0 - \tfrac{1}{2}d\theta_0 \leqslant \theta \leqslant \theta_0 + \tfrac{1}{2}d\theta_0 \mid p, H) \propto \mathrm{P}(\theta_0 - \tfrac{1}{2}d\theta_0 \leqslant \theta \leqslant \theta_0 + \tfrac{1}{2}d\theta_0 \mid H) \times \mathrm{L}(p \mid \theta_0 - \tfrac{1}{2}d\theta_0 \leqslant \theta \leqslant \theta_0 + \tfrac{1}{2}d\theta_0, H). \tag{8.7}$$

If we now require the " best " value of θ, we should, in accordance with Bayes' postulate, take the prior probability to be a constant and once again we should have to maximize L for variations of θ.

We might, however, have chosen to represent our hypotheses, not by θ, but by some quantity ϕ which is a function of θ, e.g. the standard deviation instead of the variance. In this case we should have reached equation (8.7) with ϕ written everywhere instead of θ; we should have taken the prior probability as constant; and we should have arrived at the conclusion that we should maximize L for variations of ϕ.

But are we being consistent in so doing? If we assume that the elementary intervals of θ are equiprobable we cannot assume the same of ϕ, and thus the use of Bayes' postulate appears to involve self-contradiction. The principle of maximum likelihood is free from this difficulty, for if $L(\theta)$ is maximized at $\hat{\theta}$, and $\phi(\theta)$ is a function of θ, $L(\phi)$ is maximized at $\hat{\phi} = \phi(\hat{\theta})$. Thus it makes no difference to the result which parametrization is used.

8.10 This is one of the grounds on which adherents of the frequency school have rejected Bayes' postulate in favour of the principle of maximum likelihood; but in our view the matter has been misunderstood. It seems that Bayes' postulate and the principle give the same answer in the continuous case as well as in the discrete case when proper regard is had to the limiting processes involved. We saw in **7.11** that in speaking of probability in a continuum it was essential to specify the nature of the process to the limit. If we regard θ (from the frequency viewpoint) as having emanated from a population by a process random in the limit for intervals $d\theta$ then Bayes' postulate applied to this process will clearly give a different answer from that obtained by supposing that θ emanated from a process random in the limit for $d\phi$. The two are different just as the probabilities in Example 7.4 are different, and for the same reason. Thus the apparent inconsistency is not an inconsistency at all, but a difficulty introduced by ignoring the limiting process in continuous populations.

It remains true, of course, that for many practical purposes we do not know how the actual value of θ arose. If we require a theory of inference which is unaffected by our ignorance on such points, the objection to Bayes' postulate remains and does not apply to the principle of maximum likelihood. On the other hand there is still the difficulty of adducing convincing reasons why we should adopt the principle of maximum likelihood as a principle of statistical inference. As we shall see in Volume 2, it has properties that give it a kind of posterior justification.

Example 8.4

Let us consider Example 8.2 from the likelihood viewpoint. The probability of n white balls on hypothesis (a) is unity; that on hypothesis (b) is 2^{-n}. The likelihood of (a) is then always greater than that of (b) and hence we choose (a). This is true for any n. In particular, as n increases, the ratio of the likelihoods tends to the ratio of posterior probabilities.

Example 8.5

Consider an independent sample of size n from the normal distribution

$$dF = \frac{1}{\sigma\sqrt{(2\pi)}}\exp\left\{-\frac{1}{2}\left(\frac{x-\mu}{\sigma}\right)^2\right\}dx.$$

If the observations are $x_1, x_2, \ldots, x_n$ the likelihood function may be written

$$L = \frac{1}{\sigma^n(2\pi)^{\frac{1}{2}n}}\exp\left\{-\frac{1}{2}\sum_{j=1}^{n}\left(\frac{x_j-\mu}{\sigma}\right)^2\right\}.$$

We consider a range of possible values of μ which could have generated these observations. To estimate μ we take that value which maximizes L, i.e. find the largest value of L satisfying

$$\frac{\partial L}{\partial \mu} = 0, \quad \frac{\partial^2 L}{\partial \mu^2} < 0. \tag{8.8}$$

Since L is positive this is, in general, the same as maximizing log L, sometimes (as here) a more convenient procedure. We then have

$$\frac{\partial \log L}{\partial \mu} = +\Sigma\left(\frac{x_j-\mu}{\sigma}\right) = 0, \tag{8.9}$$

and thus the estimator of μ, say $\hat{\mu}$, is given by

$$\Sigma x_j = \Sigma\hat{\mu} = n\hat{\mu}$$

or

$$\hat{\mu} = \bar{x}, \text{ the mean of the } x\text{'s}. \tag{8.10}$$

This is easily seen to be a unique maximum, and is therefore the maximum-likelihood solution.

Had we wished to estimate both μ and σ we should find, in addition to (8.9),

$$\frac{\partial \log L}{\partial \sigma} = -\frac{n}{\sigma}+\Sigma\frac{(x_j-\mu)^2}{\sigma^3} = 0 \tag{8.11}$$

giving

$$\hat{\sigma}^2 = \frac{1}{n}\Sigma(x-\mu)^2. \tag{8.12}$$

Whereas $\hat{\mu}$ does not depend on σ, $\hat{\sigma}$ does depend on μ. We choose those estimators which maximize the likelihood for simultaneous variations in μ and σ, i.e. solve (8.9) and (8.11) together. This gives us

$$\hat{\sigma}^2 = \frac{1}{n}\Sigma(x-\bar{x})^2, \tag{8.13}$$

and (8.10) and (8.13) jointly maximize the likelihood.

EXERCISES

8.1 An event of constant probability ϖ is observed r times in n independent trials. Show that by Bayes' theorem or the method of maximum likelihood the value indicated for ϖ is r/n. (For the application of Bayes' theorem it may be assumed that ϖ is distributed equally frequently from 0 to 1.)

8.2 In the previous Exercise, trials are repeated until r events are observed, the number of trials necessary being n. Show that Bayes' theorem and the method of maximum likelihood give r/n as the appropriate estimate of ϖ.

8.3 Show that r/n, averaged over all samples, equals ϖ in Exercise 8.1, but is not equal to ϖ in Exercise 8.2.

8.4 On the assumption of the previous Exercises show that if n trials are conducted and r is the number of events observed, the probability that the event occurs on the next trial is $(r+1)/(n+2)$. (Use a more general form of the rule of succession.)

8.5 An urn is known to contain balls of different colours but the number of colours M is unknown. A sample of n balls is drawn with replacement and is found to contain balls of $m(\leqslant n)$ different colours. Show that the estimator of M indicated by the method of maximum likelihood is m.

CHAPTER 9

RANDOM SAMPLING

The sampling problem

9.1 In the previous chapter we have referred incidentally to the sampling problem, which can be stated quite simply: given a sample from a population, to determine from it some or all of the properties of that population. We noted that only in exceptional cases is it possible to make assertions about the population with complete certainty, and that consequently it is necessary to fall back on statements of a less categorical kind expressible in terms of probability.

9.2 In order to be able to apply the theory of probability to this problem, it is necessary that the sampling should be random. By a random sample, we mean a sample which has been selected in such a manner that every possible sample has a calculable chance of selection. Two remarks are necessary concerning this definition. First, it contains the word " calculable " rather than " calculated," because in practice the chances of selection often need not be exhaustively calculated for every sample—the specification and control of the sampling procedure is all that is required for the application of probability theory. Second, the definition does not imply that every possible sample must have an *equal* chance of selection. If this is in fact the case, and successive drawings are independent, the sampling procedure is called *simple random sampling*, in accordance with the usage of **7.20**. For many theoretical purposes, simple random sampling is the selection process considered, but we shall see in the third volume of this book that the search for efficiency in the design of sampling inquiries often leads to the deliberate abandonment of simple random sampling. The point here is that such abandonment does not impair the application of the theory of probability, so long as the (unequal) chances of selection are uniquely determined by the sampling process, and are therefore calculable.

9.3 In actual practice, we often meet with samples that are not random, having been chosen purposively in some way. In such circumstances, it is not possible to make precise probability statements about the parent population, and where a decision has to be taken one is forced to rely on subjective judgments of an unsatisfactory kind. It is for this reason that random sampling is of primary importance in sampling investigations of a population.

Henceforth, we shall discuss only random samples, and to avoid constant repetition we shall understand the unqualified use of the words " sample " or " sampling distribution " to refer to simple random sampling conditions.

9.4 It is useful to begin a discussion of random sampling by considering the types of parent population from which samples can be chosen.

(a) In the first place, the population may be finite and existent, e.g. the population of human beings in Europe at a fixed point of time, or the population of apples on a given tree. A sampling process which extracts members one at a time from this population will evidently eventually exhaust the supply of members if continued long enough. Thus the sampling, though random, is not simple, for successive drawings are not independent.

We may, however, reduce this process to one of simple sampling by replacing the members after withdrawal. The population then remains the same at each trial. The two cases are sometimes distinguished as " sampling without replacement " and " sampling with replacement."

Furthermore, we may also in many cases regard the sampling as simple to an adequate approximation even when there is no replacement. If the population is large compared with the size of the sample, the abstraction of relatively few members will not materially affect the constitution of the remaining population, which may thus be regarded as approximately the same for subsequent samplings.

(b) Sampling with replacement from a finite population may, in fact, be regarded as sampling from an infinite population, for the process will never exhaust the supply. We may, however, have to deal with a population which is infinite in rather a different sense, namely, that of a limiting form. We may, for example, wish to consider the probability of a sample from the positive integers or the real numbers from 0 to 1. The latter case presents itself in sampling from a continuous frequency-distribution which we must necessarily regard as infinite.

Thus, if we replace an observational distribution by a conceptual continuous mathematical distribution, we replace at the same time a finite population by an infinite population. The drawing of random samples from such a population is attended by the circumstances referred to in **7.11**, namely, that the process to the limit must be taken into account.

(c) Thirdly, the population may be purely hypothetical. Consider, for example, the throws of a die. We may picture the continual throwing as a sampling process drawing members from some population. In such cases what we are really doing is constructing a conceptual population round the sample.

The concept of the hypothetical population is necessitated by ideas of frequency in probability. It is not required (and indeed has been explicitly rejected by Jeffreys) in the approach which takes probability as an undefinable measurement of attitudes of doubt. But if we take probability as a relative frequency, then to speak of the probability of a sample such as that given by throwing a die or growing wheat on a plot of soil, we must consider the sample against the background of a population. There are obvious logical difficulties in regarding such a sample as a selection—it is a selection without a choice—and still greater difficulties about supposing the selection to be random ; for to do so we must try to imagine that all the other members of the population, themselves imaginary, had an equal probability of assuming the mantle of reality, and that in some way the actual event was chosen to do so. This is a rather baffling conception. At the same time, it has to be admitted that certain events such as dice-throwing do happen as if the constituents were chosen at random from an existent population, and it accordingly seems that the concept of the hypothetical population can be justified empirically.

Randomness in sampling practice

9.5 In its colloquial use the word " random " is applied to any method of choice which lacks aim or purpose. We speak of drawing names at random out of a hat, choosing plants at random from a field of corn, selecting family budgets at random from the population, meaning thereby that the selection is completely haphazard.

Now it is found in practice that choice by a human being is not random in the stricter sense that it produces equally frequently events which we are entitled to expect to have equal prior probabilities. Some examples will make this clear.

Example 9.1

In the course of certain work at the Rothamsted Experimental Station, sets of eight wheat plants were chosen for measurement. Six of these were chosen by approved methods, referred to below, and may be taken to be truly random. The other two were chosen haphazardly by eye. If, in any set, the eight plants were ranged in order of magnitude, the two selected by eye could have any number from one to eight ; and if they, in common with the other six, were chosen at random, they should occupy these places with approximately equal frequency in a large number of sets. Table 9.1 shows what actually occurred on two different occasions (a) on May 31st, before the ears of wheat had formed, and (b) on June 28th, after the ears had formed.

Table 9.1—Distribution of plants chosen haphazardly in ranks 1 to 8 (Yates, 1935)

Date	Observation	Numbers bearing specified rank								TOTAL
		1	2	3	4	5	6	7	8	
May 31st	Shoot height . .	9	7	11	8	11	18	21	31	116
June 28th	Ear height . . .	9	19	27	23	15	10	5	4	112

The divergence of actual from expected results is quite striking. On May 31st, before the ears had formed, the observer was strongly biased towards the taller shoots ; whereas in June he was biased strongly towards the central plants and avoided short and tall plants.

Thus it is seen that bias can appear even in a trained observer, and that the bias need not be consistent in over- or under-estimation in different circumstances.

Example 9.2

Table 9.2 shows the frequencies of final digits in a number of measurements made by four different observers.

It is hard to suppose that there was any genuine difference which would lead to the appearance of certain digits at the expense of others, and we may confidently suppose that the deviations from approximate equality indicate bias on the part of the observer.

Observer A had decided preference for 0, 2, 8 and 9, avoiding the centre of the

Table 9.2—Bias in scale reading : distribution of final digits in measurements by four observers (Yule, 1927)

Final digit	Frequency of final digit per 1000			
	A	B	C	D
0	158	122	251	358
1	97	98	37	49
2	125	98	80	90
3	73	90	72	63
4	76	100	55	37
5	71	112	222	211
6	90	98	71	62
7	56	99	75	70
8	126	101	72	44
9	129	81	65	16
TOTAL	1001	999	1000	1000

scale. Observer B is quite good, his deviations from expected values being small, though he also showed some preference for 0. Observer C was poor, rounding off one measurement in two to the whole or half unit. Observer D was obviously very bad indeed, nearly 57 per cent. of his measurements being rounded off to the whole or half unit.

The observations were all made by reading a scale, those under A being on drawings to the nearest tenth of a millimetre, those under B, C and D being measurements on the heads of living subjects to the nearest millimetre. We may conclude from this that different observers may exhibit different degrees of bias even under comparable circumstances, and that even those who are aware of the existence of the possibility of bias and the necessity for taking great care (as observer A was) may nevertheless fail to avoid it.

Example 9.3

An observer was placed before a machine consisting of a circular disc divided into ten equal sections in which were inscribed the digits 0 to 9. The disc rotated at high speed and every now and then a flash occurred from a nearby electric lamp of such short duration that the disc appeared at rest. The observer had to watch the disc and write down the number occurring in the division indicated by a fixed pointer.

This was a machine designed to produce random numbers, and had been found by another observer to do so. But this particular observer produced a definite bias. The frequencies of digits in 10,000 run off by him are shown in Table 9.3.

Table 9.3—Distribution of digits obtained by an observer in using a randomizing machine (Kendall and Babington Smith, 1939)

Digit . . .	0	1	2	3	4	5	6	7	8	9	TOTAL
Frequency . .	1083	865	1053	884	1057	1007	1081	997	1025	948	10,000

If the observer was unbiased the digits should appear in approximately equal numbers ; but there is a bias in favour of all the even numbers and against the odd numbers 1, 3 and 9. The cause of this bias is obscure, for the observer did not have to estimate (as in the previous example) but merely to write down something which he saw, or thought he saw. The explanation seemed to be that he had a strong number-preference, i.e. that he actually mis-saw the numbers, or that his brain controlled his ocular impressions and censored them. We have here to deal with one of the deadliest forms of bias in psychology.

Example 9.4

Every year a number of crop reporters in England and Wales estimate the prospective yields of certain crops, forecasts being obtained at different periods of the year and final estimates when the crop is harvested. Table 9.4 shows the average estimated yield of potatoes at the various times for the years 1929–1936.

Table 9.4—Bias in crop forecasting : forecasts of yields of potatoes in England and Wales (tons per acre) (*from the official agricultural statistics*)

Year	Sept. 1st		Oct. 1st		Nov. 1st		Final estimate
	Yield	% difference from final	Yield	% difference from final	Yield	% difference from final	
1929	5·7	−17·4	6·2	−10·1	6·5	−5·8	6·9
1930	6·0	− 7·7	6·1	− 6·2	6·1	−6·2	6·5
1931	5·5	0·0	5·3	− 3·6	5·3	−3·6	5·5
1932	6·4	− 3·0	6·2	− 6·1	6·3	−4·5	6·6
1933	6·4	− 4·5	6·2	− 7·5	6·4	−4·5	6·7
1934	6·0	−15·5	6·3	−11·3	6·7	−5·6	7·1
1935	5·6	− 9·7	5·7	− 8·1	6·0	−3·2	6·2
1936	6·0	− 3·2	5·9	− 4·8	5·8	−6·5	6·2

This table exhibits very clearly an effect which has shown itself in nearly all the English crop reports (and appears also in other countries), namely, the chronic pessimism of crop forecasts. In every case but one in the table, the forecasts are below the final estimate. Nor do crop reporters seem able to learn by experience that they are underestimating. Nothing in this table indicates that the differences between forecast and final estimate diminished during the period concerned.

It should also be noticed that these estimates are the weighted average of a large number of independent observations. One of the commoner misunderstandings in this type of work is based on the supposition that, though individuals may make mistakes, their errors will cancel out in the aggregate. Our present example shows this to be untrue in general. There can appear a systematic bias affecting all the individuals making estimates.

9.6 The foregoing examples are enough to indicate that human bias is very prevalent. Trained observers may be biased even when conscious of their own imperfections ; different observers may be biased in different ways in similar circumstances ; and the same observer may be biased in different ways in different circumstances. It is abundantly clear that we must look for true randomness elsewhere than in mere lack of purpose on the part of human observers. There may be persons whose psychological processes are so finely balanced that they can deliberately select random samples, but few statisticians who have experimented in this interesting field would regard themselves as so gifted.

9.7 In Chapter 7 we saw that the primary function of randomness in probability was that it ensured that certain primitive events were equally probable. We may say that a method of selection is random for a population U if, when applied to U, it gives all members an equal probability of being chosen; or, in the language of frequency, if, when continually applied to U, it educes the members approximately equally frequently.

But this is not enough. Suppose we had a population of two members A and B, and sampled with replacement. Then a method which chooses A and B alternately and produces the series $ABAB\ldots$ educes each member approximately equally frequently; but it is not what we customarily mean by a random method. What we require of a random method is that in such circumstances it should produce a series in which no systematic arrangement is evident. Not only single characteristics, but all possible groups of characteristics should appear equally frequently.

9.8 A further point is to be noted. We may, in drawing the sample, be interested in one particular variate exhibited by the members, and it is possible that a method may give a satisfactory random sample *so far as this variate is concerned* without doing so for other variates. Suppose, for example, we are anxious to take a random sample from the inhabitants of a particular area. If we are concerned with a variate such as eye-colour it might be sufficient to choose a house every so often, say every tenth house, and select one inhabitant of that house as part of the sample. Such a method would not give every inhabitant of the area an equal chance of being chosen; but if we look back to the time when the inhabitants took up residence, we may imagine that the colour of their eyes did not influence their geographical distribution, and thus if we consider the distribution of the inhabitants into houses as independent of eye-colour, we may suppose that so far as eye-colour is concerned the sample is random. But the matter would stand differently if we were sampling for income, for we may reasonably expect poorer people to live in more crowded conditions, and a sample of one person from each house chosen would therefore under-represent the poor. Thus our sample would not be random with respect to income.

Thus a method which may properly be deemed to be random for one variate may not be so for another.

The technique of random sampling

9.9 Suppose, then, that we are given a population and a variate is specified. How are we to draw a random sample, i.e. how can we find a method which is random for that population and that variate? The answer lies partly in theory and partly in practice.

(a) In the first place we must require that there is no obvious connection between the method of selection and the properties under consideration. The method and the properties must be independent so far as our prior knowledge is concerned. In sampling a field of wheat for shoot height, for example, we must not use a method which could be influenced by that height, such as skimming a hoop over the field and selecting the plants round which it fell (for the hoop might tend to catch on the taller plants). Again, in sampling the inhabitants of a town by choosing names from a telephone directory, we should undoubtedly tend to get the more well-to-do classes and hence, if the variate under consideration is wealth or any related characteristic such as number of children, political opinion, standard of education and so on, the sample would not be random. If we were concerned with characteristics such as height, hair colour, or blood group the sample might be random, though it is not difficult in many similar cases to think of reasons why the variate might be linked with wealth.

If this matter is viewed from the standpoint of the subjective theory of probability the absence of knowledge about relationship between the method of selection and the characteristic under consideration may be sufficient to ensure randomness, for the probabilities of elementary propositions then become equal—the probabilities being measures of prior attitudes of mind.(*) But if the frequency viewpoint is adopted, it is not enough that there should be absence of knowledge of this kind, for unknown to the observer there may be relations which will prevent the elementary propositions from being true in approximately equal proportions. The presumption is that if we make as great an effort as possible to ascertain whether any relationship exists and fail to find it, there is no relationship; and hence we can assume the randomness of the method with more or less confidence. But in this approach the assumption of randomness is ultimately part of the general uncertainty of the inference from sample to population.

(b) Secondly, we may rely on previous experience of a random method of selection to justify its use on new occasions. This is evidently an extrapolation, and though most people would regard it as reasonable, the fact has to be realized. The subjective theory of probability can embrace this extrapolation within its scope, for the probabilities given by the method are assessable in terms of prior knowledge; but the frequency theory has to take the extrapolation as an additional assumption.

9.10 One of the most reliable methods of drawing random samples consists of constructing a model of the population and sampling from the model. We may, for instance, note down the characteristics of each member on a card and sample by choosing cards from the pack corresponding to the whole population. This is the method adopted in lotteries and the process is known as lottery or ticket sampling. It is moderately effective but suffers in practice from two disadvantages: the labour of constructing the card population, and the danger of bias in the drawing of cards. To be reasonably satisfied about the randomness of the shuffling entails a good deal of trouble and labour, and the same object can be attained much more simply by the use of random sampling numbers, which we now consider.

(*) At least, this is our interpretation of the position; but the writers on the subjective theory have not discussed practical sampling problems at any length, and we may be putting a gloss on their views which they would not accept.

Random sampling numbers

9.11 The easiest way of constructing a miniature population is to attach a positive integer to each member, most simply by numbering the members from 1 onwards. The set of integers so obtained is the miniature population and the problem of drawing a random sample reduces to finding a series of random numbers. The advantages of this method are obvious: no physical model population has to be constructed; the numbering can be carried out in any convenient manner; and the series of random numbers can be applied to any enumerable population so that any series of random numbers has a very wide range of application.

One point should be made clear here. If the numbering of the population is carried out in such a way as to be independent of certain characteristics of the population, any set of numbers will serve to draw a sample random with respect to those characteristics. The randomness in such a case lies, so to speak, in the allocation of integers to the population, not in deciding which ordinals to select for the sample. But in practice a procedure of this kind is of no value, since it only throws back to the difficulty of numbering the population "at random". The usual course is to number the population in any convenient way, related to the characteristics or not, and then seek for a set of numbers which are a random set from the possible integers of the population.

9.12 One of the more obvious ways of drawing random samples from an enumerated population is to use haphazard numbers taken from some totally unrelated source. Suppose, for instance, we wished to take a sample from the visible stars in the sky. We will ignore the small complications due to the existence of double stars and unresolved objects. Since the position of a star on the celestial sphere is defined by latitude and longitude, what is then required is a series of random pairs of latitudes and longitudes. At first sight it seems plausible to take an ordinary atlas and choose the figures set out in the index for place-names arranged alphabetically; for there is little reason to expect any relationship between the distribution of stars in the sky and the distribution of places on the Earth's surface. A little reflection, however, will show that the method is unsound. There are large stretches of territory and sea on the Earth which have no place-names on them—the poles, deserts and oceans; consequently no numbers will occur for these regions and there will be corresponding areas on the celestial sphere which have no chance of being included.

9.13 As a next attempt we might take a book containing a number of digits, e.g. a telephone directory, or a set of statistical tables or mathematical tables, open it at hazard and choose the digits which first strike the eye, or which occur at the top of the page, and so on. This is an improvement, but it is still open to some objection.

(a) *Telephone directories.* Table 1.4 on page 6 shows the distribution of 10,000 digits taken from a London telephone directory. Pages were chosen by opening the directory haphazardly; numbers of less than four digits and numbers in heavy type were ignored; and of the four-figure numbers remaining the two right-hand ones were taken for all numbers on the page. If the numbers were random we should expect about 1000 of each digit in the total of 10,000. Actually there are very considerable deviations from this expectation, and we shall see in a later chapter that they cannot be explained as sampling fluctuations. There are significant deficiencies in 5's and 9's, due to several causes such as the tendency to avoid these digits because they sound alike, the reservation of numbers ending in 99 for testing purposes by telephone engineers and so on. It is evident that tables of random numbers could not be constructed from directories such as this.

(b) *Mathematical tables.* Evidently care has to be exercised in using mathematical tables in constructing random series. Suppose, for instance, we take a set of logarithm tables. There are clearly relationships between successive logarithms, expressible by the fact that differences are approximately constant if the interval is small. Moreover there is a very curious theorem about digits in certain classes of table which throws theoretical doubt on the method. Consider the logarithms to base 10 of the natural numbers from 1 onwards. Suppose we choose the kth digit in each and so obtain a series of digits 0–9. Then the proportional frequency of any digit in this series does *not* tend to a limit as the length of the series increases, whatever k may be.(*) Just what does happen does not appear to be known, but it would seem that certain systematic effects begin to show themselves and these will obviously endanger the randomness of the series.

(c) *Statistical tables.* If we have a volume of statistics such as populations of towns and rural districts, there are some grounds for supposing that if the numbers are large —say, four figures or more—the final digits will be random. Here again, however, the use of such tables requires care—they may have been compiled by an observer with number preferences, and some rounding up may have taken place.

9.14 However, the necessity for the ordinary student to construct random series of his own has been obviated by the publication of various tables of Random Sampling Numbers. There are several such available:

(a) Tippett's numbers comprise 41,600 digits taken from census reports combined into fours to make 10,400 four-figure numbers (*Tracts for Computers*, No. 15, Cambridge U.P.).

(b) Kendall and Babington Smith's numbers comprise 100,000 digits grouped in twos and fours and in 100 separate thousands (*Tracts for Computers*, No. 24). These numbers were obtained from a machine specially constructed for the purpose on the lines very briefly described in Example 9.3.

(c) Fisher and Yates' numbers comprise 15,000 digits arranged in twos (*Statistical Tables for Biological, Agricultural and Medical Research*). These numbers were obtained from the 15th–19th digits in A. J. Thompson's tables of logarithms and were subsequently adjusted, it having been found that there were too many sixes.

(d) The Rand Corporation has published *A Million Random Digits* (1955) arranged in groups of five. They were constructed by a kind of electronic roulette wheel; but it is interesting to observe that even after searches for systematization had been made in the circuits, the numbers were still not quite random and had to be adjusted.

The above-mentioned tables consist of randomized digits. Wold (*Tracts for Computers*, No. 25) has compiled a table of 25,000 normal observations, by converting the Kendall–Babington Smith tables with the aid of tables of the normal integral.

(*) Cf. J. Franel, *Vierteljahrschrift der Naturforschenden Gesellschaft in Zürich* (1917), **62**, 286.

The Rand Corporation's book also contains 100,000 normal observations. Fieller and others (*Tracts for Computers*, No. 26) gave 3000 pairs of normal observations for each correlation coefficient value 0·1 (0·1) 0·9. Krishna Iyer and Sinha (1967) give 1050 pairs of normal observations for each correlation value 0·05 (0·10) 0·95.

Quenouille (1959) gives 1000 random values from each of eight distributions, these being the normal, lognormal, exponential, double exponential, rectangular, and three Edgeworth expansions. *Tables of Normal and Log-Normal Random Deviates* (Almqvist and Wiksell, 1962) have been prepared from (b) and (d) above at the University of Gothenburg. Barnett (*Tracts for Computers*, No. 27) gives 10,000 random exponential observations and 1000 random standardized normal squares.

Newman and Odell (1971) give methods for obtaining random variates from these and other (including some multivariate) distributions.

9.15 Before considering the basis of these tables it may be helpful to give some examples of their use. Here are the first 200 of the Kendall–Babington Smith tables :—

Table 9.5—Random sampling numbers (*Tracts for Computers*, No. 24)

23 15	75 48	59 01	83 72	59 93	76 24	97 08	86 95	23 03	67 44
05 54	55 50	43 10	53 74	35 08	90 61	18 37	44 10	96 22	13 43
14 87	16 03	50 32	40 43	62 23	50 05	10 03	22 11	54 38	08 34
38 97	67 49	51 94	05 17	58 53	78 80	59 01	94 32	42 87	16 95
97 31	26 17	18 99	75 53	08 70	94 25	12 58	41 54	88 21	05 13

Example 9.5

To draw a sample of 10 men from the population of 8585 men of Table 1.7.

The first process is to number the population ; and here, as in most similar cases, one numbering has already been provided by the frequency-distribution. We take numbers 1 and 2 to be those in the group 57– inches, numbers 3 to 6 those in the group 58–, and so on, those in the group 77– inches being numbers 8584 and 8585.

Now we take 10 four-figure numbers from the tables, e.g. reading across in Table 9.5 we have 2315, 7548, 5901, 8372, 5993, 7624, [9708], [8695], 2303, 6744, 0554, 5550.

The two numbers in square brackets are greater than 8585 and we ignore them. We now select the individuals corresponding to the remaining 10 numbers. They will be found to be in the intervals 65–, 70–, 68–, 72–, 68–, 70–, 65–, 69–, 63–, 68– inches respectively. The mean of these values considered as located at the centres of intervals is 68·24, as against a value in the population of 67·46.

Example 9.6

To draw a sample of 12 from the population in the following bivariate table, showing the relation between inoculation and attack in cholera.

	Not attacked	Attacked	TOTAL
Inoculated	276 (0001–3312)	3 (3313–3348)	279
Not inoculated . . .	473 (3349–9024)	66 (9025–9816)	539
TOTALS	749	69	818

There are now 818 members. We could, of course, take three-figure numbers from the tables, obtaining, e.g. from Table 9.5,

231, 575, 485, etc.

But this is rather troublesome as the numbers are not grouped in threes. It is more convenient to take four-figure numbers as before and to associate each member of the population with 12 numbers in the tables, e.g. the first would correspond to 0000–0011, the second to 0012–0023, and so on. We then get the numbers shown in brackets in the above table. We ignore numbers above 9816 as before.

The two numbers omitted in the previous Example can now be used, and, from all 12 numbers listed there, we find the following results:

	Not attacked	Attacked	Total
Inoculated	3	0	3
Not inoculated . . .	8	1	9
Totals	11	1	12

Here, for example, the member corresponding to the number 2315 falls in the not-attacked : inoculated class, and so on.

It has so happened in this example that no member in the very small inoculated : attacked class has been selected. Suppose we had had a series containing

3314, 3323, 3333, 3341.

All these fall into the group and there are four of them, as against only three members in the population. Had we been confronted with this position we should have had to decide whether the sampling was to be with or without replacement. If it was without replacement, we should have to suppose that the first three numbers in the group 3313–3348 exhausted that part of the population and ignore all numbers of the group occurring subsequently.

While on this point, we might remark that the use of random sampling numbers involves sampling with replacement as a general rule. If we wish to sample without replacement we should not use the same number twice. This involves keeping a record of what has already been used, a tedious procedure except in cases where the sample or the population is small.

Example 9.7

To construct a series of random permutations of the digits 1 to 5.

Here we are not concerned with the digits 0, 6, 7, 8 and 9 and so ignore them in the table of random numbers. We read through the table and note the digits as they occur, e.g. in Table 9.5 we have 2315, 7548, etc. The 7 is to be ignored and also the second 5, for one 5 has already occurred. We then reach the permutation 23154. Then we start again, the next series being 8, 5901, 8372, 5993, 7624, etc., giving the permutation 51324; and so on. This method is clearly rather wasteful of random digits—Plackett (1968) discusses more efficient methods.

Tables of 400 random permutations of the integers 1 to 20 are given in Kendall's *Rank Correlation Methods*, 4th edn (Griffin, 1974). Permutations of fewer than 20 can be derived by omitting unwanted integers, and there can be derived by obvious procedures 800 permutations of the integers 1 to 10, 1600 of the integers 1 to 5 and so on. Cochran and Cox, in their *Experimental Designs* (Wiley, 2nd edn, 1957), give 1000 permutations of each of 9 and 16 integers.

Moses and Oakford (*Tables of Random Permutations*, Allen and Unwin, 1963) used the Rand's million digits (**9.14**(d)) to construct 960 random permutations of 9 integers; 850 of 16; 720 of 20; 448 of 30; 400 of 50; 216 of 100; 96 of 200; 35 of 500 and 20 of 1000.

Example 9.8

To take a random sample from the normal population $dF = \frac{1}{\sqrt{(2\pi)}} e^{-\frac{1}{2}x^2} dx.$

This is a particularly interesting case, for we have to select a sample from an infinite population. Such a process, as has been seen, can only be considered as a limiting one.

Suppose that we divide the distribution into variate-ranges of width 0·1. We obtain from a table of the normal d.f. (e.g. Appendix Table 2):

Variate-value x	$F(x)$	x	$F(x)$
$-\infty$	0·00000	0·0	0·5000
.	.	.	.
.	.	.	.
.	.	.	.
−3·7	0·00011	0·1	0·5398
−3·6	0·00016	0·2	0·5793
−3·5	0·00023	0·3	0·6179
.		.	
.		.	
.		.	
−0·4	0·3085	3·5	0·99977
−0·3	0·3821	3·6	0·99984
−0·2	0·4207	3·7	0·99989
		.	.
		.	.
		.	.
−0·1	0·4602	∞	1·00000

We may now use a four-figure random number to select a value of x through the population formed by $F(x)$: e.g. the number 5461 corresponds to a variate-value +0·1− and the number 3500 to −0·4−.

Had we taken the table to n places of decimals we should have required n− figure numbers. Furthermore, we can make the approximation more exact by taking a finer variate interval. Such matters as this are to be decided in the light of the degree of approximation required.

This method is refined by avoiding the preliminary grouping. Instead, for a random number y, read off, in a table of the normal d.f., the deviate corresponding to the value y of the d.f. This is simply an application of the reverse of the transformation in **1.27**; e.g., in Appendix Table 2, the random number 7486 gives the normal deviate +0·67. A more detailed table would generally be used to avoid the need for interpolation between tabulated values of the d.f.

9.16 Random sampling numbers must obey certain conditions before they can be used. Any set of numbers whatever is random in the sense that it might arise from random sampling ; but such a set might not be suitable as a table of random sampling numbers. From the examples already given it is clear that we desire such a table to have very great flexibility. It should give random results in as many cases as possible, whether used in part or in whole.

Now it is impossible to construct a table of random sampling numbers which will satisfy this requirement entirely. Suppose, to take an extreme case, we constructed a table of $10^{10^{10}}$ digits. The chance of any digit being a zero is $\frac{1}{10}$ and thus the chance that any given block of a million digits are all zeros is 10^{-10^6}. Such a set should therefore arise fairly often in the set of $10^{10^{10}-6}$ blocks of a million. If it did not, the whole set would not be satisfactory for certain sampling experiments. Clearly, however, the set of a million zeros is not suitable for drawing samples in an experiment requiring less than a million digits.

Thus, it is to be expected that in a table of random sampling numbers there will occur patches which are not suitable for use by themselves. The unusual must be given a chance of occurring in its due proportion, however small. Kendall and Babington Smith attempted to deal with this problem by indicating the portions of their table (5 thousands out of 100) which it would be better to avoid in sampling experiments requiring fewer than 1000 digits.

9.17 If a table of random numbers is used to draw members from a population of ten, we expect the members to appear in approximately equal proportions. In other words we expect such a table to contain the ten digits 0–9 in approximately equal proportions. Similarly we expect the hundred pairs 00–99 to appear in approximately equal proportions, and so on. Various tests of this kind, based on a comparison between actual frequencies and those required to satisfy the laws of probability, can be devised. No table can satisfy them all, but if it satisfies tests which (a) ensure the randomness of the numbers for the commoner types of sampling inquiry for which it is likely to be used and (b) are capable of revealing any particular sort of bias to which the numbers are susceptible in virtue of their mode of formation, it is likely to be of general application.

9.18 The development of high-speed computers has made it possible to incorporate a " randomizing unit " into some machines which will select random samples *ad hoc*. The use of such mechanisms requires investigation ; it seems difficult to generate sets of numbers which pass the usual tests for randomness—cf. **9.14**(d). An electronic generator which passes the tests is described by Thomson (1959).

Deterministic arithmetical processes have also been used to produce " pseudo-random " numbers for use in computers. The most satisfactory seems to be the *Residue Class* method, where the numbers y_i are generated as the sequence $y_i = \rho_i c_1^{-c_2}$ where $\rho_i = c_3^{c_4} \rho_{i-1} \bmod c_1^{c_2}$, $\rho_0 \equiv 1$ and the c_j are suitably chosen constants. The method has been successfully tested for randomness. For a review, see Teichroew (1965). Sowey (1972) gives a comprehensive classified bibliography on random number generation and testing.

Sampling from a continuous population

9.19 Random sampling numbers offer the best method known at the present time of drawing random samples from an enumerable population and, as was seen in Example 9.8, may also be used to draw samples from a continuous population specified mathematically. But cases sometimes occur in which they cannot be employed. For instance, if we wish to take a sample of milk or flour, we cannot in practice number each particle and extract it from the population for examination. In such cases we are usually compelled to fall back on more intuitively founded procedure. To take a random sample from a milk churn, for instance, we might stir the contents thoroughly and scoop up a sample haphazardly. Sometimes, when the population is of manageable size, we can proceed systematically by dividing it into a number of parcels and selecting parcels by the ordinary technique of random numbers. Most sciences have their own peculiar sampling problems and no attempt can be made here to discuss them all. At this point we leave the technique of random sampling and assume hereafter, unless the contrary is stated, that the material we are discussing has been obtained by a random process. We return to the comparison of the efficiencies of different sampling methods in Vol. 3.

Sampling for attributes

9.20 As an introduction to the general sampling problems we shall consider the sampling of attributes, which raises many of the difficulties of principle but is not obscured by too much mathematics.

Suppose we have a random sample from a population whose members all exhibit either an attribute A or its negative not-A. Our sample is n in number, and a proportion p, or a number pn, exhibit the attribute; and consequently a proportion q, or number qn $(p+q = 1)$ do not. We will assume that the population is large, or that sampling is with replacement, so that the probability of obtaining an A at any drawing is not affected by other drawings and is therefore a constant, say ϖ.

The problems we have to consider are of three types :—

(a) Suppose we have some reason for supposing that the proportion of A's in the population is given by a known ϖ. Does the observed proportion p bear out this hypothesis or is it so divergent from ϖ as to lead us to doubt the hypothesis? For example, in an experiment with plants exhibiting two strains of a quality such as height in pea plants, we may wish to test whether the breeding follows the simple Mendelian law of dominant and recessive. If we begin with two pure strains tall and short, cross-breed a first generation and then produce a second generation by interbreeding, the proportional frequencies of " short " and " tall " in this generation will be $\frac{3}{4}$ and $\frac{1}{4}$ if " short " is dominant and $\frac{1}{4}$ and $\frac{3}{4}$ if " tall " is dominant, provided that the simple Mendelian law holds. Suppose we carry out such an experiment and find that for 400 plants the frequencies are 70 and 330. Can the divergence from the theoretical values 100 and 300 have arisen by chance, or is it large enough to throw doubt on the hypothesis that the simple Mendelian law is operating?

(b) In the foregoing type of problem we have some reason for testing a value of ϖ given *a priori*; but we may know nothing of ϖ, and in such a case our principal problem is to estimate it from the sample.

(c) Then, having estimated it, we wish to know the degree of reliability of the estimate. How far is the estimate likely to deviate from the real value of ϖ?

9.21 Consider the first type of problem, in which ϖ is given *a priori*. If we select repeated samples of size n from the population, the distribution of the number of A-individuals in the sample will be arrayed by the binomial expansion of $(\chi+\varpi)^n$, where $\chi = 1-\varpi$. (This is, in fact, the binomial distribution dealt with in **5.2**, but we are now using Greek symbols instead of p and q to distinguish the fact that these are population parameters.) The probability of obtaining np or fewer A-individuals is the sum of the first $(np+1)$ terms in this distribution. Call this P. We test the hypothesis by determining beforehand a (generally small) probability of error, say α, which we are prepared to tolerate in rejecting the hypothesis, and compare P with α. If P is the larger, we do not reject the hypothesis, while if α is the larger we do.

Superficially, this procedure is a reasonable one; we reject the hypothesis if the observed value of the statistic is located in the most " unlikely " portion of its sampling distribution. However, a satisfactory analysis of the logic of test inference must await a more sophisticated discussion in the second volume.

It will have been noted that P, determined as above, relates to only one of the " tails " of the sampling distribution. It is, in fact, more common in practice for tests to employ a P calculated from *both* tails of the distribution. That is to say, quite frequently we are interested in departure from the hypothesis under test which would have the effect either of decreasing or of increasing the expected number of A-individuals in the sample, while on other occasions we are interested only in departures in one direction. Once again, full discussion of the rationale of these procedures must be deferred to the second volume. Here, it is sufficient to note that in our Mendelian law example we should be following the intuitively sound procedure in regarding only low numbers of " tall " plants as providing evidence against the hypothesis that " tall " is dominant.

Example 9.9

In certain coin-tossing experiments a coin was tossed 20 times and came down " heads " 15 times. Does this conflict with the hypothesis that the coin was unbiassed?

Here we have to test the hypothesis that $\varpi = \frac{1}{2}$. Bias in the coin can operate in either direction (towards " heads " or " tails "), so we calculate P including both extremes of the sampling distribution. The probability that in 20 tosses we should get 15 or more heads is the sum of the first six terms in $(\frac{1}{2}+\frac{1}{2})^{20}$, and from Table 5.2 this is 0·0207. The probability of 15 or more tails is, by symmetry, the same. P is therefore 0·0414. If we had only been prepared to tolerate a risk of wrongly rejecting the hypothesis which is lower than this, the hypothesis would not be rejected. In the contrary case, the hypothesis would have been rejected. Commonly, a risk of 0·05 is tolerated, and in this case the hypothesis would be rejected.

9.22 In the example just given we purposely took a fairly low value of n in order that the terms of the binomial could be calculated directly. In practice n is often fairly large—100 or more—and the evaluation and summation of individual terms would be most tedious. We can, if complete accuracy is desired, use the method of summation given in **5.7**, and evaluate from the incomplete B-function or from the tables listed there. But for all ordinary purposes it is quite enough to use the normal approximation to the binomial. We saw in Example 4.6 that as $n \rightarrow \infty$ the binomial tends to the normal distribution, with mean $n\varpi$ and variance $n\varpi\chi$ in our present notation. Probabilities can therefore be evaluated from the normal integral. In fact, for many purposes, it is not even necessary to carry out the actual evaluations. From the tables of the integral (Appendix Table 2) we note that the probability of an absolute deviation as great as, or greater than, the standard deviation is 0·3173 ; for twice, the s.d. is 0·0455 ; for thrice the s.d. is 0·0027 ; for four times the s.d. is 0·00006. Thus, if we set our region of rejection outside $n\varpi \pm 2\sqrt{(n\varpi\chi)}$, we shall have a probability of 0·0455 of wrongly rejecting the hypothesis and similarly for other multiples of $\sqrt{(n\varpi\chi)}$.

It makes no difference whether we compare the actual frequencies or the proportions, since mean and standard deviation are equally affected when the variate is divided by a constant.

Example 9.10

In some dice-throwing experiments Weldon threw dice 49,152 times, and of these 25,145 yielded a 4, 5, or 6. Is this consonant with the hypothesis that the dice were unbiased ?

If the dice are unbiased the probability of a 4, 5 or 6 is $\frac{1}{2}$. Thus $n\varpi$ is 24,576 and the observed np is 569 in excess of this value.

$$\sqrt{(n\varpi\chi)} = \sqrt{(49{,}152 \times \tfrac{1}{2} \times \tfrac{1}{2})} = 110{\cdot}9.$$

The observed deviation is more than 5 times this quantity and we accordingly suspect very strongly that the dice were biased.

Standard error

9.23 The quantity $\sqrt{(n\varpi\chi)}$ is a particular case, appropriate to the binomial, of an important statistical concept known as the standard error. It is the standard deviation of the sampling distribution of the statistic, here np. It is particularly important in the class of cases, which is relatively large, wherein that sampling distribution can be taken to be normal either exactly or to an adequate degree of approximation.

9.24 Let us now turn to the case (b) in which no value of ϖ is given *a priori*. If the sample gives a proportion of A's equal to p, what shall we take as our estimator of ϖ ? The most obvious course is to take p itself ; and this is the course dictated by the more sophisticated ideas described in Chapter 8.

Consider first of all the method of maximum likelihood. The probability of obtaining np A's and nq not-A's is

$$\binom{n}{np} \varpi^{np} \chi^{nq}. \tag{9.1}$$

This is proportional to the likelihood, and neglecting constants we have to maximize

$$\mathrm{L} = k\varpi^{np}(1-\varpi)^{nq}$$

for variations of ϖ. Since L is non-negative, $\frac{\partial L}{\partial \varpi}$ and $\frac{\partial}{\partial \varpi}(\log L)\left(=\frac{1}{L}\frac{\partial L}{\partial \varpi}\right)$ vanish together and it is therefore sufficient to maximize log L. We have

$$\frac{\partial}{\partial \varpi}(\log L) = \frac{np}{\varpi} - \frac{nq}{1-\varpi} = 0$$

giving, as our estimator $\hat{\varpi}$,

$$\hat{\varpi} = p. \tag{9.2}$$

The method of Bayes will give the same result if we suppose the possible values of ϖ equally distributed between 0 and 1 for $d\varpi \to 0$. For then

$$P(\varpi \mid p) \propto P(\varpi \mid H) P(p \mid \varpi H) \tag{9.3}$$

$$\propto \varpi^{np} \chi^{nq} \, d\varpi \tag{9.4}$$

which, as before, is maximized when $\varpi = p$.

There is another way of looking at this problem of estimation. Suppose we took a large number of samples of the same size n from the population with a proportion of ϖ A's and χ not-A's. Our estimator of ϖ would be p in each case, p varying from sample to sample; and the mean value of all such estimates would be

$$\begin{aligned} E(p) &= \sum_{p=0}^{1} p \binom{n}{np} \varpi^{np} \chi^{nq} \qquad (9.5) \\ &= \frac{1}{n} \Sigma \left\{ np \binom{n}{np} \varpi^{np} \chi^{nq} \right\} \\ &= \frac{1}{n} \varpi . n \Sigma \left\{ \binom{n-1}{np-1} \varpi^{np-1} \chi^{nq} \right\} \\ &= \varpi \{\varpi + \chi\}^{n-1} = \varpi, \end{aligned}$$

so that the mean value of our estimator over all possible samples is ϖ. Such an estimator is called *unbiased*—if we follow the rule of estimation the average of our estimates in a large number of cases will be exactly the correct value ϖ. It may thus be argued that the unbiased estimator should be taken as a reliable estimator of ϖ.

9.25 In this case, therefore, all the approaches lead to the same conclusion (a happy state of affairs which, as we shall see in the sequel, does not always exist). Consider now the next stage of the problem: (c) what is the reliability of the estimator? In other words, how far is the estimate likely to differ from the true value?

We know that if the sample value p differs from ϖ by $t\sqrt{(\varpi\chi/n)}$, the probability of the difference becomes smaller as t increases. Thus, with an assigned degree of probability we can say that it is improbable that p will differ from ϖ by more than an assigned amount. But to specify this amount exactly we require to know ϖ; and this is precisely the quantity we are trying to find.

The problem as put in this form can only be solved as an approximation. If n is large the standard error of p is of the order $n^{-\frac{1}{2}}$, so that we may put

$$\varpi = p + \frac{k}{n^{\frac{1}{2}}}.$$

Thus

$$\sqrt{\frac{\varpi\chi}{n}} = \sqrt{\left\{\frac{1}{n}\left(p+\frac{k}{n^{\frac{1}{2}}}\right)\left(q-\frac{k}{n^{\frac{1}{2}}}\right)\right\}} = \sqrt{\left\{\frac{pq}{n}\left(1+\frac{k(q-p)}{pqn^{\frac{1}{2}}}\right)\right\}},$$

neglecting terms of order n^{-1},

$$= \sqrt{\frac{pq}{n}}\left\{1+\frac{1}{2}\frac{k(q-p)}{pqn^{\frac{1}{2}}}\right\}. \tag{9.6}$$

Hence for large n the standard error of p is approximately equal to $\sqrt{(pq/n)}$; and we thus reach the fundamental result that in large samples for attributes the standard error may be calculated by using the estimates of the parameters under estimate instead of the (unknown) values of those parameters themselves.

In Volume 2 we shall consider a different approach which frees us from the necessity of making this type of approximation.

Example 9.11

In a sample of 600, 240 are found to possess the attribute A. Thus $p = 0{\cdot}40$, $np = 240$, $\sqrt{(npq)} = 12$. We can therefore regard it as somewhat improbable that $n\varpi$ differs from 240 by more than twice this amount, 24, and highly improbable that it differs by more than 36. We thus can say with some assurance that $n\varpi$ lies in the range 240 ± 24 and with great assurance that it lies in the range 240 ± 36.

9.26 We now turn to a general consideration of the problems of sampling which have been exemplified above. In the first place, let us note the role of the sampling distribution in this branch of the subject. We construct from the observations some statistic t. The sampling distribution of this statistic will in general (but not always) depend on some parameters of the parent population. The probability of the observed t then permits the making of statements, by inverse probability, likelihood or otherwise, about these parameters, and so we are enabled to draw inferences about the parent population. The sampling distribution is thus fundamental to the whole subject and several subsequent chapters will be devoted entirely to the methods of finding sampling distributions when the population is specified.

If we wish to test some hypothesis about the parent which is expressible by the determination of certain parameters *a priori*, the problem is fairly simple. Given the values of the parameters, we can determine from the sampling distribution the probability of the observed value of the statistic, and use this to assess the acceptability of the hypothesis. Complications can arise even here, however, for in general, several statistics can be computed from the same sample, and they need not necessarily all lead to the same conclusion about the hypothesis; for instance, a sample might have a mean which throws doubt on the hypothesis and a variance which does not. We shall discuss this difficulty more fully in the second volume.

9.27 When the parameters of the population are not given *a priori*, we have the double problem of estimating the parameters from the sample and assigning probable limits to the estimates so obtained. We have already touched on some of the principles of estimation and shall develop the topic more systematically in Volume 2. When we have obtained an estimate—itself a statistic—we seek its sampling distribution and therefrom can assign probable limits to the population value. A special class of cases arises when we can find a statistic whose sampling distribution depends on only one parameter of the population (as in the case of attributes).

9.28 These latter types of problem permit of certain important approximations, namely in the case when the sample is large. We saw in Chapter 7 that under very general conditions the sum of n independent variables, distributed in whatever form, tends to normality as n tends to infinity. Now many of the ordinary statistics in current use can be expressed as the sum of variates, e.g. all the moments; and many others may also be shown to tend to normality for large samples. Thus we may approximate—

(a) by taking a statistic, calculated from the sample as if it were a population, to be the estimate of the corresponding parameter in that population, e.g. the variance of the sample may be taken as an estimate of the variance of the population;

(b) by calculating the mean and variance of the sampling distribution using, instead of the unknown parameter values, the statistic values calculated according to (a);

(c) by assuming that the distribution is normal and hence determining probabilities from the normal integral with the aid of the sampling mean and sampling variance (the latter being the square of the standard error).

9.29 Just how large n must be for such approximations to be valid it is not always easy to say. For some distributions, particularly that of the mean, quite a satisfactory approximation is given by low values of n, say $n > 30$. For others n has to be much higher before the approximation begins to give satisfactory results, e.g. for the product-moment correlation coefficient even values as high as 500 are not good enough in samples from a normal population. In fact, the form of the parent distribution, as well as the statistic considered, influences the rapidity of approach of the sampling distribution to the normal form.

Example 9.12

Returning to the binomial distribution, let us consider how far our procedure need be modified when sampling without replacement.

If a random sample of n is drawn without replacement from a population of N, a proportion ϖ of which bear the attribute A, the sampling distribution of the proportion p of attribute-bearers in the sample is arrayed by the hypergeometric distribution **(5.18)** and a typical term is, in our present notation,

$$\binom{n}{np}\frac{(N\varpi)^{[np]}(N\chi)^{[nq]}}{N^{[n]}}.$$

The mean value of p in this distribution, from (5.53), is known to be ϖ; and thus the sample proportion remains an unbiassed estimator of the population proportion. It is not a Bayes estimator or a maximum likelihood estimator (see Exercise 9.10)—these do not take simple forms in this case—but it differs from them only by terms of order n^{-1}.

The variance of p, from (5.55), is given by

$$\operatorname{var} p = \frac{N-n}{N-1}\cdot\frac{\varpi\chi}{n}. \tag{9.7}$$

This differs from the ordinary binomial case (sampling with replacement) by the factor $(N-n)/(N-1)$ which is less than unity for $n > 1$. The variance in sampling without replacement is therefore smaller than with replacement, as is intuitively obvious from the fact that the extreme sampling possibilities are greater in the latter case.

Example 9.13 Sequential sampling

Cases sometimes occur, especially in sampling for attributes, where the sample size is not fixed in advance. Thus, instead of drawing a sample of size n and noting the number r ($= np$) of "successes", or appearances of the attribute, we decide to go on adding to the sample until some (predetermined) number of successes r appear. In this case r is fixed and n is a random variable.

The distribution of n has been obtained at (5.36). We now re-write it in the present notation:

$$\text{Prob}\,(n = s) = \binom{s-1}{r-1}\varpi^r \chi^{s-r}, \qquad s = r, r+1, \ldots, \infty. \tag{9.8}$$

The successive terms are

$$\varpi^r\{1, r\chi, \tfrac{1}{2}r(r+1)\chi^2, \ldots\},$$

i.e. those of the negative binomial

$$\frac{\varpi^r}{(1-\chi)^r}.$$

The mean (with origin at r) and variance of the distribution are, by (5.38), $r\chi/\varpi$ and $r\chi/\varpi^2$ respectively.

For the maximum likelihood estimator of ϖ we find, as for the ordinary binomial, the value p. This, however, is no longer unbiased. In fact, we shall show that $(r-1)/(n-1)$ is an unbiased estimator of ϖ.

For

$$\begin{aligned}
E\left(\frac{r-1}{n-1}\right) &= \varpi^r \sum_{n=r}^{\infty} \frac{r-1}{n-1}\chi^{n-r}\binom{n-1}{r-1} \\
&= \varpi^r \sum_{n} \binom{n-2}{r-2}\chi^{n-r} \\
&= \varpi^r (1-\chi)^{-(r-1)} \\
&= \varpi.
\end{aligned}$$

If $r = 1$, the estimator $(r-1)/(n-1) = 0$ if $n > 1$, and if we define it to be equal to 1 if $n = 1$, it remains unbiased, as may easily be confirmed.

The biased character of r/n follows, for it cannot have the same expectation as $(r-1)/(n-1)$, being always greater. However, n/r is an unbiased estimator of $1/\varpi$, as $E(n) = r/\varpi$, the mean about zero, as remarked at the end of **5.15.**

At first sight this seems rather paradoxical. n/r is an unbiased estimator of $1/\varpi$, but r/n is a biased estimator of ϖ. The reason is, of course, our somewhat arbitrary definition of bias as a departure from a mean value. It should not occasion surprise that if $E(x) = \alpha$, $E(1/x)$ is not equal to $1/\alpha$. For a positive variate this is, in fact, inevitable (cf. Exercises 9.13 and 9.15).

The method of sampling under which we proceed with the drawing of the sample until a certain number of " successes " have turned up is sometimes called " inverse sampling ", a term to which there are some objections. It is a particular case of a method which we shall study in Chapter 34, Volume 2, and is most conveniently known as sequential sampling (for attributes), the characteristic feature of which is that we decide as we go along whether to draw further members for the sample, in the light of what has been drawn up to date.

EXERCISES

9.1 Of 10,000 babies born in a particular country 5100 are male. Taking this to be a random sample of the births in that country, show that it throws considerable doubt on the hypothesis that the sexes are born in equal proportions.

Consider how far this conclusion would be modified if the sample consisted of 1000 births, 510 of which were male.

9.2 In a sample of n_1 from a population π_1, the proportion observed with an attribute A is p_1. In another independent sample of n_2 from a population π_2 the proportion is p_2. To test whether the populations have identical proportions of A in them, it is proposed to consider the difference p_1-p_2. Show that on the hypothesis of identical proportions the difference has zero mean and a variance which may be estimated as

$$(p_1 n_1+p_2 n_2)(q_1 n_1+q_2 n_2)/\{n_1 n_2(n_1+n_2)\},$$

where $q_1 = 1-p_1$, $q_2 = 1-p_2$.

9.3 If a proportion ϖ has to be estimated from a simple random sample with proportion p, and if f is the prior probability of ϖ, then the posterior probability of ϖ is, according to Bayes' theorem, proportional to

$$f\varpi^{np}(1-\varpi)^{nq}.$$

Show that this is a maximum if

$$\frac{1}{f}\frac{\partial f}{\partial \varpi}+n\frac{p-\varpi}{\varpi(1-\varpi)} = 0.$$

Hence, in general, as n increases, the solution tends to $\varpi = p$, whatever the prior probability of ϖ.

9.4 By using the Bienaymé-Tchebycheff inequality, show that in sampling from a large population for an attribute, the probability that the observed proportion p in a sample of n differs from the population proportion ϖ by more than k is not greater than $1/(4nk^2)$.

(This is an exact result, no assumptions about the limiting normality of the binomial or the use of estimates in calculating standard errors being involved. The limits are, however, much too wide in general.)

9.5 In a large population the proportion ϖ of members bearing an attribute A is small. Show that in samples of size n the proportion of A's is a maximum likelihood estimator of ϖ; that it is unbiased; and that the variance of the number, say n_1, of A's observed may be estimated as equal to n_1.

9.6 In k different populations the proportions of an attribute are $\varpi_1, \varpi_2, \ldots, \varpi_k$. One population is chosen at random (with probability $1/k$) and from it a sample of n members is drawn with replacement at random, yielding a proportion p bearing the attribute. Show that, under repetitions of this process (including the renewed selection of a parent population), the mean value of p is $\varpi \left(\equiv \sum_{i=1}^{k} \varpi_i/k\right)$ and that its variance is $\varpi(1-\varpi)/n+(n-1)\Sigma(\varpi_i-\varpi)^2/(nk)$.

Obtain the corresponding formulae if the probability of selection of the ith population is α_i, where $\sum_{i=1}^{k} \alpha_i = 1$.

9.7 In the previous exercise, a sample of kn is taken by drawing n members from each population. Show that the proportion p in the kn sample members together has mean ϖ and variance

$$\varpi(1-\varpi)/(nk)-\Sigma(\varpi_i-\varpi)^2/(nk^2).$$

9.8 From a population in which the proportion of individuals possessing a certain attribute is ϖ, a sample of size n is drawn with replacement. From this a sub-sample of n_1 is drawn with replacement. If p, p_1 are the proportions in the sample and sub-sample respectively show that $p-p_1$ has a zero mean and variance

$$\varpi(1-\varpi)\frac{n-1}{nn_1}.$$

9.9 In the previous Exercise, if the sample of size n is selected from a population of size N without replacement and the sub-sample of n_1 is selected from the n without replacement, show that $p-p_1$ has zero mean and variance

$$\varpi(1-\varpi)\frac{N}{N-1}\cdot\frac{n-n_1}{nn_1}.$$

9.10 Show that, to order N^{-1}, the maximum likelihood estimator of the proportion ϖ in the hypergeometric distribution is

$$p+\frac{1}{2N}(p-q),$$

and that, to the same order, the variance of this estimator is

$$\frac{N-n}{N-3}\cdot\frac{\varpi(1-\varpi)}{n}.$$

9.11 Define a variable x as unity if the individual concerned possesses an attribute A and zero in the contrary case. If p is the proportion in a sample of n bearing the attribute show that, for sampling either with or without replacement, $E(x) = E(x^2) = E(x^k) = \varpi$. Hence obtain equation (9.7).

9.12 In Example 9.13 show that

$$E(r/n) = r\varpi^r\chi^{-r}\int_0^{\chi} t^{r-1}(1-t)^{-r}\,dt = \varpi\sum_{j=0}^{\infty}\frac{r!j!}{(r+j)!}\chi^j.$$

9.13 For a variate x which is defined in the range $0<x\leqslant\infty$, show that if $E(x)$ and $E(1/x)$ exist, $E(x)E(1/x)\geqslant 1$, the inequality becoming an equality only if the distribution of x is wholly concentrated at a single value. Hence show that if t is an unbiased estimator of θ, $1/t$ cannot be an unbiased estimator of $1/\theta$.

9.14 For the ordinary binomial distribution show that $E(n/r)$ does not exist.
Show also that $E\{(n+1)/(r+1)\} = (1-\chi^{n+1})/\varpi$ and hence that, asymptotically in n, $(n+1)/(r+1)$ is an unbiased estimator of $1/\varpi$.

9.15 If y is distributed independently of x in Exercise 9.13, show that

$$E\left(\frac{y}{x}\right)\geqslant\frac{E(y)}{E(x)}.$$

If s_1^2, s_2^2 are independent unbiased estimators of variances σ_1^2, σ_2^2 respectively, show that $E\left(\frac{s_i^2}{s_j^2}\right)\geqslant\frac{\sigma_i^2}{\sigma_j^2}$ for $i\neq j = 1, 2$.

CHAPTER 10

STANDARD ERRORS

10.1 Towards the close of the last chapter we discussed the estimation of statistical parameters from large samples and the type of judgement of their reliability which depends on the use of the standard error. It was remarked that, for large samples, an estimate of a parameter may be obtained by calculating from the sample values the value of the parameter in the sub-population composed by the sample; and it was established that for samples of size n the standard error gives a valid measure of precision, provided (a) that the sampling distribution of the statistic under discussion approaches normality and (b) that n is large in the sense there defined. It was also pointed out that a sufficiently accurate estimate of standard errors involving parent parameters could be obtained by using as the parameter values the corresponding statistics from the sample itself.

Since the majority of statistics in current use do tend to normality the theory of large samples is, in the main, devoted to the determination of standard errors. In this chapter we describe the principal methods available for the purpose, and incidentally derive formulae for the standard errors of the various statistics considered in previous chapters. To avoid the usual square roots associated with the standard error we shall write our results as sampling variances and covariances. Thus, for a statistic t we write the variance of its sampling distribution as $\operatorname{var} t$. The covariance of the joint distribution of two statistics t and u, that is, the first product-moment of their joint sampling distribution, is written $\operatorname{cov}(t, u)$.

10.2 By definition, the rth moment of a statistic t, that is the rth moment of its sampling distribution, is the mean value of t^r taken over all possible samples, and may be written $E(t^r)$ (cf. **2.27**). If the joint distribution of the variates $x_1, \ldots, x_n$, from which t is calculated, is $dF(x_1, \ldots, x_n)$, then the rth moment of t is the integral of $t^r\, dF$ (considered as a function of the x's) over the domain of the x's. In particular, if the sample is simple random and the parent distribution is dF, we have

$$E(t^r) = \int_{-\infty}^{\infty} \cdots \int_{-\infty}^{\infty} t^r\, dF(x_1) \ldots dF(x_n). \tag{10.1}$$

We are particularly interested in this chapter in the first and second moments of t, that is, the mean and variance of its sampling distribution. It may be recalled that the mean value of a sum is the sum of the mean values and that, if the variables are independent, the mean value of a product is the product of the mean values (**2.28**). These two results will be repeatedly required.

Standard errors of moments

10.3 In the following sections, we shall adopt the usual convention in regard to the distinction of population values and statistics by writing Greek letters to represent the former and Roman letters to represent the latter. We have, then, for the rth moment-statistic m'_r, corresponding to the rth moment μ'_r of the population,

$$m'_r = \frac{1}{n}\sum_{j=1}^{n} x_j^r \tag{10.2}$$

and for the mean-moment

$$m_r = \frac{1}{n}\sum_{j=1}^{n} (x_j - m'_1)^r. \tag{10.3}$$

10.4 Consider now the mean value of m'_r. We have

$$\begin{aligned} E(m'_r) &= \frac{1}{n}\Sigma E(x^r) = E(x^r) \\ &= \mu'_r. \end{aligned} \tag{10.4}$$

The sampling variance of m'_r is, by definition, $E(m'_r - \mu'_r)^2$ and thus assuming, as we do throughout, that the appropriate moments exist,

$$\begin{aligned} \operatorname{var} m'_r &= E\left\{\frac{1}{n}\Sigma x^r - \mu'_r\right\}^2 \\ &= \frac{1}{n^2}E[\{\Sigma x^r\}^2 - 2n\mu'_r\Sigma x^r + n^2\mu'^2_r] \\ &= \frac{1}{n^2}E[\{\Sigma x^r\}^2] - \mu'^2_r \\ &= \frac{1}{n^2}E\{\Sigma x_j^{2r} + \Sigma(x_j^r x_k^r)\} - \mu'^2_r \end{aligned}$$

the second summation extending over the $n(n-1)$ cases in which $j \neq k$ (permutations of j and k thus being allowed). Since the x's are independent the mean value of the product is the product of the mean values, and thus

$$\begin{aligned} \operatorname{var} m'_r &= \frac{1}{n^2}\{n\mu'_{2r} + n(n-1)\mu'^2_r\} - \mu'^2_r, \\ &= \frac{1}{n}(\mu'_{2r} - \mu'^2_r). \end{aligned} \tag{10.5}$$

This is an exact result.

In a similar way, if we have two moments, m'_q, m'_r, their sampling covariance is given by

$$\begin{aligned} \operatorname{cov}(m'_q, m'_r) &= E\{(m'_q - \mu'_q)(m'_r - \mu'_r)\} \\ &= E\left\{\left(\frac{1}{n}\Sigma x^q - \mu'_q\right)\left(\frac{1}{n}\Sigma x^r - \mu'_r\right)\right\} \\ &= \frac{1}{n^2}E(\Sigma x^{q+r}) + \frac{1}{n^2}E\sum_{j \neq k}(x_j^q x_k^r) - \frac{1}{n}\mu'_q E(\Sigma x^r) - \frac{1}{n}\mu'_r E(\Sigma x^q) + E(\mu'_q\mu'_r) \\ &= \frac{1}{n}(\mu'_{q+r} - \mu'_q\mu'_r), \end{aligned} \tag{10.6}$$

which reduces to (10.5) if $q = r$, as it must, for the covariance of a variable with itself is its variance.

10.5 The formulae for moments about the mean are not so simple, for the mean itself is subject to sampling fluctuations. We see from (10.5) with $r = 1$ that

$$\text{var}\, m_1' = (\mu_2' - \mu_1'^2)/n = \mu_2/n. \tag{10.7}$$

Thus fluctuations in m_1' are of order $n^{-\frac{1}{2}}$. We may then take an origin at the parent mean and neglect powers of m_1' higher than the first. We then have, neglecting terms of order $n^{-\frac{1}{2}}$,

$$\begin{aligned} E(m_r) &= E\{\Sigma(x-m_1')^r/n\} \\ &= E\{\Sigma x^r - rm_1'\Sigma x^{r-1}\}/n \\ &= E\{(1-r/n)\Sigma x^r - (r/n)\Sigma x_j x_k^{r-1}\}/n, \qquad j \neq k. \end{aligned}$$

Now the second term in the expectation on the right will involve the moments $\mu_1'\mu_{r-1}'$ and will vanish since we have chosen our origin at the mean of the parent distribution ($\mu_1' = 0$). We shall then have, neglecting terms of order $n^{-\frac{1}{2}}$,

$$E(m_r) = \mu_r, \tag{10.8}$$

a result which, unlike (10.4), is not exact but is an approximation to order $n^{-\frac{1}{2}}$. To the same order we have

$$\begin{aligned} \text{var}\, m_r &= E(m_r^2) - \{E(m_r)\}^2 \\ &= E\{\Sigma x^r - r\Sigma x_j x_k^{r-1}/n\}^2/n^2 - \{E(m_r)\}^2 \\ &= \frac{1}{n^2}E\left\{\Sigma x^{2r} + \Sigma x_j^r x_k^r + \frac{r^2}{n^2}\left(\Sigma x_j^r x_k^r + \Sigma x_j^2 x_k^{2r-2} + \Sigma x_j^2 x_k^{r-1} x_l^{r-1}\right) - \frac{2r}{n}\Sigma x_j^{r+1} x_k^{r-1}\right\} \\ &\qquad - \{E(m_r)\}^2, \end{aligned}$$

where $j \neq k \neq l$. The expectations of other terms occurring in the squaring vanish, since they contain μ_1'. The expectation of $\Sigma x_j^2 x_k^{2r-2}$ is of order $n(n-1)$ and the term involving it is therefore of order n^{-2} and may be neglected. The others give us, to the first order in n,

$$\text{var}\, m_r = \frac{1}{n}(\mu_{2r} - \mu_r^2 + r^2\mu_2\mu_{r-1}^2 - 2r\mu_{r-1}\mu_{r+1}). \tag{10.9}$$

Similarly it appears that

$$\text{cov}(m_r, m_q) = \frac{1}{n}(\mu_{r+q} - \mu_r\mu_q + rq\mu_2\mu_{r-1}\mu_{q-1} - r\mu_{r-1}\mu_{q+1} - q\mu_{r+1}\mu_{q-1}). \tag{10.10}$$

Example 10.1

For the height distribution of Table 1.7 we found (Examples 2.1 and 2.7) that $m_1' = 67{\cdot}46$, $\sqrt{m_2} = 2{\cdot}57$. Suppose we regard this distribution as a simple random sample from the adult male inhabitants of the United Kingdom living at the time when the data were collected. What can we say about the mean of the population?

The standard error of the mean depends on μ_2. This is an unknown quantity, but we may, in accordance with the general principles of large sample theory, use m_2 instead. We then find

$$\text{Standard error of } m_1' = \frac{2{\cdot}57}{\sqrt{8585}} = 0{\cdot}028 \text{ approximately.}$$

Thus we can say that the population mean probably lies in the range of twice this amount on either side of the sample mean, i.e. in the range $67{\cdot}46 \pm 0{\cdot}056$, and very probably in thrice the range, i.e. $67{\cdot}46 \pm 0{\cdot}084$. Our estimate of the mean would almost certainly be less than a tenth of an inch in error.

Example 10.2

From equation (10.9) with $r = 4$ we find

$$\operatorname{var} m_4 = \frac{1}{n}(\mu_8 - \mu_4^2 - 8\mu_5\mu_3 + 16\mu_2\mu_3^2).$$

In Chapter 12 we shall show how to obtain this result by other methods of a more exact character and confirm that it is, in fact, correct to order n^{-1}.

Example 10.3

To show that in samples from a symmetrical population the first product–moment between the mean and any mean–moment of even order vanishes to order n^{-1}.

We have, by definition, as in **10.5**,

$$\begin{aligned}\operatorname{cov}(m_1', m_r) &= \frac{1}{n^2} E\left[\left\{\Sigma x \left\{\Sigma x^r - \frac{r}{n}\Sigma x_j x_k^{r-1}\right\}\right\}\right] \\ &= \frac{1}{n^2} E\left\{\Sigma x^{r+1} - \frac{r}{n}\Sigma x_j^2 x_k^{r-1}\right\},\end{aligned}$$

the other terms vanishing, since they involve a unit power of x, if we take our origin at the mean of the parent population,

$$= \frac{1}{n}(\mu_{r+1} - r\mu_2\mu_{r-1}).$$

Now if r is even, μ_{r+1} and μ_{r-1}, being moments of odd order, will vanish for a symmetrical population and hence

$$\operatorname{cov}(m_1', m_r) = 0.$$

In the language of the theory of correlation, the mean and the even moment about the mean are uncorrelated to order n^{-1}.

Standard errors of functions of random variables

10.6 Suppose that x_i has mean θ_i, and that the variances and covariances of the k variates $x_1, x_2, \ldots, x_k$ are of order n^{-r}, $r > 0$. (In practice, r is usually equal to unity.) Consider the function $g(x_1, x_2, \ldots, x_k)$, which we write $g(x)$ for brevity. If $g_i'(\theta)$ is $\partial g(x)/\partial x_i$ evaluated at $\theta_1, \theta_2, \ldots, \theta_k$, we have the Taylor expansion

$$g(x) = g(\theta) + \sum_{i=1}^{k} g_i'(\theta)(x_i - \theta_i) + O(n^{-r}). \tag{10.11}$$

Thus, since $E(x_i) = \theta_i$,

$$E\{g(x)\} = g(\theta) + O(n^{-r}).$$

This formula is taken to a further term in Exercise 10.17.

If not all the $g'_i(\theta) = 0$, we have further

$$\begin{aligned}\operatorname{var}\{g(x)\} &= E\{[\sum_{i=1}^{k} g'_i(\theta)(x_i-\theta_i)]^2\}+o(n^{-r}) \\ &= \sum_{i=1}^{k} \{g'_i(\theta)\}^2 \operatorname{var} x_i + \mathop{\sum\sum}_{i\neq j=1}^{k} g'_i(\theta) g'_j(\theta) \operatorname{cov}(x_i, x_j) \\ &\qquad + o(n^{-r}). \end{aligned} \tag{10.12}$$

Similarly for two functions $g(x)$, $h(x)$, we find

$$\operatorname{cov}\{g(x), h(x)\} = \sum_{i=1}^{k} g'_i(\theta) h'_i(\theta) \operatorname{var}(x_i) + \mathop{\sum\sum}_{i\neq j=1}^{k} g'_i(\theta) h'_j(\theta) \operatorname{cov}(x_i, x_j) + o(n^{-r}). \tag{10.13}$$

If all the first derivatives above are zero, we must take further terms in the Taylor expansion, and the variance is more complicated—cf. Exercises 10.17–18.

10.7 Three particular specializations of (10.12) are of interest:—

(a) g is a function of a single random variable. We then find

$$\operatorname{var} g(x) = \left(\frac{dg}{dx}\right)_\theta^2 \operatorname{var} x. \tag{10.14}$$

In particular, if $g(x)$ is simply a linear function of the form $cx+e$ we find that $\operatorname{var} g = c^2 \operatorname{var} x$, as is otherwise obvious, the result in this case being exact.

(b) g is a linear function of the random variables. If

$$g(x_1, \ldots, x_k) = \sum_{i=1}^{k} a_i x \tag{10.15}$$

then (10.12) gives

$$\operatorname{var} g = \Sigma a_i^2 \operatorname{var} x_i + \sum_{i \neq j} a_i a_j \operatorname{cov}(x_i, x_j), \tag{10.16}$$

again an exact result. In particular, if the x's are independent, the variance of g is a weighted sum of their individual variances.

(c) g is a ratio of two random variables, say x_1/x_2. To avoid difficulties to be discussed in the next chapter, we assume here that $x_2 > 0$ if it is discrete and $\geqslant 0$ if it is continuous. Equation (10.12) gives

$$\begin{aligned}\operatorname{var}(x_1/x_2) &= \frac{\operatorname{var} x_1}{\theta_2^2} + \frac{\theta_1^2 \operatorname{var} x_2}{\theta_2^4} - \frac{2\theta_1 \operatorname{cov}(x_1, x_2)}{\theta_2^3} \\ &= \left\{\frac{E(x_1)}{E(x_2)}\right\}^2 \left\{\frac{\operatorname{var} x_1}{E^2(x_1)} + \frac{\operatorname{var} x_2}{E^2(x_2)} - \frac{2\operatorname{cov}(x_1, x_2)}{E(x_1)E(x_2)}\right\}. \end{aligned} \tag{10.17}$$

The second factor in (10.17) is the sum of the squares of the coefficients of variation of the two variables, minus twice the square of what may analogously be called their coefficient of covariation.

Example 10.4

To find the sampling variance of the fourth cumulant, we shall, for reasons which will become clear in Chapter 12, continue to write κ, not k, for the *sample* cumulant. Thus we require the variance of

$$\kappa_4 = m_4 - 3m_2^2.$$

From (10.12) with $x_1 = m_4$, $x_2 = m_2$ and $g = x_1 - 3x_2^2$, we obtain

$$\operatorname{var} \kappa_4 = \operatorname{var} m_4 + 36\mu_2^2 \operatorname{var} m_2 - 12\mu_2 \operatorname{cov}(m_4, m_2).$$

Substituting from (10.9) and (10.10) we find

$$\operatorname{var} \kappa_4 = \frac{1}{n}\{\mu_8 - 12\mu_6\mu_2 - 8\mu_5\mu_3 - \mu_4^2 + 48\mu_4\mu_2^2 + 64\mu_3^2\mu_2 - 36\mu_2^4\}.$$

For a normal population, $\mu_4 = 3\mu_2^2 = 3\sigma^4$, $\mu_6 = 15\sigma^6$, $\mu_8 = 105\sigma^8$ and we find in this case

$$\operatorname{var} \kappa_4 = 24\sigma^8/n.$$

Example 10.5

To find the sampling variance of the coefficient of variation defined at (2.28),

$$V = \sqrt{m_2}/m_1',$$

where we assume $m_1' > 0$ always, in order to use (10.17). Writing $\mathbf{V}$ for the population coefficient $\sqrt{\mu_2}/\mu_1'$, where $\mu_1' > 0$ also, we have

$$\operatorname{var}(V) = \mathbf{V}^2\left\{\frac{\operatorname{var}(\sqrt{m_2})}{E^2(\sqrt{m_2})} + \frac{\operatorname{var} m_1'}{E^2(m_1')} - \frac{2\operatorname{cov}(\sqrt{m_2}, m_1')}{E(\sqrt{m_2})E(m_1')}\right\}. \tag{10.18}$$

We saw in (10.7) that

$$\operatorname{var} m_1' = \mu_2/n.$$

Also, from (10.14) and (10.9),

$$\operatorname{var}(\sqrt{m_2}) = \frac{\operatorname{var} m_2}{4\mu_2} = \frac{\mu_4 - \mu_2^2}{4n\mu_2}.$$

Finally, in Example 10.3 we saw that to order n^{-1}

$$\operatorname{cov}(m_2, m_1') = \mu_3/n,$$

and from (10.13) we have

$$\operatorname{cov}(\sqrt{m_2}, m_1') = \frac{1}{2\sqrt{\mu_2}}\operatorname{cov}(m_2, m_1') = \mu_3/(2n\sqrt{\mu_2}).$$

Substituting these various values in (10.18) we find

$$\operatorname{var} V = \frac{\mathbf{V}^2}{n}\left\{\frac{\mu_4 - \mu_2^2}{4\mu_2^2} + \frac{\mu_2}{\mu_1'^2} - \frac{\mu_3}{\mu_2\mu_1'}\right\}. \tag{10.19}$$

In the normal case ($\mu_3 = 0$, $\mu_4 = 3\mu_2^2$) this becomes

$$\operatorname{var} V = \frac{\mathbf{V}^2}{n}\left\{\frac{1}{2} + \frac{\mu_2}{\mu_1'^2}\right\}$$

$$= \frac{\mathbf{V}^2}{2n}(1 + 2\mathbf{V}^2). \tag{10.20}$$

10.8 A few points on the use of standard errors may be noted.

(a) The standard error, strictly speaking, has been justified only in the case where the statistic tends to normality. In other cases it may be used, in conjunction with the Bienaymé–Chebyshev inequality, to make statements in probability, but the more exact statements which are customarily made rely on the normal integral.

(b) Where there is serious doubt, it may be necessary to conduct a separate inquiry into the limiting normality of statistics. Many of those in current use, especially those dependent on sums of variates, like the moments, tend to normality under a Central Limit effect; but certain others, especially those dependent on extreme values, do not.

(c) On a similar point, it should also be remembered that some statistics tend to normality more rapidly than others, and a given n may be large for some purposes but not for others. So far as it is possible to generalize with safety, we can usually (but not always) assume values of n greater than 500 to be "large"; values greater than 100 are often great enough to be "large" for our purposes; values below 100 are suspect in many instances; and values below 30 are very rarely "large". Some measures calculated from the moments tend to normality very slowly. $\sqrt{b_1}$ or b_1 (the sample values of $\sqrt{\beta_1}$ or β_1) are cases in point, and more refined methods which we discuss in Chapter 12 are preferable to the use of the standard error.

(d) The sampling variances must relate to the statistic under consideration. For instance, if the standard deviation of a normal distribution is estimated by taking $\sqrt{(\frac{1}{2}\pi)}$ times the mean deviation of the sample, instead of the more usual $s = \sqrt{m_2}$, the formula $\text{var}\, s = (\mu_4 - \mu_2^2)/(4n\mu_2)$ derivable from (10.9) is not applicable.

(e) From (10.5) and (10.9) it will be seen that the sampling variance of a moment depends on the population moment of twice the order, i.e. becomes very large for higher moments, even when n is large. This is the reason why such moments have very limited practical application.

(f) The order of the approximation makes it necessary to exercise care in the neighbourhood of vanishing values of standard errors. For instance, if the coefficient of variation $V = 0$ in a sample, the formula of Example 10.5 would be estimated from the sample as $\text{var}\, V = 0$. But it does not, of course, follow that there is no variation at all in the population, though none exists in the sample and the presence of variation in the parent will be unlikely if the sample is at all large.

(g) It is interesting to compare the sampling fluctuations, as expressed in the sampling variance, with Sheppard's corrections to the moments. Writing temporarily s_1^2 for the uncorrected variance in the sample, s_2^2 for the corrected variance, we have

$$\frac{s_2^2}{s_1^2} = 1 - \frac{1}{12}\frac{h^2}{s_1^2},$$

where h is the interval width. For many practical cases, if d is the number of intervals, dh is about equal to $6s_1$, and thus

$$\frac{s_2^2}{s_1^2} = 1 - \frac{3}{d^2}$$

$$\frac{s_2}{s_1} = 1 - \frac{3}{2d^2} \text{ approximately.}$$

Writing s for the square root of the sample variance m_2 we have

$$\operatorname{var} s = (\mu_4 - \mu_2^2)/(4n\mu_2).$$

For a normal population this is equal to $\sigma^2/(2n)$.

Thus if n is, say, 1000, the standard error of s is about $0{\cdot}0224\sigma = 2{\cdot}24$ per cent. of σ. Sheppard's correction in a case where $d = 20$ is only $0{\cdot}375$ per cent. of s_1, i.e. only about a sixth of the standard error. It is as well to make the corrections, even when n is smaller than 1000, in order to avoid systematic error; but the correction should not be misinterpreted as implying a higher degree of reliability in the corrected value than actually exists.

Similar considerations apply *a fortiori* to the higher moments.

Standard errors of bivariate moments

10.9 Extensions of the above formulae to the bivariate case are made without difficulty, only slightly more complicated algebra being involved. The reader will be able to verify the following formulae for himself:

$$\operatorname{var}(m'_{r,s}) = \frac{1}{n}(\mu'_{2r,2s} - \mu'^2_{r,s}) \tag{10.21}$$

$$\operatorname{cov}(m'_{r,s}, m'_{u,v}) = \frac{1}{n}(\mu'_{r+u,s+v} - \mu'_{r,s}\mu'_{u,v}) \tag{10.22}$$

$$\begin{aligned}\operatorname{var}(m_{r,s}) = \frac{1}{n}(&\mu_{2r,2s} - \mu^2_{r,s} + r^2\mu_{2,0}\mu^2_{r-1,s} + s^2\mu_{0,2}\mu^2_{r,s-1} \\ &+ 2rs\,\mu_{1,1}\mu_{r-1,s}\mu_{r,s-1} - 2r\,\mu_{r+1,s}\mu_{r-1,s} - 2s\,\mu_{r,s+1}\mu_{r,s-1})\end{aligned} \tag{10.23}$$

$$\begin{aligned}\operatorname{cov}(m_{r,s}, m_{u,v}) = \frac{1}{n}(&\mu_{r+u,s+v} - \mu_{r,s}\mu_{u,v} + ru\,\mu_{2,0}\mu_{r-1,s}\mu_{u-1,v} \\ &+ sv\,\mu_{0,2}\mu_{r,s-1}\mu_{u,v-1} + rv\,\mu_{1,1}\mu_{r-1,s}\mu_{u,v-1} \\ &+ su\,\mu_{1,1}\mu_{r,s-1}\mu_{u-1,v} - u\,\mu_{r+1,s}\mu_{u-1,v} \\ &- v\,\mu_{r,s+1}\mu_{u,v-1} - r\,\mu_{r-1,s}\mu_{u+1,v} - s\,\mu_{r,s-1}\mu_{u,v+1}).\end{aligned} \tag{10.24}$$

Where no ambiguity is involved we may write $\mu_{r,s}$ as μ_{rs}.

Example 10.6

The correlation coefficient in a sample is defined by $r = m_{11}/(m_{20}m_{02})^{\frac{1}{2}}$ and in the population by $\rho = \mu_{11}/(\mu_{20}\mu_{02})^{\frac{1}{2}}$. From (10.17), we have

$$\operatorname{var} r = \rho^2\left\{\frac{\operatorname{var} m_{11}}{\mu_{11}^2} + \frac{\operatorname{var}(m_{20}m_{02})^{\frac{1}{2}}}{\mu_{20}\mu_{02}} - 2\,\frac{\operatorname{cov}\{m_{11}, (m_{20}m_{02})^{\frac{1}{2}}\}}{\mu_{11}(\mu_{20}\mu_{02})^{\frac{1}{2}}}\right\}$$

and evaluating $\operatorname{var}(m_{20}m_{02})^{\frac{1}{2}}$ and $\operatorname{cov}\{m_{11}(m_{20}m_{02})^{\frac{1}{2}}\}$ from (10.12) and (10.13), we find

$$\operatorname{var} r = \rho^2\left\{\frac{\operatorname{var} m_{11}}{\mu_{11}^2} + \frac{1}{4}\left(\frac{\operatorname{var} m_{20}}{\mu_{20}^2} + \frac{\operatorname{var} m_{02}}{\mu_{02}^2} + \frac{2\operatorname{cov}(m_{20}m_{02})}{\mu_{20}\mu_{02}}\right) - \left(\frac{\operatorname{cov}(m_{11}m_{02})}{\mu_{11}\mu_{02}} + \frac{\operatorname{cov}(m_{11}m_{20})}{\mu_{11}\mu_{20}}\right)\right\}$$

from which, on substituting for variances and covariances from (10.23) and (10.24), we have

$$\operatorname{var} r = \frac{\rho^2}{n}\left\{\frac{\mu_{22}}{\mu_{11}^2}+\frac{1}{4}\left(\frac{\mu_{40}}{\mu_{20}^2}+\frac{\mu_{04}}{\mu_{02}^2}+\frac{2\mu_{22}}{\mu_{20}\mu_{02}}\right)-\left(\frac{\mu_{31}}{\mu_{11}\mu_{20}}+\frac{\mu_{13}}{\mu_{11}\mu_{02}}\right)\right\}.$$

For the bivariate normal distribution the substitution of values of Example 3.17 gives

$$\operatorname{var} r = \frac{1}{n}(1-\rho^2)^2.$$

The use of the standard error to test a hypothetical non-zero value of ρ is not, however, to be recommended, since the sampling distribution of r tends to normality very slowly. We return to this point in **16.23–33** below.

Standard errors of quantiles

10.10 Among the various quantities measuring location and dispersion which we considered in Chapter 2 there was one group, namely the quantiles, which are not algebraic functions of the observations and whose sampling variances cannot accordingly be determined by the above methods. We proceed to consider them now.

Suppose the parent distribution is represented by $dF(x) = f(x)\,dx$. The probability that, of a sample of n, $(l-1)$ fall below a value x_1, one falls in the range $x_1 \pm \frac{1}{2}dx_1$ and the remaining $(n-l)$ fall above x_1 is proportional to

$$\{F(x_1)\}^{l-1}f(x_1)\,dx_1\,\{1-F(x_1)\}^{n-l} = F_1^{l-1}(1-F_1)^{n-l}\,dF_1, \tag{10.25}$$

where $F_1 = F(x_1)$. This expression is accordingly the frequency function of the value not exceeded by a proportion l/n of the sample, i.e. the lth quantile.

Put

$$l = nq$$

so that

$$n-l = n(1-q)$$
$$= np, \text{ say.}$$

The distribution (10.25) has a modal value given by differentiating the frequency function with respect to x_1, i.e. (taking logarithms first) by

$$(l-1)\frac{f_1}{F_1}-(n-l)\frac{f_1}{(1-F_1)}+\frac{f_1'}{f_1} = 0, \tag{10.26}$$

this equation being satisfied by the modal value $\check{x}$. Now for large n, the factor f_1'/f_1 will in general be small compared with the other terms in (10.26), l and $n-l$ being large. We may therefore neglect it, and (10.26) becomes, to order n^{-1},

$$\frac{q}{F}-\frac{p}{1-F} = 0$$

or

$$F(\check{x}) = q.$$

This is in accordance with our general assumptions. To order n^{-1} the quantile of the sample equals the quantile of the parent.

Now let us investigate the distribution (10.25) in the neighbourhood of the modal value. Put

$$F_1 = q+\xi.$$

(10.25) becomes (neglecting constants)

$$(q+\xi)^{nq}(p-\xi)^{np}.$$

Taking logarithms and expanding we have, except for constants,

$$\begin{aligned} & nq\log\left(1+\frac{\xi}{q}\right)+np\log\left(1-\frac{\xi}{p}\right) \\ &= nq\left(+\frac{\xi}{q}-\frac{1}{2}\frac{\xi^2}{q^2}\cdots\right)+np\left(-\frac{\xi}{p}-\frac{1}{2}\frac{\xi^2}{p^2}\cdots\right) \\ &= -\frac{n\xi^2}{2pq}+O(\xi^3). \end{aligned}$$

Now for large samples ξ will be small compared with q, and we neglect the terms of higher order. Thus the distribution of ξ is

$$dF \propto \exp\left(\frac{-n\xi^2}{2pq}\right)d\xi$$

or, evaluating the necessary constant by integration,

$$dF = \frac{1}{\sqrt{(2\pi)}\sqrt{(pq/n)}}\exp\left(\frac{-n\xi^2}{2pq}\right)d\xi, \tag{10.27}$$

showing that ξ is in the limit distributed normally with variance

$$\operatorname{var}\xi = \frac{pq}{n}. \tag{10.28}$$

This is the variance of ξ, which is a proportion. To find the variance of x_1 we note that $d\xi = dF_1 = f_1\,dx_1$ and hence that

$$\operatorname{var}x_1 = \frac{pq}{nf_1^2}. \tag{10.29}$$

In practice this formula is often applied to grouped frequency distributions, and in such applications it is to be remembered that f_1, the ordinate of the parent, is to be taken as the frequency *per unit interval* at x_1, this being the best estimate of the ordinate.

Example 10.7

If x_1 is the median, $p = q = \frac{1}{2}$ and we have var (median) $= 1/(4nf_1^2)$, where f_1 is the median ordinate. For instance, if the parent population is normal with variance σ^2, the median ordinate is (from Appendix Table 1) $0{\cdot}39894/\sigma$. Hence the standard error of the median is

$$\begin{aligned} & \frac{\sigma}{\sqrt{n}}\cdot\frac{1}{2\times 0{\cdot}39894} \\ &= 1{\cdot}2533\frac{\sigma}{\sqrt{n}}. \end{aligned}$$

The standard error of the mean in samples of n from a normal population is $\frac{\sigma}{\sqrt{n}}$, which is thus considerably smaller than the standard error of the median.

10.11 To find the covariance of two quantiles we generalize equation (10.25). If we have a random sample of n individuals the probability that $(l-1)$ lie below x_1, one lies at $x_1 \pm \frac{1}{2}dx_1$, $(n-l-m)$ lie between x_1 and x_2, one at $x_2 \pm \frac{1}{2}dx_2$, and the remaining $(m-1)$ above x_2 is

$$dF \propto F_1^{l-1}(F_2-F_1)^{n-l-m}(1-F_2)^{m-1}\,dF_1\,dF_2, \tag{10.30}$$

where $F_1 = F(x_1)$, $F_2 = F(x_2)$.

We put

$$l = q_1 n$$
$$m = p_2 n$$

and find, for the equations giving the modal values corresponding to (10.26),

$$\frac{q_1}{F_1} - \frac{(q_2-q_1)}{F_2-F_1} = 0$$
$$\frac{q_2-q_1}{F_2-F_1} - \frac{p_2}{1-F_2} = 0$$

giving, for the limiting modal values,

$$\left.\begin{aligned} F(\check{x}_1) &= q_1 \\ F(\check{x}_2) &= q_2 \end{aligned}\right\} \tag{10.31}$$

The conditions as to the relative smallness of $f'(\check{x})/f(\check{x})$ are satisfied in any ordinary case. Now put

$$F_1 = q_1+\xi_1$$
$$F_2 = q_2+\xi_2.$$

The joint distribution of ξ_1 and ξ_2 then becomes

$$dF \propto (q_1+\xi_1)^{q_1 n}(q_2-q_1+\xi_2-\xi_1)^{(q_2-q_1)n}(p_2-\xi_2)^{p_2 n}\,d\xi_1\,d\xi_2.$$

On proceeding as in the previous section, taking logarithms, expanding and neglecting terms in ξ^3 and higher, we find ultimately

$$dF \propto \exp\left\{-\frac{n}{2(q_2-q_1)}\left(\frac{q_2}{q_1}\xi_1^2 - 2\xi_1\xi_2 + \frac{p_1}{p_2}\xi_2^2\right)\right\}d\xi_1\,d\xi_2. \tag{10.32}$$

Thus the joint distribution of ξ_1 and ξ_2 tends to the bivariate normal form, and on comparing (10.32) with the standard form of Example 3.17, we see that

$$\frac{1}{(1-\rho^2)\operatorname{var}\xi_1} = \frac{nq_2}{(q_2-q_1)q_1}$$
$$\frac{1}{(1-\rho^2)\operatorname{var}\xi_2} = \frac{np_1}{(q_2-q_1)p_2}$$
$$\frac{\rho^2}{(1-\rho^2)\operatorname{cov}(\xi_1,\xi_2)} = \frac{\rho}{(1-\rho^2)\sqrt{\{\operatorname{var}\xi_1\operatorname{var}\xi_2\}}} = \frac{n}{(q_2-q_1)},$$

whence it is easy to find

$$\left.\begin{aligned} \operatorname{var}\xi_1 &= \frac{p_1 q_1}{n} \\ \operatorname{var}\xi_2 &= \frac{p_2 q_2}{n} \\ \operatorname{cov}(\xi_1,\xi_2) &= \frac{p_2 q_1}{n} \end{aligned}\right\} \tag{10.33}$$

The asymmetry of the result for the covariance is due to the fact that p_2 relates necessarily to the *upper* quantile. For the corresponding expression in x_1 and x_2 we have

$$\text{cov}(x_1, x_2) = \frac{p_2 q_1}{n f_2 f_1}. \tag{10.34}$$

With equations (10.33) and (10.34) we can find expressions for the variances of the interquartile range and similar statistics.

Example 10.8

The variance of the difference δ of two quantiles at x_1 and x_2 is given, from (10.16), by

$$d\delta = dx_1 - dx_2,$$

$$\text{var}\,\delta = \text{var}\,x_1 + \text{var}\,x_2 - 2\,\text{cov}(x_1, x_2)$$

$$= \frac{1}{n}\left\{\frac{p_1 q_1}{f_1^2} + \frac{p_2 q_2}{f_2^2} - \frac{2p_2 q_1}{f_1 f_2}\right\}.$$

When the quantiles are the two quartiles, $p_2 = q_1 = \frac{1}{4}$, $p_1 = q_2 = \frac{3}{4}$, and for the variance of the *semi*-interquartile range we have

$$\text{var}\,(\text{s.i.q.}) = \frac{1}{64n}\left(\frac{3}{f_1^2} + \frac{3}{f_2^2} - \frac{2}{f_1 f_2}\right),$$

where f_1, f_2 are the frequencies per unit interval at the two quartiles, f_2 relating to the upper quartile.

For instance, if the parent distribution is normal, $f_1 = f_2$ and we find

$$\text{var}\,(\text{s.i.q.}) = \frac{1}{16 n f_1^2}.$$

From the tables the deviate corresponding to the quartile is 0·6745 and the ordinate at this point, $f_1 = 0{\cdot}3178/\sigma$, so that the standard error of the semi-interquartile range is

$$\frac{\sigma}{\sqrt{n}(4 \times 0{\cdot}3178)}$$

$$= 0{\cdot}7867\frac{\sigma}{\sqrt{n}}.$$

10.12 In amplification of the point mentioned in **10.8** (d), it is worth while stressing again the fact that a standard error is related to the way in which a parameter is estimated. For instance, the standard deviation of a normal distribution can be estimated from a sample in several ways : from the second moment ; by taking $\sqrt{(\frac{1}{2}\pi)}$ times the mean deviation ; by taking $1/0{\cdot}6745$ times the semi-interquartile range ; and so on. Each method will have its appropriate standard error, that for the first (**10.8**) being $\sigma/\sqrt{(2n)}$, and that for the third $1{\cdot}6495\sigma/\sqrt{(2n)}$, as is easily deduced from Example 10.8. At a later stage considerations such as this will lead us to the inquiry, what is the estimate, if any, with the *minimum* sampling variance ? For present purposes it is enough to note the importance of not using a quoted formula without reference to the method of estimation of the parameter concerned.

Standard error of the mean deviation

10.13 Suppose that we are sampling from a population with mean μ and variance σ^2. Let d_α be the sample mean deviation measured about the *fixed* point α, i.e.

$$d_\alpha = \frac{1}{n}\sum_{i=1}^{n} |x_i - \alpha|.$$

Then

$$E(d_\alpha) = E|x_i - \alpha| = \delta_\alpha, \tag{10.35}$$

where δ_α is the corresponding population mean deviation, and from (10.16)

$$\text{var}\, d_\alpha = \frac{1}{n^2}\left\{\sum_{i=1}^{n} \text{var}(|x_i - \alpha|) + \sum_{i \neq j}\sum \text{cov}(|x_i - \alpha|, |x_j - \alpha|)\right\}.$$

Since the $|x_i - \alpha|$ are completely independent, their covariances are zero and we have

$$\text{var}\, d_\alpha = \frac{1}{n}\{E(|x-\alpha|^2) - E^2(|x-\alpha|)\}$$

$$= \frac{1}{n}\{\sigma^2 + (\alpha-\mu)^2 - \delta_\alpha^2\}. \tag{10.36}$$

If $\alpha = \mu$, the parent mean, this becomes

$$\text{var}\, d_\mu = \frac{1}{n}(\sigma^2 - \delta_\mu^2). \tag{10.37}$$

We denoted δ_μ by δ_1 in Chapter 2.

Incidentally, (10.37) establishes that $\delta_\mu \leqslant \sigma$—cf. the stronger inequality given at (2.22).

(10.35–7) are exact results. Consider now the sample mean deviation d_a about a value a, itself subject to sampling variation, which has the properties

$$E(a) = \alpha, \qquad \text{var}\, a = O(n^{-1}).$$

The $|x_i - a|$ are not independent, since a is common to them all, and their covariances will not disappear as above. Their sum may, however, be $o(n)$, and therefore asymptotically negligible compared with the sum of the variances. Since $d_a = d_\alpha\{1 + O(n^{-\frac{1}{2}})\}$, we then find as in **10.5** that *approximately*

$$E(d_a) = \delta_\alpha, \qquad \text{var}\, d_a = \text{var}\, d_\alpha.$$

Exercise 10.14 shows that these approximations hold for any symmetrical population when $a = \bar{x}$, the sample mean, and gives the general asymptotic expression for var $d_{\bar{x}}$. We now write simply d for $d_{\bar{x}}$, so that

$$\text{var}\, d = \frac{1}{n}(\sigma^2 - \delta_\mu^2). \tag{10.38}$$

For a normal population we have seen at (5.75) that $\delta_\mu = \sigma\sqrt{(2/\pi)}$ and hence

$$\text{var}\, d = \sigma^2(1 - 2/\pi)/n. \tag{10.39}$$

In this normal case, the mean and variance of d can be obtained exactly. Since

$$E(d) = E|x_i - \bar{x}|$$

and $(x_i - \bar{x})$ has zero mean and exact variance $\sigma^2(n-1)/n$ by (10.16), (5.75) gives

$$E(d) = \left\{\frac{2\sigma^2(n-1)}{\pi n}\right\}^{\frac{1}{2}}.$$

For the sampling variance, Helmert (1876) and Fisher (1920) found

$$\text{var}\, d = \frac{2\sigma^2(n-1)}{\pi n^2}[\tfrac{1}{2}\pi + \sqrt{\{n(n-2)\}} - n + \arcsin\{1/(n-1)\}], \tag{10.40}$$

a proof of which is indicated in Exercise 15.6 below. For large n, this expression tends to (10.39). Godwin (1945, 1948) gives a recurrence relation for the moments, an approximation to the distribution, and an exact expression for the d.f. of d, which is tabulated for $n = 2(1)10$ by Hartley (1945) and reproduced in Vol. II of the *Biometrika Tables*, Vol. I of which gives small-sample tables of the percentage points and the moments of d.

Standard error of the mean difference

10.14 Let us write g for the sample value of Gini's mean difference without repetition, for which the parental value is Δ (**2.21**). We write μ and σ^2 for the population mean and variance. We have

$$g = \frac{1}{n(n-1)} \sum_{i,j=1}^{n} |x_i - x_j|, \tag{10.41}$$

and

$$E(g) = E|x_i - x_j| = \Delta. \tag{10.42}$$

Also

$$E(g^2) = \frac{1}{n^2(n-1)^2} E\{\Sigma |x_i - x_j|\}^2$$

$$= \frac{1}{n^2(n-1)^2} E\{\Sigma(x_i - x_j)^2 + \Sigma' |x_i - x_j||x_i - x_k| + \Sigma'' |x_i - x_j||x_k - x_l|\},$$

$$i \neq j \neq k \neq l,$$

where there are $2n(n-1)$ terms in the summation Σ, $4n(n-1)(n-2)$ in Σ' and $n(n-1)(n-2)(n-3)$ in Σ''. We then find

$$E(g^2) = \frac{1}{n(n-1)}\{2E(x_i - x_j)^2 + 4(n-2)E(|x_i - x_j||x_i - x_k|) + (n-2)(n-3)E(|x_i - x_j||x_k - x_l|)\}.$$

The first expected value is $2\sigma^2$, by (2.27), and the third is Δ^2 because the x's are all independent there. Writing $\mathcal{J}$ for the second, we have

$$E(g^2) = \frac{1}{n(n-1)}\{4\sigma^2 + 4(n-2)\mathcal{J} + (n-2)(n-3)\Delta^2\}$$

and hence

$$\text{var}\, g = \frac{1}{n(n-1)}\{4\sigma^2 + 4(n-2)\mathcal{J} - 2(2n-3)\Delta^2\}, \tag{10.43}$$

where

$$\mathcal{J} = \int_{-\infty}^{\infty}\int_{-\infty}^{\infty}\int_{-\infty}^{\infty} |x-y||x-z| f(x)f(y)f(z)\,dx\,dy\,dz$$

$$= \int_{-\infty}^{\infty}\left\{\int_{-\infty}^{x}\int_{-\infty}^{x}(x-y)(x-z) + \int_{-\infty}^{x}\int_{x}^{\infty}(x-y)(z-x) + \int_{x}^{\infty}\int_{-\infty}^{x}(y-x)(x-z) + \int_{x}^{\infty}\int_{x}^{\infty}(y-x)(z-x)\right\} dF(y)\,dF(z)\,dF(x),$$

and if we write $G(x) = \int_{-\infty}^{t} tf(t)\,dt$, integration with respect to y and z gives

$$\mathcal{J} = \int_{-\infty}^{\infty} \{(xF-G)^2+2(xF-G)(xF-G+\mu-x)+(xF-G+\mu-x)^2\}\,dF(x)$$

$$= 4\int_{-\infty}^{\infty} \{(xF-G)^2+(\mu-x)(xF-G)\}\,dF(x)+\sigma^2. \tag{10.44}$$

Writing I for the integral in (10.44), we finally have from (10.43)

$$\operatorname{var} g = \frac{2}{n(n-1)}\{2(n-1)\sigma^2+8(n-2)I-(2n-3)\Delta^2\}. \tag{10.45}$$

This derivation is due to Lomnicki (1952), a more complicated proof having been first given by Nair (1936).

Sillitto (1969) gives a method of estimating var g from the sample.

Three special cases are of interest:

Normal population : any mean, variance σ^2.

$$E(g) = 2\sigma/\sqrt{\pi} = 1{\cdot}1284\sigma,$$

$$\operatorname{var} g = \frac{4\sigma^2}{n(n-1)}\{\tfrac{1}{3}(n+1)+2(n-2)\sqrt{3}/\pi-2(2n-3)/\pi\}\sim\sigma^2(0{\cdot}8068)^2/n. \tag{10.46}$$

Third and fourth moments, and an approximation to the distribution, of g in the normal case are given by Kamat (1959).

Exponential population : $dF = \exp(-x/\sigma)\,dx/\sigma.\quad (0\leqslant x\leqslant\infty)$

$$E(g) = \sigma,$$

$$\operatorname{var} g = \sigma^2\frac{2(2n-1)}{3n(n-1)}\sim 4\sigma^2/(3n). \tag{10.47}$$

Rectangular population : $dF = dx/k.\quad (0\leqslant x\leqslant k)$

$$E(g) = \tfrac{1}{3}k,$$

$$\operatorname{var} g = \frac{k^2(n+3)}{45n(n-1)}\sim k^2/(45n). \tag{10.48}$$

Some commonly required standard errors

10.15 It may be useful to list here for reference (see table overleaf) some of the standard errors commonly required. Certain of the results have been obtained already in this chapter; others are direct consequences of the methods we have developed; some relate to statistics which we shall meet at a later stage.

For convenience in writing, the variance is usually given, not the standard error; but sometimes the numerical coefficient of a variance is expressed as a square so that the square root is obtainable at sight.

Statistic	Variance multiplied by n	Notes
Mean, m_1'	$\mu_2\ (= \sigma^2)$	True for any population with finite second moment μ_2.
Sample variance, m_2 . . .	$\mu_4-\mu_2^2$	For normal parent, $= 2\sigma^4$.
Sample s.d., s	$(\mu_4-\mu_2^2)/(4\mu_2)$	For normal parent, $= \sigma^2/2$.
Third moment, m_3 . . .	$\mu_6-\mu_3^2-6\mu_4\mu_2+9\mu_2^3$	For normal parent, $= 6\sigma^6$.
Fourth moment, m_4 . . .	$\mu_8-\mu_4^2-8\mu_5\mu_3+16\mu_2\mu_3^2$	For normal parent, $= 96\sigma^8$.
$\sqrt{b_1} = m_3/m_2^{3/2}$ . . .	6	For normal parent only. See **12.18** and Exercise 12.9.
$b_2 = m_4/m_2^2$	24	For normal parent only. See Exercise 12.10.
Coefficient of variation, V .	See Example 10.5.	For normal parent only, $= V^2(1+2V^2)/2$.
Pearson mode (cf. **6.2**) .	See Yasukawa (1926) for formulae and tables.	Distribution skew for moderate n.
Mean deviation	See **10.13** and Exercise 10.14.	For normal parent, $= \sigma^2(1-2/\pi)$.
Gini's mean difference . .	See **10.14**.	For normal parent, $= (0{\cdot}8068)^2\sigma^2$.
Median	$1/(4y_0^2)$ when y_0 is ordinate at median.	For normal parent and small samples see Hojo (1931) and K. Pearson (1931). For large normal samples equals $(1{\cdot}2533)^2\sigma^2$.
Quartile	$3/(16y^2)$ where y is the ordinate at the quartile.	For normal parent, $= (1{\cdot}3626)^2\sigma^2$. See also Hojo (1931).
Deciles	See **10.10**.	For normal parent, (deciles 4, 6) $= (1{\cdot}2680)^2\sigma^2$; (deciles 3, 7) $= (1{\cdot}3180)^2\sigma^2$; (deciles 2, 8) $= (1{\cdot}4288)^2\sigma^2$; (deciles 1, 9) $= (1{\cdot}7094)^2\sigma^2$.
Semi-interquartile range .	$\frac{1}{4}\{3/16y_1^2+3/16y_2^2 -1/8y_1y_2\}\sigma^2$, where y_1, y_2 are the quartile ordinates.	For normal parent, $= (0{\cdot}7867)^2\sigma^2$.
Correlation coefficient r (product-moment)	See Example 10.6.	For bivariate normal parent, $= (1-\rho^2)^2$. Not to be recommended except for very large samples. See Chapter 16.

EXERCISES

10.1 In the height distribution of Table 1.7 it has been found that

$$m_2 = 6{\cdot}616$$
$$m_3 = -0{\cdot}207$$
$$m_4 = 137{\cdot}689.$$

Regarding the distribution as a random sample from a population which is approximately normal, show that m_3 does not differ significantly from zero (which, of course, must be so if the assumption of normality is to be maintained) and that m_4 has a standard error of about 4 per cent. of its value.

10.2 A multinomial distribution has k classes with probabilities $\varpi_1, \varpi_2, \ldots, \varpi_k$. In a sample of size n the observed frequencies in these classes are $f_1, f_2, \ldots, f_k$, and $p_i = f_i/n$. Show that a function $T(np_1, np_2, \ldots, np_k)$, where T does not depend on n except as shown, has asymptotic variance

$$n \operatorname{var} T = \sum_{i=1}^{k} \varpi_i \left(\frac{\partial T}{\partial p_i}\right)^2 - \left(n\frac{\partial T}{\partial n}\right)^2,$$

the derivatives being taken at the values $p_i = \varpi_i$.

(Fisher, *Statistical Methods for Research Workers*)

10.3 With a random sample of 100 it is found that the odds are 99 to 1 against a certain statistic differing from the parent value by more than two units. Find how big the sample would have to be to increase these odds to 199 to 1; and find the odds that with the sample of 100 the statistic would differ from the parent value by one unit.

10.4 If k random samples of different sizes are drawn from a population in which the proportion of members bearing an attribute A is $\bar{\omega}$, show that the standard error of the mean proportion of A in the samples is $\sqrt{\dfrac{\bar{\omega}(1-\bar{\omega})}{kH}}$ where H is the harmonic mean of the numbers in the samples.

10.5 Using the results of **10.6**, show that the correlation in large samples between any regular functions $y(x_1)$ and $z(x_2)$ is the same as that between x_1 and x_2.

10.6 Show that for large samples from a symmetrical distribution any sample central moment of odd order has zero covariance to order n^{-1} with any sample central moment of even order.

10.7 In the notation of this chapter show that, to order n^{-1},

$$\operatorname{cov}(m_r, m_q') = (\mu_{q+r} - \mu_q \mu_r - r\mu_{q+1}\mu_{r-1})/n.$$

10.8 Show that, for a symmetrical distribution, the covariance in large samples of two sample central moments of even order cannot be negative to order n^{-1}.

10.9 The variance of m_2 in large samples is $(\mu_4 - \mu_2^2)/n$. If this is estimated as $(m_4 - m_2^2)/n$, show that the variance of the estimated $\operatorname{var} m_2$ in samples from a normal population is $56\sigma^8/n^2$. Compare with the value of $\operatorname{var} m_2$ in normal variation, $2\sigma^4/n$.

10.10 $x_1, x_2, \ldots, x_p, \ldots, x_n$ are independent, identically distributed variates with zero mean, and $\bar{x}_p = \sum_{i=1}^{p} x_i/p$, $\bar{x}_n = \sum_{i=1}^{n} x_i/n$. Show that the correlation coefficient (i.e. the covariance divided by the product of the standard errors)

(a) between $\bar{x}_p$ and $\bar{x}_n$ is $(p/n)^{\frac{1}{2}}$;
(b) between $\bar{x}_p-\bar{x}_n$ and $\bar{x}_n$ is zero; and
(c) between $\bar{x}_p$ and $\bar{x}_p-\bar{x}_n$ is $\left(1-\frac{p}{n}\right)^{\frac{1}{2}}$.

10.11 Verify equation (10.46).

10.12 Show that the sampling variances of the first four cumulants, as calculated from the moments, are given to order n^{-1} by

$$\text{var}\,\kappa_1 = \frac{1}{n}\kappa_2, \qquad \text{var}\,\kappa_2 = \frac{1}{n}(\kappa_4+2\kappa_2^2), \qquad \text{var}\,\kappa_3 = \frac{1}{n}(\kappa_6+9\kappa_4\kappa_2+9\kappa_3^2+6\kappa_2^3),$$

$$\text{var}\,\kappa_4 = \frac{1}{n}(\kappa_8+16\kappa_6\kappa_2+48\kappa_5\kappa_3+34\kappa_4^2+72\kappa_4\kappa_2^2+144\kappa_3^2\kappa_2+24\kappa_2^4).$$

(*Note.* The sample values of the κ's on the left have been written with Greek letters; to replace them in the customary way by k's might produce confusion with the k-statistics to be introduced in Chapter 12.)

10.13 Show that the mean value of the variance is given exactly by

$$E(m_2) = \frac{n-1}{n}\mu_2$$

and that its variance is given exactly by

$$\text{var}\,m_2 = \left(\frac{n-1}{n}\right)^2\frac{\mu_4-\mu_2^2}{n}+\frac{2(n-1)}{n^3}\mu_2^2.$$

Hence verify that (10.9) is accurate to order n^{-1}.

10.14 In **10.13**, show as in **2.18** that the mean deviation

$$d_{\bar{x}} = \frac{2}{n}\sum_{x_i<\bar{x}}(\bar{x}-x_i) = \frac{2}{n}\left\{p\sum_{x_i>\bar{x}}x_i-(1-p)\sum_{x_i<\bar{x}}x_i\right\}$$

where there are np observations $\leqslant x$ in value. Hence show that

$$\text{var}\,d_{\bar{x}} \sim \frac{4}{n}\left\{P^2\int_{\mu}^{\infty}(x-\mu)^2\,dF+(1-P)^2\int_{-\infty}^{\mu}(x-\mu)^2\,dF\right\}-\delta_{\mu}^2,$$

where $P = F(\mu)$. Show that when the population is symmetrical, this reduces to (10.38).

(Gastwirth, 1974)

10.15 Show that in odd samples of n from a population with frequency-function uniform over a unit range the sampling variance of the distribution of the median is given exactly by $1/\{4(n+2)\}$.

10.16 In the notation of this chapter, if two variates are independent, if the distribution of each is symmetric, and if $r+s$ is an even number, show that

$$n\,\text{cov}\,(m_{r,s}, m_{u,v}) = \mu_{r+u,0}\,\mu_{0,s+v}-\mu_{r0}\,\mu_{u0}\,\mu_{0s}\,\mu_{0v}.$$

10.17 In **10.6** show that to the next order of approximation

$$E\{g(x)\} = g(\theta)+\tfrac{1}{2}\left[\sum_{i=1}^{k}g_i''(\theta)\,\text{var}\,x_i+\sum\sum_{i\neq j}g_{ij}''(\theta)\,\text{cov}\,(x_i,x_j)\right]+o(n^{-r})$$

and that if all the $g_i'(\theta) = 0$, we also have to the next order

$$E\{[g(x)-g(\theta)]^2\} = \tfrac{1}{4}E\left\{\left[\sum_{i=1}^{k} g_i''(\theta)(x_i-\theta_i)^2+\sum\sum_{i\neq j} g_{ij}''(\theta)(x_i-\theta_i)(x_j-\theta_j)\right]^2\right\}$$

and
$$\text{var}\{g(x)\} = E\{[g(x)-g(\theta)]^2\}-[E\{g(x)-g(\theta)\}]^2.$$

10.18 In binomial sampling for attributes (see **9.21**) with population parameter ϖ and sample proportion p, use the results of (10.14) and Exercise 10.17 to show that to the first orders

$$\begin{aligned}\text{var}\{p(1-p)\} &= \varpi(1-\varpi)(1-2\varpi)^2/n, & \varpi \neq \tfrac{1}{2},\\ &= (n-1)/(8n^3), & \varpi = \tfrac{1}{2}.\end{aligned}$$

Verify directly from the moments of the binomial that

$$\text{var}\{p(1-p)\} = \frac{\varpi(1-\varpi)(n-1)^2}{n^3}\left\{(1-2\varpi)^2+\frac{2\varpi(1-\varpi)}{n-1}\right\}$$

exactly, agreeing with the above first-order results. (The first-order result for $\varpi = \frac{1}{2}$ is exact here, because $p(1-p)$ is exactly represented by the second-order Taylor expansion.)

10.19 Show that in samples of size n from the rectangular population $dF = k\,dx$, $0 \leqslant x \leqslant 1/k$, the variance of the mean deviation is approximately $1/(48k^2n)$.

10.20 Using the first result of Exercise 10.17, and the results of Exercise 10.13, show that, writing $s^2 = m_2n/(n-1)$, the standard deviation of a population is estimated with bias of order less than that of s itself by

$$s\left(1+\frac{\dfrac{m_4}{s^4}-1}{8n}\right),$$

reducing to $s\{1+(4n)^{-1}\}$ in the normal case.

10.21 Two lists of M and N items have D items in common. Independent random samples of sizes m and n respectively ($m \geqslant n \geqslant 2$) are taken without replacement and found to have d items in common. Show that an unbiassed estimate of D is given by

$$\hat{D} = \frac{MN}{mn}d$$

and that

$$\text{var}\,\hat{D} = \frac{MND}{mn}\left\{1+\frac{m-1}{M-1}\cdot\frac{n-1}{N-1}(D-1)\right\}-D^2.$$

Show that as d approaches n, $\hat{D}$ exceeds N and thus that the estimate takes " impossible " values.

(Cf. Goodman (1952); also Deming and Glasser (1959))

10.22 In Exercise 10.21, show that an unbiassed estimate of var $\hat{D}$ is given by

$$\hat{V} = \hat{D}\left(\frac{M-1}{m-1}\cdot\frac{N-1}{n-1}-1\right)+\hat{D}^2\left(1-\frac{mn}{MN}\cdot\frac{M-1}{m-1}\cdot\frac{N-1}{n-1}\right)$$

and hence that $\hat{V}$ may be negative, and thus an " impossible " estimate of var $\hat{D}$, when m, n, M and N are all small.

10.23 x and y are jointly distributed with moments μ_{rs}. Show that

$$\text{var}(xy) = \mu_{10}'^2\mu_{02}+\mu_{20}\mu_{01}'^2+\mu_{20}\mu_{02}+\text{cov}(x^2, y^2)-\mu_{11}^2-2\mu_{10}'\mu_{11}\mu_{01}' \qquad \text{(A)}$$

exactly, and that when x is independent of y, this reduces to

$$\text{var}(xy) = \mu_{10}'^2\mu_{02}+\mu_{20}\mu_{01}'^2+\mu_{20}\mu_{02}. \qquad \text{(B)}$$

Show that (B) agrees to order n^{-1} with the large-sample result obtained from (10.12) if μ_{20} and μ_{02} are of order n^{-1}.

If m_{20}, m_{02} are unbiased estimates of μ_{20}, μ_{02}, show that an unbiased estimate of (B) is

$$x^2 m_{02}+y^2 m_{20}-m_{20} m_{02}.$$

(Cf. Goodman (1960); Goodman (1962) generalizes to the product of more than two variates, and Bohrnstedt and Goldberger (1969) give the covariance of two products xy, uv.)

CHAPTER 11

EXACT SAMPLING DISTRIBUTIONS

11.1 The role of the sampling distribution in statistical inference has been indicated in Chapter 7. In the present chapter we propose to give an account of the main methods of finding such distributions when the population from which the sample was derived is specified. It will, as usual, be assumed that the sampling is simple and random. Thus, if the parent distribution is $F(x)$ the joint distribution of n values $x_1, \ldots, x_n$ is $F(x_1)\,F(x_2)\ldots F(x_n)$; and if z is a statistic

$$z = z(x_1, \ldots, x_n) \tag{11.1}$$

the distribution function of z is given by

$$G(z_0) = \int \ldots \int dF(x_1) \ldots dF(x_n), \tag{11.2}$$

the integration being taken over the domain of the x's such that $z(x_1, \ldots, x_n) \leqslant z_0$.

Formally, (11.2) is the solution of our problem, which thus reduces to the purely mathematical one of evaluating certain multiple integrals or sums. The methods with which we are here concerned are fundamentally devices of various kinds to facilitate the integrative process. They fall into three main groups :—

(a) straightforward evaluation of the integral (11.2) by ordinary analytical processes such as a convenient change of variable;
(b) the use of geometrical terminology to effect the same object and to avoid cumbrous analytical formulae; and
(c) the use of characteristic functions.

11.2 As an illustration of the straightforward analytical approach, let us find the distribution of the sum of squares of n independent variables, each of which is distributed normally with zero mean and unit variance. The joint distribution of the n variables is then the product of n quantities of type $\frac{1}{\sqrt{(2\pi)}} e^{-\frac{1}{2}x^2}$, that is to say

$$dF = \frac{1}{(2\pi)^{\frac{1}{2}n}} \exp\{-\tfrac{1}{2}(x_1^2+x_2^2+\ldots+x_n^2)\}\, dx_1 \ldots dx_n. \tag{11.3}$$

We require the sampling distribution of

$$z = x_1^2+x_2^2+\ldots+x_n^2. \tag{11.4}$$

We have thus to evaluate the multiple integral

$$G(z_0) = \int \ldots \int \frac{1}{(2\pi)^{\frac{1}{2}n}} \exp(-\tfrac{1}{2}\Sigma x^2)\, dx_1 \ldots dx_n$$

over the domain of x's conditioned by $z \leqslant z_0$.

Make the polar transformation to variables $z, \theta_1, \theta_2, \ldots, \theta_{n-1}$

$$\left.\begin{aligned} x_1 &= z^{\frac{1}{2}} \cos\theta_1 \cos\theta_2 \ldots \cos\theta_{n-1}, \\ x_j &= z^{\frac{1}{2}} \cos\theta_1 \cos\theta_2 \ldots \cos\theta_{n-j} \sin\theta_{n-j+1}, \\ & \quad j = 2, 3, \ldots, n-1 \; ; \; n > 2, \\ x_n &= z^{\frac{1}{2}} \sin\theta_1. \end{aligned}\right\} \tag{11.5}$$

The Jacobian of this transformation is given by

$$\frac{\partial(x_1, \ldots, x_n)}{\partial(z, \theta_1, \ldots, \theta_{n-1})},$$

which is equal to $\frac{1}{2}z^{\frac{1}{2}n-1}$ times the determinant

$$\begin{vmatrix} \cos\theta_1\cos\theta_2\ldots\cos\theta_{n-1} & \cos\theta_1\cos\theta_2\ldots\cos\theta_{n-2}\sin\theta_{n-1}\ldots & \sin\theta_1 \\ -\sin\theta_1\cos\theta_2\ldots\cos\theta_{n-1} & -\sin\theta_1\cos\theta_2\ldots\cos\theta_{n-2}\sin\theta_{n-1}\ldots & \cos\theta_1 \\ -\cos\theta_1\sin\theta_2\ldots\cos\theta_{n-1} & -\cos\theta_1\sin\theta_2\ldots\cos\theta_{n-2}\sin\theta_{n-1}\ldots & 0 \\ \ldots \quad \ldots & \ldots \quad \ldots \quad \ldots & \ldots \\ -\cos\theta_1\cos\theta_2\ldots\sin\theta_{n-1} & +\cos\theta_1\cos\theta_2\ldots\cos\theta_{n-1} \quad \ldots & 0 \end{vmatrix}.$$

Taking out common factors in columns, we find that this determinant is equal to

$$\cos^{n-1}\theta_1\cos^{n-2}\theta_2\ldots\cos\theta_{n-1}\sin\theta_1\;\sin\theta_2\ldots\sin\theta_{n-1}\text{ times}$$

$$\begin{vmatrix} 1 & 1 & 1 & \ldots & 1 \\ -\tan\theta_1 & -\tan\theta_1 & -\tan\theta_1 & \ldots & \cot\theta_1 \\ -\tan\theta_2 & -\tan\theta_2 & -\tan\theta_2 & \ldots & 0 \\ \ldots & \ldots & \ldots & \ldots & \ldots \\ -\tan\theta_{n-2} & -\tan\theta_{n-2} & \cot\theta_{n-2} & \ldots & 0 \\ -\tan\theta_{n-1} & \cot\theta_{n-1} & 0 & \ldots & 0 \end{vmatrix}$$

and, on subtracting each column from the preceding one, the determinant is found to reduce to $\cos^{n-2}\theta_1\cos^{n-3}\theta_2 \;\ldots\; \cos\theta_{n-2}$.

Thus our integral becomes

$$\int\ldots\int\frac{1}{(2\pi)^{\frac{1}{2}n}}e^{-\frac{1}{2}z}\,\tfrac{1}{2}z^{\frac{1}{2}n-1}\cos^{n-2}\theta_1\ldots\cos\theta_{n-2}\,dz\,d\theta_1\ldots d\theta_{n-1}. \tag{11.6}$$

The advantage of the transformation is that the limits of the variables are now much simpler. z itself can vary from 0 upwards, θ_{n-1} from 0 to 2π and the other θ's from $-\frac{1}{2}\pi$ to $\frac{1}{2}\pi$. Thus the integral (11.6) divides into a product of integrals, those in θ_j being constants, and we find for our distribution function of z

$$G(z_0) = k\int_0^{z_0} e^{-\frac{1}{2}z}\,z^{\frac{1}{2}n-1}\,dz. \tag{11.7}$$

The constant k may be evaluated by integration between 0 and ∞ and hence the distribution sought is

$$dG = \frac{1}{2^{\frac{1}{2}n}\Gamma(\frac{1}{2}n)}e^{-\frac{1}{2}z}\,z^{\frac{1}{2}n-1}\,dz, \qquad 0 \leqslant z \leqslant \infty, \tag{11.8}$$

a Pearson Type III distribution, usually known as the χ^2 distribution, of which the parameter n in (11.8) is called the number of " degrees of freedom ".

Exercise 11.26 deals with the case $n = 2$.

11.3 The essential feature of the change of variables is the simplification of the domain of integration as defined by the limits of the new variables. In general, we take the statistic whose sampling distribution is being sought as one of the new variables and choose $n-1$ others in any way which may be convenient to the particular problem. Then, if J is the Jacobian of the transformation, namely

$$J = \frac{\partial(x_1, \ldots, x_n)}{\partial(z, \theta_1, \ldots, \theta_{n-1})},$$

the integral (11.2) becomes

$$G(z) = \int \ldots \int f(x_1) \ldots f(x_n) \frac{\partial(x_1, \ldots, x_n)}{\partial(z, \theta_1, \ldots, \theta_{n-1})} dz\, d\theta_1 \ldots d\theta_{n-1}, \tag{11.9}$$

$f(x_j)$ being the frequency function of the population and x_j being expressed in terms of z and the θ's. The integration now takes place with respect to the θ's, which can usually be chosen so as to vary between limits which are independent of z; and thus the integral (11.2) is replaced by more easily calculable definite integrals.

As always in such cases J is subject to an ambiguity of sign which must be determined so as to make the transformed integral positive. The validity of the variate-transformation depends on the familiar conditions governing the change of variable in a multiple integral. For example, it is a sufficient condition that the new variables and their first derivatives shall be continuous in the x's and that J does not change sign in the domain of integration. Some further examples will make the general type of investigation clear.

Example 11.1

To find the distribution of the mean of a sample of n values $x_1, \ldots, x_n$ from the Cauchy distribution

$$dF = \frac{dx}{\pi(1+x^2)}, \qquad -\infty \leqslant x \leqslant \infty.$$

The joint distribution is

$$\frac{1}{\pi^n} \prod_{j=1} \frac{dx_j}{(1+x_j^2)}, \tag{11.10}$$

and the statistic z is given by

$$nz = \sum_{j=1}^{n} x_j. \tag{11.11}$$

We have to integrate (11.10) over a domain of x's subject to $\Sigma x \leqslant nz$. Let us take new variables $u_1 = x_1$, $u_2 = x_2, \ldots, u_{n-1} = x_{n-1}$, and $u_n = \frac{1}{n}\sum_1^n x = z$. Here J is evidently equal to the constant n. Our new variables $u_1, \ldots, u_{n-1}$ may extend from $-\infty$ to $+\infty$ and the new variable u_n from $-\infty$ to z. We then have

$$G(z) = \int_{-\infty}^{z} du_n \int_{-\infty}^{\infty} \ldots \int_{-\infty}^{\infty} \frac{n}{\pi^n} \prod_{j=1}^{n-1} \frac{du_j}{(1+u_j^2)\{1+(nu_n - u_1 - \ldots - u_{n-1})^2\}} \tag{11.12}$$

and the *frequency* function of z is given by the $(n-1)$-fold multiple integral in $u_1, \ldots, u_{n-1}$ in (11.12). This integral may be evaluated by step-by-step integration. We have

$$\frac{1}{(1+x^2)\{r^2+(a-x)^2\}} = \frac{1}{\{a^2+(r+1)^2\}\{a^2+(r-1)^2\}}$$
$$\left[\frac{2ax}{x^2+1}+\frac{a^2+r^2-1}{x^2+1}+\frac{2a^2-2ax}{r^2+(a-x)^2}+\frac{a^2-r^2+1}{r^2+(a-x)^2}\right]$$

whence, integrating with respect to x from $-\infty$ to $+\infty$, we find on the right

$$\frac{1}{\{a^2+(r+1)^2\}\{a^2+(r-1)^2\}}\left[a\log(x^2+1)-a\log\{r^2+(a-x)^2\}\right.$$
$$\left.+(a^2+r^2-1)\arctan x+\frac{a^2-r^2+1}{r}\arctan\frac{x-a}{r}\right]_{-\infty}^{\infty}$$

reducing to
$$\pi\left(\frac{r+1}{r}\right)\left\{\frac{1}{a^2+(r+1)^2}\right\}. \tag{11.13}$$

Thus in (11.12), taking $x = u_{n-1}$, $r = 1$, $a = nu_n-u_1- \ . \ . \ . \ -u_{n-2}$, we find that the $(n-1)$-fold integral reduces to

$$\int_{-\infty}^{\infty}\cdots\int_{-\infty}^{\infty}\frac{n}{\pi^{n-1}}\prod_{j=1}^{n-2}\frac{du_j}{(1+u_j^2)}\frac{2}{\{2^2+(nu_n-u_1-\ldots-u_{n-2})^2\}}.$$

Integrating with respect to $u_{n-2}, u_{n-3}, \ldots$ successively, we reduce this eventually to

$$\frac{n^2}{\pi\{n^2+(nu_n)^2\}} = \frac{1}{\pi(1+u_n^2)}. \tag{11.14}$$

Thus the distribution of z is given by

$$dG = \frac{dz}{\pi(1+z^2)}, \qquad -\infty \leqslant z \leqslant \infty, \tag{11.15}$$

and is the same as that of a single observation.

This is an interesting example of the failure of the Central Limit theorem, the mean of samples of n observations failing to tend to normality for large n since the moments of the distribution do not exist.

Exercise 11.25 gives an even more extreme instance of the failure of the Central Limit effect.

Example 11.2

To find the distribution of a linear function of n independent variables $x_1, \ldots, x_n$ where x_j is distributed normally with zero mean and variance σ_j^2.

Let the linear function be

$$z = a_1x_1+\ldots+a_nx_n. \tag{11.16}$$

Then by a transformation $u_j = x_j/\sigma_j$ we have

$$z = \Sigma a_j\sigma_j u_j \tag{11.17}$$

and u_j is now distributed with zero mean and unit variance. Our problem is thus equivalent to finding the distribution of a linear function of variables each of which is normally distributed with zero mean and unit variance.

Consider a linear transformation of type

$$v_i = \sum_{j=1}^{n} c_{ij}u \tag{11.18}$$

and let us determine the c's so that they are orthogonal, i.e. so that

$$\left.\begin{aligned}\sum_{i=1}^{n} c_{ij}c_{ik} &= 0, \quad j \neq k\\ &= 1, \quad j = k\end{aligned}\right\}. \tag{11.19}$$

This can be done in an infinity of ways, for (11.19) imposes $\frac{1}{2}n(n-1)+n$ conditions on n^2 constants. We then find

$$\sum_{i=1}^{n} v_i^2 = \Sigma(\Sigma\, c_{ij}u_j)^2 = \sum_{=1}^{n} u_j^2.$$

Also, for the Jacobian of the transformation

$$J = \left|\frac{\partial v}{\partial u}\right| = |c_{ij}|.$$

If we multiply $|c_{ij}|$ by its transpose we find, in virtue of (11.19), a determinant whose elements are all zero except for units in the diagonals, namely a determinant which is equal to unity. Thus $J^2 = 1$ and $J = 1$. The joint distribution of the u's is, by hypothesis,

$$dF = \frac{1}{(2\pi)^{\frac{1}{2}n}}\exp\left(-\tfrac{1}{2}\sum_{j=1}^{n} u_j^2\right)\Pi\, du \tag{11.20}$$

and hence that of the v's is of the same form:—

$$dF = \frac{1}{(2\pi)^{\frac{1}{2}n}}\exp\left(-\tfrac{1}{2}\sum_{j=1}^{n} v_j^2\right)\prod_j dv. \tag{11.21}$$

The distribution of the other variables v is then, from (11.21), seen to be independent of v_1 and we have, for the (marginal) distribution of v_1,

$$dF_1 = \frac{1}{\sqrt{(2\pi)}}\exp(-\tfrac{1}{2}v_1^2)\,dv_1. \tag{11.22}$$

Thus v_1 is distributed normally with unit variance and zero mean, independently of the other v_i. By symmetry, the same is true of any v_j. Now v_1 is an arbitrary linear function of the x_j. If we identify it with z of (11.16), we have the result that z is normally distributed with zero mean and variance obtained from (11.17) as

$$\operatorname{var} z = \Sigma\, a_j^2\sigma_j^2. \tag{11.23}$$

Example 11.3 Helmert's transformation

The orthogonal transformation used in the previous example is especially useful for samples from a normal population, for the Jacobian is unity and the density function remains of the same form. There is one such orthogonal transformation which has a particular interest. Let us put

$$\left.\begin{aligned}u_1 &= (x_1-x_2)/\sqrt{2}\\ u_2 &= (x_1+x_2-2x_3)/\sqrt{6}\\ u_3 &= (x_1+x_2+x_3-3x_4)/\sqrt{12}\\ \cdots &\quad \cdots \quad \cdots\\ u_{n-1} &= \{x_1+x_2+\ldots+x_{n-1}-(n-1)x_n\}/\sqrt{\{n(n-1)\}}\\ u_n &= (x_1+x_2+\ldots+x_n)/\sqrt{n}.\end{aligned}\right\} \tag{11.24}$$

It is readily verified that (11.19) is satisfied and the Jacobian of the transformation is unity. Thus, if the x's are all independent and normally distributed about zero mean with unit variance, so also are the u's.

Now, for *any* orthogonal transformation with u_n as in (11.24),

$$\begin{aligned}\Sigma(x-\bar{x})^2 &= \Sigma x^2 - n\bar{x}^2\\ &= \sum_{i=1}^{n} u_i^2 - u_n^2\\ &= \sum_{i=1}^{n-1} u_i^2.\end{aligned}$$

Thus we see that the sum of squares of n standardized variates *measured from their mean* is distributed like the sum of squares of $n-1$ normal variates with zero mean. It follows from (11.8) that the distribution of

$$w = \Sigma(x-\bar{x})^2$$

is given by

$$dG = \frac{1}{2^{\frac{1}{2}(n-1)}\Gamma\{\frac{1}{2}(n-1)\}} e^{-\frac{1}{2}w} w^{\frac{1}{2}(n-3)}\, dw. \tag{11.25}$$

Moreover, this quantity w (a simple multiple of the sample variance) is independent of $\bar{x}$, for the distribution of the u's is $k\exp\{-\frac{1}{2}(w+u_n^2)\}$, where $u_n = n^{-\frac{1}{2}}\bar{x}$. Thus in normal samples the mean is distributed independently of the variance. An alternative proof using c.f.'s is indicated in Exercise 11.18.

It makes no essential difference if the x's have common mean μ and variance σ^2; $\bar{x}$ is then replaced by $(\bar{x}-\mu)/\sigma$ and w by w/σ^2.

This interesting and important result characterizes the normal distribution; that is to say, if the mean and variance of random samples from a population are independent, the population must be normal. (Cf. Exercise 11.19.)

Order-statistics

11.4 It may not always be possible to find variate-transformations in which z, the statistic of interest, is one new variate and there are $(n-1)$ others whose domains of integration are not dependent on z. A case of particular interest is provided by the so-called "order-statistics", the theory of which we shall consider in detail in Chapter 14 and in Volume 2. If we have a sample of n observations, $y_1, y_2, \ldots, y_n$, and rearrange them in ascending order of magnitude, renumbering them so that the suffix 1 is attached to the smallest, y_r is called the rth order-statistic. The median and the quantiles are order-statistics in this sense (the median, for example, being the $\frac{1}{2}(n+1)$th observation or midway between the $\frac{1}{2}n$th and the $\frac{1}{2}(n+2)$th); so also would be the lowest value of the sample ($r = 1$) or the highest value ($r = n$).

If we have a sample of n values $x_1, x_2, \ldots, x_n$ from a distribution function $F(x)$, their distribution is given by

$$dG = dF(x_1)\,dF(x_2)\ldots dF(x_n).$$

A transformation to

$$y_r = \int_{-\infty}^{x_r} dF(t) = F(x_r) \tag{11.26}$$

gives for the distribution of the y's

$$dH = dy_1 dy_2 \dots dy_n, \qquad (0 \leqslant y_i \leqslant 1, \text{ all } i). \tag{11.27}$$

The y's and the x's are in the same order since y is a non-decreasing function of x.

If we now order the y's so that y_1 is the smallest, y_2 the next smallest and so on, the domain of variation is given by

$$0 \leqslant y_1 \leqslant y_2 \leqslant \dots \leqslant y_n \leqslant 1. \tag{11.28}$$

If the d.f. is continuous, as we now assume, there is probability zero that any two y_i are equal.

To integrate (11.27) over this domain we require

$$\int_0^1 \dots \left[\int_0^{y_4} \left\{\int_0^{y_3} \left(\int_0^{y_2} dy_1\right) dy_2\right\} dy_3\right] \dots dy_n. \tag{11.29}$$

This reduces, on step-by-step integration, to $1/n!$. For the *ordered* variates, then, the joint distribution of the y's is

$$dK = n! \prod_{i=1}^{n} dy_i, \tag{11.30}$$

the constant obviously being the number of ways of ordering n observations.

Suppose now that we are interested in the (marginal) distribution of y_r, the rth smallest of the y's. To simplify the derivation slightly we make the further transformation

$$\left.\begin{aligned} z_i &= y_i, & i &= 1, 2, \dots, r \\ &= 1 - y_i, & i &= r+1, r+2, \dots, n \end{aligned}\right\} \tag{11.31}$$

The Jacobian is unity and the distribution of the z's is accordingly

$$dL = n! \prod_{i=1}^{n} dz_i \tag{11.32}$$

over the domain

$$0 \leqslant z_1 \leqslant z_2 \leqslant \dots \leqslant z_r \leqslant 1 - z_{r+1} \leqslant \dots \leqslant 1 - z_n \leqslant 1$$

which may be re-written

$$0 \leqslant z_1 \leqslant z_2 \leqslant \dots \leqslant z_r; \quad 0 \leqslant z_n \leqslant z_{n-1} \leqslant \dots \leqslant z_{r+1} \leqslant 1 - z_r.$$

To find the distribution of z_r we integrate out the other variables in two separate blocks, as follows:

$$\begin{aligned} dM &= dz_r n! \int_0^{z_r} \dots \int_0^{z_3} \int_0^{z_2} dz_1 dz_2 \dots dz_{r-1} \times \int_0^{1-z_r} \int_0^{z_{r+1}} \dots \int_0^{z_{n-1}} dz_n dz_{n-1} \dots dz_{r+1} \\ &= dz_r n! \frac{z_r^{r-1}}{(r-1)!} \frac{(1-z_r)^{n-r}}{(n-r)!} \\ &= \frac{n!}{(r-1)!(n-r)!} z_r^{r-1} (1-z_r)^{n-r} dz_r, \qquad 0 \leqslant z_r \leqslant 1. \end{aligned} \tag{11.33}$$

Thus the distribution of z_r is a Beta-distribution of the first kind as at (6.14). To obtain that of the rth order-statistic x_r we simply put $F(x_r) = z_r$, obtaining

$$dN = \frac{n!}{(r-1)!(n-r)!} \{F(x_r)\}^{r-1} \{1 - F(x_r)\}^{n-r} f(x_r) \, dx_r. \tag{11.34}$$

This expression could have been obtained more directly (cf. (10.25)), but we have given a full derivation to illustrate the processes involved.

Example 11.4

From (11.33) and Example 2.8, we have the exact result

$$\text{var}(y_r) = r(n-r+1)/\{(n+1)^2(n+2)\},$$

since $y_r = z_r$ by (11.31). Writing np for r, this becomes

$$\text{var}(y_{np}) = np\{n(1-p)+1\}/\{(n+1)^2(n+2)\}.$$

From (11.26) and (10.14) we see that, as $n \to \infty$ with p fixed,

$$\text{var}(y_{np}) \sim \{F'(x_{np})\}^2 \text{var}(x_{np}),$$

where $F'(x_{np})$ $(\neq 0)$ is to be evaluated as the expectation of x_{np}, which is then the corresponding population quantile ξ_p defined by $F(\xi_p) = p$. Thus

$$\text{var}(x_{np}) \sim \{F'(\xi_p)\}^{-2} p(1-p)/n,$$

agreeing with (10.29). The median is the special case $p = \frac{1}{2}$.

The geometrical method

11.5 A considerable amount of cumbrous analysis may often be avoided by the use of geometrical representation of the domain of integration. We may imagine the values $x_1, \ldots, x_n$ of any given sample as the co-ordinates of a point in an n-dimensional Euclidean hyperspace. The function $dF(x) \ldots dF(x_n)$ may then be regarded as the *density* at the point and the total frequency between z_1 and z_2 will be the integral of this density (the *weight*) in a region lying between the two loci $z(x_1, \ldots, x_n) = z_1$ and $z(x_1, \ldots, x_n) = z_2$, which in general will be hypersurfaces in the n-fold space, i.e. will themselves be spaces of $(n-1)$ dimensions. The distribution function of z will be the total weight between the hypersurface corresponding to $z = -\infty$ and that corresponding to z; and the frequency function will be the element of weight between the hypersurfaces $z-\frac{1}{2}dz$ and $z+\frac{1}{2}dz$.

Example 11.5

Consider again the problem of Example 11.2. In the n-fold u-space the density is given by

$$\frac{1}{(2\pi)^{\frac{1}{2}n}} \exp(-\tfrac{1}{2}\Sigma u^2).$$

The statistic z $(= \Sigma a_j x_j)$ determines a hyperplane

$$z = \Sigma a_j \sigma_j u_j \tag{11.35}$$

and we have to find the total weight between this hyperplane and the corresponding hyperplane at $-\infty$, i.e. the weight on one side—the "lower" side—of the hyperplane (11.35).

Now Σu^2 is the square of the distance of the point $u_1, \ldots, u_n$ from the origin and is therefore unchanged by any rotation of the co-ordinate axes. Choose such a rotation which brings the axis of one variable perpendicular to the hyperplane (11.35), meeting it in Q. Let P be the sample point $u_1, \ldots, u_n$ and O the origin. Then

$$\Sigma u^2 = OP^2 = OQ^2 + QP^2,$$

so that the density at P is

$$\frac{1}{(2\pi)^{\frac{1}{2}n}} \exp(-\tfrac{1}{2}OQ^2) \exp(-\tfrac{1}{2}QP^2).$$

For variation over the hyperplane, OQ^2 is constant and the integral of $\exp(-\frac{1}{2}QP^2)$ is thus a constant independent of OQ. Hence the frequency function of z is given by

$$f(z) = k \exp(-\tfrac{1}{2}OQ^2),$$

k being some constant.

But OQ is the distance from O to the hyperplane and is given by

$$OQ^2 = \frac{z^2}{\Sigma a_j^2 \sigma_j^2}.$$

Hence

$$f(z) = k \exp\left\{-\frac{1}{2} \frac{z^2}{\Sigma a_j^2 \sigma_j^2}\right\},$$

i.e. z is distributed normally with variance $\Sigma a_j^2 \sigma_j^2$ about zero mean.

The reader will find it instructive to compare this example with Example 11.2. They are, in effect, the same thing expressed in different language.

Example 11.6

Consider again the illustration of **11.2**. The elegance of the geometrical approach is well brought out by the analogous derivation of the result there obtained.

In fact, our density function, as before, is given by $k \exp(-\frac{1}{2}OP^2)$. We require the distribution of the statistic $z = OP^2$, and the density is obviously constant over the surface $z =$ constant, that is to say the surface of an n-dimensional hypersphere. The frequency function of z is then the integral of this constant density between the hyperspheres z and $z + dz$, i.e. is proportional to $\exp(-\frac{1}{2}OP^2)$ times the element of the volume of the hypersphere, which itself is proportional to the nth power of the radius OP. Thus we have

$$dF = k \exp(-\tfrac{1}{2}OP^2) \frac{d}{dz} OP^n \, dz$$

$$\propto e^{-\frac{1}{2}z} z^{\frac{1}{2}(n-2)} \, dz,$$

giving, on evaluation of the constant,

$$dF = \frac{1}{2^{\frac{1}{2}n} \Gamma(\frac{1}{2}n)} e^{-\frac{1}{2}z} z^{\frac{1}{2}(n-2)} \, dz$$

as at (11.8).

Now suppose that the quantities $x_1, \ldots, x_n$, while still being normally distributed with unit variance, are subject to p independent homogeneous linear restrictions of type

$$a_1 x_1 + a_2 x_2 + \ldots + a_n x_n = 0.$$

In thc n-space the variables x will then be constrained to lie on p hyperplanes. The first will cut the hypersphere of constant density in a hypersphere of one lower dimension, also, of course, of constant density; the second will cut this in a hypersphere of one lower dimension still, and so on. The result of the linear restrictions will be to constrain the variables to a hypersphere of p fewer dimensions, and thus the distribution of z in these circumstances will be as before, but with $n-p$ instead of n, i.e.

$$dF = \frac{1}{2^{\frac{1}{2}(n-p)}\Gamma\{\frac{1}{2}(n-p)\}} e^{-\frac{1}{2}z} z^{\frac{1}{2}(n-p-2)} dz. \tag{11.36}$$

Exercise 11.20 indicates a proof of (11.36) using c.f.'s.

Example 11.7 The sampling distribution of the mean and variance in normal samples

Writing $\bar{x}$ for the mean of a sample, we have, for the variance s^2,

$$s^2 = \frac{1}{n}\Sigma(x-\bar{x})^2$$
$$= \frac{1}{n}\Sigma x^2 - \bar{x}^2.$$

In samples from a normal population with zero mean and unit variance the density at the point $x_1, \ldots, x_n$ is proportional to

$$\exp(-\tfrac{1}{2}\Sigma x^2) = \exp\{-\tfrac{1}{2}(ns^2+n\bar{x}^2)\}. \tag{11.37}$$

Let us find the sampling distributions of s and $\bar{x}$. From (11.37) it is seen that the density function can be expressed simply in terms of those quantities, and we then have to find some transformation of the volume element $dx_1 \ldots dx_n$. We have considered this from the analytical viewpoint in Example 11.3. Let us now consider it geometrically.

In the n-space consider the unit vector whose direction cosines are $\frac{1}{\sqrt{n}}, \frac{1}{\sqrt{n}}, \ldots, \frac{1}{\sqrt{n}}$, say OQ where O is the origin. If P is the sample point, let PM be the perpendicular from P on to OQ. Then the length of OM is

$$\frac{x_1}{\sqrt{n}} + \frac{x_2}{\sqrt{n}} + \ldots + \frac{x_n}{\sqrt{n}} = \bar{x}\sqrt{n}.$$

The length of OP is $\sqrt{\Sigma x^2}$. Thus the length of PM is $(\Sigma x^2 - n\bar{x}^2)^{\frac{1}{2}} = s\sqrt{n}$.

The element of volume at P may be regarded as the product of an elemental increment in OM, equal to $d\bar{x}\sqrt{n}$, and the elemental volume in the perpendicular hyperplane through M. In the hyperplane the contours of equal density, as in the last example, are hyperspheres of radius $s\sqrt{n}$ centred at M, and consequently the element of volume is equal to $k\,d\bar{x}s^{n-2}\,ds$ multiplied by other elements which need not concern us since they are independent of $\bar{x}$ and s. We have then for the element of frequency

$$dF \propto \exp\{-\tfrac{1}{2}(ns^2+n\bar{x}^2)\}s^{n-2}\,d\bar{x}\,ds, \tag{11.38}$$

and this splits into two factors

$$dF_1 \propto \exp(-\tfrac{1}{2}n\bar{x}^2)\,d\bar{x}, \tag{11.39}$$
$$dF_2 \propto \exp(-\tfrac{1}{2}ns^2)\,s^{n-2}\,ds. \tag{11.40}$$

Thus, as we have seen, in samples from a normal population the distributions of mean and variance are independent. We have from (11.40), for the distribution of s^2,

$$dF \propto \exp\left(-\tfrac{1}{2}ns^2\right) s^{n-3}\, ds^2$$

and, on evaluation of the constant,

$$dF = \frac{n^{\frac{1}{2}(n-1}}{2^{\frac{1}{2}(n-1)}\Gamma\left\{\frac{1}{2}(n-1)\right\}} \exp\left(-\tfrac{1}{2}ns^2\right)(s^2)^{\frac{1}{2}(n-3)}\, d(s^2), \qquad 0 \leqslant s^2 \leqslant \infty, \qquad (11.41)$$

which is (11.25) again, with $w = ns^2$.

It is interesting to compare this with the distribution of the previous example. In the latter case we found the distribution of the sum of squares of the variables *measured from the population mean.* In this case we have found the distribution of $1/n$ times the sum of the squares *measured from the sample mean.* In its equivalent form (11.25), (11.41) is identical with (11.36) with a single linear constraint ($p = 1$), as we observed in Example 11.3.

Example 11.8 " Student's " distribution

In the previous example we have

$$\frac{\bar{x}\sqrt{n}}{s\sqrt{n}} = \frac{OM}{PM} = \cot\phi,$$

where ϕ is the angle POM.

If, then, we define a statistic $z = \bar{x}/s$, z will be constant over the cone obtained by rotating OP about the unit vector, keeping the angle ϕ constant. The distribution of z will then be given by determining the weight between the cones defined by ϕ and $\phi + d\phi$.

Consider the intersection of these cones with the hypersphere of radius OP. They will cut off an annulus on the sphere whose " content " (the n-dimensional analogue of volume) will be proportional to $OP\, d\phi . PM^{n-2}$

$$= OP^{n-1} \sin^{n-2}\phi\, d\phi.$$

The density function is constant and proportional to $\exp\left(-\frac{1}{2}OP^2\right)$ on the hypersphere, and thus the total frequency between the cones will be proportional to

$$\int_0^\infty e^{-\frac{1}{2}OP^2} OP^{n-1} \sin^{n-2}\phi\, d\phi\, d(OP)$$

$$\propto \sin^{n-2}\phi\, d\phi, \qquad 0 \leqslant \phi \leqslant \pi.$$

The distribution of z ($= \cot\phi$) is then given by

$$dF \propto \frac{k\, dz}{(1+z^2)^{\frac{1}{2}n}},$$

or, on evaluation of the constant,

$$dF = \frac{1}{B\left\{\frac{1}{2}(n-1), \frac{1}{2}\right\}} \frac{dz}{(1+z^2)^{\frac{1}{2}n}}.$$

Since z is the ratio of two functions of unit dimension in the variables, this distribution holds for samples from a normal population irrespective of the scale, that is to say, irrespective of the variance of the parent population.

The distribution is usually put in a slightly different form. Put

$$t = \frac{\bar{x}\sqrt{n}}{\left\{\frac{1}{n-1}\Sigma(x-\bar{x})^2\right\}^{\frac{1}{2}}} = \sqrt{(n-1)}\,z. \tag{11.42}$$

In terms of t, we have

$$dF = \frac{1}{\sqrt{(n-1)}\,B\{\frac{1}{2}(n-1),\frac{1}{2}\}}\frac{dt}{\left(1+\frac{t^2}{n-1}\right)^{\frac{1}{2}n}} \tag{11.43}$$

$$= \frac{\Gamma\{\frac{1}{2}(\nu+1)\}}{\sqrt{(\nu\pi)}\,\Gamma(\frac{1}{2}\nu)}\frac{dt}{(1+t^2/\nu)^{\frac{1}{2}(\nu+1)}} \tag{11.44}$$

where $\nu = n-1$.

This celebrated expression is known as " Student's " distribution after the pen-name of its discoverer (1908a). It will be discussed in detail in Chapter 16.

Example 11.9 *Distribution of the mean of samples from a rectangular population* (Hall (1927))

Consider now a sample of n values from the rectangular distribution

$$dF = dx, \quad 0 \leqslant x \leqslant 1.$$

In the n-space the density function will be a constant everywhere inside a hypercube

$$0 \leqslant x_j \leqslant 1,\ j = 1,\dots,n \tag{11.45}$$

and zero elsewhere. The unit vector will be along the diagonal of this cube. If P is the sample point $(x_1, \dots, x_n)$ and PM the perpendicular on to this diagonal, then, as shown in Example 11.7, $OM = \bar{x}\sqrt{n}$. Thus, for the distribution of $\bar{x}$ we require the element of weight (which in this case is proportional to the element of volume) between the hyperplanes $\bar{x}$ and $\bar{x}+d\bar{x}$; and this is equivalent to finding the content of the hyperplane (its " area ") cut off by the various faces of the hypercube. The complication of the problem arises from the fact that as $\bar{x}$ increases this region changes its shape according to the number of edges of the hypercube cut by the hyperplane.

Consider the " quadrants "

$$\left.\begin{array}{l} x_j \geqslant r_j \\ r_j = 0 \text{ or } 1 \end{array}\right\} j = 1, 2, \dots, n, \tag{11.46}$$

whose corners are the corners of the hypercube. Any one of the corners may have 0 or 1 or 2 . . . or n of its co-ordinates equal to unity and the rest zero. We divide the quadrants into $(n+1)$ sets according as the corner has $0, 1, \dots, n$ of its co-ordinates equal to unity, that is, according as

$$r = \sum_{j=1}^{n} r_j$$

is equal to $0, 1, \dots, n$. A quadrant of the tth set may be called Q_t. There will be $\binom{n}{t}$ different Q_t's. Let S be any point of Q_0, i.e. any point whose co-ordinates are all $\geqslant 0$, and let just s of its co-ordinates be $\geqslant 1$. Then S will belong to just $\binom{s}{0}, = 1, Q_0$; $\binom{s}{1} Q_1$'s; $\binom{s}{2} Q_2$'s, and so on. Now if $s > 0$,

$$\sum_{t=0}^{s} (-1)^t \binom{s}{t} = (1-1)^s = 0. \tag{11.47}$$

Hence, if whenever a point belongs to a Q_t we give it a density $(-1)^t$ and then sum over all Q, the resultant density will be 1 or 0 according as the point belongs to the hypercube or not.

Let the segment of the hyperplane

$$z = \Sigma x \tag{11.48}$$

lying in Q_0 have content $V_n(z)$. Then the segment lying in any member of (11.46) will have content $V_n(z-r)$ which is zero if $r \geqslant z$. Further, the segment of (11.48) lying in any member of (11.46) will have the content

$$\sum_{r=0}^{k} (-1)^r \binom{n}{r} V_n(z-r) \tag{11.49}$$

where $k = [z]$, = the greatest integer less than z.

To find $V_n(z)$, let $\bar{V}_{n-1}(z)$ be the projection of $V_n(z)$ perpendicular to one of the axes, so that

$$V_n(z) = \sqrt{n}\bar{V}_{n-1}(z).$$

Now $\bar{V}_n(z)$ is the content of the n-dimensional region bounded by (11.48) and the coordinate hyperplanes—a region whose base is therefore of content $V_n(z)$. The perpendicular from O to this base is $\frac{z}{\sqrt{n}}$. Hence

$$\bar{V}_n(z) = \frac{1}{n}\left(\frac{z}{\sqrt{n}}\right) V_n(z)$$

and

$$\bar{V}_{n-1}(z) = \frac{1}{n-1}\left(\frac{z}{\sqrt{(n-1)}}\right) V_{n-1}(z)$$

or

$$V_n(z) = \frac{1}{n-1}\sqrt{\left(\frac{n}{n-1}\right)}\, z V_{n-1}(z). \tag{11.50}$$

Since $V_2(z) = z\sqrt{2}$ repeated applications of this formula give

$$V_n(z) = \frac{\sqrt{n}}{(n-1)!} z^{n-1}.$$

Substituting in (11.49) we find for the content of the region common to the hypercube and the hyperplane

$$f(z) = \frac{\sqrt{n}}{(n-1)!} \sum_{r=0}^{k} (-1)^r \binom{n}{r} (z-r)^{n-1} \tag{11.51}$$

for values of z between k and $k+1$.

Since

$$\int_0^n f(z) \frac{dz}{\sqrt{n}} = 1$$

the distribution of the mean $m = \frac{z}{n}$ is given by

$$g(m) = \frac{n^n}{(n-1)!} \sum_{=0}^{k} (-1)^r \binom{n}{r} \left(m - \frac{r}{n}\right)^{n-1}, \quad \frac{k}{n} \leqslant m \leqslant \frac{k+1}{n}. \tag{11.52}$$

This is the required distribution. It is unusual in consisting of n arcs of degree $(n-1)$ in m, having $(n-1)$-point contact at their joins, that is at the points $\frac{k}{n}$ $(k = 1, 2, \ldots, n)$.

The distribution is symmetrical since the hyperplane z = constant is perpendicular to the unit vector, which itself is an axis of symmetry of the hypercube.

For particular values $n = 2, 3, 4$, (11.52) gives the following results for the frequency function :—

$$n=2: \quad 4m, \qquad 0 \leqslant m \leqslant \tfrac{1}{2},$$
$$4(1-m), \qquad \tfrac{1}{2} \leqslant m \leqslant 1.$$
$$n=3: \quad \frac{27m^2}{2}, \qquad 0 \leqslant m < \tfrac{1}{3},$$
$$\frac{27}{2}\{m^2 - 3(m-\tfrac{1}{3})^2\}, \qquad \tfrac{1}{3} \leqslant m \leqslant \tfrac{2}{3},$$
$$\frac{27}{2}(1-m)^2, \qquad \tfrac{2}{3} \leqslant m \leqslant 1.$$
$$n=4: \quad \frac{128}{3}m^3, \qquad 0 \leqslant m \leqslant \tfrac{1}{4},$$
$$\frac{128}{3}\{m^3 - 4(m-\tfrac{1}{4})^3\}, \qquad \tfrac{1}{4} \leqslant m \leqslant \tfrac{1}{2},$$
$$\frac{128}{3}\{(1-m)^3 - 4(\tfrac{3}{4}-m)^3\}, \qquad \tfrac{1}{2} \leqslant m \leqslant \tfrac{3}{4},$$
$$\frac{128}{3}(1-m)^3, \qquad \tfrac{3}{4} \leqslant m \leqslant 1.$$

For $n = 2$ the distribution is "triangular." For larger n it approaches normality in virtue of the Central Limit theorem, and the approximation is good even at $n = 3$, as the reader should verify graphically. Stephens (1966) and Buckle *et al.* (1969) give percentiles of the distribution (11.52) for $n \leqslant 30$.

The method of characteristic functions

11.6 It has already been noted in **7.18** that the characteristic function of the sum of n independent variables is the product of their characteristic functions. This simple property enables us to find the sampling distribution of a wide class of statistics which are expressible as sums, and particularly of the mean.

If we have a sample of n values from a population whose characteristic function is $\phi(t)$, the characteristic function of their sum is ϕ^n. Thus the distribution function of their sum z is given by $F(z)$ where (cf. (4.4) and (4.33))

$$F(z) - F(0) = \frac{1}{2\pi}\int_{-\infty}^{\infty} \frac{1-e^{-izt}}{it}\phi^n\,dt \tag{11.53}$$

and the frequency function is

$$f(z) = \frac{1}{2\pi}\int_{-\infty}^{\infty} e^{-itz}\phi^n\,dt. \tag{11.54}$$

The following examples will illustrate the power of these results.

Example 11.10 Distribution of the mean for the binomial distribution

The characteristic function of the binomial distribution arrayed by $(q+p)^r$ is

$$(q+pe^{it})^r.$$

The c.f. of the sampling distribution of the sum of n values is then

$$(q+pe^{it})^{rn}$$

and that of the distribution of the mean ($1/n$ of that sum) is $(q+pe^{it/n})^{rn}$. But this is the c.f. of the binomial distribution arrayed by

$$(q+p)^{rn}, \tag{11.55}$$

the interval being $1/n$ instead of unity; and hence this distribution is that of the mean.

Example 11.11 Distribution of the mean for the Poisson distribution

The characteristic function of the Poisson distribution whose general term is $e^{-\lambda}\lambda^r/r!$ is

$$\exp\{\lambda(e^{it}-1)\}.$$

The c.f. of the mean is then $\exp\{n\lambda(e^{it/n}-1)\}$ and hence the distribution of the mean is the Poisson distribution whose general term is

$$e^{-n\lambda}\frac{(n\lambda)^r}{r!}, \tag{11.56}$$

the interval being $1/n$ instead of unity.

Example 11.12 Distribution of the mean for the normal population

The characteristic function of the normal distribution

$$dF = \frac{1}{\sigma\sqrt{(2\pi)}}\exp\left\{-\tfrac{1}{2}\left(\frac{x-\mu}{\sigma}\right)^2\right\}dx$$

is seen from Example 3.4 to be

$$\exp(-\tfrac{1}{2}t^2\sigma^2+it\mu).$$

The c.f. of the distribution of the mean of n values is then

$$\exp\left\{n\left(-\tfrac{1}{2}\frac{t^2\sigma^2}{n^2}+\frac{it\mu}{n}\right)\right\} = \exp\left\{-\tfrac{1}{2}\frac{t^2\sigma^2}{n}+it\mu\right\}. \tag{11.57}$$

This is the c.f. of a normal distribution with mean μ and variance σ^2/n, which is therefore the distribution required.

Example 11.13 Distribution of the mean for the Gamma population

The characteristic function of the distribution

$$dF = \frac{1}{\Gamma(\gamma)}e^{-x/a}\left(\frac{x}{a}\right)^{\gamma-1}\frac{dx}{a}, \qquad a>0,$$

is (Example 3.6)

$$\phi(t) = \frac{1}{(1-ita)^\gamma}.$$

The c.f. of the distribution of the mean of n values is then

$$\frac{1}{\left(1-\frac{ita}{n}\right)^{n\gamma}}.$$

This is the c.f. of the distribution

$$dF = \frac{1}{\Gamma(\gamma n)}e^{-nx/a}\left(\frac{nx}{a}\right)^{n\gamma-1}\frac{n\,dx}{a}. \tag{11.58}$$

Example 11.14 Distribution of the mean for the rectangular population

The characteristic function of the distribution $dF = dx$ $(0 \leqslant x \leqslant 1)$ is

$$\int_0^1 e^{itx}\, dx = \frac{e^{it}-1}{it}.$$

The c.f. of the mean of n values is then $\left(\frac{e^{it/n}-1}{it/n}\right)^n$, and the frequency function is thus

$$f(x) = \frac{1}{2\pi}\int_{-\infty}^{\infty} e^{-itx}\left(\frac{e^{it/n}-1}{it/n}\right)^n dt. \tag{11.59}$$

This integral is everywhere analytic and the range of integration may be changed to the contour Γ consisting of the real axis from $-\infty$ to $-c$, the small semicircle of radius c and centre at the origin, and the real axis from c to ∞. Thus

$$f(x) = \frac{1}{2\pi}\int_\Gamma e^{-itx}\left(\frac{e^{it/n}-1}{it/n}\right)^n dt = \frac{(-1)^n}{2\pi}\int_\Gamma e^{-itx}\sum_{j=0}^{n}(-1)^j\binom{n}{j}\frac{e^{itj/n}}{(it/n)^n}dt. \tag{11.60}$$

Now

$$\int_\Gamma \frac{e^{igz}}{z^n}dz = 0 \text{ if } g>0$$

$$= -2\pi i^n \frac{g^{n-1}}{(n-1)!} \text{ if } g \leqslant 0.$$

This may be seen by integrating along a contour consisting of Γ and the infinite semicircle *above* the real axis if $g>0$ and *below* it if $g \leqslant 0$.

Substituting in (11.60) we find

$$f(x) = \frac{(-1)^n}{2\pi}\int_\Gamma \Sigma(-1)^j\binom{n}{j}\frac{e^{itj/n-itx}}{(it/n)^n}dt$$

$$= \frac{(-1)^{n-1}}{(n-1)!}\sum_{j<nx}(-1)^j\binom{n}{j}\frac{\left(\frac{j}{n}-x\right)^{n-1}}{\left(\frac{1}{n}\right)^n}$$

$$= \frac{n^n}{(n-1)!}\sum_{j=0}^{[nx]}(-1)^j\binom{n}{j}\left(x-\frac{j}{n}\right)^{n-1}$$

This, with a few changes of notation, is the same as (11.52). The result was given by Irwin (1927) in this form, but is traceable as far back as Lagrange.

Gray and Odell (1966) obtain the distribution of any linear function of rectangular variates with different ranges. Marsaglia (1965) gives the d.f. of the ratio of sums of rectangular variates.

11.7 General expressions may also be derived for the distributions of geometric means and of the moments about fixed points.

In fact, if $y = \log x$, the characteristic function of y is

$$\alpha(t) = \int_{-\infty}^{\infty} e^{it\log x}\, dF = \int_{-\infty}^{\infty} x^{it}\, dF.$$

The distribution of the sum of n independent values of y, say nz, is then given by

$$F(nz)-F(0) = \frac{1}{2\pi}\int_{-\infty}^{\infty}\frac{1-e^{-itnz}}{it}\alpha^n\, dt, \tag{11.61}$$

and the distribution of the mean is that of z. But $z = \log u$, where u is the geometric mean, and hence the distribution of u may be found.

The frequency function, when it exists, is

$$f(nz) = \frac{1}{2\pi}\int_{-\infty}^{\infty} e^{-itnz}\alpha^n\, dt.$$

Similarly the characteristic function of a power of the variate, say x^r, is given by

$$\beta(t) = \int_{-\infty}^{\infty} \exp(itx^r)\, dF$$

and thus the distribution of the rth moment, say z, by

$$F(nz) - F(0) = \int_{-\infty}^{\infty} \frac{e^{-itnz}-1}{it}\beta^n\, dt. \tag{11.62}$$

Example 11.15 Distribution of the geometric mean in samples from a rectangular population

If the population is $\quad dF = dx/a, \quad 0 \leqslant x \leqslant a,$

the characteristic function of $\log x$ is

$$\int_0^a x^{it}\frac{dx}{a} = \frac{a^{it}}{1+it}.$$

The *frequency* function of $u = \Sigma \log x$ is then given by

$$f(u) = \frac{1}{2\pi}\int_{-\infty}^{\infty} \frac{e^{-itu}\, a^{nit}}{(1+it)^n} dt$$
$$= \frac{1}{2\pi}\int_{-\infty}^{\infty} \frac{e^{it(n\log a - u)}}{(1+it)^n} dt, \qquad n\log a - u \geqslant 0.$$

This integral may be evaluated by contour integration and we find

$$f(u) = \frac{(n\log a - u)^{n-1} e^{-(n\log a - u)}}{\Gamma(n)},$$

whence, putting $z = e^{u/n}$, we find for the distribution of the geometric mean z

$$f(z) = \frac{n^n z^{n-1}}{a^n\Gamma(n)}\left(\log\frac{a}{z}\right)^{n-1}. \tag{11.63}$$

Example 11.16 Distribution of the second-order moment about the population mean in samples from a normal population

If the distribution is

$$dF = \frac{1}{\sigma\sqrt{(2\pi)}}\exp(-\tfrac{1}{2}x^2/\sigma^2)\, dx$$

the characteristic function of x^2 is

$$\frac{1}{\sigma\sqrt{(2\pi)}}\int_{-\infty}^{\infty} \exp(itx^2)\exp(-\tfrac{1}{2}x^2/\sigma^2)\, dx$$
$$= \frac{1}{(1-2\sigma^2 it)^{\frac{1}{2}}}.$$

The c.f. of the mean of n values, say m_2, is then

$$\frac{1}{(1-2\sigma^2 it/n)^{\frac{1}{2}n}} \tag{11.64}$$

and the frequency function of this is

$$\frac{1}{2\pi}\int_{-\infty}^{\infty}\frac{\exp(-itm_2)}{(1-2\sigma^2 it/n)^{\frac{1}{2}n}}dt.$$

This may be integrated by a contour integration, or the result written down directly from the consideration that (11.64) is the characteristic function of the distribution

$$dF = \frac{n^{\frac{1}{2}n}}{(2\sigma^2)^{\frac{1}{2}n}\Gamma(\frac{1}{2}n)}\exp(-nm_2/2\sigma^2)\, m_2^{\frac{1}{2}n-1}\, dm_2, \tag{11.65}$$

a result which may be compared with that of (11.8), to which it reduces on writing $z = nm_2/\sigma^2$.

The distribution of a sum or difference of random variables

11.8 The distribution of the sum of two independent variates may be obtained directly as follows. Their joint distribution is

$$dF(x_1, x_2) = dF_1(x_1)\,dF_2(x_2) = f_1(x_1)f_2(x_2)\,dx_1\,dx_2.$$

The transformation $z = x_1+x_2$, $y = x_2$, has Jacobian equal to 1, and gives

$$dG(z,y) = f_1(z-y)f_2(y)\,dz\,dy. \tag{11.66}$$

Integrating out y, we obtain, for the marginal distribution of z,

$$dH(z) = \left\{\int_{-\infty}^{\infty} f_1(z-y)f_2(y)\,dy\right\}dz$$

or, for the frequency function of z,

$$h(z) = \int_{-\infty}^{\infty} f_1(z-y)f_2(y)\,dy. \tag{11.67}$$

The distribution function of z is obtained by integrating $h(t)$ from $-\infty$ to z, giving,

$$H(z) = \int_{-\infty}^{\infty} F_1(z-y)f_2(y)\,dy. \tag{11.68}$$

Note that if $x_1 \geqslant 0$, y is to be integrated from $-\infty$ to z; if x_2 also is $\geqslant 0$, y is integrated from 0 to z.

If we are interested in the difference $z = x_1 - x_2$, rather than their sum, the reader will see by retracing the argument that (11.67) is changed only in that $(z-y)$ is replaced by $(z+y)$. In this case, even if x_1 and x_2 are non-negative, z is not.

(11.67) holds for discrete variates (with integration replaced by summation as usual) although no Jacobian enters into the transformation to (11.66), as Example 11.17 shows.

Example 11.17

Consider the distribution of the sum of two independent Poisson variates. Their joint distribution is

$$f(x_1, x_2) = e^{-\lambda_1}\frac{\lambda_1^{x_1}}{x_1!}\cdot e^{-\lambda_2}\frac{\lambda_2^{x_2}}{x_2!}, \quad x_1, x_2 = 0, 1, 2, \ldots$$

(11.67) becomes
$$h(z) = \sum_{y=0}^{z} e^{-\lambda_1}\frac{\lambda_1^{z-y}}{(z-y)!}\cdot\frac{e^{-\lambda_2}\lambda_2^{y}}{y!}$$
$$= e^{-(\lambda_1+\lambda_2)}\lambda_1^{z}\sum_{y=0}^{z}\frac{(\lambda_2/\lambda_1)^y}{(z-y)!\,y!}$$
$$= \frac{e^{-(\lambda_1+\lambda_2)}\lambda_1^{z}}{z!}\sum_{y=0}^{z}\binom{z}{y}\left(\frac{\lambda_2}{\lambda_1}\right)^y$$
$$= e^{-(\lambda_1+\lambda_2)}\lambda_1^{z}\left(1+\frac{\lambda_2}{\lambda_1}\right)^z\bigg/ z! = e^{-(\lambda_1+\lambda_2)}(\lambda_1+\lambda_2)^z/z!.$$

Thus the sum is also distributed in the Poisson form, with parameter equal to the sum of the two components' parameters. It follows at once that the sum of n independent Poisson variates is a Poisson, parameters being additive. Cf. the less general result of Example 11.11.

The difference between two independent Poisson variates cannot be a Poisson, since it can take negative values. If the two Poissons have the same parameter $\lambda = a$, say, their difference will have all odd cumulants zero and even cumulants equal to $2a$. Its distribution was given in Example 4.5.

(11.67) can be used to obtain successively the distribution of the sum of any number of variables whose individual distributions are known. If all the variables have the same distribution, the general form may be suggested when the results for two or three variates have been worked out. Its correctness can then be verified by induction. The following example illustrates the method.

Example 11.18

In Example 11.6 we found that the distribution of the sum of squares of n independent normal variates is given by

$$dF = \frac{1}{2^{\frac{1}{2}n}\Gamma(\frac{1}{2}n)}e^{-\frac{1}{2}z}z^{\frac{1}{2}n-1}\,dz. \tag{11.69}$$

Suppose we had surmised this form from an examination of a few cases for low n. Let x be another variate distributed normally about zero mean with unit variance, and $x^2 = v$. Then v has the distribution

$$dF = \frac{1}{2^{\frac{1}{2}}\Gamma(\frac{1}{2})}e^{-\frac{1}{2}v}v^{-\frac{1}{2}}\,dv, \qquad 0 \leqslant v \leqslant \infty.$$

Then, from (11.67) the frequency function of the distribution of $u = z+v$ is given, since z and v are both non-negative, by

$$\int_0^u \frac{1}{2^{\frac{1}{2}}\Gamma(\frac{1}{2})}e^{-\frac{1}{2}(u-y)}(u-y)^{-\frac{1}{2}}\frac{1}{2^{\frac{1}{2}n}\Gamma(\frac{1}{2}n)}e^{-\frac{1}{2}y}y^{\frac{1}{2}(n-2)}\,dy$$
$$= \frac{e^{-\frac{1}{2}u}}{2^{\frac{1}{2}(n+1)}\Gamma(\frac{1}{2})\Gamma(\frac{1}{2}n)}\int_0^u (u-y)^{-\frac{1}{2}}y^{\frac{1}{2}(n-2)}\,dy$$
$$= \frac{B(\frac{1}{2},\frac{1}{2}n)}{2^{\frac{1}{2}(n+1)}\Gamma(\frac{1}{2})\Gamma(\frac{1}{2}n)}e^{-\frac{1}{2}u}u^{\frac{1}{2}(n-1)}$$
$$= \frac{1}{2^{\frac{1}{2}(n+1)}\Gamma\{\frac{1}{2}(n+1)\}}e^{-\frac{1}{2}u}u^{\frac{1}{2}(n-1)}$$

which is the same as (11.69) with $n+1$ for n. Hence the distribution holds generally.

The distribution of a ratio of random variables

11.9 Cases sometimes arise in which we require the sampling distribution of the ratio of two independent variates, x_1 and x_2. Their joint distribution is, as before,

$$dF(x_1, x_2) = dF_1(x_1)\,dF_2(x_2) = f_1(x_1)f_2(x_2)\,dx_1\,dx_2. \tag{11.70}$$

We now transform to new variables

$$u = x_1/x_2,$$
$$v = x_2,$$

and the Jacobian of the transformation is found to be

$$J = v. \tag{11.71}$$

Care is necessary here to ensure that the change of variable obeys the validity conditions outlined in **11.3**. Essentially, we must ensure that $J \neq 0$ within the domain of integration, although zero values on the boundary of the domain do not affect the validity of the transformation. From (11.71) we see that this implies

$$x_2 = v \neq 0 \tag{11.72}$$

except on the boundary. Thus we must in the first place restrict our discussion to ratios of random variables such that the *denominator* can range *either* from 0 to ∞, *or* from $-\infty$ to 0, but not both. In this case we have, from (11.70) and (11.71),

$$dG(u, v) = f_1(uv)f_2(v)\,v\,du\,dv, \tag{11.73}$$

and thus the marginal distribution of u is

$$dH(u) = \left\{\int_{-\infty}^{\infty} f_1(uv)f_2(v)\,v\,dv\right\}du, \tag{11.74}$$

and its distribution function

$$H(u) = \int_{-\infty}^{\infty} F_1(uv)f_2(v)\,dv. \tag{11.75}$$

Example 11.19

Consider again the distribution of the ratio $\bar{x}/s$ discussed in Example 11.8. Here $\bar{x}$ is the mean of samples of n from a normal population and is thus distributed as

$$dF_1 \propto \exp(-n\bar{x}^2/2\sigma^2)\,d\bar{x}.$$

s is distributed as

$$dF_2 \propto \exp(-ns^2/2\sigma^2)\,s^{n-2}\,ds,$$

as we have seen at (11.39) and (11.40), where we had $\sigma^2 = 1$.

Then the distribution of z $(= \bar{x}/s)$ is, from (11.74), a constant times

$$\int_0^{\infty} \exp(-nz^2s^2/2\sigma^2)\exp(-ns^2/2\sigma^2)\,s^{n-2}.s.ds$$

$$\propto \frac{1}{(1+z^2)^{\frac{1}{2}n}},$$

as we found in Example 11.8.

11.10 The distribution of the ratio u of **11.9** can also be obtained in terms of characteristic functions if x_2 is non-negative with finite mean. From (11.75), we have for non-negative v

$$H(u) = \int_0^\infty F_1(uv)\,dF_2(v),$$

which on using (4.4) becomes, writing $\phi_1(t)$, $\phi_2(t)$ for the c.f.'s of x_1, x_2 respectively,

$$\begin{aligned} H(u) &= \int_0^\infty \left\{F_1(0)+\frac{1}{2\pi}\int_{-\infty}^{\infty}\frac{1-e^{-ituv}}{it}\cdot\phi_1(t)\,dt\right\}dF_2(v) \\ &= F_1(0)+\frac{1}{2\pi i}\int_{-\infty}^{\infty}\frac{\phi_1(t)}{t}\left\{\int_0^\infty (1-e^{-ituv})\,dF_2(v)\right\}dt \\ &= F_1(0)+\frac{1}{2\pi i}\int_{-\infty}^{\infty}\frac{\phi_1(t)}{t}\{1-\phi_2(-tu)\}\,dt. \end{aligned} \tag{11.76}$$

Differentiating with respect to u, we find that the f.f. of u, if it exists, is given by

$$h(u) = \frac{1}{2\pi i}\int_{-\infty}^{\infty}\phi_1(t)\phi_2'(-tu)\,dt, \tag{11.77}$$

provided that the integral converges, a result originally due to Cramér (1937). Daniels (1954) pointed out that the finiteness of the mean of x_2 is a sufficient condition for the integral on the right of (11.77) to converge absolutely for all u.

Geary (1944) generalized (11.77) to the case of non-independent x_1, x_2 with joint c.f. $\phi(t_1, t_2)$, where x_2 is non-negative with finite mean as before. His result is

$$h(u) = \frac{1}{2\pi i}\int_{-\infty}^{\infty}\left[\frac{\partial\phi(t_1,t_2)}{\partial t_2}\right]_{t_2=-t_1 u} dt_1. \tag{11.78}$$

The proof of (11.78) sketched in Exercise 11.24 below indicates the importance of the condition that $E(x_2)$ be finite.

The c.f. of u, $\phi_u(t)$, may sometimes more simply be obtained from

$$\phi_u(t) = \int_{-\infty}^{\infty}\left\{\int_{-\infty}^{\infty} e^{itx_1/x_2}\,dF_1(x_1)\right\}dF_2(x_2) = \int_{-\infty}^{\infty}\phi_1\left(\frac{t}{x_2}\right)dF_2(x_2). \tag{11.79}$$

The same method holds for a *product* of variables—see Example 11.22.

Example 11.20 Fisher's variance-ratio distribution

Suppose that we have two independent samples of n_1, n_2 members from normal populations with variances σ_1^2 and σ_2^2 respectively. We define the sample variances as of type $\Sigma(x-\bar{x})^2/n$ and call them s_1^2 and s_2^2. Their distributions are then, by (11.40),

$$dF_1 \propto \exp\left(-\frac{n_1 s_1^2}{2\sigma_1^2}\right)s_1^{n_1-3}\,ds_1^2, \qquad dF_2 \propto \exp\left(-\frac{n_2 s_2^2}{2\sigma_2^2}\right)s_2^{n_2-3}\,ds_2^2.$$

The distribution of the ratio $t^2 = s_1^2/s_2^2$ is then, from (11.74), given by

$$\begin{aligned} f(t^2) &\propto \int_0^\infty \exp\left(-\frac{n_1 t^2 s_2^2}{2\sigma_1^2}\right)(s_2 t)^{n_1-3}\exp\left(-\frac{n_2 s_2^2}{2\sigma_2^2}\right)s_2^{n_2-3}.s_2^2.ds_2^2 \\ &\propto \int_0^\infty \exp\left\{-\frac{s_2^2}{2}\left(\frac{n_1 t^2}{\sigma_1^2}+\frac{n_2}{\sigma_2^2}\right)\right\}s_2^{n_1+n_2-4}t^{n_1-3}\,ds_2^2 \end{aligned}$$

$$\propto \frac{(t^2)^{\frac{1}{2}(n_1-3)}}{\left(\frac{n_1 t^2}{\sigma_1^2}+\frac{n_2}{\sigma_2^2}\right)^{\frac{1}{2}(n_1+n_2-2)}}, \qquad 0 \leqslant t^2 \leqslant \infty.$$

This is a Beta distribution of the second kind (cf. (6.17)). It is usually put in different form. Writing

$$z = \tfrac{1}{2}\log\frac{n_1(n_2-1)s_1^2/\sigma_1^2}{n_2(n_1-1)s_2^2/\sigma_2^2} = \tfrac{1}{2}\log\frac{n_1(n_2-1)}{n_2(n_1-1)}t^2\Big/\left(\frac{\sigma_1^2}{\sigma_2^2}\right), \tag{11.80}$$

we find for the frequency function of z

$$f(z) \propto \frac{e^{(n_1-1)z}}{\{(n_1-1)e^{2z}+(n_2-1)\}^{\frac{1}{2}(n_1+n_2-2)}}, \qquad -\infty \leqslant z \leqslant \infty, \tag{11.81}$$

or, writing $\nu_1 = n_1-1$ and $\nu_2 = n_2-1$ and evaluating the constant term,

$$f(z) = \frac{2\nu_1^{\frac{1}{2}\nu_1}\nu_2^{\frac{1}{2}\nu_2}}{B(\frac{1}{2}\nu_1, \frac{1}{2}\nu_2)}\frac{e^{\nu_1 z}}{(\nu_1 e^{2z}+\nu_2)^{\frac{1}{2}(\nu_1+\nu_2)}}. \tag{11.82}$$

This form was chosen by R. A. Fisher because interpolation in tables of z is easier than in tables of t. The distribution is the basis of many tests of hypotheses, and at a later stage (Chapter 16) we shall study its properties at some length.

Let us obtain the distribution by Cramér's result at (11.77). For the respective c.f.'s of s_1^2 and s_2^2 we have (cf. Examples 11.3 and 11.13)

$$\phi_1(t) = \left(1-\frac{2\sigma_1^2 it_1}{n_1}\right)^{-\frac{1}{2}(n_1-1)} \qquad \phi_2(t) = \left(1-\frac{2\sigma_2^2 it_2}{n_2}\right)^{-\frac{1}{2}(n_2-1)}$$

Formula (11.77) then gives us for the ratio $v = \dfrac{s_1^2/\sigma_1^2}{s_2^2/\sigma_2^2}$

$$f(v) \propto \frac{1}{2\pi i}\int_{-\infty}^{\infty}\left(1-\frac{2it}{n_1}\right)^{-\frac{1}{2}(n_1-1)}\left(1+\frac{2itv}{n_2}\right)^{-\frac{1}{2}(n_2+1)} dt.$$

A contour integration over the imaginary axis and the infinite semicircle to the left of it then gives us $\quad f(v) \propto \left[\dfrac{d^\alpha}{dt^\alpha}\left(1-\dfrac{2tv}{n_2}\right)^{-\frac{1}{2}(n_2+1)}\right]_{t=-\frac{1}{2}n_1}$

where $\alpha = \frac{1}{2}(n_1-3)$, giving

$$f(v) \propto \frac{v^{\frac{1}{2}(n_1-3)}}{\left(1+\frac{n_1 v}{n_2}\right)^{\frac{1}{2}(n_1+n_2-2)}}.$$

This, on the substitution $v = \dfrac{t^2}{\sigma_1^2/\sigma_2^2}$, reduces to the form for t^2 above. It will be noticed that this proof applies only for odd n_1, since α must be integral.

There are other (not necessarily independent) random variables whose ratio is distributed in this form—cf. Kotlarski (1964).

11.11 In **11.9** we referred to the problem of the distribution of a ratio when the denominator has a range overlapping zero. One way of handling the difficulty is to split the range into two parts at zero. We assume the denominator random variable to be continuous at zero, since otherwise there is a finite probability that the ratio is infinite. We obtain from (11.73), separately for the two parts of the range,

$$dG(u,v)\begin{cases} = f_1(uv)f_2(v)\,|\,v\,|\,du\,dv, & v<0, \\ = f_1(uv)f_2(v)\,v\,du\,dv, & v\geqslant 0, \end{cases} \tag{11.83}$$

so that (11.74) is replaced by

$$dH(u) = \left\{\int_0^\infty f_1(uv)f_2(v)\,v\,dv - \int_{-\infty}^0 f_1(uv)f_2(v)\,v\,dv\right\}du, \tag{11.84}$$

and the distribution function follows as before.

In practice, we are usually interested in a ratio whose denominator is non-negative, so that the methods of **11.10** apply, as in Examples 11.19 and 11.20. But even when the denominator can take values on either side of zero, the problem can sometimes be simplified by considering the square of the ratio, thus getting back to **11.10**.

Example 11.21

Consider the ratio of two independent normal variables with zero mean and unit variance. From (11.84), its frequency function is

$$dH(u) = \frac{du}{2\pi}\left\{\int_0^\infty e^{-\frac{1}{2}(uv)^2}e^{-\frac{1}{2}v^2}\,v\,dv - \int_{-\infty}^0 e^{-\frac{1}{2}(uv)^2}e^{-\frac{1}{2}v^2}\,v\,dv\right\}.$$

Simplifying, and making use of the symmetry of the normal distribution, this becomes

$$dH(u) = \frac{du}{\pi}\int_0^\infty \exp\{-\tfrac{1}{2}v^2(u^2+1)\}\,v\,dv.$$

This integral is immediately seen to be (cf. the mean deviation of a normal distribution at (5.26)) equal to $1/(1+u^2)$. Thus

$$dH(u) = \frac{du}{\pi(1+u^2)}, \qquad -\infty\leqslant u\leqslant\infty, \tag{11.85}$$

which is a Cauchy distribution.

Here, as suggested in **11.11**, consideration of the square of the ratio is fruitful, leading to the distribution of the ratio of two squares of standardized normal variables, since each of these is distributed in the Type III form of (11.8) with $n = 1$. Thus the distribution of the squared ratio is equivalent to Fisher's distribution of t in Example 11.20. It is easily confirmed that the latter reduces to our result above if we put $n_1 = n_2 = 2$,(*) $\sigma_1 = \sigma_2 = 1$.

There are other identical (but not necessarily independent) variables whose ratio has the Cauchy distribution—cf. Laha (1959) and Kotlarski (1964) and a specific instance in Exercise 11.23.

Example 11.22

Now consider the product, instead of the ratio, of the variables in Example 11.21. As at (11.79), the c.f. of $z = x_1x_2$ is

$$\phi(t) = \int_{-\infty}^\infty \phi_1(tx_2)\,dF_2(x_2)$$

(*) Not 1, because we are squaring about the *population* means here, not the sample means, and thus have an extra degree of freedom. Cf. Example 11.7 on this point.

$$= \int_{-\infty}^{\infty} e^{-\frac{1}{2}(tx_2)^2}(2\pi)^{-\frac{1}{2}}e^{-\frac{1}{2}x_2^2}\,dx_2$$
$$= (2\pi)^{-\frac{1}{2}}\int_{-\infty}^{\infty} e^{-\frac{1}{2}x_2^2(1+t^2)}\,dx_2$$
$$= (1+t^2)^{-\frac{1}{2}}.$$

Inverting this c.f., using (4.5), we have

$$2\pi f(z) = \int_{-\infty}^{\infty}(1+t^2)^{-\frac{1}{2}}e^{-izt}\,dt$$
$$= 2\int_{0}^{\infty}(1+t^2)^{-\frac{1}{2}}\cos zt\,dt$$

using symmetries. Thus $f(z) = \frac{1}{\pi}\int_0^{\infty}(1+t^2)^{-\frac{1}{2}}\cos zt\,dt = \frac{1}{\pi}K_0(|z|)$,

where K_0 is the modified Bessel function of the third kind. Thus $f(z)$ is symmetrical about zero, as is intuitively obvious.

Exercise 11.21 contains related results. Lomnicki (1967) gives results for products of n independent identical variates, which need not be normal.

It follows at once from the independence of x_1 and x_2 that $E(z^r) = E(x_1^r)E(x_2^r)$, whatever the distributions of x_1 and of x_2. Here, in the normal case, we find for the skewness and kurtosis of z, from (5.60), that $\gamma_1 = 0$ and $\gamma_2 = 6$, just as for "Student's" distribution (11.44) with $\nu = 5$, as can be verified from Example 3.3.

11.12 Up to this point we have been mainly concerned with the distribution of a single statistic calculated from the members of a simple random sample. The methods may, however, readily be generalized to obtain the joint distribution of several statistics. For example, if there are several statistics $z_1, z_2, \ldots, z_p$, and the joint distribution of the sample values $x_1, \ldots, x_n$ is represented by $dF(x_1, \ldots, x_n)$, the characteristic function of the z's is given by

$$\phi(t_1, \ldots, t_p) = \int_{-\infty}^{\infty}\cdots\int_{-\infty}^{\infty}\exp(it_1z_1+\ldots+it_pz_p)\,dF(x_1, \ldots, x_n) \quad (11.86)$$

and the frequency function of the z's (if it exists) by

$$f(z_1, \ldots, z_p) = \frac{1}{(2\pi)^p}\int_{-\infty}^{\infty}\cdots\int_{-\infty}^{\infty}\exp(-it_1z_1-\ldots-it_pz_p)\,\phi(t_1, \ldots, t_p)\,dt_1\ldots dt_p. \quad (11.87)$$

Examples of the use of these results will occur in the sequel.

Steepest descents

11.13 Although, in this chapter, we have been concerned with the derivation of exact results for sampling distributions, it is convenient to refer here to an approximative method developed by Daniels (1954), which differs radically from the methods to be discussed in Chapter 12. The frequency function of a mean is expressed as an inversion of the nth power of the characteristic function. The resultant integral is evaluated approximately by choosing a path of integration through a saddlepoint of the integrand in such a way that the integrand is negligible outside the immediate neighbourhood of the saddlepoint. The approximation which results is, in simpler cases at least, remarkably good. The method results in expansions of the frequency function of the Edgeworth type (**6.18**) but is more powerful than the Edgeworth expansion itself. It will be used in **48.19–22**, Vol. 3, in connexion with the sampling theory of serial correlations.

EXERCISES

11.1 Find the distribution of means of samples of n independent observations from

$$dF = e^{-x}\,dx, \quad 0 \leqslant x \leqslant \infty,$$

(a) by induction; (b) by the use of characteristic functions.

11.2 Derive by the method of characteristic functions the sampling distribution of Example 11.1 for the mean of samples from the Cauchy distribution

$$dF = \frac{dx}{\pi(1+x^2)}, \quad -\infty \leqslant x \leqslant \infty.$$

11.3 Show that if $\bar{x}$ is the arithmetic mean and g the geometric mean in samples of n from the Gamma distribution

$$dF = \frac{e^{-x}\,x^{p-1}}{\Gamma(p)}\,dx, \quad 0 \leqslant x \leqslant \infty,$$

the c.f. of $S = \log(\bar{x}/g)$ is

$$\phi(t) = \{\Gamma(p-it/n)/\Gamma(p)\}^n\,\Gamma(np)/\{\Gamma(np-it)\,n^{it}\},$$

with mean

$$\kappa_1 = \frac{\Gamma'(np)}{\Gamma(np)} - \frac{\Gamma'(p)}{\Gamma(p)} - \log n.$$

(Bain and Engelhardt (1975) approximate the d.f. of S.)
Glaser (1976) obtains the exact distribution of $g/\bar{x}$.

11.4 $x_1, \ldots, x_n$ are n independent normal variates with zero mean and unit variance. Show that $x_i^2 \Big/ \left(\sum_{j=1}^{n} x_j^2 \right)$ is distributed independently of $\sum_{j=1}^{n} x_j^2$. Show also that if $\bar{x}$ is the mean of the x's, $(x_i-\bar{x})^2/\Sigma(x_j-\bar{x})^2$ is independent of $\Sigma(x_j-\bar{x})^2$.

11.5 Use the result of the previous Exercise to show that if m_2, m_3 and m_4 are the sample central moments in a normal sample

$$\mathrm{E}\left(\frac{m_3}{m_2^{3/2}}\right)^k = \frac{\mathrm{E}m_3^k}{\mathrm{E}m_2^{3k/2}},$$

$$\mathrm{E}\left(\frac{m_4}{m_2^2}\right)^k = \frac{\mathrm{E}m_4^k}{\mathrm{E}m_2^{2k}}.$$

11.6 Show that the conditional distribution of " Student's " t (defined at (11.42)) for fixed s^2 is exactly normal with mean equal to zero (the assumed population mean) and variance

$$(n-1)\,\sigma^2/(ns^2)$$

11.7 Show that the joint distribution of two order-statistics x_r, x_s $(r<s)$ is given by

$$dF(x_r, x_s) = \frac{n!}{(r-1)!\,(s-r-1)!\,(n-s)!} F_r^{r-1}(F_s-F_r)^{s-r-1}(1-F_s)^{n-s}\,dF_r\,dF_s$$

where F_j is the distribution function of x with $x = x_j$, $j = r, s$. Integrate out x_s to obtain the distribution of x_r at (11.34).

11.8 $x_1, x_2, \ldots, x_n$ are independent Beta variables, distributed in the form (6.14), x_i having parameters (p_i, q_i). Show that if $p_2 = p_1+q_1$, the product $x_1 x_2$ is distributed in the same form with parameters (p_1, q_1+q_2); and hence that if $p_r = p_1 + \sum_{s=1}^{r-1} q_s$ for $r = 2, 3, \ldots, n$, the product $x_1 x_2 \ldots x_r$ has the same distribution with parameters $(p_1, \sum_{s=1}^{r} q_s)$.

(The product of other sets of independent variables can have this distribution—cf. Kotlarski (1962) and also I. R. James (1972).)

11.9 Show that the distribution of the geometric mean of n variables, one from each of the distributions with frequency functions

$$\frac{x^{p-1}e^{-x}}{\Gamma(p)}, \frac{x^{p+1/n-1}e^{-x}}{\Gamma(p+1/n)}, \ldots, \frac{x^{p+(n-1)/n-1}e^{-x}}{\Gamma\{p+(n-1)/n\}},$$

is the same as the distribution of the arithmetic mean of n independent variables distributed in the first of these forms. Show that for $n = 2$, $p = \frac{1}{4}$, this gives a pair of independent random variables whose product is exactly normally distributed.

(cf. Kullback, 1934)

11.10 If $x_1, x_2, \ldots, x_n$ are independent Gamma variables

$$f(x_i) = \frac{1}{\Gamma(p_i)} e^{-x_i} x_i^{p_i-1}, \quad p_i>0, \; 0 \leqslant x \leqslant \infty,$$

show that the variables

$$y_1 = \sum_{i=1}^{n} x_i,$$

$$y_r = x_r \Big/ \sum_{s=1}^{r-1} x_s, \quad r = 2, 3, \ldots, n$$

are also independent, y_1 being a Gamma variable with parameter $\sum_{i=1}^{n} p_i$ and y_r $(r>1)$ being distributed in the second kind of Beta distribution (6.17) with parameters p_r and $\sum_{s=1}^{r-1} p_s$. Show that the converse result also holds: if the y's are thus independently distributed, then the x's are.

(For more general results, cf. Aitchison (1963).)

11.11 Show that the ratio $v = x/y$ of two independent normal variables has frequency function

$$f(v) = \frac{1}{\sqrt{(2\pi)}} \frac{\mu_y \sigma_x^2 + \mu_x \sigma_y^2 v}{(\sigma_x^2+\sigma_y^2 v^2)^{\frac{3}{2}}} \exp\left\{-\tfrac{1}{2}\frac{(\mu_x-\mu_y v)^2}{\sigma_x^2+\sigma_y^2 v^2}\right\}$$

where the μ_j and σ_j^2 are the means and variances of the variables, and it is assumed that μ_y is so large compared with σ_y that the range of y is effectively positive.

Hence show that $(\mu_x - \mu_y v)(\sigma_x^2+\sigma_y^2 v^2)^{-\frac{1}{2}}$ is normally distributed about zero mean with unit variance.

(Geary, 1930)

11.12 Generalizing **11.2**, suppose that the x_i have unit variances but means μ_i, which may differ. Writing $\lambda = \sum_{i=1}^{n} \mu_i^2$, show that the distribution of z at (11.4) is now

$$f(z) = \frac{e^{-\frac{1}{2}(z+\lambda)} z^{\frac{1}{2}n-1}}{2^{\frac{1}{2}n}\Gamma\{\frac{1}{2}(n-1)\}\Gamma(\frac{1}{2})} \sum_{r=0}^{\infty} \frac{\lambda^r z^r}{(2r)!} B\{\tfrac{1}{2}(n-1), \tfrac{1}{2}+r\},$$

reducing to (11.8) when $\lambda = 0$. (The c.f. of this f.f. was given in Exercise 7.21.)

11.13 Let z_1, z_2 be independently distributed as in Exercise 11.12, with constants (n, λ) equal to (ν_1, λ) for z_1 and to $(\nu_2, 0)$ for z_2. Show that the distribution of $\dfrac{(z_1/\nu_1)}{(z_2/\nu_2)} = w$ is

$$g(w) = e^{-\frac{1}{2}\lambda} \sum_{r=0}^{\infty} \frac{(\frac{1}{2}\lambda)^r}{r!} \left(\frac{\nu_1}{\nu_2}\right)^{\frac{1}{2}\nu_1+r} \frac{1}{B(\frac{1}{2}\nu_1+r, \frac{1}{2}\nu_2)} \frac{w^{\frac{1}{2}\nu_1-1+r}}{\left(1+\frac{\nu_1}{\nu_2}w\right)^{\frac{1}{2}(\nu_1+\nu_2)+r}}$$

which reduces to Example 11.20 when $\lambda = 0$, $\nu_1 = n_1-1$, $\nu_2 = n_2-1$.

(The distributions of this and the previous exercises are called *non-central*. We shall study them in Chapter 24, Vol. 2.)

11.14 x_1 and x_2 are independently distributed in the form (11.8) with $2n$ and $2m$ degrees of freedom respectively. Show that the c.f. of $y = cx_1 - bx_2$ $(c, b > 0)$ is

$$\phi(t) = (1-2ict)^{-n}(1+2ibt)^{-m}.$$

Show from the negative binomial distribution (5.36) that

$$p^m \sum_{u=0}^{n-1} \binom{m+u-1}{m-1} q^u + q^n \sum_{v=0}^{m-1} \binom{n+v-1}{n-1} p^v \equiv 1$$

and hence, substituting $p = 1-q = (1+2ibt)c/(c+b)$, that

$$\phi(t) = \sum_{j=1}^{n} \binom{n+m-j-1}{m-1} \left(\frac{c}{c+b}\right)^m \left(\frac{b}{c+b}\right)^{n-j} (1-2ict)^{-j}$$
$$+ \sum_{k=1}^{m} \binom{n+m-k-1}{n-1} \left(\frac{b}{c+b}\right)^n \left(\frac{c}{c+b}\right)^{m-k} (1+2ibt)^{-k},$$

expressing the c.f. as that of a mixture of $(n+m)$ distributions of the same forms as cx_1 and $(-bx_2)$, with negative binomial probabilities as the weights.

(Jayachandran and Barr, 1970)

11.15 z_1, z_2 are independent variates distributed in the generalized Gamma form

$$dF(z_i) = \frac{\exp\left(-\frac{z_i}{\theta_i}\right)\left(\frac{z_i}{\theta_i}\right)^{a_i}}{\Gamma(a_i+1)} d\left(\frac{z_i}{\theta_i}\right), \quad z_i > 0,\ \theta_i > 0$$

where a_i is zero or a positive integer. Writing $a_1 + a_2 = a$, $\phi = \frac{1}{\theta_1} + \frac{1}{\theta_2}$, and using the identity

$$\int_0^\infty t^{a_1}(t+s)^{a_2} e^{-\lambda t}\,dt \equiv \lambda^{-(a+1)} \sum_{r=0}^{a_2} \binom{a_2}{r} (a-r)!\,(\lambda s)^r,$$

show that the distribution of $u = z_1 - z_2$ is

$$h(u) \propto \exp\left(\frac{u}{\theta_2}\right) A(u, \phi), \quad -\infty \leqslant u \leqslant \infty,$$

where

$$A(u, \phi) = \begin{cases} \exp(-\phi u)\,\phi^{-(a+1)} \sum_{r=0}^{a_1} \binom{a_1}{r} (a-r)!\,(\phi u)^r, & u \geqslant 0, \\ \phi^{-(a+1)} \sum_{r=0}^{a_2} \binom{a_2}{r} (a-r)!\,(\phi u)^r, & u < 0. \end{cases}$$

Show that if $\theta_1 = 2c$, $\theta_2 = 2b$, $a_1 = n-1$ and $a_2 = m-1$, $h(u)$ is the f.f. corresponding to the c.f. in Exercise 11.14.

The special case $\theta_1 = \theta_2 = 1$, $\phi = 2$, $a_1 = a_2 = \frac{1}{2}a$ yields the distribution of the difference between two identical Gamma variates with parameter an integer.

(Cf. Lentner and Buehler (1963). For the case where $a_1 = a_2$ is non-integral, cf. K. Pearson *et al.* (1932).)

11.16 In Exercise 11.15, show that the distribution of $v = z_1+z_2$ is (for any, not necessarily integral, $a_i>0$)

$$g(v)=\sum_{r=0}^{\infty}\binom{a_1+r}{r}\left(\frac{\theta_2}{\theta_1}\right)^{a_1+1}\left(1-\frac{\theta_2}{\theta_1}\right)^r \frac{e^{-\frac{v}{\theta_2}}\left(\frac{v}{\theta_2}\right)^{a+r+1}}{\Gamma(a+r+2)}\, d\left(\frac{v}{\theta_2}\right).$$

Show also that the conditional distribution of z_1 given u is, for integral a_i,

$$g(z_1|u) = z_1^{a_1}(z_1-u)^{a_2}\exp(-\phi z_1)/A(u,\phi), \quad \begin{cases} z_1>0, \\ z_1>u. \end{cases}$$

(Lentner and Buehler, 1963)

11.17 x is a standardized normal variate and, independently, z has the f.f. (11.8) with n an even integer $n = 2p>0$. Use the method of Example 11.22 to show that $w = xz^{\frac{1}{2}}$ is distributed like the sum of p independent variates w_j with f.f. $\frac{1}{2}\exp(-|w_j|)$.

11.18 Derive the joint characteristic function of mean and variance in samples of n observations from a normal distribution, and show that it factorizes into their marginal c.f.'s, and that therefore they are independently distributed.

11.19 If the c.f. of sample mean and variance in samples of size n is $\phi(t_1, t_2)$ and that of the frequency distribution $f(x)$ is $\alpha(t)$, show that

$$\left[\frac{\partial}{\partial t_2}\phi(t_1,t_2)\right]_{t_2=0} = \frac{n-1}{n}i\left[\alpha^{n-1}(t_1/n)\int_{-\infty}^{\infty}x^2 e^{it_1x/n}f(x)\,dx-\alpha^{n-2}(t_1/n)\left\{\int_{-\infty}^{\infty}x\,e^{it_1x/n}f(x)\,dx\right\}^2\right].$$

Hence, if mean and variance are independent, show that

$$-\alpha\frac{d^2\alpha}{dt^2}+\left(\frac{d\alpha}{dt}\right)^2 = \sigma^2\alpha^2$$

where σ^2 is the population variance. Hence show that the distribution must be normal.
(Lukacs (1942). Geary (1936) established the result for distributions possessing finite cumulants of all orders.)

11.20 From Exercise 7.15, show that if there are n independent standardized normal variates x_j, the joint c.f. of

$$x = \sum_{j=1}^{n} a_j x_j \text{ and } y = \sum_{j=1}^{n} x_j^2 \text{ is}$$

$$\phi_{x,y}(t_1,t_2) = (1-2it_2)^{-\frac{1}{2}n}\exp\left\{\frac{-\frac{1}{2}a^2}{1-2it_2}\right\}, \quad \text{where } a^2 = t_1^2\sum_{j=1}^{n}a_j^2.$$

Use this result and (4.34) to show that the conditional c.f. of y given that $x = 0$ is

$$\phi_{21}(t_2) = (1-2it_2)^{-\frac{1}{2}(n-1)}$$

corresponding to (11.36) with $p = 1$. If there are p linearly independent functions $x^{(r)} = \sum_{j=1}^{n} a_{jr}x_j$, $r = 1, 2, \ldots, p$, show that the $(p+1)$-variate c.f. $\phi_{x,y}(t_{11},\ldots,t_{1p},t_2)$ is as above with $a^2 = \sum_{j=1}^{n}\left(\sum_{r=1}^{p}a_{jr}t_{1r}\right)^2$, and use the generalization of (4.34) to verify (11.36) for any p.

11.21 x_r, y_r, $r = 1, 2, \ldots,$ are independent standardized normal variates. Show that $z = x_1y_1+x_2y_2$ is distributed exactly as w_1-w_2, where the w_j are independent with f.f. e^{-w_j} and hence or otherwise that z is distributed in the form $g(z) = \frac{1}{2}\exp(-|z|)$ and that $|z|$ is again exponentially distributed. More generally, show that $\sum_{r=1}^{2n} x_r y_r$ is distributed as the difference between two Gamma variables each with parameter n, i.e. as in Exercise 11.15 with $a_1 = a_2 = n-1$ and $\theta_1 = \theta_2 = 1$. (The property that $|w_1 - w_2|$ is distributed exactly as w_j characterizes the exponential distribution $\theta \exp(-\theta w)$ among continuous distributions—cf. Puri and Rubin (1970).)

11.22 Generalizing Example 11.21, show that if x and y are bivariate normally distributed with correlation ρ, the statistic

$$z = \frac{x-\mu_x}{\sigma_x} \Big/ \frac{y-\mu_y}{\sigma_y}$$

has the frequency function

$$g(z) = \frac{(1-\rho^2)^{\frac{1}{2}}\, dz}{\pi(1-2\rho z+z^2)}, \qquad -\infty \leqslant z \leqslant \infty.$$

Generalizing Exercise 11.11, show that if $v = x/y$ and $\mu_y/\sigma_y \longrightarrow \infty$ so that $\text{Prob}(y>0) \longrightarrow 1$, the variable $(\mu_x - \mu_y v)/(\sigma_x^2 - 2\rho\sigma_x\sigma_y v + \sigma_y^2 v^2)^{\frac{1}{2}}$ is approximately standardized normal.

(Cf. Geary (1930), Hinkley (1969) and Marsaglia (1965) for the distribution of x/y.)

11.23 In Example 11.21, show that if instead of normal variates we consider two independent variates distributed as in Exercise 2.12, their ratio is still distributed in the Cauchy form (11.85).

(Steck, 1958a)

11.24 If x_1, x_2 ($x_2 \geqslant 0$, $E(x_2) = \mu$) have the joint f.f. $f(x_1, x_2)$ and c.f. $\phi(t_1, t_2)$, show that the f.f. of $u = x_1/x_2$ is $h(u) = \int_0^\infty f(uv, v)\, v\, dv$, generalizing (11.74). Now consider the joint f.f.

$$g(x_1, x_2) = \frac{x_2}{\mu} f(x_1, x_2).$$

Show that its c.f. is

$$\psi(t_1, t_2) = \frac{1}{i\mu}\frac{\partial\phi(t_1, t_2)}{\partial t_2},$$

and that the c.f. of the variate $w = x_1 - ux_2$ is $[\psi(t_1, t_2)]_{t_2 = -ut_1}$. By showing that the f.f. of w at $w = 0$ is equal to $h(u)/\mu$, use the Inversion Theorem to establish (11.78).

(Daniels, 1954)

11.25 If $x_1, x_2, \ldots, x_n$ are independent standardized normal variates, and $y_j = 1/x_j^2$, with c.f. $\phi_y(t)$, show using (11.79) that the c.f. of x_2/x_1 is $\phi_y(\frac{1}{2}it^2)$, so that from Examples 11.21 and 4.2,

$$\phi_y(t) = \exp\{-(-2it)^{\frac{1}{2}}\}.$$

Verify that the f.f. of y is given by (4.70).

Hence show that the c.f. of the mean $\bar{y}$ of the y's is

$$\phi_{\bar{y}}(t) = \phi_y(nt) = \phi_{ny}(t),$$

so that the distribution of the mean of n observations is a multiple n of that of a single observation, becoming more widely dispersed as n increases.

By putting $n = 2$, show that $v = 2x_1x_2/(x_1^2+x_2^2)^{\frac{1}{2}}$ is a standardized normal variate.

11.26 In **11.2**, show that when $n = 2$, the polar variables z and θ_1 are independently distributed, the former as χ^2 with 2 d.fr. and the latter uniformly on the interval $(0, 2\pi)$. Hence show that $w = z^{\frac{1}{2}}\cos(p\theta_1)$ and $v = z^{\frac{1}{2}}\sin(p\theta_1)$ are independent standardized normal variables for any positive integer p. Putting $p = 2$ in v, derive the final result of Exercise 11.25.

11.27 If x_1 has the symmetric stable c.f. (4.69) with $a = 0$ and exponent α_1, and the non-negative variate x_2 has the stable c.f. (4.68) with $a = 0$ and exponent $\alpha_2 < 1$, show using the method of Example 11.22 that $z = x_1 x_2^{1/\alpha_1}$ is symmetric stable with exponent $\alpha_1\alpha_2$.

If $y_1, y_2, \ldots, y_n$ are independent standardized normal variates, show, using Exercise 11.25, that

$$z = y_1/\{y_2\, y_3^2\, y_4^{2^2} \ldots y_n^{2^{n-2}}\}$$

is symmetric stable with exponent $\alpha = 1/2^{n-2}$.

(Example 11.21 gives the case $n = 2$; cf. Brown and Tukey, 1946.)

CHAPTER 12

CUMULANTS OF SAMPLING DISTRIBUTIONS—(1)

12.1 In the previous chapter we have considered methods of deriving sampling distributions in an exact form when the parent population is completely specified. Those methods are not applicable when the population is not completely known, and they may in any case lead to results which are difficult to apply in practice, e.g. by yielding an integral which has not been tabulated. In such cases we can frequently deal with the problem by finding an approximate form for the sampling distribution, particularly by ascertaining its lower moments, and then fitting a tractable type of curve such as one of the Pearson system.

A procedure of this kind has, in fact, already been considered in Chapter 10, wherein it was seen that approximate expressions could be derived for the first and second moments of sampling distributions in terms of the low moments of the parent. When the sampling distribution tends to normality this, in effect, solves our problem, for its first and second moments determine a normal distribution. The methods of this chapter are really developments of this idea. We shall discuss exact methods of finding the moments of sampling distributions in terms of population moments. Our results are important not only on their own account, but in giving an accurate method of judging the degree of approximation of the expressions for large n discussed in Chapter 10. In particular, we shall be able to take up some points which had to be left on one side in that chapter—e.g. the rapidity with which some functions of the moments such as $\sqrt{b_1}$ approach normality.

12.2 The statistics with whose distributions we are customarily concerned may be usefully classified into three groups. The largest group comprises statistics which are symmetric functions of the observations; that is to say, if the observations are $x_1, x_2, \ldots, x_n$, the statistic depends explicitly on every x and its value is unchanged if we interchange any two x's. Moreover, our statistics are nearly always algebraic. This class includes the arithmetic mean, moments, quantities such as b_1 and b_2 which are derived from moments and, in the bivariate domain, product-moments. The second group comprises statistics based on order-properties of the sample values—the median, the quantiles, the range, extreme values and so forth. The third group is best defined as including those statistics, such as the sample mode, which are not included in the first two groups.

The methods we shall develop in this and the following chapters apply to the first group only. We may recall that there are three different types of moment concerned in the investigation: (a) the moments of the parent population, (b) the moments of the sample and (c) the moments of the sampling distribution. They will be referred to as parent moments, sample moments (moment-statistics) and sampling moments respectively. Similarly we shall consider parent cumulants, sample cumulants and sampling cumulants.

12.3 In Chapter 10 we obtained the exact results

$$\left.\begin{aligned} \mathrm{E}\, m'_r &= \mu'_r \\ \operatorname{var} m'_r &= \mathrm{E}(m'_r - \mu'_r)^2 = \frac{1}{n}(\mu'_{2r} - \mu'^2_r) \end{aligned}\right\} \tag{12.1}$$

and noted that formulae for sampling moments about the mean were more difficult to obtain. Although we shall later reject this approach in favour of another, it is instructive to consider what happens if we try to generalize the procedure of that chapter to our present problem. Suppose, for example, that we are interested in the sampling distribution of the variance. The above equations give us the first two sampling moments of the second moment about an arbitrary point. For the first sampling moment of the variance we have

$$\begin{aligned} \mathrm{E}\, m_2 &= \mathrm{E}\,\{\Sigma x^2/n - (\Sigma x/n)^2\} \\ &= \mathrm{E}\left\{\frac{n-1}{n^2}\Sigma x^2 - \frac{1}{n^2}\Sigma x_i x_j\right\}, \qquad i \neq j \\ &= \frac{n-1}{n}\mathrm{E}\, x^2 - \frac{n-1}{n}\mathrm{E}\, x_i x_j. \end{aligned} \tag{12.2}$$

Since x_i and x_j are independent, $\mathrm{E}\, x_i x_j$ is equal to μ'^2_1. Equation (12.2) then gives us

$$\mathrm{E}\, m_2 = \frac{n-1}{n}(\mu'_2 - \mu'^2_1) = \frac{n-1}{n}\mu_2. \tag{12.3}$$

This may be compared with the approximate expression given by the methods of Chapter 10, namely

$$\mathrm{E}\, m_2 = \mu_2. \tag{12.4}$$

12.4 It will be clear that the same method can be used to derive the sampling moment of any order of a statistic which can be expressed as a symmetric function (rational and integral) of the observations. For instance, to find the fourth moment of the sample variance we expand $\{\Sigma x^2/n - (\Sigma x/n)^2\}^4$ in terms of sums of products of type $\Sigma x_i^\alpha x_j^\beta \ldots x_k^\gamma$ and, on taking expectations, replace them by $n(n-1)\ldots(n-t+1)\mu'_\alpha \mu'_\beta \ldots \mu'_\gamma$ where t is the number of different suffixes $i, j \ldots k$ in the sum. This gives us the desired sampling moment in terms of parent moments.

12.5 The method is straightforward enough. Its execution, however, leads to some tedious algebra and some cumbrous expressions, except in very simple cases. We shall therefore find it convenient to systematize the working in terms of the notation of symmetric functions.

Suppose that we have a set of x's, $x_1, x_2, \ldots, x_n$. When we write an expression such as $\Sigma x_i^2 x_j x_l^3 x_k$ we shall suppose that all the suffixes are different and that, subject to this, the summation takes place over all values of x. Thus, in the example just given, there are $n(n-1)(n-2)(n-3)$ terms in the summation. It is to be noted that in an expression such as $\Sigma x_i x_j$ every pair occurs twice, for example $x_1 x_2$ and $x_2 x_1$; whereas in $\Sigma x_i^2 x_j$, $x_1^2 x_2$ and $x_2^2 x_1$ occur, but each does not occur twice.

The *augmented* symmetric functions are defined by

$$[p_1^{\pi_1} p_2^{\pi_2} \dots p_s^{\pi_s}] = \Sigma x_i^{p_1} x_j^{p_1} \dots x_q^{p_2} x_r^{p_2} \dots x_u^{p_s} x_v^{p_s} \tag{12.5}$$

where there are π_1 powers p_1, π_2 powers p_2, and so on on the right of (12.5). For example

$$\Sigma x_i^2 x_j x_l^3 x_k = [1^2 2\, 3]\,; \quad \Sigma x_i^2 x_j^2 x_k^2 = [2^3].$$

More usual functions are the *monomial* symmetric functions defined by

$$(p_1^{\pi_1} p_2^{\pi_2} \dots p_s^{\pi_s}) = [p_1^{\pi_1} p_2^{\pi_2} \dots p_s^{\pi_s}]/\{\pi_1!\,\pi_2! \dots \pi_s!\}. \tag{12.6}$$

Two particular cases are of special importance: the *unitary* functions

$$a_r = (1^r) = \Sigma x_i x_j \dots x_l / r! \tag{12.7}$$

and the one-part functions or power-sums

$$s_r = (r) = \Sigma x_i^r = [r]. \tag{12.8}$$

Tables exist giving these functions in terms of one another. The most useful are the tables giving the power-sums in terms of the augmented symmetrics and vice versa (David and Kendall, 1949). From (12.5) we have the fundamental result

$$\mathrm{E}[p_1^{\pi_1} p_2^{\pi_2} \dots p_s^{\pi_s}] = n(n-1)\dots(n-\rho+1)(\mu'_{p_1})^{\pi_1}(\mu'_{p_2})^{\pi_2}\dots(\mu'_{p_s})^{\pi_s} \tag{12.9}$$

where $\rho = \sum_{i=1}^{s} \pi_i$ and $p = \sum_{i=1}^{s} p_i \pi_i$ is the *weight* of the symmetric function. Appendix Table 10 gives the relationships up to weight 6.

Example 12.1

For the sample variance we have, in terms of the power sums,

$$m_2 = \frac{(2)}{n} - \frac{(1)^2}{n^2}.$$

From Appendix Table 10 or directly we have

$$(2) = [2]\,; \quad (1)^2 = [2] + [1^2].$$

Hence

$$m_2 = \frac{[2]}{n} - \frac{[2]+[1^2]}{n^2}$$

$$= \frac{n-1}{n^2}[2] - \frac{1}{n^2}[1^2]. \tag{12.10}$$

Taking expectations and using (12.9) we have

$$\mathrm{E}\,m_2 = \frac{n-1}{n^2} n\mu'_2 - \frac{1}{n^2} n(n-1)\mu_1'^2$$

$$= \frac{n-1}{n}\mu_2. \tag{12.11}$$

Let us note here a very useful abbreviation of the working. The statistic m_2 is independent of the origin of calculation and hence its sampling moments cannot depend on μ'_1. Without loss of generality, therefore, we can take this parent mean to be zero, the other moments then becoming moments about the mean. This implies that we can ignore any augmented symmetric function containing a unit. Thus from (12.10) we have at once

$$\mathrm{E}\, m_2 = \frac{n-1}{n}\mu_2.$$

In a similar way we find

$$m_2^2 = \frac{(2)^2}{n^2} - \frac{2(2)(1)^2}{n^3} + \frac{(1)^4}{n^4}. \tag{12.12}$$

From the tables,

$$\begin{aligned}(2)^2 &= [4]+[2^2]\\ (2)(1)^2 &= [4]+2[31]+[2^2]+[21^2]\\ (1)^4 &= [4]+4[31]+3[2^2]+6[21^2]+[1^4].\end{aligned}$$

We can ignore those augmented symmetrics containing a unit, and on substitution in (12.12) we find

$$\begin{aligned}m_2^2 &= \frac{[4]+[2^2]}{n^2} - 2\frac{[4]+[2^2]}{n^3} + \frac{[4]+3[2^2]}{n^4}\\ &= \frac{(n-1)^2}{n^4}[4] + \frac{n^2-2n+3}{n^4}[2^2],\end{aligned}$$

whence, immediately,

$$\mathrm{E}\, m_2^2 = \frac{(n-1)^2}{n^3}\mu_4 + \frac{(n-1)(n^2-2n+3)}{n^3}\mu_2^2. \tag{12.13}$$

We then find, using (12.11),

$$\begin{aligned}\operatorname{var} m_2 &= \mathrm{E}\, m_2^2 - (\mathrm{E}\, m_2)^2\\ &= \frac{(n-1)^2}{n^3}\mu_4 - \frac{(n-1)(n-3)}{n^3}\mu_2^2. \end{aligned}\tag{12.14}$$

For large n this becomes approximately

$$\operatorname{var} m_2 = \frac{\mu_4 - \mu_2^2}{n},$$

confirming the result we reached in (10.9) with $r = 2$.

If, in (12.14), we put $\kappa_4 = \mu_4 - 3\mu_2^2$, $\kappa_2 = \mu_2$ we find

$$\operatorname{var} m_2 = \left(\frac{n-1}{n}\right)^2 \left\{\frac{\kappa_4}{n} + \frac{2\kappa_2^2}{n-1}\right\}. \tag{12.15}$$

k-statistics

12.6 The algebraic complexity of the results obtained by this straightforward approach and the amount of work required to reach them, especially before the tables of symmetric functions were available, led to a search for simpler methods. Fisher (1929) revolutionized the subject in two ways: by proposing new symmetric functions to characterize the distributions, the so-called *k*-statistics; and by showing how their sampling cumulants could be obtained by combinatorial methods.

We consider a family of statistics $k_1, k_2, \ldots, k_p, \ldots$ which are symmetric functions of the observations and are such that the mean value of k_p is the pth cumulant κ_p :—

$$\mathrm{E}k_p = \kappa_p. \tag{12.16}$$

There is a possible source of confusion here due to notation. Whereas the moment-statistic m_p is the same function of the sample values as μ_p is of the parent values, the same relation does not hold between k_p and κ_p. Conversely, $\mathrm{E}m_p$ is not equal to μ_p.

12.7 Note first of all that k_p is uniquely determined by this definition; for if there were two functions k_p and k'_p obeying (12.16) their difference $k_p - k'_p$ would have a zero mean value. But this difference is itself a symmetric function and can therefore be expressed as the sum of terms Σx^p, $\Sigma x_j x_k^{p-1}$, etc., and hence its mean value is a series of terms each of which is a product of moments. The vanishing of this series would imply a relationship among the moments, which is impossible except perhaps for particular parent populations. Hence $k_p - k'_p$ must vanish identically and thus $k_p = k'_p$.

Secondly, note that the k's are, like the central moments, independent of the origin of measurement, except for k_1, which is equal to the mean itself. In fact, we have by Taylor's theorem

$$k_p(x_1+h, x_2+h, \ldots, x_n+h)$$
$$= k_p(x_1, x_2, \ldots, x_n) + \frac{h}{1!}Dk_p(x_1, x_2, \ldots, x_n) + \frac{h^2}{2!}D^2 k_p(x_1, x_2, \ldots, x_n) + \ldots \tag{12.17}$$

where

$$D = \frac{\partial}{\partial x_1} + \frac{\partial}{\partial x_2} + \ldots + \frac{\partial}{\partial x_n}.$$

Taking mean values, and remembering that κ_p itself is independent of the origin, except for κ_1, we have

$$\kappa_p = \kappa_p + \frac{h}{1!}\mathrm{E}(Dk_p) + \ldots \tag{12.18}$$

Thus $\mathrm{E}(Dk_p)$ and other terms on the right vanish separately, for (12.18) is an identity in h. In virtue of the remark above, this implies that $Dk_p = 0$, $D^2 k_p = 0$, and so on; and hence, from (12.17),

$$k_p(x_1+h, x_2+h, \ldots, x_n+h) = k_p(x_1, x_2, \ldots, x_n),$$

i.e. k_p is independent of h. The exception to this rule is k_1 which has as its mean value $\kappa_1 = \mu'_1$ and thus

$$k_1 = \frac{1}{n}\Sigma x. \tag{12.19}$$

12.8 We now proceed to find explicit expressions for the k-statistics in terms of the observations $x_1, \ldots, x_n$. By definition k_p is of degree p in these observations (for κ_p is of order p in the moments, that is, the sum of the orders of the moments comprising any term in κ_p is p). We may then write

$$k_p = \Sigma\Sigma(x_1^{p_1} x_2^{p_1} \ldots x_{\pi_1}^{p_1} x_{\pi_1+1}^{p_2} \ldots x_{\pi_1+\pi_2}^{p_2} \ldots x_{\pi_1+\ldots+\pi_s}^{p_s}) A(p_1^{\pi_1} \ldots p_s^{\pi_s}) \tag{12.20}$$

where the second summation extends over all the ways of assigning the $\pi_1 + \pi_2 + \ldots + \pi_s$

subscripts (including permutations) from the n available and the first summation extends over all partitions of the number p, $(p_1^{\pi_1} p_2^{\pi_2} \ldots p_s^{\pi_s})$. $A(p_1^{\pi_1} \ldots p_s^{\pi_s})$ is a number depending on the partition.

We have
$$p_1\pi_1 + p_2\pi_2 + \ldots + p_s\pi_s = p \tag{12.21}$$
and define ρ by
$$\pi_1 + \pi_2 + \ldots + \pi_s = \rho. \tag{12.22}$$

We assume $n \geqslant p$. On taking mean values of (12.20) we have, since the x's are independent,
$$\kappa_p = \Sigma \{(\mu_{p_1}^{\pi_1} \mu_{p_2}^{\pi_2} \ldots \mu_{p_s}^{\pi_s}) AB\}, \tag{12.23}$$
where B is the number of ways of picking out the ρ subscripts from n, permutations allowed, and is therefore equal to $n(n-1)\ldots(n-\rho+1) = n^{[\rho]}$.

Now from equation (3.39), we have
$$\kappa_p = p!\Sigma\Sigma\left(\frac{\mu_{p_1}}{p_1!}\right)^{\pi_1} \ldots \left(\frac{\mu_{p_s}}{p_s!}\right)^{\pi_s} \frac{(-1)^{\rho-1}(\rho-1)!}{\pi_1! \ldots \pi_s!}, \tag{12.24}$$
the summation extending over all partitions subject to (12.21) and (12.22). On identifying corresponding terms in (12.23) and (12.24) we find the values of the A's and on substituting in (12.20) obtain finally
$$k_p = \frac{p!\Sigma(-1)^{\rho-1}(\rho-1)!}{n^{[\rho]}} \Sigma \frac{x_1^{p_1} \ldots x_\rho^{p_s}}{(p_1!)^{\pi_1} \ldots (p_s!)^{\pi_s} \pi_1! \ldots \pi_s!}, \tag{12.25}$$
the explicit expression of k_p in terms of the x's.

We may notice an important simplification of this expression which is crucial in a discussion of the sampling properties of the k's. Apart from factors in ρ and n a typical term in (12.25) may be written
$$p!\left(\frac{x_1^{p_1}}{p_1!}\frac{x_2^{p_1}}{p_1!} \ldots \frac{x_{\pi_1}^{p_1}}{p_1!} \ldots \frac{x_\rho^{p_s}}{p_s!}\right) \cdot \frac{1}{\pi_1! \ldots \pi_s!}$$
where, it is to be remembered, permutations of the subscripts are allowed. There will be a term of this type corresponding to every partition of ρ into π's and of p into p's. Consequently we may write
$$k_p = \Sigma \frac{(-1)^{\rho-1}(\rho-1)!}{n^{[\rho]}} \Sigma (x_{\gamma_1} x_{\gamma_2} \ldots x_{\gamma_p}), \tag{12.26}$$
where there is a term in the second summation corresponding to every possible way of assigning the subscripts. In this assignment, subscripts are regarded as distinct entities. For example, if from the n subscripts we choose p_1 to be 1, p_1 to be 2, ... p_2 to be π_1+1, and so on, there will be as many different terms as there are ways of choosing p_1 from the 1's, and so on, i.e.
$$\frac{p!}{(p_1!)^{\pi_1} \ldots (p_s!)^{\pi_s} \pi_1! \ldots \pi_s!}. \tag{12.27}$$

In fact, (12.25) is a condensed form of (12.26) in which all the terms leading to the same x-product are added together, their number being given by (12.27).

Expression of k-statistics in terms of symmetric products and sums

12.9 We can write down the k's in terms of the augmented symmetric functions at once from the expressions for cumulants in terms of moments. For example, we have

$$\kappa_3 = \mu_3' - 3\mu_2'\mu_1' + 2\mu_1'^3.$$

Hence, comparing (12.24) and (12.25), we have

$$k_3 = \frac{[3]}{n} - \frac{3\,[21]}{n(n-1)} + \frac{2\,[1^3]}{n(n-1)(n-2)}.$$

Substituting for the augmented symmetrics in terms of power sums,

$$[3] = (3), [21] = -(3)+(2)(1), [1^3] = 2(3)-3(2)(1)+(1)^3$$

we find

$$k_3 = \frac{1}{n(n-1)(n-2)}\{n^2(3)-3n(2)(1)+2(1)^3\}.$$

Writing $s_r = (r)$ we may put this in the form

$$k_3 = \frac{1}{n(n-1)(n-2)}(n^2 s_3 - 3n s_2 s_1 + 2s_1^3).$$

12.10 The first eight k-statistics in terms of the power sums are as follows :—

$$\left.\begin{aligned}
k_1 &= \frac{1}{n}s_1\\
k_2 &= \frac{1}{n^{[2]}}(ns_2 - s_1^2)\\
k_3 &= \frac{1}{n^{[3]}}(n^2 s_3 - 3n s_2 s_1 + 2s_1^3)\\
k_4 &= \frac{1}{n^{[4]}}\{(n^3+n^2)s_4 - 4(n^2+n)s_3 s_1 - 3(n^2-n)s_2^2 + 12n s_2 s_1^2 - 6s_1^4\}\\
k_5 &= \frac{1}{n^{[5]}}\{(n^4+5n^3)s_5 - 5(n^3+5n^2)s_4 s_1 - 10(n^3-n^2)s_3 s_2 + 20(n^2+2n)s_3 s_1^2\\
&\quad +30(n^2-n)s_2^2 s_1 - 60n s_2 s_1^3 + 24s_1^5\}\\
k_6 &= \frac{1}{n^{[6]}}\{(n^5+16n^4+11n^3-4n^2)s_6 - 6(n^4+16n^3+11n^2-4n)s_5 s_1\\
&\quad -15n(n-1)^2(n+4)s_4 s_2 - 10(n^4-2n^3+5n^2-4n)s_3^2\\
&\quad +30(n^3+9n^2+2n)s_4 s_1^2 + 120(n^3-n)s_3 s_2 s_1 + 30(n^3-3n^2+2n)s_2^3\\
&\quad -120(n^2+3n)s_3 s_1^3 - 270(n^2-n)s_2^2 s_1^2 + 360n s_2 s_1^4 - 120s_1^6\}\\
k_7 &= \frac{1}{n^{[7]}}\{(n^6+42n^5+119n^4-42n^3)s_7 - 7(n^5+42n^4+119n^3-42n^2)s_6 s_1\\
&\quad -21(n^5+12n^4-31n^3+18n^2)s_5 s_2 - 35(n^5+5n^3-6n^2)s_4 s_3\\
&\quad +42(n^4+27n^3+44n^2-12n)s_5 s_1^2 + 210(n^4+6n^3-13n^2+6n)s_4 s_2 s_1\\
&\quad +140(n^4+5n^2-6n)s_3^2 s_1 + 210(n^4-3n^3+2n^2)s_3 s_2^2\\
&\quad -210(n^3+13n^2+6n)s_4 s_1^3 - 1260(n^3+n^2-2n)s_3 s_2 s_1^2\\
&\quad -630(n^3-3n^2+2n)s_2^3 s_1 + 840(n^2+4n)s_3 s_1^4 + 2520(n^2-n)s_2^2 s_1^3\\
&\quad -2520n s_2 s_1^5 + 720s_1^7\}
\end{aligned}\right\} \quad (12.28)$$

$$\left.\begin{aligned}k_8 = \frac{1}{n^{[8]}}\{&(n^7+99n^6+757n^5+141n^4-398n^3+120n^2)s_8-8(n^6+99n^5+757n^4\\&+141n^3-398n^2+120n)s_7s_1-28(n^6+37n^5-39n^4-157n^3\\&+278n^2-120n)s_6s_2-56(n^6+9n^5-23n^4+111n^3-218n^2+120n)s_5s_3\\&-35(n^6+n^5+33n^4-121n^3+206n^2-120n)s_4^2+56(n^5+68n^4+359n^3\\&-8n^2-60n)s_6s_1^2+336(n^5+23n^4-31n^3-23n^2+30n)s_5s_2s_1\\&+560(n^5+5n^4+5n^3-5n^2-6n)s_4s_3s_1+420(n^5+2n^4-25n^3\\&+46n^2-24n)s_4s_2^2+560(n^5-4n^4+11n^3-20n^2+12n)s_3^2s_2\\&-336(n^4+38n^3+99n^2-18n)s_5s_1^3-2520(n^4+10n^3-17n^2+6n)s_4s_2s_1^2\\&-1680(n^4+2n^3+7n^2-10n)s_3^2s_1^2-5040(n^4-2n^3-n^2+2n)s_3s_2^2s_1\\&-630(n^4-6n^3+11n^2-6n)s_2^4+1680(n^3+17n^2+12n)s_4s_1^4\\&+13{,}440(n^3+2n^2-3n)s_3s_2s_1^3+10{,}080(n^3-3n^2+2n)s_2^3s_1^2\\&-6720(n^2+5n)s_3s_1^5-25{,}200(n^2-n)s_2^2s_1^4+20{,}160n\,s_2s_1^6-5040s_1^8\}\end{aligned}\right\} \quad (12.28)$$

In particular, taking the origin at zero ($s_1 = 0$), we have

$$\left.\begin{aligned}k_1 &= m_1'\\k_2 &= \frac{n}{n-1}m_2\\k_3 &= \frac{n^2}{(n-1)(n-2)}m_3\\k_4 &= \frac{n^2}{(n-1)(n-2)(n-3)}\{(n+1)m_4-3(n-1)m_2^2\}\end{aligned}\right\} \quad (12.29)$$

expressing the k's in terms of the moment-statistics.

Formulae for k_9, k_{10} and k_{11} are given explicitly by Zia Ud-Din (1954, 1959). The last of these occupies two pages.

12.11 There is a well-known theorem of symmetric functions which states that any rational integral algebraic symmetric function of $x_1, \ldots, x_n$ can be expressed uniquely, rationally, integrally and algebraically in terms of the symmetric sums s_r. It can thus be so expressed in terms of the k's, for from equations such as (12.28) the s's can be so expressed in terms of the k's. Thus an investigation of the sampling constants of any symmetric function expressible in terms of rational integral algebraic symmetric functions can be translated into an investigation concerning the k's.

The reader who is prepared to take the algebra for granted may prefer to pass over the rest of this chapter and the next rather lightly, noting the main results without following the proofs. We are about to embark on a combinatorial method for deriving systematically the sampling cumulants of k-statistics in terms of parent cumulants. We shall find that the results are substantially simpler than equivalent results stated in terms of moments.

Sampling cumulants of k-statistics

12.12 The problem of determining the sampling-moments or the sampling cumulants of k-statistics is that of finding mean values of powers and products of those statistics. To any number a with partition $(a_1^{\alpha_1} a_2^{\alpha_2} \dots a_s^{\alpha_s})$ there will correspond a moment

$$\mu'(a_1^{\alpha_1} \dots a_s^{\alpha_s}) = \mathrm{E}(k_{a_1}^{\alpha_1} \dots k_{a_s}^{\alpha_s}) \tag{12.30}$$

and a cumulant $\kappa(a_1^{\alpha_1} \dots a_s^{\alpha_s})$ related to the moments by the identity (cf. (3.30))

$$\Sigma\left\{\kappa(a_1^{\alpha_1} \dots a_s^{\alpha_s})\frac{t_{a_1}^{\alpha_1}}{\alpha_1!} \dots \frac{t_{a_s}^{\alpha_s}}{\alpha_s!}\right\} = \log\left\{\Sigma\mu'(b_1^{\beta_1} \dots b_m^{\beta_m})\frac{t_{b_1}^{\beta_1}}{\beta_1!} \dots \frac{t_{b_m}^{\beta_m}}{\beta_m!}\right\}. \tag{12.31}$$

For example, the fourth cumulant of k_2 will be expressible in terms of the fourth moment of k_2 about the origin and moments of lower order. These quantities will be written $\kappa(2^4)$ and $\mu'(2^4)$, in accordance with (12.30). Again, the cumulant $\kappa(32)$ corresponds to the moment $\mu'(32)$, the mean value of $k_3 k_2$. Generally, in the simultaneous distribution of the k's there will be a separate formula of degree a for every partition of a.

Now the product $k_{a_1}^{\alpha_1} \dots k_{a_s}^{\alpha_s}$ is homogeneous and of total degree a in the x's. Hence, when mean values are taken $\mu'(a_1^{\alpha_1} \dots a_s^{\alpha_s})$ will be homogeneous and of total order a in the parent μ's. Since the κ's themselves are of homogeneous order in the μ's it follows that $\kappa(a_1^{\alpha_1} \dots a_s^{\alpha_s})$ is of homogeneous order in the κ's. Hence we get the first rule for the sampling of k-statistics :—

Rule 1. $\kappa(a_1^{\alpha_1} \dots a_s^{\alpha_s})$ consists of the sum of terms each of which, except for constants, is a product of parent κ's of order a.

For instance, $\kappa(2^4)$ is of total order 8 and is therefore the sum of terms in κ_8, $\kappa_6\kappa_2$, $\kappa_5\kappa_3$, κ_4^2, $\kappa_4\kappa_2^2$, $\kappa_3^2\kappa_2$ and κ_2^4. Similarly $\kappa(32)$ will contain a term in κ_5 and one in $\kappa_3\kappa_2$ and no others. As seen in the next rule, no terms in κ_1 appear.

Rule 2. No term in $\kappa(a_1^{\alpha_1} \dots a_s^{\alpha_s})$ contains κ_1, except $\kappa(1)$ itself.

This follows as in Example 12.1. The k-statistics are independent of the origin and hence their sampling distribution cannot depend on the variable quantity κ_1. The exception occurs when we are dealing with the only statistic which is dependent on the origin, namely k_1, and here $\kappa(1) = \kappa_1$ as is evident from the definitions.

12.13 We now enunciate and illustrate the rules by which the terms in $\kappa(a_1^{\alpha_1} \dots a_s^{\alpha_s})$ can be found. As the proof of the validity of the rules is difficult to grasp until their nature has been comprehended we defer a proof until the next chapter.

To find the term in $\kappa_{b_1}^{\beta_1} \dots \kappa_{b_m}^{\beta_m}$ in $\kappa(a_1^{\alpha_1} \dots a_s^{\alpha_s})$ consider the two-way table

$$\begin{array}{ccccccc|c}
 & & & & & & & b_1 \\
 & & & & & & & b_1 \\
 & & & & & & & \cdot \\
 & & & & & & & \cdot \\
 & & & & & & & \cdot \\
 & & & & & & & b_2 \\
 & & & & & & & \cdot \\
 & & & & & & & \cdot \\
 & & & & & & & \cdot \\
 & & & & & & & \cdot \\
\hline
a_1 & a_1 & \dots & a_2 & \dots & & & a
\end{array} \tag{12.32}$$

where there is a row corresponding to every κ in the term $\kappa_{b_1}^{\beta_1} \ldots \kappa_{b_m}^{\beta_m}$ and a column corresponding to every part in $\kappa(a_1^{\alpha_1} \ldots a_s^{\alpha_s})$. Consider the various possible arrays that can complete the body of the table by the insertion of positive integers whose row and column sums are the respective b and a numbers; e.g. if we are seeking the coefficient of $\kappa_6 \kappa_2^2$ in $\kappa(4^2 2)$ we shall consider such arrays as

$$\begin{array}{ccc|c} 2 & 2 & 2 & 6 \\ 1 & 1 & . & 2 \\ 1 & 1 & . & 2 \\ \hline 4 & 4 & 2 & 10 \end{array} \qquad \begin{array}{ccc|c} 2 & 3 & 1 & 6 \\ 1 & 1 & . & 2 \\ 1 & . & 1 & 2 \\ \hline 4 & 4 & 2 & 10 \end{array} \qquad \begin{array}{ccc|c} 3 & 3 & . & 6 \\ 1 & . & 1 & 2 \\ . & 1 & 1 & 2 \\ \hline 4 & 4 & 2 & 10 \end{array} \tag{12.33}$$

Then the rules by which these arrays give the coefficients of $\kappa_{b_1}^{\beta_1} \ldots \kappa_{b_m}^{\beta_m}$ are as follows:

Rule 3. Every array in which the numbers in the body of the array fall into two or more blocks, each confined to separate rows and columns, is to be ignored.

For instance, in the foregoing example

$$\begin{array}{ccc|c} 4 & 2 & . & 6 \\ . & 2 & . & 2 \\ . & . & 2 & 2 \\ \hline 4 & 4 & 2 & 10 \end{array}$$

is to be ignored, since the 2×2 block in the top left-hand corner has no row or column number in common with the entry in the bottom right-hand corner.

Rule 4. There will be a contribution to the coefficient of $\kappa_{b_1}^{\beta_1} \ldots \kappa_{b_m}^{\beta_m}$ in $\kappa(a_1^{\alpha_1} \ldots a_s^{\alpha_s})$ corresponding to each array that can complete (12.32) without becoming subject to Rule 3. Each contribution consists of a numerical coefficient multiplied by a function of n called the pattern function.

Rule 5. The numerical coefficient is the number of ways in which the column totals, considered as composed of distinct individuals, can be allocated to form the array concerned, divided by $\beta_1!\beta_2!\ldots\beta_m!$.

Rule 6. The pattern function depends only on the configuration of zeros in the array, not on the actual numbers composing it or on the row and column totals, and is given by considering the separations of the rows into distinct groups or separates.

(i) With a separation into one separate, there is associated the number n; with a separation into two separates, $n(n-1)$; . . . ; with a separation into q separates, $n(n-1) \ldots (n-q+1)$.

(ii) In each separation we count the number of separates in which a particular column is represented by a non-zero entry. If in ρ separates, we assign the factor

$$\frac{(-1)^{\rho-1}(\rho-1)!}{n(n-1)\ldots(n-\rho+1)} \text{ to that column.}$$

This is done for each column and the factors multiplied together.

(iii) For each separation, the number in (i) is multiplied by the product in (ii).

(iv) The result, summed over all separations, gives the pattern function.

Rule 7. Any array containing a row which consists of a single non-zero entry has a vanishing pattern function and is to be ignored.

Rule 8. Any array containing a column which consists of a single non-zero entry has a pattern function $1/n$ times that of the array obtained by omitting that column.

Rule 9. Any array the non-zero elements of which consist of two groups connected only by a single column has a vanishing pattern function and is to be ignored.

Example 12.2

As an illustration of these rules (which are not as difficult as they look), suppose that we seek the coefficient of $\kappa_6 \kappa_2^2$ in $\kappa(4^2 2)$. If the reader will write down the thirty or so possible arrays with column totals 4, 4, 2 and row totals 6, 2, 2, he will find that the only ones which do not vanish are those of (12.33) and permutations of rows and columns with the same sum, namely

$$
\begin{array}{ccc|c} 2 & 2 & 2 & 6 \\ 1 & 1 & . & 2 \\ 1 & 1 & . & 2 \\ \hline 4 & 4 & 2 & 10 \end{array}
\qquad
\begin{array}{ccc|c} 2 & 3 & 1 & 6 \\ 1 & 1 & . & 2 \\ 1 & . & 1 & 2 \\ \hline 4 & 4 & 2 & 10 \end{array}
\qquad
\begin{array}{ccc|c} 3 & 2 & 1 & 6 \\ 1 & 1 & . & 2 \\ . & 1 & 1 & 2 \\ \hline 4 & 4 & 2 & 10 \end{array}
\qquad
\begin{array}{ccc|c} 2 & 3 & 1 & 6 \\ 1 & . & 1 & 2 \\ 1 & 1 & . & 2 \\ \hline 4 & 4 & 2 & 10 \end{array}
$$

(a) (b) (c) (d)

$$
\begin{array}{ccc|c} 3 & 2 & 1 & 6 \\ . & 1 & 1 & 2 \\ 1 & 1 & . & 2 \\ \hline 4 & 4 & 2 & 10 \end{array}
\qquad
\begin{array}{ccc|c} 3 & 3 & . & 6 \\ 1 & . & 1 & 2 \\ . & 1 & 1 & 2 \\ \hline 4 & 4 & 2 & 10 \end{array}
\qquad
\begin{array}{ccc|c} 3 & 3 & . & 6 \\ . & 1 & 1 & 2 \\ 1 & . & 1 & 2 \\ \hline 4 & 4 & 2 & 10 \end{array}
\qquad (12.34)
$$

(e) (f) (g)

With practice the reader will find it unnecessary to write down arrays such as (c), (d) and (e), which are merely obtained from (b) by permuting rows and columns, but for clarity at this stage they have been set out in full. There is one trap here to be particularly noticed. In array (b) the two columns summing to 4 and the two rows summing to 2 are different, and their permutations result in 4 different arrays. But in array (f), though the rows and columns are different, permutations produce only 2 different arrays.

Each of these arrays contributes to the coefficient required. Consider first of all that from (a). The numerical coefficient is $\left(\frac{4!}{2!1!1!}\right)\left(\frac{4!}{2!1!1!}\right)\cdot\frac{1}{2!} = 72$. The first factor in brackets is the number of ways of allocating 4 individuals in the partition 2, 1, 1, similarly for the second, and we divide by 2! since there are 2 members of the row totals the same, this being the only β factor not equal to unity.

Under Rule 8, the pattern function is $1/n$ times that of

$$
\begin{array}{cc} \times & \times \\ \times & \times \\ \times & \times \end{array}
$$

The separations, five in number, are as follows :—

$$\begin{array}{ccccc}
\times\ \times & \times\ \times & \times\ \times & \times\ \times & \times\ \times \\
 & \cdots\cdots & \left.\begin{array}{c}\cdots\cdots \\ \times\ \times \\ \cdots\cdots\end{array}\right\} & & \cdots\cdots \\
\times\ \times & \times\ \times & & \times\ \times & \times\ \times \\
 & & & \cdots\cdots & \cdots\cdots \\
\times\ \times & \times\ \times & \times\ \times & \times\ \times & \times\ \times
\end{array}$$

The first is, of course, the original pattern array (one separate). The next three consist each of two separates, obtained by taking alone the first, second and third row. The last consists of three separates, one for each row.

The contributions respectively under Rule 6 will be found to be

$$n.\left(\frac{1}{n}\right)\left(\frac{1}{n}\right) = \frac{1}{n}$$

$$3n(n-1)\left\{\frac{-1}{n(n-1)}\right\}\left\{\frac{-1}{n(n-1)}\right\} = \frac{3}{n(n-1)}$$

$$n(n-1)(n-2)\left\{\frac{(-1)^2 2!}{n(n-1)(n-2)}\right\}\left\{\frac{(-1)^2 2!}{n(n-1)(n-2)}\right\} = \frac{4}{n(n-1)(n-2)}.$$

The sum of these is $\frac{n}{(n-1)(n-2)}$ and hence the contribution from array (a) in (12.34) is $\frac{72}{(n-1)(n-2)}$.

Now for arrays (b) to (e), which have all the same numerical factor and the same pattern function and can therefore be considered together. For any one the numerical factor is

$$\left(\frac{4!}{2!\,1!\,1!}\right)\left(\frac{4!}{3!\,1!}\right)\left(\frac{2!}{1!\,1!}\right)\cdot\frac{1}{2!} = 48$$

and that of the four together is thus 192.

Under Rule 6 the pattern function will depend on the configuration

$$\begin{array}{ccc}
\times & \times & \times \\
\times & \times & . \\
\times & . & \times
\end{array}$$

where $\times$ stands for a non-zero entry and a period for a zero entry. There are five separations of this, one of one separate, three of two separates, and one of three separates. The contribution from the first is

$$n\frac{1}{n}\frac{1}{n}\frac{1}{n} = \frac{1}{n^2}$$

for each column has a non-zero entry in the separate. The contribution from the three separations given respectively by isolating the first, second and third row will be found to be

$$n(n-1)\left[\frac{-1}{n^3(n-1)^3}+\frac{1}{n^3(n-1)^2}+\frac{1}{n^3(n-1)^2}\right] = \frac{2n-3}{n^2(n-1)^2}.$$

The contribution from the separation of three separates is

$$n(n-1)(n-2)\left[\frac{2!}{n(n-1)(n-2)}\cdot\frac{-1}{n(n-1)}\cdot\frac{-1}{n(n-1)}\right] = \frac{2}{n^2(n-1)^2}.$$

The pattern function is the sum of these three contributions and is thus $\frac{1}{(n-1)^2}$.

The contribution from arrays (f) and (g) in (12.34) will be found to be $\frac{32}{(n-1)^2}$.

Hence, adding all the contributions together, we find that the coefficient of $\kappa_6\kappa_2^2$ in $\kappa(4^2 2)$ is

$$\frac{72}{(n-1)(n-2)}+\frac{192}{(n-1)^2}+\frac{32}{(n-1)^2} = \frac{8(37n-65)}{(n-1)^2(n-2)},$$

as shown in equation (12.66) below.

Rule 10. The expression for any $\kappa(a_1^{\alpha_1}\ldots)$ which contains a unit part may be obtained from that without the part by (1) dividing throughout by n and (2) increasing the suffix of one of the κ's by unity in every possible way.

For example, it may be shown that

$$\kappa(2^2) = \frac{\kappa_4}{n}+\frac{2\kappa_2^2}{n-1}.$$

Hence

$$\kappa(2^2 1) = \frac{\kappa_5}{n^2}+\frac{4\kappa_3\kappa_2}{n(n-1)}$$

$$\kappa(2^2 1^2) = \frac{\kappa_6}{n^3}+\frac{4\kappa_3^2}{n^2(n-1)}+\frac{4\kappa_4\kappa_2}{n^2(n-1)},$$

and so on.

12.14 The reader cannot be blamed for doubting whether this rather elaborate combinatorial procedure represents much of an advance on the straightforward algebraical approach considered earlier in the chapter. He should try a few of the more complicated cases both ways. In point of fact, it is so easy to make algebraical slips with either method that most formulae in current use have been checked by being calculated by both methods. The combinatorial method, however, is something more than an algebraical short cut. We shall find in the sequel that several results of importance are derivable directly from it, whereas the straightforward evaluation of expectations yields them only by non-straightforward manœuvres.

Example 12.3

For comparison with Example 12.1 let us find the variance of m_2 by the combinatorial method.

From (12.29) we have

$$k_2 = \frac{n}{n-1}m_2.$$

Hence
$$\operatorname{var} m_2 = \left(\frac{n-1}{n}\right)^2 \operatorname{var} k_2$$
$$= \left(\frac{n-1}{n}\right)^2 \kappa(2^2).$$

$\kappa(2^2)$ consists of two terms, one in κ_4 and one in κ_2^2. The only array contributing to the first is

$$\begin{array}{cc|c} 2 & 2 & 4 \\ \hline 2 & 2 & 4 \end{array}$$

with a numerical factor unity and a pattern function $1/n$. The arrays giving the second are of type

$$\begin{array}{cc|c} & & 2 \\ & & 2 \\ \hline 2 & 2 & 4 \end{array}$$

If any entry in this were a 2 the row in which it appeared would contain only a single entry and hence the array would vanish. The only contributing array is therefore

$$\begin{array}{cc|c} 1 & 1 & 2 \\ 1 & 1 & 2 \\ \hline 2 & 2 & 4 \end{array}$$

The numerical coefficient is $\left(\dfrac{2!}{1!1!}\right)^2 \cdot \dfrac{1}{2!} = 2$. The pattern function will be found to be $1/(n-1)$. Hence

$$\kappa(2^2) = \frac{\kappa_4}{n} + \frac{2\kappa_2^2}{n-1}$$

is given in (12.35), and

$$\operatorname{var} m_2 = \left(\frac{n-1}{n}\right)^2 \kappa(2^2)$$
$$= \left(\frac{n-1}{n}\right)^2 \left\{\frac{1}{n}(\mu_4 - 3\mu_2^2) + \frac{2}{n-1}\mu_2^2\right\}$$
$$= \frac{(n-1)^2}{n^3}\mu_4 + \frac{(3-n)(n-1)}{n^3}\mu_2^2,$$

agreeing with (12.14).

Example 12.4

To find the third moment of k_2 we require $\kappa(2^3)$. This will be the sum of factors in κ_6, $\kappa_4\kappa_2$, κ_3^2 and κ_2^3.

The coefficient of the first is $1/n^2$. For the second we have to consider the array

$$\begin{array}{ccc|c} 1 & 1 & 2 & 4 \\ 1 & 1 & . & 2 \\ \hline 2 & 2 & 2 & 6 \end{array}$$

all others vanishing except the two equivalent partitions obtained when the column with the single entry appears in the first or second place. The numerical factor is then

$$3.\left(\frac{2!}{1!1!}\right)^2 = 12.$$

The pattern function is $1/n$ times that of

$$\begin{matrix} \times & \times \\ \times & \times \end{matrix}$$

i.e. is $1/\{n(n-1)\}$. The coefficient of $\kappa_4\kappa_2$ is then $12/\{n(n-1)\}$.

For the term in κ_3^2 the only contributory array is

$$\begin{array}{ccc|c} 1 & 1 & 1 & 3 \\ 1 & 1 & 1 & 3 \\ \hline 2 & 2 & 2 & 6 \end{array}$$

with a factor $\left(\frac{2!}{1!1!}\right)^3 \cdot \frac{1}{2!} = 4$ and pattern function $\frac{n-2}{n(n-1)^2}$.

For the last term we have to consider the array

$$\begin{array}{ccc|c} 1 & 1 & . & 2 \\ 1 & . & 1 & 2 \\ . & 1 & 1 & 2 \\ \hline 2 & 2 & 2 & 6 \end{array}$$

with a numerical coefficient 8 and a pattern function $\frac{1}{(n-1)^2}$. Collecting terms together we get

$$\kappa(2^3) = \frac{\kappa_6}{n^2} + \frac{12\kappa_4\kappa_2}{n(n-1)} + \frac{4(n-2)}{n(n-1)^2}\kappa_3^2 + \frac{8}{(n-1)^2}\kappa_2^3.$$

This is also the value of the third moment $\mu(2^3)$ measured about the mean of the sampling distribution κ_2. We see that if the parent is normal the third moment reduces to $\frac{8\kappa_2^3}{(n-1)^2}$, i.e. is of order n^{-2}, indicating a rapid tendency towards symmetry.

12.15 In consequence of Rules 1 and 2 and Exercise 3.5, any $\kappa(a_1^{\alpha_1} \ldots a_s^{\alpha_s}) = 0$ if a is odd and the population is symmetrical.

Few things illustrate the usefulness of expressing the formulae in terms of cumulants and the power of the combinatorial method better than the simplification imported when the parent population is normal. In this case only terms in κ_2 survive, all higher cumulants vanishing.

Example 12.5

As an illustration let us prove that $\kappa(pq) = 0$ for normal samples unless $p = q$. The only term which can appear in $\kappa(pq)$ is $\kappa_2^{\frac{1}{2}(p+q)}$ and evidently, if $p+q$ is odd, this itself is impossible and $\kappa(pq)$ must vanish. If $p+q$ is even we have to consider

the array

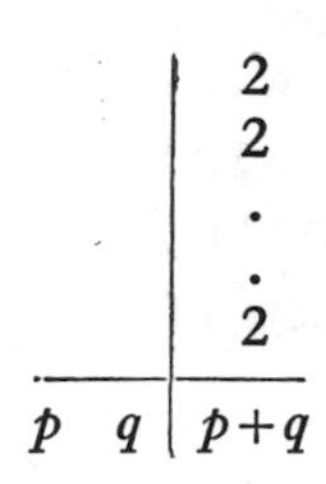

Now if any entry in this array is 2 the array vanishes since the row concerned will contain only one entry. The reverse can only happen if all the entries are unity, in which case the sums p and q must be equal. This establishes the result.

Example 12.6

If $\kappa(a_1^{\pi_1} \ldots a_s^{\pi_s})$ contains r parts, every term in it is of order $n^{-(r-1)}$. For example every term in $\kappa(3^2 2^2)$ is of order n^{-3}.

By Rule 7, every array with non-vanishing pattern function (p.f.) has at least two non-zero entries (n.z.e.) in each row. By Rule 8, we may confine ourselves to arrays with at least two n.z.e. in each column. We have to show that the p.f. of every such r-column array is of order $n^{-(r-1)}$. Every array has a single-separate separation; by Rule 6, this contributes $n\left(\frac{1}{n}\right)^r = n^{-(r-1)}$. We now need only show that no separation with more than one separate has a contribution of higher order. Consider any 2-separate separation, S_2; if the p.f. is not to vanish by Rule 9, at least 2 columns must have a n.z.e. in each separate, and by Rule 6 the contribution of S_2 is of order $n^2\left(\frac{1}{n^2}\right)^2\left(\frac{1}{n}\right)^{r-2}$ or less, i.e. it is $o(n^{-(r-1)})$. Any 3-separate separation, S_3, may be derived by a sub-separation of an S_2, and using Rule 3 it follows that the contribution of S_3 is of order $n^3\left(\frac{1}{n^2}\right)^3\left(\frac{1}{n}\right)^{r-3}$, i.e. $o(n^{-(r-1)})$. The same argument, step-by-step, shows that no separation can contribute a term of order as high as $n^{-(r-1)}$. This completes the proof.

Example 12.7

We may use the properties of sampling cumulants to prove a characterizing property of the normal distribution that was discussed analytically in Example 11.3 and geometrically in Example 11.7: namely, that the mean and variance are independent in normal samples.

First, we recall from **4.16–17** that two variates are independent if and only if their joint c.f. factorizes into their univariate c.f.'s, i.e.,

$$\phi(t_1, t_2) = \phi_1(t_1)\phi_2(t_2).$$

As in the univariate case (cf. **4.21**) any finite product-moment may be obtained by differentiating the joint c.f. and putting $t_1 = t_2 = 0$, while if the joint distribution is uniquely determined by its moments we may expand the c.f. as an infinite series. Expanding all three c.f.'s and taking logarithms, we have on using (3.74)

$$\sum_{r,s=0}^{\infty} \kappa_{rs} \frac{(it_1)^r}{r!}\frac{(it_2)^s}{s!} = \sum_{r=1}^{\infty} \kappa_{r0}\frac{(it_1)^r}{r!} + \sum_{s=1}^{\infty} \kappa_{0s}\frac{(it_2)^s}{s!}$$

which implies that $\kappa_{rs} = 0$ if the product $rs \neq 0$. Even if the infinite series expansion is impermissible, it is clear that any finite $\kappa_{rs} = 0$ if $rs \neq 0$.

Conversely, if all $\kappa_{rs} = 0$, $rs \neq 0$, and the c.f. may be thus expanded, independence of the variates follows, so that this is then a necessary and sufficient condition for independence. Its stringency may be gauged from Exercise 7.15, which gave a bivariate distribution for which $\kappa_{rs} = 0$ for any $r \neq 2$ and any s, with the variables clearly dependent.

Now consider the joint distribution in normal samples of mean and variance, or equivalently of k_1 and k_2. Since $\kappa_p = 0, p > 2$, we have from Rule 10 that $\kappa(2^s\,1^r) = 0$ (any $r, s \neq 0$). Hence k_1 and k_2 are independent, as we have already seen.

We may now establish the converse, due to Geary (1936), that only in normal variation will mean and variance be independent. For, by Rule 10,

$$\kappa(21^r) = \kappa_{r+2}/n^r, \qquad r > 0.$$

If k_1 and k_2 are independent this is zero for all $r > 0$. This implies normality on the part of the parent.

It is rather remarkable that we have not had to use the relations $\kappa(2^s\,1^r) = 0$ for $s > 1$ to establish the converse; and that the converse requires independence for only one value of n.

Later writers, e.g. Lukacs (1942), have proved Geary's theorem under less restrictive conditions on the existence of cumulants—cf. Exercise 11.19.

12.16 The following formulae give results for statistics of degree not greater than 10, with some of the 12th degree. They hold for n not less than the order of the k-statistic considered.

Second k-statistic

$$\kappa(2^2) = \frac{\kappa_4}{n} + \frac{2\kappa_2^2}{n-1} \tag{12.35}$$

$$\kappa(2^3) = \frac{\kappa_6}{n^2} + \frac{12\kappa_4\kappa_2}{n(n-1)} + \frac{4(n-2)}{n(n-1)^2}\kappa_3^2 + \frac{8}{(n-1)^2}\kappa_2^3 \tag{12.36}$$

$$\begin{aligned}\kappa(2^4) = {} & \frac{\kappa_8}{n^3} + \frac{24}{n^2(n-1)}\kappa_6\kappa_2 + \frac{32(n-2)}{n^2(n-1)^2}\kappa_5\kappa_3 + \frac{8(4n^2-9n+6)}{n^2(n-1)^3}\kappa_4^2 \\ & + \frac{144}{n(n-1)^2}\kappa_4\kappa_2^2 + \frac{96(n-2)}{n(n-1)^3}\kappa_3^2\kappa_2 + \frac{48}{(n-1)^3}\kappa_2^4 \end{aligned} \tag{12.37}$$

$$\begin{aligned}\kappa(2^5) = {} & \frac{\kappa_{10}}{n^4} + \frac{40\kappa_8\kappa_2}{n^3(n-1)} + \frac{80(n-2)}{n^3(n-1)^2}\kappa_7\kappa_3 + \frac{40(5n^2-12n+9)}{n^3(n-1)^3}\kappa_6\kappa_4 \\ & + \frac{16(n-2)(6n^2-12n+7}{n^3(n-1)^4}\kappa_5^2 + \frac{480}{n^2(n-1)^2}\kappa_6\kappa_2^2 + \frac{1280(n-2)}{n^2(n-1)^3}\kappa_5\kappa_3\kappa_2\end{aligned}$$

$$+\frac{320(4n^2-9n+6)}{n^2(n-1)^4}\kappa_4^2\kappa_2+\frac{480(2n^2-7n+6)}{n^2(n-1)^4}\kappa_4\kappa_3^2+\frac{1920}{n(n-1)^3}\kappa_4\kappa_2^3$$
$$+\frac{1920(n-2)}{n(n-1)^4}\kappa_3^2\kappa_2^2+\frac{384}{(n-1)^4}\kappa_2^5 \qquad (12.38)$$

$$\kappa(2^6) = \frac{1}{n^5}\kappa_{12}+\frac{60}{n^4(n-1)}\kappa_{10}\kappa_2+\frac{160(n-2)}{n^4(n-1)^2}\kappa_9\kappa_3+\frac{240(2n^2-5n+4)}{n^4(n-1)^3}\kappa_8\kappa_4$$
$$+\frac{96(n-2)(7n^2-14n+9)}{n^4(n-1)^4}\kappa_7\kappa_5+\frac{4(113n^4-520n^3+950n^2-800n+265)}{n^4(n-1)^5}\kappa_6^2$$
$$+\frac{1200}{n^3(n-1)^2}\kappa_8\kappa_2^2+\frac{4800(n-2)}{n^3(n-1)^3}\kappa_7\kappa_3\kappa_2+\frac{2400(5n^2-12n+9)}{n^3(n-1)^4}\kappa_6\kappa_4\kappa_2$$
$$+\frac{160(n-2)(31n-53)}{n^3(n-1)^4}\kappa_6\kappa_3^2+\frac{960(n-2)(6n^2-12n+7)}{n^3(n-1)^5}\kappa_5^2\kappa_2$$
$$+\frac{1920(n-2)(9n^2-23n+16)}{n^3(n-1)^5}\kappa_5\kappa_4\kappa_3+\frac{480(11n^3-41n^2+59n-31)}{n^3(n-1)^5}\kappa_4^3$$
$$+\frac{9600}{n^2(n-1)^3}\kappa_6\kappa_2^3+\frac{38400(n-2)}{n^2(n-1)^4}\kappa_5\kappa_3\kappa_2^2+\frac{9600(4n^2-9n+6)}{n^2(n-1)^5}\kappa_4^2\kappa_2^2$$
$$+\frac{28800(2n^2-7n+6)}{n^2(n-1)^5}\kappa_4\kappa_3^2\kappa_2+\frac{960(n-2)(5n-12)}{n^2(n-1)^5}\kappa_3^4+\frac{28800}{n(n-1)^4}\kappa_4\kappa_2^4$$
$$+\frac{38400(n-2)}{n(n-1)^5}\kappa_3^2\kappa_2^3+\frac{3840}{(n-1)^5}\kappa_2^6 \qquad (12.39)$$

Third k-statistic

$$\kappa(3^2) = \frac{1}{n}\kappa_6+\frac{9}{n-1}\kappa_4\kappa_2+\frac{9}{n-1}\kappa_3^2+\frac{6n}{(n-1)(n-2)}\kappa_2^3 \qquad (12.40)$$

$$\kappa(3^3) = \frac{1}{n^2}\kappa_9+\frac{27}{n(n-1)}\kappa_7\kappa_2+\frac{27(3n-4)}{n(n-1)^2}\kappa_6\kappa_3+\frac{27(4n-7)}{n(n-1)^2}\kappa_5\kappa_4$$
$$+\frac{54(4n-7)}{(n-1)^2(n-2)}\kappa_5\kappa_2^2+\frac{162(5n-12)}{(n-1)^2(n-2)}\kappa_4\kappa_3\kappa_2+\frac{36(7n^2-30n+34)}{(n-1)^2(n-2)^2}\kappa_3^3$$
$$+\frac{108n(5n-12)}{(n-1)^2(n-2)^2}\kappa_3\kappa_2^3 \qquad (12.41)$$

$$\kappa(3^4) = \frac{1}{n^3}\kappa_{12}+\frac{54}{n^2(n-1)}\kappa_{10}\kappa_2+\frac{108(2n-3)}{n^2(n-1)^2}\kappa_9\kappa_3+\frac{27(17n^2-49n+35)}{n^2(n-1)^3}\kappa_8\kappa_4$$
$$+\frac{108(7n^2-20n+16)}{n^2(n-1)^3}\kappa_7\kappa_5+\frac{27(17n^2-47n+39)}{n^2(n-1)^3}\kappa_6^2+\frac{27(37n-70)}{n(n-1)^2(n-2)}\kappa_8\kappa_2^2$$
$$+\frac{324(19n^2-67n+54)}{n(n-1)^3(n-2)}\kappa_7\kappa_3\kappa_2+\frac{162(65n^2-245n+234)}{n(n-1)^3(n-2)}\kappa_6\kappa_4\kappa_2$$
$$+\frac{108(82n^3-481n^2+958n-640)}{n(n-1)^3(n-2)^2}\kappa_6\kappa_3^2+\frac{108(59n^2-220n+224)}{n(n-1)^3(n-2)}\kappa_5^2\kappa_2$$
$$+\frac{324(75n^3-473n^2+1016n-756)}{n(n-1)^3(n-2)^2}\kappa_5\kappa_4\kappa_3$$

$$+\frac{27(173n^4-1503n^3+4962n^2-7380n+4200)}{n(n-1)^3(n-2)^3}\kappa_4^3$$
$$+\frac{108(71n^2-263n+234)}{(n-1)^3(n-2)^2}\kappa_6\kappa_2^3+\frac{648(79n^2-343n+378)}{(n-1)^3(n-2)^2}\kappa_5\kappa_3\kappa_2^2$$
$$+\frac{486(63n^2-290n+352)}{(n-1)^3(n-2)^2}\kappa_4^2\kappa_2^2+\frac{972(99n^3-688n^2+1612n-1280)}{(n-1)^3(n-2)^3}\kappa_4\kappa_3^2\kappa_2$$
$$+\frac{162(87n^3-594n^2+1420n-1176)}{(n-1)^3(n-2)^3}\kappa_3^4+\frac{972n(23n^2-103n+118)}{(n-1)^3(n-2)^3}\kappa_4\kappa_2^4$$
$$+\frac{648n(103n^2-510n+640)}{(n-1)^3(n-2)^3}\kappa_3^2\kappa_2^3+\frac{648n^2(5n-12)}{(n-1)^3(n-2)^3}\kappa_2^6 \qquad (12.42)$$

Fourth *k*-statistic

$$\kappa(4^2) = \frac{1}{n}\kappa_8+\frac{16}{n-1}\kappa_6\kappa_2+\frac{48}{n-1}\kappa_5\kappa_3+\frac{34}{n-1}\kappa_4^2+\frac{72n}{(n-1)(n-2)}\kappa_4\kappa_2^2$$
$$+\frac{144n}{(n-1)(n-2)}\kappa_3^2\kappa_2+\frac{24n(n+1)}{(n-1)(n-2)(n-3)}\kappa_2^4 \qquad (12.43)$$

$$\kappa(4^3) = \frac{1}{n^2}\kappa_{12}+\frac{48}{n(n-1)}\kappa_{10}\kappa_2+\frac{16(13n-17)}{n(n-1)^2}\kappa_9\kappa_3+\frac{12(41n-65)}{n(n-1)^2}\kappa_8\kappa_4$$
$$+\frac{48(16n-29)}{n(n-1)^2}\kappa_7\kappa_5+\frac{12(37n-70)}{n(n-1)^2}\kappa_6^2+\frac{72(11n-19)}{(n-1)^2(n-2)}\kappa_8\kappa_2^2$$
$$+\frac{288(19n-41)}{(n-1)^2(n-2)}\kappa_7\kappa_3\kappa_2+\frac{48(203n-523)}{(n-1)^2(n-2)}\kappa_6\kappa_4\kappa_2$$
$$+\frac{144(56n^2-257n+302)}{(n-1)^2(n-2)^2}\kappa_6\kappa_3^2+\frac{1440(4n-11)}{(n-1)^2(n-2)}\kappa_5^2\kappa_2$$
$$+\frac{1152(22n^2-106n+133)}{(n-1)^2(n-2)^2}\kappa_5\kappa_4\kappa_3+\frac{8(709n^2-3430n+4456)}{(n-1)^2(n-2)^2}\kappa_4^3$$
$$+\frac{288(19n^3-98n^2+125n+2)}{(n-1)^2(n-2)^2(n-3)}\kappa_6\kappa_2^3+\frac{1728(24n^3-140n^2+200n+4)}{(n-1)^2(n-2)^2(n-3)}\kappa_5\kappa_3\kappa_2^2$$
$$+\frac{432(61n^3-371n^2+552n+12)}{(n-1)^2(n-2)^2(n-3)}\kappa_4^2\kappa_2^2+\frac{864(103n^3-629n^2+948n+24)}{(n-1)^2(n-2)^2(n-3)}\kappa_4\kappa_3^2\kappa_2$$
$$+\frac{288(41n^4-384n^3+1209n^2-1282n-36)}{(n-1)^2(n-2)^2(n-3)^2}\kappa_3^4+\frac{288n(53n^2-179n-52)}{(n-1)^2(n-2)^2(n-3)}\kappa_4\kappa_2^4$$
$$+\frac{1728n(29n^3-196n^2+317n+62)}{(n-1)^2(n-2)^2(n-3)^2}\kappa_3^2\kappa_2^3+\frac{1728n(n+1)(n^2-5n+2)}{(n-1)^2(n-2)^2(n-3)^2}\kappa_2^6 \qquad (12.44)$$

Fifth *k*-statistic

$$\kappa(5^2) = \frac{1}{n}\kappa_{10}+\frac{25}{n-1}\kappa_8\kappa_2+\frac{100}{n-1}\kappa_7\kappa_3+\frac{200}{n-1}\kappa_6\kappa_4+\frac{125}{n-1}\kappa_5^2$$
$$+\frac{200n}{(n-1)(n-2)}\kappa_6\kappa_2^2+\frac{1200n}{(n-1)(n-2)}\kappa_5\kappa_3\kappa_2+\frac{850n}{(n-1)(n-2)}\kappa_4^2\kappa_2$$

$$+\frac{1500n}{(n-1)(n-2)}\kappa_4\kappa+\frac{600n(n+1)}{(n-1)(n-2)(n-3)}\kappa_4\kappa_2^3$$
$$+\frac{1800n(n+1)}{(n-1)(n-2)(n-3)}\kappa_3^2\kappa_2^2+\frac{120n^2(n+5)}{(n-1)(n-2)(n-3)(n-4)}\kappa_2^5 \quad (12.45)$$

Sixth *k*-statistic

$$\kappa(6^2) = \frac{1}{n}\kappa_{12}+\frac{1}{n-1}(36\kappa_{10}\kappa_2+180\kappa_9\kappa_3+465\kappa_8\kappa_4+780\kappa_7\kappa_5+461\kappa_6^2)$$
$$+\frac{n}{(n-1)(n-2)}(450\kappa_8\kappa_2^2+3600\kappa_7\kappa_3\kappa_2+7200\kappa_6\kappa_4\kappa_2+6300\kappa_6\kappa_3^2$$
$$+4500\kappa_5^2\kappa_2+21600\kappa_5\kappa_4\kappa_3+4950\kappa_4^3)$$
$$+\frac{n(n+1)}{(n-1)(n-2)(n-3)}(2400\kappa_6\kappa_2^3+21600\kappa_5\kappa_3\kappa_2^2+15300\kappa_4^2\kappa_2^2$$
$$+54000\kappa_4\kappa_3^2\kappa_2+8100\kappa_3^4)$$
$$+\frac{n^2(n+5)}{(n-1)(n-2)(n-3)(n-4)}(5400\kappa_4\kappa_2^4+21600\kappa_3^2\kappa_2^3)$$
$$+\frac{n(n+1)(n^2+15n-4)}{(n-1)(n-2)(n-3)(n-4)(n-5)}720\kappa_2^6 \quad (12.46)$$

Product–cumulant formulae

$$\kappa(32) = \frac{1}{n}\kappa_5+\frac{6}{n-1}\kappa_3\kappa_2 \quad (12.47)$$

$$\kappa(42) = \frac{1}{n}\kappa_6+\frac{8}{n-1}\kappa_4\kappa_2+\frac{6}{n-1}\kappa_3^2 \quad (12.48)$$

$$\kappa(52) = \frac{1}{n}\kappa_7+\frac{10}{n-1}\kappa_5\kappa_2+\frac{20}{n-1}\kappa_4\kappa_3 \quad (12.49)$$

$$\kappa(62) = \frac{1}{n}\kappa_8+\frac{12}{n-1}\kappa_6\kappa_2+\frac{30}{n-1}\kappa_5\kappa_3+\frac{20}{n-1}\kappa_4^2 \quad (12.50)$$

$$\kappa(72) = \frac{1}{n}\kappa_9+\frac{14}{n-1}\kappa_7\kappa_2+\frac{42}{n-1}\kappa_6\kappa_3+\frac{70}{n-1}\kappa_5\kappa_4 \quad (12.51)$$

$$\kappa(82) = \frac{1}{n}\kappa_{10}+\frac{16}{n-1}\kappa_8\kappa_2+\frac{56}{n-1}\kappa_7\kappa_3+\frac{112}{n-1}\kappa_6\kappa_4+\frac{70}{n-1}\kappa_5^2 \quad (12.52)$$

$$\kappa(43) = \frac{1}{n}\kappa_7+\frac{12}{n-1}\kappa_5\kappa_2+\frac{30}{n-1}\kappa_4\kappa_3+\frac{36n}{(n-1)(n-2)}\kappa_3\kappa_2^2 \quad (12.53)$$

$$\kappa(53) = \frac{1}{n}\kappa_8+\frac{1}{n-1}(15\kappa_6\kappa_2+45\kappa_5\kappa_3+30\kappa_4^2)+\frac{n}{(n-1)(n-2)}(60\kappa_4\kappa_2^2+90\kappa_3^2\kappa_2) \quad (12.54)$$

$$\kappa(63) = \frac{1}{n}\kappa_9+\frac{1}{n-1}(18\kappa_7\kappa_2+63\kappa_6\kappa_3+105\kappa_5\kappa_4)$$
$$+\frac{n}{(n-1)(n-2)}(90\kappa_5\kappa_2^2+360\kappa_4\kappa_3\kappa_2+90\kappa_3^3) \quad (12.55)$$

$$\kappa(73) = \frac{1}{n}\kappa_{10}+\frac{1}{n-1}(21\kappa_8\kappa_2+84\kappa_7\kappa_3+168\kappa_6\kappa_4+105\kappa_5^2)$$
$$+\frac{n}{(n-1)(n-2)}(126\kappa_6\kappa_2^2+630\kappa_5\kappa_3\kappa_2+420\kappa_4^2\kappa_2+630\kappa_4\kappa_3^2) \quad (12.56)$$

$$\kappa(54) = \frac{1}{n}\kappa_9+\frac{1}{n-1}(20\kappa_7\kappa_2+70\kappa_6\kappa_3+120\kappa_5\kappa_4)$$
$$+\frac{n}{(n-1)(n-2)}(120\kappa_5\kappa_2^2+600\kappa_4\kappa_3\kappa_2+180\kappa_3^3)$$
$$+\frac{n(n+1)}{(n-1)(n-2)(n-3)}240\kappa_3\kappa_2^3 \quad (12.57)$$

$$\kappa(64) = \frac{1}{n}\kappa_{10}+\frac{1}{n-1}(24\kappa_8\kappa_2+96\kappa_7\kappa_3+194\kappa_6\kappa_4+120\kappa_5^2)$$
$$+\frac{n}{(n-1)(n-2)}(180\kappa_6\kappa_2^2+1080\kappa_5\kappa_3\kappa_2+720\kappa_4^2\kappa_2+1260\kappa_4\kappa_3^2)$$
$$+\frac{n(n+1)}{(n-1)(n-2)(n-3)}(480\kappa_4\kappa_2^3+1080\kappa_3^2\kappa_2^2) \quad (12.58)$$

$$\kappa(32^2) = \frac{1}{n^2}\kappa_7+\frac{16}{n(n-1)}\kappa_5\kappa_2+\frac{12(2n-3)}{n(n-1)^2}\kappa_4\kappa_3+\frac{48}{(n-1)^2}\kappa_2^2\kappa_3 \quad (12.59)$$

$$\kappa(42^2) = \frac{1}{n^2}\kappa_8+\frac{20}{n(n-1)}\kappa_6\kappa_2+\frac{8(5n-7)}{n(n-1)^2}\kappa_5\kappa_3+\frac{4(7n-10)}{n(n-1)^2}\kappa_4^2$$
$$+\frac{80}{(n-1)^2}\kappa_4\kappa_2^2+\frac{120}{(n-1)^2}\kappa_3^2\kappa_2 \quad (12.60)$$

$$\kappa(52^2) = \frac{1}{n^2}\kappa_9+\frac{24}{n(n-1)}\kappa_7\kappa_2+\frac{20(3n-4)}{n(n-1)^2}\kappa_6\kappa_3+\frac{20(5n-7)}{n(n-1)^2}\kappa_5\kappa_4$$
$$+\frac{120}{(n-1)^2}\kappa_5\kappa_2^2+\frac{480}{(n-1)^2}\kappa_4\kappa_3\kappa_2+\frac{120}{(n-1)^2}\kappa_3^3 \quad (12.61)$$

$$\kappa(62^2) = \frac{1}{n^2}\kappa_{10}+\frac{28}{n(n-1)}\kappa_8\kappa_2+\frac{12(7n-9)}{n(n-1)^2}\kappa_7\kappa_3+\frac{4(41n-56)}{n(n-1)^2}\kappa_6\kappa_4$$
$$+\frac{20(5n-7)}{n(n-1)^2}\kappa_5^2+\frac{168}{(n-1)^2}\kappa_6\kappa_2^2+\frac{840}{(n-1)^2}\kappa_5\kappa_3\kappa_2+\frac{560}{(n-1)^2}\kappa_4^2\kappa_2$$
$$+\frac{840}{(n-1)^2}\kappa_4\kappa_3^2 \quad (12.62)$$

$$\kappa(3^22) = \frac{1}{n^2}\kappa_8+\frac{21}{n(n-1)}\kappa_6\kappa_2+\frac{6(8n-11)}{n(n-1)^2}\kappa_5\kappa_3+\frac{9(3n-5)}{n(n-1)^2}\kappa_4^2$$
$$+\frac{18(6n-11)}{(n-1)^2(n-2)}\kappa_4\kappa_2^2+\frac{18(9n-20)}{(n-1)^2(n-2)}\kappa_3^2\kappa_2+\frac{36n}{(n-1)^2(n-2)}\kappa_2^4 \quad (12.63)$$

$$\kappa(432) = \frac{1}{n^2}\kappa_9+\frac{26}{n(n-1)}\kappa_7\kappa_2+\frac{24(3n-4)}{n(n-1)^2}\kappa_6\kappa_3+\frac{10(11n-17)}{n(n-1)^2}\kappa_5\kappa_4$$

$$+\frac{36(5n-9)}{(n-1)^2(n-2)}\kappa_5\kappa_2^2+\frac{12(61n-128)}{(n-1)^2(n-2)}\kappa_4\kappa_3\kappa_2+\frac{36(5n-12)}{(n-1)^2(n-2)}\kappa_3^3$$
$$+\frac{360n}{(n-1)^2(n-2)}\kappa_3\kappa_2^3 \quad (12.64)$$

$$\kappa(532) = \frac{1}{n^2}\kappa_{10}+\frac{31}{n(n-1)}\kappa_8\kappa_2+\frac{101n-131}{n(n-1)^2}\kappa_7\kappa_3+\frac{5(37n-55)}{n(n-1)^2}\kappa_6\kappa_4$$
$$+\frac{5(23n-35)}{n(n-1)^2}\kappa_5^2+\frac{30(9n-16)}{(n-1)^2(n-2)}\kappa_6\kappa_2^2+\frac{30(45n-92)}{(n-1)^2(n-2)}\kappa_5\kappa_3\kappa_2$$
$$+\frac{60(15n-31)}{(n-1)^2(n-2)}\kappa_4^2\kappa_2+\frac{30(45n-103)}{(n-1)^2(n-2)}\kappa_4\kappa_3^2+\frac{720n}{(n-1)^2(n-2)}\kappa_4\kappa_2^3$$
$$+\frac{1620n}{(n-1)^2(n-2)}\kappa_3^2\kappa_2^2 \quad (12.65)$$

$$\kappa(4^22) = \frac{1}{n^2}\kappa_{10}+\frac{32}{n(n-1)}\kappa_8\kappa_2+\frac{8(13n-37)}{n(n-1)^2}\kappa_7\kappa_3+\frac{4(49n-73)}{n(n-1)^2}\kappa_6\kappa_4$$
$$+\frac{4(29n-46)}{n(n-1)^2}\kappa_5^2+\frac{8(37n-65)}{(n-1)^2(n-2)}\kappa_6\kappa_2^2+\frac{1536}{(n-1)^2}\kappa_5\kappa_3\kappa_2$$
$$+\frac{144(7n-15)}{(n-1)^2(n-2)}\kappa_4^2\kappa_2+\frac{72(21n-50)}{(n-1)^2(n-2)}\kappa_4\kappa_3^2+\frac{96(10n^2-27n-1)}{(n-1)^2(n-2)(n-3)}\kappa_4\kappa_2^3$$
$$+\frac{144(17n^2-53n-2)}{(n-1)^2(n-2)(n-3)}\kappa_3^2\kappa_2^2+\frac{192n(n+1)}{(n-1)^2(n-2)(n-3)}\kappa_2^5 \quad (12.66)$$

$$\kappa(43^2) = \frac{1}{n^2}\kappa_{10}+\frac{33}{n(n-1)}\kappa_8\kappa_2+\frac{6(19n-25)}{n(n-1)^2}\kappa_7\kappa_3+\frac{3(65n-107)}{n(n-1)^2}\kappa_6\kappa_4$$
$$+\frac{6(19n-34)}{n(n-1)^2}\kappa_5^2+\frac{18(19n-33)}{(n-1)^2(n-2)}\kappa_6\kappa_2^2+\frac{72(23n-52)}{(n-1)^2(n-2)}\kappa_5\kappa_3\kappa_2$$
$$+\frac{54(19n-48)}{(n-1)^2(n-2)}\kappa_4^2\kappa_2+\frac{54(33n^2-148n+172)}{(n-1)^2(n-2)^2}\kappa_4\kappa_3^2$$
$$+\frac{72n(17n-40)}{(n-1)^2(n-2)^2}\kappa_4\kappa_2^3+\frac{108n(27n-70)}{(n-1)^2(n-2)^2}\kappa_3^2\kappa_2^2+\frac{216n^2}{(n-1)^2(n-2)^2}\kappa_2^5 \quad (12.67)$$

$$\kappa(32^3) = \frac{1}{n^3}\kappa_9+\frac{30}{n^2(n-1)}\kappa_7\kappa_2+\frac{2(31n-53)}{n^2(n-1)^2}\kappa_6\kappa_3+\frac{12(9n^2-23n+16)}{n^2(n-1)^3}\kappa_5\kappa_4$$
$$+\frac{240}{n(n-1)^2}\kappa_5\kappa_2^2+\frac{360(2n-3)}{n(n-1)^3}\kappa_4\kappa_3\kappa_2+\frac{24(5n-12)}{n(n-1)^3}\kappa_3^3+\frac{480}{(n-1)^3}\kappa_3\kappa_2^3 \quad (12.68)$$

$$\kappa(42^3) = \frac{1}{n^3}\kappa_{10}+\frac{36}{n^2(n-1)}\kappa_8\kappa_2+\frac{4(23n-37)}{n^2(n-1)^2}\kappa_7\kappa_3+\frac{4(47n^2-120n+81)}{n^2(n-1)^3}\kappa_6\kappa_4$$
$$+\frac{12(9n^2-24n+17)}{n^2(n-1)^3}\kappa_5^2+\frac{360}{n(n-1)^2}\kappa_6\kappa_2^2+\frac{288(5n-7)}{n(n-1)^3}\kappa_5\kappa_3\kappa_2$$
$$+\frac{144(7n-10)}{n(n-1)^3}\kappa_4^2\kappa_2+\frac{24(49n-95)}{n(n-1)^3}\kappa_4\kappa_3^2+\frac{960}{(n-1)^3}\kappa_4\kappa_2^3+\frac{2160}{(n-1)^3}\kappa_3^2\kappa_2^2 \quad (12.69)$$

$$\kappa(3^2 2^2) = \frac{1}{n^3}\kappa_{10} + \frac{37}{n^2(n-1)}\kappa_8\kappa_2 + \frac{6(17n-27)}{n^2(n-1)^2}\kappa_7\kappa_3 + \frac{3(61n^2-166n+117)}{n^2(n-1)^3}\kappa_6\kappa_4$$
$$+\frac{2(59n^2-154n+113)}{n^2(n-1)^3}\kappa_5^2 + \frac{6(67n-131)}{n(n-1)^2(n-2)}\kappa_6\kappa_2^2$$
$$+\frac{24(71n^2-246n+202)}{n(n-1)^3(n-2)}\kappa_5\kappa_3\kappa_2 + \frac{36(29n^2-103n+93)}{n(n-1)^3(n-2)}\kappa_4^2\kappa_2$$
$$+\frac{36(38n^2-155n+160)}{n(n-1)^3(n-2)}\kappa_4\kappa_3^2 + \frac{72(14n-23)}{(n-1)^3(n-2)}\kappa_4\kappa_2^3$$
$$+\frac{144(19n-44)}{(n-1)^3(n-2)}\kappa_3^2\kappa_2^2 + \frac{288n}{(n-1)^3(n-2)}\kappa_2^5 \qquad (12.70)$$

12.17 Additional formulae for the case of a normal parent population are as follows. There are two general formulae :—

$$\kappa(2^r) = \frac{2^{r-1}(r-1)!}{(n-1)^{r-1}}\kappa_2^r \qquad (12.71)$$

$$\kappa(p^q 2^r) = \frac{2^r(r+\frac{1}{2}pq-1)!}{(\frac{1}{2}pq-1)!(n-1)^r}\kappa_2^r\kappa(p^q) \qquad (12.72)$$

and the following specific formulae of degree 12 and upwards (those of lower degree, of course, being derivable from equations (12.34) to (12.70) by putting all κ's higher than the second equal to zero).

$$\kappa(3^2 2^4) = \frac{34{,}560n}{(n-1)^5(n-2)}\kappa_2^7 \qquad (12.73)$$

$$\kappa(3^4 2) = \frac{7776n^2(5n-12)}{(n-1)^4(n-2)^3}\kappa_2^7 \qquad (12.74)$$

$$\kappa(3^4 2^2) = \frac{108{,}864n^2(5n-12)}{(n-1)^5(n-2)^3}\kappa_2^8 \qquad (12.75)$$

$$\kappa(3^4 2^3) = \frac{1{,}741{,}824n^2(5n-12)}{(n-1)^6(n-2)^3}\kappa_2^9 \qquad (12.76)$$

$$\kappa(3^6) = \frac{466{,}560n^3(22n^2-111n+142)}{(n-1)^5(n-2)^5}\kappa_2^9 \qquad (12.77)$$

$$\kappa(3^6 2) = \frac{18}{n-1}\kappa_2\kappa(3^6) \qquad (12.78)$$

$$\kappa(3^6 2^2) = \frac{360}{(n-1)^2}\kappa_2^2\kappa(3^6) \qquad (12.79)$$

$$\kappa(4^2 2^2) = \frac{1920n(n+1)}{(n-1)^3(n-2)(n-3)}\kappa_2^6 \qquad (12.80)$$

$$\kappa(4^2 2^3) = \frac{23{,}040n(n+1)}{(n-1)^4(n-2)(n-3)}\kappa_2^7 \qquad (12.81)$$

$$\kappa(4^2 2^4) = \frac{322{,}560n(n+1)}{(n-1)^5(n-2)(n-3)}\kappa_2^8 \qquad (12.82)$$

$$\kappa(4^3 2) = \frac{20{,}736n(n+1)(n^2-5n+2)}{(n-1)^3(n-2)^2(n-3)^2}\kappa_2^7 \tag{12.83}$$

$$\kappa(4^3 2^2) = \frac{290{,}304n(n+1)(n^2-5n+2)}{(n-1)^4(n-2)^2(n-3)^2}\kappa_2^8 \tag{12.84}$$

$$\kappa(4^3 2^3) = \frac{4{,}644{,}864n(n+1)(n^2-5n+2)}{(n-1)^5(n-2)^2(n-3)^2}\kappa_2^9 \tag{12.85}$$

$$\kappa(4^4) = \frac{6912n(n+1)}{(n-1)^3(n-2)^3(n-3)^3}\{53n^4-428n^3+1025n^2-474n+180\}\kappa_2^8 \tag{12.86}$$

$$\kappa(4^4 2) = \frac{16}{n-1}\kappa_2\kappa(4^4) \tag{12.87}$$

$$\kappa(4^4 2^2) = \frac{288}{(n-1)^2}\kappa_2^2\kappa(4^4) \tag{12.88}$$

$$\kappa(4^5) = \frac{484.12^5}{n^4}\kappa_2^{10} \text{ approximately} \tag{12.89}$$

It should be noted that (12.73) to (12.89) are all expressions of even degree. Those of odd degree are all zero, because $\kappa_r = 0$ for $r > 2$ in normal variation. In fact, (12.72) implies that $\kappa(p^q 2^r) = 0$ if pq is odd or if $q = 1$ and $p > 2$. Methods of proof of (12.71) and (12.72) are suggested in Exercises 13.10 and 13.11. Hsu and Lawley (1939) gave exact results for $\kappa(4^5)$ and $\kappa(4^6)$.

12.18 As an illustration of the way in which the sampling formulae can be used to approximate to a sampling distribution, let us consider the distribution of $\sqrt{b_1}$, the sample value of the measure of skewness at (3.89), in samples from a normal population. We have, in terms of the sample moments,

$$\sqrt{b_1} = \frac{m_3}{m_2^{3/2}} = \frac{n-2}{\sqrt{\{n(n-1)\}}}\frac{k_3}{k_2^{3/2}}.$$

For a normal distribution the variance of k_3, $\kappa(3^2)$ is, by (12.40), equal to

$$\frac{6n\kappa_2^3}{(n-1)(n-2)}.$$

We therefore consider the statistic

$$x = \sqrt{\frac{(n-1)(n-2)}{6n}}\,k_3 k_2^{-3/2} \tag{12.90}$$

$$= \frac{n-1}{\sqrt{\{6(n-2)\}}}\sqrt{b_1} \tag{12.91}$$

which will, to order n^{-1}, have unit variance. We have

$$x = \sqrt{\frac{(n-1)(n-2)}{6n}}\frac{k_3}{\kappa_2^{3/2}}\left\{1+\frac{k_2-\kappa_2}{\kappa_2}\right\}^{-3/2} \tag{12.92}$$

Since the population is symmetrical, the mean value of x is zero. We then have, expanding (12.92),

$$x^2 = \frac{(n-1)(n-2)}{6n}\cdot\frac{1}{\kappa_2^3}\left\{k_3^2-\frac{3}{\kappa_2}k_3^2(k_2-\kappa_2)+\frac{6}{\kappa_2^2}k_3^2(k_2-\kappa_2)^2-\frac{10}{\kappa_2^3}k_3^2(k_2-\kappa_2)^3\right.$$
$$\left.+\frac{15}{\kappa_2^4}k_3^2(k_2-\kappa_2)^4-\frac{21}{\kappa_2^5}k_3^2(k_2-\kappa_2)^5+\frac{28}{\kappa_2^6}k_3^2(k_2-\kappa_2)^6+\dots\right\}. \quad (12.93)$$

The variance may be obtained by taking mean values of both sides, and since κ_2 is the mean value of k_2 we have

$$\text{var}\, x = \frac{(n-1)(n-2)}{6n}\cdot\frac{1}{\kappa_2^3}\left\{\mu(3^2)-\frac{3}{\kappa_2}\mu(3^2 2)+\frac{6}{\kappa_2^2}\mu(3^2 2^2)-\frac{10}{\kappa_2^3}\mu(3^2 2^3)\right.$$
$$\left.+\frac{15}{\kappa_2^4}\mu(3^2 2^4)-\frac{21}{\kappa_2^5}\mu(3^2 2^5)+\frac{28}{\kappa_2^6}\mu(3^2 2^6)+\dots\right\}. \quad (12.94)$$

We now express the product μ's in terms of product κ's by using equation (12.31) and identifying coefficients. For a normal distribution $\kappa(32^r) = 0$ and we will take our approximation to order n^{-4}, so that κ's of five parts or more may be neglected. We then find

$$\text{var}\, x = \frac{(n-1)(n-2)}{6n}\cdot\frac{1}{\kappa_2^3}\left[\kappa(3^2)-\frac{3}{\kappa_2}\kappa(3^2 2)+\frac{6}{\kappa_2^2}\{\kappa(3^2 2^2)+\kappa(3^2)\kappa(2^2)\}\right.$$
$$-\frac{10}{\kappa_2^3}\{\kappa(3^2 2^3)+3\kappa(3^2 2)\kappa(2^2)+\kappa(3^2)\kappa(2^3)\}+\frac{15}{\kappa_2^4}\{6\kappa(3^2 2^2)\kappa(2^2)$$
$$+4\kappa(3^2 2)\kappa(2^3)+\kappa(3^2)\kappa(2^4)+3\kappa(3^2)\kappa^2(2^2)\}-\frac{21}{\kappa_2^5}\{15\kappa(3^2 2)\kappa^2(2^2)$$
$$\left.+10\kappa(3^2)\kappa(2^3)\kappa(2^2)\}+\frac{28}{\kappa_2^6}.15\kappa(3^2)\kappa^3(2^2)\right]. \quad (12.95)$$

Substituting the values of equations (12.35) to (12.89) we find, after some purely algebraic reduction,

$$\text{var}\, x = 1-\frac{6}{n-1}+\frac{28}{(n-1)^2}-\frac{120}{(n-1)^3}+\dots$$
$$= 1-\frac{6}{n}+\frac{22}{n^2}-\frac{70}{n^3}+\dots \quad (12.96)$$

In a similar way it may be shown that

$$\mu_4(x) = 3-\frac{1056}{n^2}+\frac{24{,}132}{n^3}-\dots \quad (12.97)$$

$\mu_3(x)$ is zero, for the distribution is symmetrical.

In point of fact, the exact results for this case are known (cf. Exercise 12.9) and it follows from them that

$$\left\{\frac{(n-2)(n+1)(n+3)}{6n(n-1)}\right\}^{\frac{1}{2}}k_3/k_2^{3/2} = \frac{\{(n+1)(n+3)\}^{\frac{1}{2}}}{n-1}x = \left\{\frac{(n+1)(n+3)}{6(n-2)}\right\}^{\frac{1}{2}}\sqrt{b_1}$$

has unit variance exactly and fourth moment equal to $3\left(1+\frac{12}{n}-\dots\right)$, indicating a fairly rapid approach to normality.

There are two ways of improving on the first approximation that x is normally distributed with unit variance. In the first place we may consider a transformation to a new variate ξ, chosen so that ξ is normally distributed to order n^{-2}. Secondly, we may fit a Pearson curve to the distribution of x, using the values of moments given by (12.96) and (12.97). The appropriate curve is the Type VII

$$dF \propto \left(1+\frac{x^2}{a^2}\right)^{-m} dx. \tag{12.98}$$

The first line was adopted by Fisher (1929), who obtained the following transformation:

$$\xi = x\left(1+\frac{3}{n}+\frac{91}{4n^2}\right)-\frac{3}{2n}\left(1-\frac{111}{2n}\right)(x^3-3x)-\frac{33}{8n^2}(x^5-10x^3+15x). \tag{12.99}$$

The second was adopted by E. S. Pearson (1930), who tabulated the percentiles of the fitted distribution. The *Biometrika Tables* give the 95% and 99% points (which with changed sign are also the 5% and 1% points, by symmetry) of the distribution of $\sqrt{b_1}$ for $n \geqslant 25$, and D'Agostino and Pearson (1973) give the S_U approximation (cf. **6.35**) to the distribution.

12.19 By such methods approximations can be obtained to the sampling distributions of any statistics which are expressible as symmetric functions. The reader may care to refer to investigations on these lines by F. N. David (1949*a*, *b*) into the sampling moments of the coefficient of variation, the logarithm of the variance-ratio (Fisher's z) and the variance-ratio itself. Difficulties associated with the expansion of denominators in ratio-statistics may sometimes be overcome by the use of the following device.

Consider a statistic t which is in the form of a ratio a/b, where b is always positive. Then

$$\begin{aligned} F(t_0) &= \mathrm{P}\{t \leqslant t_0\} = \mathrm{P}\{a/b \leqslant t_0\} \\ &= \mathrm{P}\{a \leqslant b\,t_0\} = \mathrm{P}\{a-b\,t_0 \leqslant 0\}. \end{aligned} \tag{12.100}$$

We then consider the statistic $a-b\,t_0$, which does not suffer from the disadvantage of being a ratio. For example, suppose our statistic is "Student's" t, given by

$$t = \frac{(\bar{x}-\mu)\sqrt{(n-1)}}{s},$$

where μ is the parent mean and $s^2 = \Sigma(x-\bar{x})^2/n$. The distribution is symmetrical and hence

$$\begin{aligned} \mathrm{P}\{|t| \leqslant |t_0|\} &= \mathrm{P}\{t^2 \leqslant t_0^2\} \\ &= \mathrm{P}\{(\bar{x}-\mu)^2(n-1)-s^2 t_0^2 \leqslant 0\}. \end{aligned} \tag{12.101}$$

We then consider a statistic

$$u = (n-1)(\bar{x}-\mu)^2 - t_0^2 s^2 \tag{12.102}$$

and can obtain as many moments or cumulants as we like for given μ and t_0^2.

In this particular instance, for normal variation we know that $\bar{x}$ and s are independent and can find the cumulants of u directly. For the characteristic function of u, writing it as θ, we have, from (11.39) and (11.40),

$$\phi(t) \propto \iint \exp[\{(n-1)(\bar{x}-\mu)^2 - t_0^2 s^2\}\theta] \exp\{-\tfrac{1}{2}n(\bar{x}-\mu)^2\}\, d\bar{x} \exp\{-\tfrac{1}{2}ns^2\} s^{n-2}\, ds$$

$$= \left\{1-\frac{2(n-1)}{n}\theta\right\}^{-\frac{1}{2}} \left\{1+\frac{2t_0^2\theta}{n}\right\}^{-\frac{1}{2}(n-1)}. \quad (12.103)$$

Hence we find, for u,

$$\kappa_1 = \frac{n-1}{n}(1-t_0^2) \quad (12.104)$$

$$\kappa_2 = \frac{2(n-1)}{n^2}(n-1+t_0^4). \quad (12.105)$$

As n tends to infinity ϕ tends to $(1-2\theta)^{-\frac{1}{2}} \exp(-\theta t_0^2)$, which apart from the second term is the c.f. of a χ^2 distribution with one degree of freedom.

Now we require only the probability that $u \leqslant 0$, which is therefore the probability that a deviate with mean (12.104) and variance (12.105) is less than zero. For $n = 20$, $t_0 = 2$ we have $\kappa_1 = -2{\cdot}85$, $\kappa_2 = 3{\cdot}42$. This is equivalent to a standardized deviate of $-2{\cdot}85/\sqrt{3{\cdot}42} = -1{\cdot}563$. The probability that this is exceeded in a normal distribution is 0·941, which is accordingly our approximation to the probability that t will not exceed $t_0 = 2$ in absolute value. The true value is 0·940.

Although we have considered a normal case, the method is particularly useful in judging the effect of departures from normality on certain statistics. Reference may be made to David and Johnson (1951) for some work on these lines concerning the variance-ratio.

Sampling from finite populations

12.20 In sampling without replacement from finite populations the algebra of expectations becomes much more complex in virtue of the fact that successive drawings are no longer independent. Let us follow through the work of **12.3** in the finite case to illustrate the point. Equation (12.2) still holds, i.e.

$$\mathrm{E}\, m_2 = \frac{n-1}{n}\mathrm{E}\, x^2 - \frac{n-1}{n}\mathrm{E}\, x_i x_j, \qquad (i \neq j). \quad (12.106)$$

If capital M's are written to denote the moments of the parent population of N members, we now have, by definition,

$$\mathrm{E}\, x^2 = M_2'$$

$$\mathrm{E}\, x_i x_j = \sum_{i,j=1}^{N} x_i x_j / \{N(N-1)\} \qquad (i \neq j)$$

$$= \frac{\left(\sum_1^N x_i\right)^2 - \sum_1^N x_i^2}{N(N-1)} = \frac{N^2 M_1'^2 - N M_2'}{N(N-1)}.$$

Substituting in (12.106) we then find

$$\mathrm{E}\, m_2 = \frac{n-1}{n}\left\{\frac{N}{N-1}(M_2' - M_1'^2)\right\}$$

$$= \frac{n-1}{n} \cdot \frac{N}{N-1} M_2. \quad (12.107)$$

Again, with sufficient pertinacity, we can find all the moments of symmetric-function statistics by this kind of approach, but the algebra rapidly becomes unmanageable. A simpler method is as follows.

12.21 Let us first of all note that the expectation of any augmented symmetric function for a sample bears a very simple relation to the corresponding function for the parent. Denoting the two functions by the subscripts n and N respectively, we have from (12.9)

$$E[p_1^{\pi_1}\dots p_s^{\pi_s}]_n/n^{[\rho]} = [p_1^{\pi_1}\dots p_s^{\pi_s}]_N/N^{[\rho]} \tag{12.108}$$

where $\rho = \sum_i \pi_i$. It follows that if K_p is the pth k-statistic for the parent, we have

$$E_N k_p = K_p \tag{12.109}$$

where E_N means an expectation in the set of N values forming the parent. This attractive property of k-statistics is one of their most useful features. We have, for example, $k_2 = nm_2/(n-1)$ and hence

$$E_N \frac{nm_2}{n-1} = K_2 = \frac{NM_2}{N-1}$$

from which (12.107) follows immediately.

Secondly, let us imagine the population of N as itself a sample from an infinite population with cumulants κ_p. If E refers to expectation in this infinite population we shall have

$$E k_p = E K_p = \kappa_p.$$

The sample of n observations may now be considered as having been drawn from the infinite population in two stages, the first of which consists in drawing the population of N, and the second in selecting the sample from the latter. Thus we may regard the expectation operation E as consisting of two stages, the second of which is E_N over the finite population.

Now if a symmetric function f has expectation *linear* in the cumulants,

$$E(f) = \sum a_j \kappa_j, \tag{12.110}$$

we may write down the analogous relation

$$E_N(f) = \sum a_j K_j. \tag{12.111}$$

For if this were not true, $E_N(f)$ could be expressed as some other function of K's whose expectation E would be the same as that of $\sum a_j K_j$, and this would imply a relationship among the K's. This is sometimes known as the Irwin–Kendall principle from the authors who introduced it (1944).

From (12.109), we see that

$$E(K_p^s k_p) = E\{K_p^s E_N(k_p)\} = E(K_p^{s+1})$$

and using this with $s = 1$, we find that

$$\left.\begin{aligned} E(k_p-K_p)^2 = E(k_p^2)+E(K_p^2) - 2E(K_p k_p) &= E(k_p^2)-E(K_p^2) \\ \text{or} \qquad \text{var}_N(k_p) = \text{var}(k_p)-\text{var}(K_p).& \end{aligned}\right\} \tag{12.112}$$

Example 12.8 Moments of the mean in the finite population case

The mean-statistic m_1 is equal to k_1. Thus, since $E(k_1) = \kappa_1$, (12.110–11) give

$$E_N(k_1) = K_1.$$

From the known result for the variance of the mean we have

$$E(k_1^2) = \frac{\kappa_2}{n}+\kappa_1^2, \quad E(K_1^2) = \frac{\kappa_2}{N}+\kappa_1^2.$$

Hence, using (12.112),

$$E(k_1-K_1)^2 = E k_1^2 - E K_1^2 = \kappa_2\left(\frac{1}{n}-\frac{1}{N}\right).$$

It then follows at once that

$$E_N(k_1-K_1)^2 = K_2\left(\frac{1}{n}-\frac{1}{N}\right), \tag{12.113}$$

giving us the variance of the mean. If we prefer to express this in terms of parent moments we have

$$E_N(k_1-K_1)^2 = \frac{N-n}{(N-1)n}M_2. \tag{12.114}$$

For the third moment we have

$$E(k_1-\kappa_1)^3 = E(k_1-K_1+K_1-\kappa_1)^3$$

$$= E(k_1-K_1)^3+3E(k_1-K_1)^2(K_1-\kappa_1)+3E(k_1-K_1)(K_1-\kappa_1)^2+E(K_1-\kappa_1)^3. \tag{12.115}$$

The third term on the right vanishes, since $E_N(k_1-K_1) = 0$. Now

$$E(k_1-\kappa_1)^3 = \frac{\kappa_3}{n^2}, \quad E(K_1-\kappa_1)^3 = \frac{\kappa_3}{N^2}.$$

Hence, from (12.115) we find

$$E(k_1-K_1)^3 = \kappa_3\left(\frac{1}{n^2}-\frac{1}{N^2}\right)-3E(K_1-\kappa_1)(k_1-K_1)^2.$$

Writing

$$\alpha_r = \frac{1}{n^r}-\frac{1}{N^r} \tag{12.116}$$

we then find

$$\begin{aligned} E(k_1-K_1)^3 &= \alpha_2\kappa_3-3E(K_1-\kappa_1)E_N(k_1-K_1)^2 \\ &= \alpha_2\kappa_3-3E(K_1-\kappa_1)\alpha_1 K_2 \\ &= \alpha_2\kappa_3-3\alpha_1 E(K_1-\kappa_1)(K_2-\kappa_2) \\ &= \alpha_2\kappa_3-3\alpha_1\kappa_3/N \\ &= \left(\alpha_2-\frac{3\alpha_1}{N}\right)\kappa_3. \end{aligned}$$

Hence, for the third moment,

$$E_N(k_1-K_1)^3 = \left(\alpha_2-\frac{3\alpha_1}{N}\right)K_3. \tag{12.117}$$

Substituting for α_2, α_1 and

$$K_3 = \frac{N^2}{(N-1)(N-2)}M_3$$

we find

$$E_N(m_1-M_1)^3 = \frac{(N-n)(N-2n)}{n^2(N-1)(N-2)}M_3. \tag{12.118}$$

Likewise we find

$$\mathrm{E}_N(k_1-K_1)^4 = K_4\left\{\alpha_3-4\alpha_2/N+6\alpha_1/N^2-\frac{3(N-1)}{N(N+1)}\alpha_1^2\right\}+3K_2^2\frac{N-1}{N+1}\alpha_1^2 \qquad (12.119)$$

which is equivalent to

$$\mathrm{E}_N(m_1-M_1)^4 = \frac{N-n}{n^3(N-1)(N-2)(N-3)}\{(N^2-6nN+N+6n^2)M_4 + 3N(n-1)(N-n-1)M_2^2\}. \qquad (12.120)$$

The reader should check that the results (5.55) which we obtained for the hypergeometric distribution are special cases of (12.114), (12.118) and (12.120), given by putting M_r equal to the corresponding binomial moments in (5.4–5) with $n = 1$, and remembering that there the variable is a sum and not a mean as here.

Example 12.9

We have, generalizing (12.112),

$$\left.\begin{aligned} \mathrm{E}(k_r-K_r)(k_p-K_p) &= \mathrm{E}(k_r k_p)-\mathrm{E}(K_r K_p),\\ \text{or}\qquad \mathrm{cov}_N(k_r, k_p) &= \mathrm{cov}(k_r, k_p)-\mathrm{cov}(K_r, K_p). \end{aligned}\right\} \qquad (12.121)$$

From (12.121) with $p = 1$ and Rule 10, we find

$$\mathrm{E}(k_r-K_r)(k_1-K_1) = \frac{1}{n}\kappa_{r+1}-\frac{1}{N}\kappa_{r+1} = \alpha_1\kappa_{r+1}$$

from which $$\mathrm{E}_N(k_r-K_r)(k_1-K_1) = \alpha_1 K_{r+1}.$$

Thus, if the finite population is symmetrical, any k-statistic of even order is uncorrelated with k_1, for K_{r+1} then vanishes since it is homogeneous of odd degree in the central moments, by **12.7–8.** Cf. the large-sample result for even-order central moments in Example 10.3.

12.22 In developing our next point, we encounter a difficulty of notation. We have already used the symbols κ, k and K. In the next chapter, when we proceed to multivariate extensions, we shall require corresponding quantities with several suffixes, e.g. κ_{rs}, k_{rs}, K_{rs}. We now require a notation for a quantity whose mean value is the product of κ's, e.g. $\kappa_r\kappa_s$. Previous writers (Tukey (1950), Wishart (1952), Kendall (1952)) notwithstanding, we shall write these quantities with the letter l and thus we define them by

$$\mathrm{E}\, l_{rs\ldots u} = \kappa_r\kappa_s\ldots\kappa_u. \qquad (12.122)$$

The l-statistics are called "polykays" by some authors, but we feel that there are limits to linguistic miscegenation that should not be exceeded.

It will, as before, follow that if $L_{rs\ldots u}$ is the corresponding statistic for a finite population

$$\mathrm{E}_N\, l_{rs\ldots u} = L_{rs\ldots u}. \qquad (12.123)$$

The use of these functions derives from the fact that we can express them in terms of augmented symmetrics once and for all and hence derive *non-linear* functions of them as linear functions to which the argument of **12.21** can be applied. The necessary table is given by Abdel-Aty (1954) up to weight 12. Wishart (1952) gave expressions up to order 8 for these multiple k-statistics in terms of products of simple k-statistics. For example we have, from his table,

$$k_2 k_1 = l_2 l_1 = l_3/n + l_{21} = k_3/n + l_{21} \tag{12.124}$$

and it follows at once that

$$\mathrm{E}(k_2 k_1) = \kappa_3/n + \kappa_2 \kappa_1, \tag{12.125}$$

and also that

$$\mathrm{E}_N(k_2 k_1) = K_3/n + K_2 K_1. \tag{12.126}$$

Similarly we have

$$k_1^4 = k_4/n^3 + 4l_{31}/n^2 + 3l_{22}/n^2 + 6l_{211}/n + l_{1111} \tag{12.127}$$

whence
$$\mathrm{E}(k_1^4) = \kappa_4/n^3 + 4\kappa_3 \kappa_1/n^2 + 3\kappa_2^2/n^2 + 6\kappa_2 \kappa_1^2/n + \kappa_1^4, \tag{12.128}$$

again with a parallel expression for $\mathrm{E}_N(k_1^4)$.

Example 12.10

One of the most useful practical applications of these results is that they enable us to write down unbiased estimators of products of cumulants. We found in Examples 12.1 and 12.3 that

$$\operatorname{var} m_2 = \frac{(n-1)^2}{n^2}\left\{\frac{\kappa_4}{n} + \frac{2\kappa_2^2}{n-1}\right\}. \tag{12.129}$$

In some practical applications of this result we may require to calculate the quantity on the right, but if we have only a sample available we are ignorant of κ_4 and κ_2 and have to estimate them. Thus we require a quantity whose mean value is the function on the right in (12.129). In terms of the multiple k-statistics this is written down at sight as

$$\frac{(n-1)^2}{n^2}\left\{\frac{k_4}{n} + \frac{2l_{22}}{n-1}\right\}. \tag{12.130}$$

From Wishart's table we then read off

$$l_{22} = \frac{n-1}{n+1}\left(k_2^2 - \frac{k_4}{n}\right)$$

and, on substitution in (12.130), for the unbiased estimator we find

$$\frac{(n-1)^2}{n^2(n+1)}\left\{\frac{(n-1)k_4}{n} + 2k_2^2\right\}. \tag{12.131}$$

Example 12.11

To find an unbiassed estimator of $\mu_3 \mu_1'^2$.

This is equal to $\kappa_3 \kappa_1^2$ and the estimator is l_{311}. From Appendix Table 11 we write down the estimator directly in terms of augmented symmetric functions as

$$l_{311} = \frac{2}{n^{[5]}}[1^5] - \frac{3}{n^{[4]}}[21^3] + \frac{[31^2]}{n^{[3]}}.$$

From Appendix Table 10 we can then substitute for the augmented symmetrics in terms of the power-sums (r). We find

$$l_{311} = \frac{2}{n^{[5]}}\{24(5)-30(4)(1)-20(3)(2)+20(3)(1)^2+15(2)^2(1)-10(2)(1)^3+(1)^5\}$$
$$-\frac{3}{n^{[4]}}\{-6(5)+6(4)(1)+5(3)(2)-3(3)(1)^2-3(2)^2(1)+(2)(1)^3\}$$
$$+\frac{1}{n^{[3]}}\{2(5)-2(4)(1)-(3)(2)+(3)(1)^2\}$$

which reduces, on putting $(r) = s_r$, to

$$\frac{1}{n^{[5]}}\{2n(n+2)s_5-2(n^2+2n+6)s_4s_1-(n^2+8n-8)s_3s_2+(n^2+2n+16)s_3s_1^2$$
$$+(9n-6)s_2^2s_1-(3n+8)s_2s_1^3+2s_1^5\}.$$

Dwyer (1964) considers the l-statistics of deviations from the sample mean, leading to general moment formulae. Cf. also Dwyer and Tracy (1964), Tracy (1968, 1969), Nagambal and Tracy (1970) and Tracy and Gupta (1973, 1974). See also the collection of papers edited by Tracy (1972).

EXERCISES

12.1 By using (12.25) directly, write down k_3 and k_4 in terms of the augmented symmetric functions and hence verify (12.29).

12.2 Assuming that

$$k_3 = a_0 s_3 + a_1 s_2 s_1 + a_2 s_1^3$$

obtain the coefficients a by taking expectations of both sides and hence derive k_3 of (12.28).

12.3 Use the approach of Example 12.1 to obtain $\kappa(2^3)$ in terms of the parent cumulants.

12.4 Show that the pattern functions of the following patterns :—

$$\begin{matrix} \times & \times & \times \\ \times & . & \times \\ \times & \times & . \end{matrix} \qquad\qquad \begin{matrix} \times & \times & \times & \times \\ . & . & \times & \times \\ \times & \times & . & . \end{matrix}$$

are $\dfrac{1}{(n-1)^2}$ and $\dfrac{1}{n(n-1)^2}$ respectively.

(Fisher, 1929)

12.5 Show that the pattern function of the pattern

$$\begin{matrix} \times & \times & \times & \times \\ \times & \times & \times & \times \end{matrix} \quad \text{etc.}$$

with p columns is

$$\frac{1}{n^{p-1}}\left\{1-\frac{(-1)^{p-1}}{(n-1)^{p-1}}\right\}.$$

(Fisher, 1929)

12.6 Verify the formulae of equations (12.37) and (12.43).

12.7 If a pattern contains a column with three entries ; if the patterns obtained by suppressing this column and (1) amalgamating the three rows, (2) amalgamating the pairs of three rows, and (3) leaving the rows unamalgamated are A, B_1, B_2, B_3 and C respectively, show that the pattern function for the original pattern is

$$\frac{n}{(n-1)(n-2)}A-\frac{1}{(n-1)(n-2)}(B_1+B_2+B_3)+\frac{2}{n(n-1)(n-2)}C.$$

Deduce that the function of the pattern

$$\begin{matrix} \times & \times & \times \\ \times & \times & \times \\ \times & \times & \times \\ \times & \times & . \end{matrix}$$

is

$$\frac{n^3-8n^2+17n+2}{(n-1)^2(n-2)^2(n-3)}.$$

(Fisher and Wishart, 1931)

12.8 Use the combinatorial method to verify that, to order n^{-1},

$$\text{var}\, m_4 = (\mu_8-\mu_4^2+16\mu_3^2\mu_2-8\mu_5\mu_3)/n.$$

12.9 Show that in normal samples k_2 is independent of $k_p/k_2^{\frac{1}{2}p}$ for $p = 3, 4, \ldots$ (cf. Exercises 11.4–5). Hence show that this ratio has mean zero for all p, and all odd-order moments zero for p odd. For $p = 3$, show that

$$\text{var}\,(k_3/k_2^{3/2}) = \frac{6n(n-1)}{(n-2)(n+1)(n+3)}$$

$$\mu_4(k_3/k_2^{3/2}) = \frac{108n^2(n-1)^2(n^2+27n-70)}{(n-2)^3(n+1)(n+3)(n+5)(n+7)(n+9)}$$

and verify equations (12.96) and (12.97).

(Cf. Fisher, 1930; μ_6 and μ_8 are listed by P. Williams (1935).)

12.10 Defining y by the relation

$$y = \left\{\frac{(n-1)(n-2)(n-3)}{24n(n+1)}\right\}^{\frac{1}{2}}\frac{k_4}{k_2^2},$$

show by the method of **12.18** that the moments of the distribution of y in samples from a normal population are

$$\mu_1 = 0,$$

$$\mu_2 = 1-\frac{12}{n}+\frac{88}{n^2}-\frac{532}{n^3}\ldots,$$

$$\mu_3 = 6\sqrt{\frac{6}{n}}\left(1-\frac{65}{2n}+\frac{4811}{8n^2}-\frac{136,605}{16n^3}\ldots\right),$$

$$\mu_4 = 3+\frac{468}{n}-\frac{32,196}{n^2}+\frac{1,118,388}{n^3}\ldots.$$

Show as in Exercise 12.9 that

$$\text{var}(k_4/k_2^2) = 24n(n-1)^2/\{(n-3)(n-2)(n+3)(n+5)\}$$

exactly, and hence verify $\mu_2(y)$.

(E. S. Pearson, 1930; Fisher, 1930; the *Biometrika Tables* give percentage points of $b_2 = \frac{m_4}{m_2^2} = \frac{(n-2)(n-3)}{(n^2-1)}\frac{k_4}{k_2^2}+\frac{3(n-1)}{(n+1)}$ for $n \geq 200$ and D'Agostino and Pearson (1973) give them for $20 \leq n \leq 200$.)

12.11 Show that, in samples from a finite population,

$$E_N(k_2-K_2)^2 = \frac{(N-n)(Nn-n-N-1)}{n(n-1)N(N+1)}K_4+\frac{2(N-n)}{(n-1)(N+1)}K_2^2$$

which is equivalent to

$$E_N\left(m_2-\frac{n-1}{n}\frac{N}{N-1}M_2\right)^2 = \frac{(n-1)N(N-n)}{n^3(N-1)^2(N-2)(N-3)}\{(nN-N-n-1)(N-1)M_4-(nN^2-3N^2+6N-3n-3)M_2^2\}.$$

12.12 Show likewise that

$$E_N(k_3-K_3)(k_2-K_2) = \frac{(N-n)\{(n-1)(N-1)-6\}}{n(n-1)N(N+5)}K_5+\frac{6(N-n)}{(n-1)(N+5)}K_3K_2.$$

12.13 Show that in samples from a symmetrical finite population k_{2r} is uncorrelated with any k-statistic of odd order.

12.14 Show that if $M(z)$ is the moment-generating function of x, the m.g.f., say $M(\xi)$, of a function $y = f(x)$, may be written

$$M(\xi) = \left[\exp\left\{\xi f\left(\frac{d}{dz}\right)\right\}M(z)\right]_{z=0}$$

Hence show that the generating function of the moments of the k-statistics,

$$\Sigma\left\{\mu(p_1^{\pi_1}\ldots p_s^{\pi_s})\frac{t_{p_1}^{\pi_1}}{\pi_1!}\ldots\frac{t_{p_s}^{\pi_s}}{\pi_s!}\right\}$$

is given by

$$\left[\exp\{t_1\mathrm{K}_1+t_2\mathrm{K}_2+t_3\mathrm{K}_3+\ldots\}\exp\left\{\kappa_1 s_1+\frac{\kappa_2 s_2}{2!}+\ldots\right\}\right]_{x=0}$$

where K_r is the same function of the operators $\frac{\partial}{\partial x}$ as k_r is of the observations x and $s_r = \Sigma(x^r)$.

Deduce that

$$\mathrm{K}_p(s_p) = p!$$
$$\mathrm{K}_p(s_{p_1}s_{p_2}\ldots) = 0,$$

where $(p_1p_2\ldots)$ is any partition of p.

(Fisher, 1929)

12.15 Show that

$$E\{k_p(x_1+h, x_2, \ldots, x_n)\} = \kappa_p+\frac{1}{n}h^p$$

and hence that

$$\mathrm{S}_p k_p = p!,$$
$$\mathrm{S}_q k_p = 0, \qquad q \neq p,$$

where S_p is the same function of the operators $\frac{\partial}{\partial x}$ as s_p is of the observations x.

(Kendall, 1940a)

12.16 Show that the generating function of the moments of the k-statistics is given by

$$\left[\exp\left\{\frac{\mathrm{S}_1\kappa_1}{1!}+\frac{\mathrm{S}_2\kappa_2}{2!}+\ldots\right\}\exp\{k_1t_1+k_2t_2+\ldots\}\right]_{x=0}$$

and hence derive the result $\mathrm{S}_p k_p = p!$ of the previous Exercise.

(Kendall, 1942)

12.17 Use Exercise 12.16 to show that, in the expression of k_p in terms of the symmetric sums s, the sum of the coefficients is $1/n$.

12.18 Show that an unbiassed estimator of $\mu_1'^4$ is

$$\frac{1}{n^{[4]}}\{s_1^4-6s_2s_1^2+3s_2^2+8s_3s_1-6s_4\}.$$

12.19 Show that

$$l_{222} = -\frac{[1^6]}{n^{[6]}}+\frac{3[21^4]}{n^{[5]}}-\frac{3[2^2 1^2]}{n^{[4]}}+\frac{[2^3]}{n^{[3}}$$

as given in Appendix Table 11.

12.20 By the method of Example 12.7 show that if k_1 and k_p are uncorrelated, for all $p>1$, the parent distribution is normal; and that if k_1 is independent of any $k_p(p>1)$ the parent is likewise normal. (Use Exercise 4.5.)

12.21 Show that the joint distribution of k_1 and k_2 in samples of size n has correlation coefficient

$$\rho = \frac{\mathrm{cov}(k_1, k_2)}{\{\mathrm{var}(k_1)\,\mathrm{var}(k_2)\}^{\frac{1}{2}}} = \frac{\gamma_1}{\left(\gamma_2+\frac{2n}{n-1}\right)^{\frac{1}{2}}},$$

where γ_1, γ_2 are the measures of skewness and kurtosis (cf. (3.89–90)) of the parent population, so that $\rho = 0$ for any symmetrical parent, although Example 12.7 shows that k_1 and k_2 are independent only in the normal case. Use the fact that $\rho^2 < 1$ to verify the result in Exercise 3.19 that $\beta_2 > 1 + \beta_1$.

Show using (5.20) that for the Poisson distribution (5.18) as $n \to \infty$, $\rho \to (1+2\lambda)^{-\frac{1}{2}}$, arbitrarily close to 1 as $\lambda \to 0$.

(Cf p. 219)

12.22 Generalizing Exercises 12.9–10, show that in normal samples

$$E(k_p^a k_r^b / k_2^s) = E(k_p^a k_r^b)/E(k_2^s) = \frac{(n-1)^{s-1}}{(n+1)(n+3)\ldots(n+2s-3)} \frac{E(k_p^a k_r^b)}{\kappa_2^s}$$

for any non-negative integers p, r with $2s = ap + br$ a positive integer. Derive $\text{var}(k_3/k_2^{3/2})$ from (12.40) and $\text{var}(k_4/k_2^2)$ from (12.43) and show that these variables are uncorrelated. Show from (12.67) that

$$E(k_4 k_3^2 / k_2^5) = 216n^2(n-1)^2/\{(n-2)^2(n+1)(n+3)(n+5)(n+7)\}$$

and hence show that the correlation coefficient between k_4/k_2^2 and k_3^2/k_2^3 in normal samples is asymptotically $(27/n)^{\frac{1}{2}}$.

(Cf. Fisher, 1930)

CHAPTER 13

CUMULANTS OF SAMPLING DISTRIBUTIONS—(2)

13.1 In this chapter we shall discuss two main topics. The first is the extension to bivariate and multivariate statistics of the methods developed in the last chapter for the univariate case. The second is the structure of k-statistics and cumulants, with particular reference to the proof of the combinatorial rules for developing their sampling properties.

Bivariate k-statistics

13.2 Given any bipartite number pp' we shall have for any partition $\{(p_1p'_1)^{\pi_1}(p_2p'_2)^{\pi_2}\ldots\}$ and the bivariate cumulant $\kappa_{pp'}$ a k-statistic $k_{pp'}$ whose mean value is $\kappa_{pp'}$. Explicitly

$$k_{pp'} = p!p'!\Sigma\frac{(-1)^{\rho-1}(\rho-1)!}{n^{[\rho]}}\Sigma\frac{(x_1^{p_1}y_1^{p_1'}x_2^{p_1}y_2^{p_1'}\ldots x_\rho^{ps}y_\rho^{p's})}{(p_1!)^{\pi_1}(p'_1!)^{\pi_1}\ldots\pi_1!\pi_2!\ldots}. \tag{13.1}$$

In particular, corresponding to (12.28) we have

$$\left.\begin{aligned}
k_{11} &= \frac{1}{n^{[2]}}(ns_{11}-s_{10}s_{01})\\
k_{21} &= \frac{1}{n^{[3]}}(n^2s_{21}-2ns_{10}s_{11}-ns_{20}s_{01}+2s_{10}^2s_{01})\\
k_{31} &= \frac{1}{n^{[4]}}\{n^2(n+1)s_{31}-n(n+1)s_{30}s_{01}-3n(n-1)s_{11}s_{20}\\
&\qquad -3n(n+1)s_{21}s_{10}+6ns_{11}s_{10}^2+6ns_{20}s_{10}s_{01}-6s_{01}s_{10}^3\}\\
k_{22} &= \frac{n}{(n-1)(n-2)(n-3)}\left\{(n+1)s_{22}-\frac{2(n+1)}{n}s_{21}s_{01}-\frac{2(n+1)}{n}s_{12}s_{10}\right.\\
&\qquad -\frac{(n-1)}{n}s_{20}s_{02}-\frac{2(n-1)}{n}s_{11}^2+\frac{8}{n}s_{11}s_{10}s_{01}+\frac{2}{n}s_{02}s_{10}^2+\frac{2}{n}s_{20}s_{01}^2\\
&\qquad \left.-\frac{6}{n^2}s_{10}^2s_{01}^2\right\}
\end{aligned}\right\} \tag{13.2}$$

In generalization of the mean value functions of the k's we may write, for example,

$$\mathrm{E}(k_{20}k_{11}) = \mu'\begin{pmatrix}2 & 1\\ 0 & 1\end{pmatrix}$$

$$\mathrm{E}(k_{20}k_{11}k_{02}) = \mu'\begin{pmatrix}2 & 1 & 0\\ 0 & 1 & 2\end{pmatrix}$$

with corresponding κ's. The latter may be expressed in terms of the cumulants of the bivariate distribution as in the univariate; and the coefficients will now depend on partitions of bipartite numbers. Our rules still apply (and in particular the pattern functions corresponding to particular arrays are unchanged); but the numerical coefficients associated with arrays are modified, for we now have to consider the number of ways of allocating two different sorts of individuals in a two-way partition of a bipartite number. An example will make the modification clear.

Suppose we wish to find the coefficient of $\kappa_{33}\kappa_{11}^2$ in $\kappa\begin{pmatrix}2 & 2 & 1\\ 2 & 2 & 1\end{pmatrix}$. The total degree is 10 and, the orders of the product being 6, 2, 2, we have to consider arrays that complete the table

$$\begin{array}{ccc|c} & & & 6\\ & & & 2\\ & & & 2\\ \hline 4 & 4 & 2 & 10\end{array}$$

i.e. those we discussed in Example 12.2. The pattern functions are those we have already found. For the numerical coefficients we have to regard the column totals as consisting of the pairs (2, 2), (2, 2), and (1, 1) and the row totals as consisting of (3, 3), (1, 1) and (1, 1). For instance, the array

$$\begin{array}{ccc|c} 2 & 2 & 2 & 6\\ 1 & 1 & . & 2\\ 1 & 1 & . & 2\\ \hline 4 & 4 & 2 & 10\end{array}$$

may be written either as

$$\begin{array}{ccc|c} (1,1) & (1,1) & (1,1) & (3,3)\\ (0,1) & (1,0) & . & (1,1)\\ (1,0) & (0,1) & . & (1,1)\\ \hline (2,2) & (2,2) & (1,1) & (5,5)\end{array} \qquad (13.3)$$

or as

$$\begin{array}{ccc|c} (2,0) & (0,2) & (1,1) & (3,3)\\ (0,1) & (1,0) & . & (1,1)\\ (0,1) & (1,0) & . & (1,1)\\ \hline (2,2) & (2,2) & (1,1) & (5,5)\end{array} \qquad (13.4)$$

each of which will make a contribution to the numerical coefficient. It will be found that no other arrays are possible except those obtained by permuting the first two columns.

The numerical coefficient in (13.3) and the permuted array together is

$$2\left(\frac{2!}{1!1!}\right)\left(\frac{2!}{1!1!}\right)\left(\frac{2!}{1!1!}\right)\left(\frac{2!}{1!1!}\right)\cdot\frac{1}{2!} = 16.$$

That in (13.4) and the permuted array is

$$2\left(\frac{2!}{2!}\right)\left(\frac{2!}{1!1!}\right)\left(\frac{2!}{1!1!}\right)\left(\frac{2!}{2!}\right)\cdot\frac{1}{2!} = 4.$$

The total contribution is thus 20. The pattern function is $\dfrac{1}{(n-1)(n-2)}$ as in Example 12.2.

In the same way it will be found that for the arrays

$$\begin{array}{ccc|c} 2 & 3 & 1 & 6 \\ 1 & 1 & . & 2 \\ 1 & . & 1 & 2 \\ \hline 4 & 4 & 2 & 10 \end{array} \qquad \begin{array}{ccc|c} 3 & 3 & . & 6 \\ 1 & . & 1 & 2 \\ . & 1 & 1 & 2 \\ \hline 4 & 4 & 2 & 10 \end{array}$$

the coefficients are 48 and 8. Thus, using the pattern functions in Example 12.2, the desired coefficient of $\kappa_{33}\kappa_{11}^2$ is

$$\frac{20}{(n-1)(n-2)}+\frac{48}{(n-1)^2}+\frac{8}{(n-1)^2} = \frac{4(19n-33)}{(n-1)^2(n-2)}.$$

Example 13.1

To find an exact expression for the covariance of the estimates of variance of two correlated variables, i.e.

$$\kappa\begin{pmatrix}2 & 0\\ 0 & 2\end{pmatrix}.$$

This will clearly consist of three terms, in κ_{22}, $\kappa_{20}\kappa_{02}$ and κ_{11}^2. For the first, we have the array

$$\begin{array}{cc|c} (2,0) & (0,2) & (2,2) \\ \hline (2,0) & (0,2) & (2,2) \end{array}$$

with pattern function $1/n$ and numerical coefficient unity. For the second, the contribution is zero, the only array being

$$\begin{array}{cc|c} (2,0) & . & (2,0) \\ . & (0,2) & (0,2) \\ \hline (2,0) & (0,2) & (2,2) \end{array}$$

which has a vanishing pattern function. For the third term we have

$$\begin{array}{cc|c} (1,0) & (0,1) & (1,1) \\ (1,0) & (0,1) & (1,1) \\ \hline (2,0) & (0,2) & (2,2) \end{array}$$

the pattern function for which is $\frac{1}{(n-1)}$ and numerical coefficient 2. Hence

$$\kappa\begin{pmatrix}2 & 0\\ 0 & 2\end{pmatrix} = \frac{1}{n}\kappa_{22}+\frac{2}{n-1}\kappa_{11}^2.$$

We note the analogy between this expression and the variance of k_{20}, which may be written (cf. (12.35))

$$\kappa\begin{pmatrix}2 & 2\\ 0 & 0\end{pmatrix} = \frac{\kappa_{40}}{n}+\frac{2\kappa_{20}^2}{n-1}.$$

In particular, if the population is normal all κ's except those of the second order vanish. We then find

$$\text{correlation } (k_{20}, k_{02}) = \frac{\kappa_{11}^2}{\kappa_{20}\kappa_{02}} = \rho^2,$$

where ρ is the correlation parameter of the bivariate normal form.

13.3 It is possible to derive the bivariate formulae from the univariate formulae by the symbolic process, given by Kendall (1940c), which has been illustrated in **3.29** above. This was used by Cook (1951) to derive bivariate formulae up to those of order 4 in each variate. We quote here some of the simpler results.

$$\kappa\begin{pmatrix}2 & 1\\ 0 & 1\end{pmatrix} = \frac{1}{n}\kappa_{31}+\frac{2}{n-1}\kappa_{20}\kappa_{11} \tag{13.5}$$

$$\kappa\begin{pmatrix}2 & 0\\ 0 & 2\end{pmatrix} = \frac{1}{n}\kappa_{22}+\frac{2}{n-1}\kappa_{11}^2 \tag{13.6}$$

$$\kappa\begin{pmatrix}1 & 1\\ 1 & 1\end{pmatrix} = \frac{1}{n}\kappa_{22}+\frac{1}{n-1}\kappa_{20}\kappa_{02}+\frac{1}{n-1}\kappa_{11}^2 \tag{13.7}$$

$$\kappa\begin{pmatrix}3 & 3\\ 0 & 0\end{pmatrix} = \frac{1}{n}\kappa_{60}+\frac{9}{n-1}\kappa_{40}\kappa_{20}+\frac{9}{n-1}\kappa_{30}^2+\frac{6n}{(n-1)(n-2)}\kappa_{20}^3 \tag{13.8}$$

(which is, of course, a univariate result)

$$\kappa\begin{pmatrix}3 & 2\\ 0 & 1\end{pmatrix} = \frac{1}{n}\kappa_{51}+\frac{6}{n-1}\kappa_{31}\kappa_{20}+\frac{3}{n-1}\kappa_{40}\kappa_{11}+\frac{9}{n-1}\kappa_{30}\kappa_{21}+\frac{6n}{(n-1)(n-2)}\kappa_{20}^2\kappa_{11} \tag{13.9}$$

$$\kappa\begin{pmatrix}3 & 1\\ 0 & 2\end{pmatrix} = \frac{1}{n}\kappa_{42}+\frac{3}{n-1}\kappa_{22}\kappa_{20}+\frac{6}{n-1}\kappa_{31}\kappa_{11}+\frac{3}{n-1}\kappa_{30}\kappa_{12}+\frac{6}{n-1}\kappa_{21}^2 + \frac{6n}{(n-1)(n-2)}\kappa_{20}\kappa_{11}^2 \tag{13.10}$$

$$\kappa\begin{pmatrix}3 & 0\\ 0 & 3\end{pmatrix} = \frac{1}{n}\kappa_{33}+\frac{9}{n-1}\kappa_{22}\kappa_{11}+\frac{9}{n-1}\kappa_{12}\kappa_{21}+\frac{6n}{(n-1)(n-2)}\kappa_{11}^3 \tag{13.11}$$

$$\kappa\begin{pmatrix}2 & 2\\ 1 & 1\end{pmatrix} = \frac{1}{n}\kappa_{42}+\frac{4}{n-1}\kappa_{22}\kappa_{20}+\frac{4}{n-1}\kappa_{31}\kappa_{11}+\frac{1}{n-1}\kappa_{40}\kappa_{02}+\frac{4}{n-1}\kappa_{30}\kappa_{12} + \frac{5}{n-1}\kappa_{21}^2+\frac{4n}{(n-1)(n-2)}\kappa_{20}\kappa_{11}^2+\frac{2n}{(n-1)(n-2)}\kappa_{20}^2\kappa_{02} \tag{13.12}$$

$$\kappa\begin{pmatrix}2 & 1\\ 1 & 2\end{pmatrix} = \frac{1}{n}\kappa_{33}+\frac{5}{n-1}\kappa_{22}\kappa_{11}+\frac{2}{n-1}\kappa_{13}\kappa_{20}+\frac{2}{n-1}\kappa_{31}\kappa_{02}+\frac{1}{n-1}\kappa_{30}\kappa_{03} + \frac{8}{n-1}\kappa_{21}\kappa_{12}+\frac{4n}{(n-1)(n-2)}\kappa_{11}\kappa_{20}\kappa_{02}+\frac{2n}{(n-1)(n-2)}\kappa_{11}^3 \tag{13.13}$$

There is obviously a strong family resemblance among these formulae and we shall see later how they can be written in a concise form.

Example 13.2

The methods may similarly be extended to deal with bivariate finite populations. Formulae in this field are not required often enough to make it worth while listing them, but it may be useful to illustrate the method of deriving them. Let us then derive the variance of k_{11} in samples of n from a finite population of size N. We have, after the manner of **12.21**,

$$\mathrm{E}(k_{11}-K_{11})^2 = \mathrm{E}(k_{11}^2)-\mathrm{E}(K_{11}^2). \tag{13.14}$$

From (13.7) we have

$$\mathrm{E}(k_{11}^2) = \mu'\begin{pmatrix}1 & 1\\ 1 & 1\end{pmatrix} = \kappa\begin{pmatrix}1 & 1\\ 1 & 1\end{pmatrix}+\kappa_{11}^2$$

$$= \frac{1}{n}\kappa_{22}+\frac{1}{n-1}\kappa_{20}\kappa_{02}+\frac{1}{n-1}\kappa_{11}^2+\kappa_{11}^2, \tag{13.15}$$

$$\mathrm{E}(K_{11}^2) = \frac{1}{N}\kappa_{22}+\frac{1}{N-1}\kappa_{20}\kappa_{02}+\frac{1}{N-1}\kappa_{11}^2+\kappa_{11}^2. \tag{13.16}$$

Hence (13.14) becomes

$$\operatorname{var} k_{11} = \left(\frac{1}{n}-\frac{1}{N}\right)\kappa_{22}+\left(\frac{1}{n-1}-\frac{1}{N-1}\right)\kappa_{20}\kappa_{02}+\left(\frac{1}{n-1}-\frac{1}{N-1}\right)\kappa_{11}^2,$$

which reduces to

$$\frac{\operatorname{var} k_{11}}{N-n}-\frac{1}{nN}\kappa_{22} = \frac{1}{(n-1)(N-1)}\kappa_{20}\kappa_{02}+\frac{1}{(n-1)(N-1)}\kappa_{11}^2. \tag{13.17}$$

We now require to eliminate the non-linear terms on the right. From (13.16) we have

$$\mathrm{E}(K_{11}^2)-\frac{1}{N}\kappa_{22} = \frac{1}{N-1}\kappa_{20}\kappa_{02}+\frac{N}{N-1}\kappa_{11}^2. \tag{13.18}$$

From (13.6) in terms of K's, we likewise find

$$\mathrm{E}(K_{20}K_{02})-\frac{1}{N}\kappa_{22} = \kappa_{20}\kappa_{0}+{}_2\frac{2}{N-1}\kappa_{11}^2. \tag{13.19}$$

Hence, from (13.17), (13.18) and (13.19),

$$\begin{vmatrix} \frac{\operatorname{var} k_{11}}{N-n}-\frac{1}{nN}\kappa_{22} & \frac{1}{(n-1)(N-1)} & \frac{1}{(n-1)(N-1)} \\ \mathrm{E}(K_{11}^2)-\frac{1}{N}\kappa_{22} & \frac{1}{N-1} & \frac{N}{N-1} \\ \mathrm{E}(K_{20}K_{02})-\frac{1}{N}\kappa_{22} & 1 & \frac{2}{N-1} \end{vmatrix} = 0. \tag{13.20}$$

This reduces to the expression

$$\operatorname{var} k_{11} = \frac{N-n}{(n-1)(N+1)(N-2)}\left\{\frac{N-2}{Nn}(Nn-N-n-1)K_{22}\right.$$

$$\left.+(N-1)K_{20}K_{02}+(N-3)K_{11}^2\right\}. \tag{13.21}$$

In deriving this we have, in accordance with the Irwin–Kendall principle of **12.21**, replaced the expectations over the infinite population by those over the finite population of N; terms such as $\mathrm{E}(K_{11}^2)$ being then replaced by their constant value such as K_{11}^2. Equation (13.21) should be compared with the result of Exercise 12.11.

Proof of the combinatorial rules

13.4 We now proceed to prove the validity of the rules enunciated and exemplified in **12.12–14.** Rules 1 and 2 have already been proved in **12.12.**

As a preliminary let us define an operator ∂_p such that

$$\left.\begin{aligned} \partial_p \mu'_r &= r(r-1)\ldots(r-p+1)\mu'_{r-p}, \quad r>p \\ \partial_p \mu'_p &= p! \\ \partial_p \mu'_r &= 0, \qquad r<p, \end{aligned}\right\} \tag{13.22}$$

and
$$\partial_p(AB) = (\partial_p A)B + A(\partial_p B), \tag{13.23}$$
so that ∂ acting on a product is distributive.

In virtue of (13.23) we have

$$\begin{aligned} \partial_p(\mu'_r)^m &= m(\mu'_r)^{m-1}\partial_p\mu'_r \\ &= \frac{\partial}{\partial\mu'_r}(\mu'_r)^m\,\partial_p\mu'_r. \end{aligned}$$

It follows that if f is a polynomial function in the μ's

$$\partial_p f = \frac{\partial f}{\partial\mu'_1}\partial_p\mu'_1 + \frac{\partial f}{\partial\mu'_2}\partial_p\mu'_2 + \ldots \tag{13.24}$$

and this also holds if f can be expanded in a series of polynomials in the μ's.

Now consider the expression defining the cumulants in terms of the moments, (3.30),

$$\exp\left(\kappa_1 t + \ldots + \kappa_p\frac{t^p}{p!} + \ldots\right) = 1 + \mu'_1 t + \ldots \frac{\mu'_p t^p}{p!} + \ldots$$

On operating on both sides by ∂_p there results

$$\begin{aligned} \exp\left(\kappa_1 t + \ldots + \kappa_p\frac{t^p}{p!} + \ldots\right)\left(\partial_p\kappa_1 t + \ldots + \partial_p\kappa_p\frac{t^p}{p!} + \ldots\right) &= t^p + \mu'_1 t^{p+1} + \ldots \\ &= t^p\left(1 + \mu'_1 t + \ldots + \mu'_p\frac{t^p}{p!} + \ldots\right), \end{aligned}$$

and hence

$$\partial_p\kappa_1 t + \ldots + \partial_p\kappa_p\frac{t^p}{p!} + \ldots = t^p.$$

This is an identity in t and hence

$$\left.\begin{aligned} \partial_p\kappa_p &= p! \\ \partial_q\kappa_p &= 0, \qquad q \neq p \end{aligned}\right\}. \tag{13.25}$$

For example,

$$\begin{aligned} \kappa_4 &= \mu'_4 - 4\mu'_3\mu'_1 - 3\mu'^2_2 + 12\mu'_2\mu'^2_1 - 6\mu'^4_1. \\ \partial_1\kappa_4 &= 4\mu'_3 - 4\mu'_3 - 12\mu'_2\mu'_1 - 12\mu'_2\mu'_1 + 24\mu'^3_1 + 24\mu'_2\mu'_1 - 24\mu'^3_1 \\ &= 0. \\ \partial_2\kappa_4 &= 12\mu'_2 - 24\mu'^2_1 - 12\mu'_2 + 24\mu'^2_1 \\ &= 0. \\ \partial_3\kappa_4 &= 24\mu'_1 - 24\mu'_1 \\ &= 0. \\ \partial_4\kappa_4 &= 4!. \end{aligned}$$

13.5 Now in accordance with Rule 1, which we established in **12.12**, $\kappa(a_1^{\alpha_1} \ldots a_s^{\alpha_s})$ and hence $\mu(a_1^{\alpha_1} \ldots a_s^{\alpha_s})$ may be expressed in terms of parent κ's by an equation of the form

$$\mu(a_1^{\alpha_1} a_2^{\alpha_2} \ldots) = \Sigma \{A(\kappa_{b_1}^{\beta_1} \kappa_{b_2}^{\beta_2} \ldots)\} \qquad (13.26)$$

where A is a factor which it is our object to find. To save printing we omit the primes on the μ's for the rest of this proof. Operate on both sides of (13.26) by $(\partial_{b_1}^{\beta_1} \partial_{b_2}^{\beta_2} \ldots)$. Every term on the right is annihilated except that in $(\kappa_{b_1}^{\beta_1} \kappa_{b_2}^{\beta_2} \ldots)$ and we have

$$A.(b_1!)^{\beta_1}(b_2!)^{\beta_2} \ldots \beta_1!\beta_2! \ldots = (\partial_{b_1}^{\beta_1} \partial_{b_2}^{\beta_2} \ldots)\mu(a_1^{\alpha_1} a_2^{\alpha_2} \ldots). \qquad (13.27)$$

We now consider an operator θ_p, analogous to ∂_p, which, when acting on a power of x (of any suffix), reduces the exponent by p and multiplies by $r(r-1) \ldots (r-p+1)$; and we will suppose the operator to be distributive.(*) Regarding $\mu(a_1^{\alpha_1} a_2^{\alpha_2} \ldots)$ as the mean value of $(k_{a_1}^{\alpha_1} k_{a_2}^{\alpha_2} \ldots)$ we see that the result of operating by the ∂'s on the mean value expressed in terms of μ's is the same as that given by taking the mean value of the operation of the θ's. But this latter operation results in a constant, which is equal to its mean value; and we thus have

$$A = \frac{(\theta_{b_1}^{\beta_1} \theta_{b_2}^{\beta_2} \ldots)}{(b_1!)^{\beta_1}(b_2!)^{\beta_2} \ldots \beta_1!\beta_2! \ldots}(k_{a_1}^{\alpha_1} k_{a_2}^{\alpha_2} \ldots). \qquad (13.28)$$

Our rules are concerned with the evaluation of this operation.

13.6 Consider now a completed array of type (12.32). A little reflection will show that there is one such array for every term in (13.28) which does not vanish by operation, and that every term in (13.28) will have its corresponding completed array. The numbers in the body of the array are the powers of x occurring in the k-product; added horizontally they compose the orders of the operators; added vertically they compose the orders of the corresponding k's. A completed array is, so to speak, a chart of part of the operation; and the whole operation is the sum of all possible completed arrays.

The operation (13.28) gives us the coefficients in $\mu(a_1^{\alpha_1} a_2^{\alpha_2} \ldots)$, but we wish to find those in the corresponding $\kappa(a_1^{\alpha_1} a_2^{\alpha_2} \ldots)$. The necessary allowance is made by Rule 3, which we now prove; that is, the coefficient of $(\kappa_{b_1}^{\beta_1} \kappa_{b_2}^{\beta_2} \ldots)$ in $\kappa(a_1^{\alpha_1} a_2^{\alpha_2} \ldots)$ is given by all completed arrays, *ignoring those which are resolvable into separate blocks each confined to separate rows and columns.*

Referring to equation (12.31), expressing the relation between multivariate moments and cumulants, we see that $\kappa(a_1^{\alpha_1} a_2^{\alpha_2} \ldots)$ is the sum of terms composed of products of one, two, three . . . multivariate moments. The first term is $\mu(a_1^{\alpha_1} a_2^{\alpha_2} \ldots)$ itself. Consider a two-part term such as $\mu(a_1^{\alpha_1'} a_2^{\alpha_2'} \ldots)\mu(a_1^{\alpha_1''} a_2^{\alpha_2''} \ldots)$, where $\alpha_1' + \alpha_1'' = \alpha_1$, etc. Its coefficient in the expansion on the right-hand side of (12.31) is

$$-\tfrac{1}{2}.\frac{2!}{1!1!}\frac{t_{a_1}^{\alpha_1'}}{\alpha_1'!\alpha_1''!}\frac{t_{a_2}^{\alpha_2'}}{\alpha_2'!\alpha_2''!} \ldots$$

(*) θ_p may be regarded as equivalent to $\left(\frac{\partial^p}{\partial x_1^p} + \frac{\partial^p}{\partial x_2^p} + \ldots + \frac{\partial^p}{\partial x_n^p}\right)$.

and hence the coefficient with which it appears in the formula for $\kappa(a_1^{\alpha_1} a_2^{\alpha_2}\dots)$ is

$$-\frac{\alpha_1!}{\alpha_1'!\,\alpha_1''!}\,\frac{\alpha_2!}{\alpha_2'!\,\alpha_2''!}\dots = -\binom{\alpha_1}{\alpha_1'}\binom{\alpha_2}{\alpha_2'}\dots \tag{13.29}$$

Now $\mu(a_1^{\alpha'_1} a_2^{\alpha'_2}\dots)$ will itself have an array of type (12.32) with column totals $(a_1^{\alpha'_1} a_2^{\alpha'_2}\dots)$ and row totals, say $(b_1^{\beta'_1} b_2^{\beta'_2}\dots)$; and similarly for $\mu(a_1^{\alpha''_1} a_2^{\alpha''_2}\dots)$. Provided that $\beta_1'+\beta_1'' = \beta_1$ these arrays will correspond to terms in the κ's which, when multiplied, will give a term in $(\kappa_{b_1}^{\beta_1}\kappa_{b_2}^{\beta_2}\dots)$. Thus the product of these terms may be considered as an array of type (12.32) with column totals $(a_1^{\alpha_1} a_2^{\alpha_2}\dots)$ and row totals $(b_1^{\beta_1} b_2^{\beta_2}\dots)$ and with the body of the table resolvable into two separate blocks. Since there are α_1 columns of total a_1, there will be $\binom{\alpha_1}{\alpha_1'}\binom{\alpha_2}{\alpha_2'}$ products of this type in the expression which gives $\mu(a_1^{\alpha_1} a_2^{\alpha_2}\dots)$. This factor is the same as (13.29) but of opposite sign. Hence, if we ignore the separate two-part blocks in the array for μ we shall have allowed for the products of two moments which must be subtracted from μ to give κ.

Now some of these separate blocks will themselves be separable into two blocks, and in subtracting them all from $\mu(a_1^{\alpha_1} a_2^{\alpha_2}\dots)$ we subtract too much. For example, if there are three separate blocks, L, M, N, we shall, by considering L and $(M+N)$ as two blocks, have subtracted L, M, N. We shall have done the same by considering M and $(L+N)$, and N and $(L+M)$ as two blocks. That is, we have subtracted $2L$, $2M$, $2N$ too much. We must restore these blocks to the array for μ. Such additions, summed over all blocks of three, will be found to equal the terms in the expansion of (12.31) which result from the product of three moments.

In restoring these blocks we restore too many of the cases where there are four separate blocks. These must be subtracted again, and correspond to the negative term in (12.31) involving the product of four moments. Proceeding in this way we establish Rule 3.(*)

13.7 Now we proceed to Rules 4, 5 and 6, which are the fundamental rules of the whole process. Consider again the array of type (12.32), to fix the ideas, say,

$$\begin{array}{ccc|c} 2 & 3 & 1 & 6 \\ 1 & 1 & . & 2 \\ 1 & . & 1 & 2 \\ \hline 4 & 4 & 2 & 10 \end{array} \tag{13.30}$$

This array will represent a number of terms in the operation, each of which consists of the operation of θ_6 on a term $x^2.x^3.x$ (the first row), θ_2 on $x.x$ (the second row), and so on. Provided that the suffixes of the x's in any row are alike, every suffix of the x's will provide a term, for k_p contains terms with every distribution of powers (adding to p) and suffixes. There will, for instance, be terms of the following kind :—

$$\begin{array}{ccc} x_1^2 & x_1^3 & x_1 \\ x_2 & x_2 & . \\ x_3 & . & x_3 \end{array} \qquad \begin{array}{ccc} x_1^2 & x_1^3 & x_1 \\ x_1 & x_1 & . \\ x_2 & . & x_2, \end{array} \qquad \begin{array}{ccc} x_1^2 & x_1^3 & x_1 \\ x_1 & x_1 & . \\ x_1 & . & x_1. \end{array}$$

(*) Kaplan (1952) has adduced a statistical argument in support of Rule 3.

In fact, for any completed array, we have terms in which

(i) all the x's have the same suffix (n in number, one for each suffix),
(ii) all the x's but one row have the same suffix ($n(n-1)$ in number),
(iii) all the x's but two rows have the same suffix and the remaining two are the same ($n(n-1)$ in number),

and so on. These cases correspond to the various separations dealt with in Rule 5.

Now in case (i) the term in any column arises from the term in x^p in k_p and (apart from numerical factors which are considered presently) is n^{-1}, from equation (12.25). Hence any column which contains an entry contributes a factor n^{-1} and the total function of n arising from case (i) is the product of n and of (n^{-1}) to the power of the number of columns containing a non-zero entry.

Similarly in cases (ii) and (iii) the n-function for each separation is the product of $n(n-1)$ and, for each column, a factor in n^{-1} or $\frac{-1}{n(n-1)}$ according as the column contains non-zero entries in one or in both parts of the separation; and so on.

This explains the origin of the pattern function as described in Rule 6. But in order to establish that rule completely (and incidentally to establish Rules 4 and 5) we have to show that the numerical coefficients arising from each separation are the same. When this is done the validity of Rule 6 is demonstrated, for the separate contributions in n may be added together to give the pattern function and the whole multiplied by the numerical coefficient.

θ_1 may be considered as the operation of picking out an x from the operand in all possible ways and replacing it by unity. Similarly $\frac{\theta_p}{p!}$ may be regarded as picking out p x's with the same suffix and replacing them by unity. It is thus evident that operating on a k-product by a θ-product $\frac{\theta_{b_1}\theta_{b_2}\ldots}{b_1!b_2!\ldots}$ of the same degree will yield a result which is the number of ways in which sets of x's can be picked out of the k-product so that each set contains b_1 of one suffix, b_2 of a second suffix (which may be the same as the first), and so on.

Now consider the operation (13.28) in which the k's are expressed in the simplified form (12.26). The operations θ being distributive, we shall emerge from the operation with a sum of terms comprising all the possible ways in which the individual x's can be picked out of the k-product such that the row and column totals of the two-way array are satisfied. Consider the sets corresponding to a particular array, such as (13.30). The contribution to the total will consist of the ways of picking out individuals such that

(i) from the individuals in the first k_4 are chosen four in the partition (2, 1, 1),
(ii) from the second k_4 are chosen four in the partition (3, 1),
(iii) from the k_2 are chosen two in the partition (1, 1),
(iv) these are associated in all possible ways such that individuals in a row arise from the same suffix.

On consideration it will be seen that the total number of ways of doing this is the number of ways of allocating the individuals from column totals as required by Rule 5; *and this is true whether sets of rows have the same suffix or not.*

Rules 5 and 6, and hence Rule 4, follow at once.

13.8 The remaining rules are ancillary.

Rule 7 follows from Rule 2. In fact, the pattern function is independent of the numbers composing the array, and the pattern with a row containing one element can therefore form the skeleton of an array in which that element is unity; and this would entail the appearance of κ_1, which by Rule 2 is impossible.

Rule 8 follows from Rule 6. The column containing the single element appears in just one separate of all the separations, and the contributions to the pattern function are thus all multiplied by n^{-1} owing to its presence.

Rule 10 follows from Rule 8. The addition of a unit part is equivalent to the addition of an extra column containing unity. This multiplies all pattern functions by n^{-1}, leaves numerical coefficients unchanged and increases the suffix of every κ according to the row in which the unit appears.

13.9 It only remains to prove Rule 9. Note that any pattern function can be evaluated linearly in terms of the functions of the pattern obtained by omitting one of the columns. For example, consider the right-hand column of

$$\begin{array}{cccc} \cdot & \times & \times & \times \\ \times & \cdot & \times & \times \\ \times & \times & \cdot & \cdot \\ \times & \times & \cdot & \cdot \end{array} \qquad (13.31)$$

and the contributions to the pattern function from it. The 15 separations which are possible with four rows can be divided into two classes, that in which the two rows in the fourth column lie in the same separate and that in which they do not. In separations of the first type the contributions from the first three columns will be the contributions of all separations of

$$\begin{array}{ccc} \times & \times & \times \\ \times & \times & \cdot \\ \times & \times & \cdot \end{array} \qquad (13.32)$$

in which the first two rows are amalgamated. Considering the function of the first three rows

$$\begin{array}{ccc} \cdot & \times & \times \\ \times & \cdot & \times \\ \times & \times & \cdot \\ \times & \times & \cdot \end{array} \qquad (13.33)$$

in which amalgamation has not taken place, we see that the contribution consists of all contributions which do not occur in the first. Calling the first A and the second B, we see that the contribution is

$$\frac{1}{n}A - \frac{1}{n(n-1)}(B-A) = \frac{1}{n-1}A - \frac{1}{n(n-1)}B,$$

i.e. a linear function of the derived patterns A and B. The proof of the general result follows exactly the same lines.

Now if a pattern may be divided into two groups connected only by a single column we can reduce it step by step by omitting the other columns. We end up with this single column, and the pattern function of this column must vanish; for the column total a corresponds to k_a, whose mean value the one-column array expresses, and since by definition this mean value is κ_a no composite terms such as would be given by two rows or more can appear.

13.10 The adaptation of the argument to the bivariate case is fairly straightforward. The only rule requiring examination is the one giving the numerical coefficients attaching to the pattern function; and on examination it should be clear, from the structure of the bivariate k-statistics, that the procedure sketched in **13.2** applies. The use of an annihilating operator

$$\partial_{pq}\mu_{rs} = r^{[p]}s^{[q]}\mu_{r-p,\,s-q} \tag{13.34}$$

gives the required result.

We may also state without proof that similar methods apply to multivariate statistics of higher order than the bivariate. For trivariate statistics, for example, we should have to consider partitions into three-part members. Only the simpler results in such cases are required in practice and they can usually be written down from the univariate or bivariate cases from considerations of the type which we develop in the following paragraphs.

The above proofs of the rules are due to Kendall (1940b). James (1958) gives alternative proofs of the combinatorial rules, covering the general multivariate case, and also including results on moments as well as k-statistics.

James and Mayne (1962) give a combinatorial method of obtaining the joint cumulants of any number of functions of any number of variables with finite cumulants. If each variable has its κ_r of order $n^{-(r-1)}$ they give the joint cumulants of order $\leqslant 6$ explicitly to terms of order n^{-4} or better.

Summary of results

13.11 Let us review the results of this and the previous chapter. We have seen that, by the use of simple algebraic methods, we can obtain as many moments or cumulants as we please of any statistic which is expressible as a symmetric function of the observations. We have also noted that the working is greatly simplified by the use of k-statistics and cumulants.

It is reasonable to inquire how this simplification comes about. What are the structural properties of k-statistics and cumulants which give them these advantages, and do they represent the limit of achievement in sampling simplicity? The answer to the last question seems to be affirmative and it is instructive to consider why this should be so.

13.12 The definition of univariate cumulants in terms of moments by the identity in t

$$\exp\{\Sigma \kappa_p t^p/p!\} = \Sigma \mu_p' t^p/p! \tag{13.35}$$

tends to obscure the essential structure of the cumulant-moment relationship. Let us, instead, consider the relation between multivariate cumulants and moments, as defined formally by (12.31). To fix the ideas, suppose we have four variates and require their product-moment μ_{1111}'. From (12.31) we find, on examining the coefficient in $t_1 t_2 t_3 t_4$,

$$\begin{aligned}\mu_{1111}' = {} & \kappa_{1111}+(\kappa_{1110}\kappa_{0001}+\kappa_{1101}\kappa_{0010}+\kappa_{1011}\kappa_{0100}+\kappa_{1000}\kappa_{1000})\\ & +(\kappa_{1100}\kappa_{0011}+\kappa_{1010}\kappa_{0101}+\kappa_{0110}\kappa_{1001})\\ & +(\kappa_{1100}\kappa_{0010}\kappa_{0001}+\kappa_{1010}\kappa_{0100}\kappa_{0001}+\kappa_{1001}\kappa_{0100}\kappa_{0010}+\kappa_{0110}\kappa_{1000}\kappa_{0001}\\ & +\kappa_{0101}\kappa_{1000}\kappa_{0010}+\kappa_{0011}\kappa_{1000}\kappa_{0100})\\ & +\kappa_{1000}\kappa_{0100}\kappa_{0010}\kappa_{0001}.\end{aligned} \tag{13.36}$$

Such formulae exhibit the essential simplicity of the relationship. In the expression of μ's in terms of κ's every possibility on the right occurs exactly once. When we invert the relationship to give κ's in terms of μ's we have, analogously to (13.36),

$$\begin{aligned}\kappa_{1111} = {} & \mu_{1111}'-(\mu_{1110}'\mu_{0001}'+\mu_{1101}'\mu_{0010}'+\mu_{1011}'\mu_{0100}'+\mu_{0111}'\mu_{1000}')\\ & -(\mu_{1100}'\mu_{0011}'+\mu_{1010}'\mu_{0101}'+\mu_{0110}'\mu_{1001}')\\ & +2(\mu_{1100}'\mu_{0010}'\mu_{0001}'+\text{ etc.})-6\mu_{1000}'\mu_{0100}'\mu_{0010}'\mu_{0001}'.\end{aligned} \tag{13.37}$$

This type of formula differs from (13.36) only in the factor $(\rho-1)!(-1)^{\rho-1}$ attached to sets which contain ρ parts.

13.13 In particular, these formulae remain valid if two or more of the four variates are identical. If, for example, the first and second are identical, expressions such as κ_{1111} are written κ_{211} and become trivariate; and if in addition the third and fourth are identical the cumulant becomes κ_{22} and is bivariate. In the extreme case when all are identical it becomes κ_4. Thus we can obtain all the relations of order 4 for any number of variates by condensing (13.36) or (13.37) by amalgamating suffixes. Certain terms which are distinct in the originals then become the same and can be added. For example, from (13.36) we find

$$\begin{aligned}\mu_{22}' = {} & \kappa_{22}+2\kappa_{21}\kappa_{01}+2\kappa_{12}\kappa_{10}+\kappa_{20}\kappa_{02}+2\kappa_{11}^2+\kappa_{20}\kappa_{01}^2+\kappa_{02}\kappa_{10}^2+4\kappa_{11}\kappa_{10}\kappa_{01}\\ & +\kappa_{10}^2\kappa_{01}^2.\end{aligned} \tag{13.38}$$

If we now add the suffixes we reach the known univariate form

$$\mu_4' = \kappa_4+4\kappa_3\kappa_1+3\kappa_2^2+6\kappa_2\kappa_1^2+\kappa_1^4. \tag{13.39}$$

Similarly, in the finite population case, (13.21), on adding its suffixes, reduces to the result of Exercise 12.11. The origin of the numerical coefficients in (13.39) is now clear. The symbolic methods of deriving multivariate from univariate formulae (cf. **3.28**) are merely convenient shorthand ways of working the process of amalgamation in reverse, a procedure which is possible in virtue of the symmetry of the terms in (13.36) among the suffixes.

13.14 It thus appears that the formal definitions of type (12.31) and (13.35) involving the exponential function derive their utility from the enumerative algebraic property of the exponential as illustrated in (13.36); not, for instance, from the analytical properties of the exponential such as its being equal to its own derivative or its being the inverse of the logarithmic function. The numerical coefficients appearing in our relations between cumulants and moments are convenient summaries of relations which are structurally much simpler than they appear. Similar considerations apply to k-statistics. Apart from factors in $(-1)^{\rho-1}(\rho-1)!/n^{[\rho]}$ they are, as we have noted in **12.8**, sums of products of the contributory variates x taken in all possible ways once and only once.

13.15 We can now summarize our formulae. For instance (13.36) can be written

$$\mu'_{ijkl} = \kappa_{ijkl}+\overset{4}{\Sigma}\kappa_i\kappa_{jkl}+\overset{3}{\Sigma}\kappa_{ij}\kappa_{kl}+\overset{6}{\Sigma}\kappa_i\kappa_j\kappa_{kl}+\kappa_i\kappa_j\kappa_k\kappa_l, \tag{13.40}$$

where the summations occur over all ways of grouping the subscripts, which are shown in the numbers above the summation signs. (There is a slight change of notation here: μ'_i refers to the mean value of the ith variate, not the mean of the ith power of some variate x.)

Likewise, if $s_{ij\ldots l}$ refers to the products $x_i x_j \ldots x_l$ summed over the sample, we have for the k-statistics

$$\left.\begin{aligned} k_i &= s_i/n \\ k_{ij} &= (n s_{ij}-s_i s_j)/n^{[2]} \\ k_{ijk} &= (n^2 s_{ijk}-n\overset{3}{\Sigma} s_i s_{jk}+2s_i s_j s_k)/n^{[3]} \end{aligned}\right\} \tag{13.41}$$

and so on.

For example, if the ith and jth variates are identical we find, in the more familiar notation,

$$k_2 = (n s_2-s_1^2)/\{n(n-1)\} \tag{13.42}$$

whereas the bivariate k_{11} is given by

$$k_{11} = (n s_{11}-s_{10}s_{01})/\{n(n-1)\}. \tag{13.43}$$

If we require the bivariate k_{21} we amalgamate i and j and obtain from the third formula of (13.41)

$$k_{21} = \{n^2 s_{21}-2n s_{10}s_{11}-n s_{20}s_{01}+2s_{10}^2 s_{01}\}/n^{[3]}$$

as given in (13.2).

13.16 This notation, which was suggested by Kaplan (1952) as a tensor notation, enables us to summarize the multivariate formulae in little more space than is required for the univariate results. For instance, for four variates we have for the covariance of two pairs

$$\begin{aligned} \kappa(ab, ij) &= \mathrm{E}\,\{(k_{ab}-\kappa_{ab})(k_{ij}-\kappa_{ij})\} \\ &= \frac{1}{n}\kappa_{abij}+\frac{1}{n-1}(\kappa_{ai}\kappa_{bj}+\kappa_{aj}\kappa_{bi}). \end{aligned} \tag{13.44}$$

This embodies seven types of formulae. If all variates are the same (a, b, i, j referring to one variate) we have

$$\operatorname{var} k_2 = \kappa(2^2) = \frac{1}{n}\kappa_4 + \frac{2}{n-1}\kappa_2^2. \tag{13.45}$$

If variate a is the same as i and variate b the same as j, we have

$$\operatorname{var} k_{11} = \kappa\begin{pmatrix}1 & 1\\ 1 & 1\end{pmatrix} = \frac{1}{n}\kappa_{22} + \frac{1}{n-1}\kappa_{20}\kappa_{02} + \frac{1}{n-1}\kappa_{11}^2. \tag{13.46}$$

If a and b are the same, i and j the same, we find

$$\operatorname{cov}(k_{20}, k_{02}) = \kappa\begin{pmatrix}2 & 0\\ 0 & 2\end{pmatrix} = \frac{1}{n}\kappa_{22} + \frac{2}{n-1}\kappa_{11}^2. \tag{13.47}$$

If a, b and i are the same and j is different,

$$\operatorname{cov}(k_{20}, k_{11}) = \kappa\begin{pmatrix}2 & 1\\ 0 & 1\end{pmatrix} = \frac{1}{n}\kappa_{31} + \frac{2}{n-1}\kappa_{11}\kappa_{20}. \tag{13.48}$$

If a and b are the same, i, j are different,

$$\operatorname{cov}(k_{200}, k_{011}) = \kappa\begin{pmatrix}2 & 0\\ 0 & 1\\ 0 & 1\end{pmatrix} = \frac{1}{n}\kappa_{211} + \frac{2}{n-1}\kappa_{110}\kappa_{101}. \tag{13.49}$$

If a and i are the same, b, j are different,

$$\operatorname{cov}(k_{110}, k_{101}) = \kappa\begin{pmatrix}1 & 1\\ 1 & 0\\ 0 & 1\end{pmatrix} = \frac{1}{n}\kappa_{211} + \frac{1}{n-1}(\kappa_{200}\kappa_{011} + \kappa_{110}\kappa_{101}). \tag{13.50}$$

Finally, if all four are different,

$$\operatorname{cov}(k_{1100}, k_{0011}) = \kappa\begin{pmatrix}1 & 0\\ 1 & 0\\ 0 & 1\\ 0 & 1\end{pmatrix} = \frac{1}{n}\kappa_{1111} + \frac{1}{n-1}(\kappa_{1010}\kappa_{0101} + \kappa_{1001}\kappa_{0110}). \tag{13.51}$$

13.17 The following formulae summarize further multivariate results.

$$\kappa(ab, ijk) = \kappa_{abijk}/n + \sum^{6} \kappa_{ai}\kappa_{bjk}/(n-1). \tag{13.52}$$

$$\kappa(ab, ijkl) = \kappa_{abijkl}/n + \left(\sum^{8} \kappa_{ai}\kappa_{bjkl} + \sum^{6} \kappa_{aij}\kappa_{bkl}\right)/(n-1). \tag{13.53}$$

$$\begin{aligned}\kappa(ab, ij, pq) = {} & \kappa_{abijpq}/n^2 + \sum^{12} \kappa_{abip}\kappa_{jq}/\{n(n-1)\} + \sum^{4} \kappa_{aip}\kappa_{bjq}(n-)/\{n(n-1)^2\} \\ & + \sum^{8} \kappa_{ai}\kappa_{bp}\kappa_{jq}/(n-1)^2.\end{aligned} \tag{13.54}$$

$$\begin{aligned}\kappa(abc, ijk) = {} & \kappa_{abcijk}/n + \left(\sum^{9} \kappa_{ai}\kappa_{bcjk} + \sum^{9} \kappa_{abi}\kappa_{cjk}\right)/(n-12) \\ & + \sum^{6} \kappa_{ai}\kappa_{bj}\kappa_{ck}\, n/\{(n-1)(n-2)\}.\end{aligned} \tag{13.55}$$

$$\begin{aligned}\kappa(abc, ijkl) = {} & \kappa_{abcijkl}/n + \sum^{12} \kappa_{ai}\kappa_{bcjkl}/(n-1) + \left(\sum^{12} \kappa_{abi}\kappa_{cjkl} + \sum^{18} \kappa_{aij}\kappa_{bckl}\right)/(n-1) \\ & + \sum^{36} \kappa_{ai}\kappa_{bj}\kappa_{ckl}\, n/\{(n-1)(n-2)\}.\end{aligned} \tag{13.56}$$

$$\kappa(ab, ij, pq, uv) = \kappa_{abijpquv}/n + \sum^{24} \kappa_{ai}\kappa_{bjpquv}/\{n^2(n-1)\} + \sum^{32} \kappa_{aip}\kappa_{bjquv}(n-2)/\{n^2(n-1)^2\}$$
$$+\sum^{8} \kappa_{aipu}\kappa_{bjqv}(n^2-3n+3)/\{n^2(n-1)^3\} + 3\sum^{24} \kappa_{abpu}\kappa_{ijqv}/\{n^2(n-1)\}$$
$$+(\sum^{96} \kappa_{ai}\kappa_{bp}\kappa_{jquv} + \sum^{48} \kappa_{ai}\kappa_{pu}\kappa_{bjqv})/\{n(n-1)^2\}$$
$$+\sum^{96}\kappa_{ai}\kappa_{bpu}\kappa_{jqv}(n-2)/\{n(n-1)^3\} + \sum^{48} \kappa_{bi}\kappa_{jp}\kappa_{qu}\kappa_{va}/(n-1)^3. \tag{13.57}$$

$$\kappa(abcd, ijkl) = \kappa_{abcdijkl}/n + \sum^{16} \kappa_{ai}\kappa_{bcdjkl}/(n-1) + \sum^{48} \kappa_{abi}\kappa_{cdjkl}/(n-1)$$
$$+\sum^{72} \kappa_{ai}\kappa_{bj}\kappa_{cdkl}\, n/\{(n-1)(n-2)\} + (\sum^{16} \kappa_{aijk}\kappa_{bcdl}$$
$$+\sum^{18} \kappa_{abij}\kappa_{cdkl})/(n-1) + \sum^{144} \kappa_{ai}\kappa_{bcj}\kappa_{dkl}\, n/\{(n-1)(n-2)\}$$
$$+\sum^{24} \kappa_{ai}\kappa_{bj}\kappa_{ck}\kappa_{dl}\, n(n+1)/\{(n-1)(n-2)(n-3)\}. \tag{13.58}$$

These formulae embrace 41 formulae given by Cook (1951) and a number of others besides, but as Kaplan remarks, some care is required in deriving particular cases.

Example 13.3

For the variance of k_{11} from an infinite population we have, as in Example 13.2,

$$\operatorname{var} k_{11} = \frac{1}{n}\kappa_{22} + \frac{1}{n-1}\kappa_{20}\kappa_{02} + \frac{1}{n-1}\kappa_{11}^2.$$

Let us specialize this to the case of a normal parent in which the variances are unity. We then find

$$\operatorname{var} k_{11} = \frac{1}{n-1}(1+\rho^2). \tag{13.59}$$

Let us find the third and fourth cumulants of k_{11}.

The third cumulant is given by (13.54) with a, i, p the same and b, j, q the same. We find

$$\kappa\begin{pmatrix}1 & 1 & 1\\ 1 & 1 & 1\end{pmatrix} = \frac{1}{n^2}\kappa_{33} + \frac{1}{n(n-1)}\{6\kappa_{22}\kappa_{11} + 3\kappa_{20}\kappa_{13} + 3\kappa_{31}\kappa_{02}\}$$
$$+\frac{n-2}{n(n-1)^2}\{3\kappa_{21}\kappa_{12} + \kappa_{30}\kappa_{03}\} + \frac{1}{(n-1)^2}\{2\kappa_{11}^3 + 6\kappa_{11}\kappa_{20}\kappa_{02}\}. \tag{13.60}$$

For the normal case the only surviving terms yield

$$\kappa_3(k_{11}) = \frac{1}{(n-1)^2}(2\rho^3 + 6\rho). \tag{13.61}$$

Likewise we find from (13.57), taking a, i, p, u the same and b, j, q, v the same, the surviving terms, which can only arise from the concluding term,

$$\kappa_4(k_{11}) = \frac{1}{(n-1)^3}\{36\rho^2 + 6 + 6\rho^4\}. \tag{13.62}$$

Thus, for instance, the kurtosis of the distribution of k_{11} is given by (13.59) and (13.62) as

$$\beta_2 - 3 = \frac{6}{n-1}\left(1+\frac{4\rho^2}{(1+\rho^2)^2}\right). \qquad (13.63)$$

Tables of symmetric functions

13.18 *Symmetric Function and Allied Tables*, by F. N. David, M. G. Kendall, and D. E. Barton (C.U.P., 1966), contains full versions, corrected where necessary, of the univariate and bivariate tables discussed in this and the previous chapter, with many others. Mikhail (1968) gives trivariate tables up to weight 6.

EXERCISES

13.1 Show that the correlation between k_{30} and k_{03} in samples of n from a bivariate normal population is ρ^3, where ρ is the correlation parameter of the population.

13.2 Show generally that, in the circumstances of Exercise 13.1, the correlation between k_{ro} and k_{or} is ρ^r.

13.3 A four-variate normal population has variates x_1, x_2, x_3, x_4 with unit variances and the correlation of x_i and x_j is ρ_{ij}. Show that the correlation between the sample covariance of x_1, x_2 and that of x_3, x_4 is given by

$$\frac{\rho_{13}\rho_{24}+\rho_{14}\rho_{23}}{\sqrt{(1+\rho_{12}^2)(1+\rho_{34}^2)}}.$$

13.4 Derive the formulae (13.5–13) from the general formulae in **13.16–17.**

13.5 Use the methods of this chapter to derive the large-sample result for the variance of the correlation coefficient in bivariate normal samples

$$\text{var } r = (1-\rho^2)^2/n.$$

13.6 If x is a random variable with finite cumulants of all orders satisfying $\kappa_r = O(n^{-r+1})$ show that the transformed variate $y = f(x)$ has cumulants which also satisfy this condition. (This does not preclude the cumulants from being of lower order, i.e. $o(n^{-r+1})$.)

(James, 1955)

13.7 Show that for a bivariate normal population k_{tu} and k_{vw} have zero covariance unless $t+u = v+w$.

(Wishart, 1929)

13.8 Show that for a bivariate normal population

$$dF = \frac{1}{2\pi\sigma_1\sigma_2(1-\rho^2)^{\frac{1}{2}}}\exp\left\{\frac{-1}{2(1-\rho^2)}\left(\frac{x_1^2}{\sigma_1^2}-\frac{2\rho x_1 x_2}{\sigma_1\sigma_2}+\frac{x_2^2}{\sigma_2^2}\right)\right\}dx_1\,dx_2,$$

$$\text{var}(k_{tu}) = \kappa\begin{pmatrix} t & t \\ u & u\end{pmatrix} = t!\,u!\sum_{j=1}^{t+u}\frac{(j-1)!}{j}\frac{\Delta^j 0^{t+u}}{n^{[j]}}\sigma_1^{2t}\sigma_2^{2u}F(-t,-u,1,\rho^2),$$

where $\Delta^j 0^k$ is the jth difference of the kth power of zero and F refers to the hypergeometric function.

(Wishart, 1929)

13.9 In the notation of **13.15**, show that

$$k_{ijkl} = \{n^2(n+1)s_{ijkl}-n(n+1)\sum^{4} s_i s_{jkl}-n(n-1)\sum^{3} s_{ij}s_{kl}+2n\sum^{6} s_i s_j s_{kl}-6s_i s_j s_k s_l\}/n^{[4]},$$

and hence verify the formulae for k_{31} and k_{22} of (13.2).

(Kaplan, 1952)

13.10 Referring to the result of **13.9**, show that for a normal parent population the effect of adding a new part 2 to $\kappa(a_1^{\alpha_1}\ldots a_s^{\alpha_s})$ is to give pattern functions $1/(n-1)$ times those of the original. Show also that the effect on the numerical coefficient in an array is to multiply by twice the number of rows in the array. Deduce that the effect of adding a new part 2 is equivalent to operating by

$$\frac{2\kappa_2^2}{n-1}\frac{d}{d\kappa_2}.$$

(Fisher and Wishart, 1931)

13.11 Use the previous exercise to establish equations (12.71) and (12.72).

13.12 In generalization of Exercise 13.10 show that for a multivariate normal parent the effect of adding a covariance k_{pq} (p, q referring to the pth and qth variates) is equivalent to operating by

$$\sum_{rs} \frac{1}{(n-1)}(\kappa_{pr}\kappa_{qs}+\kappa_{ps}\kappa_{qr})\frac{d}{d\kappa_{rs}},$$

where κ_{pq} is the covaıiance of the variates p and q.

(Fisher and Wishart, 1931)

13.13 Show that in normal samples

$$\text{var}(k_r/k_2^{\frac{1}{2}r}) \sim r!/n, \qquad r>2.$$

(Cf. the exact results for r = 3, 4 in Exercises 12.9–10.)

CHAPTER 14

ORDER-STATISTICS

14.1 In Chapters 10 and 11 we have briefly discussed some of the sampling properties of medians, quantiles and, more generally, the order-statistics of a sample of observations. It will be remembered that the rth order-statistic of a sample of n is simply the rth smallest variate-value in the sample. We shall denote it by a suffix in parentheses: $x_{(r)}$. Unlike statistics which are based on symmetric functions of the observations, the order-statistics do not lend themselves to investigation by the methods we have been using in Chapters 12 and 13. They have the distinctive property that if we take an ordinary random sample of values $x_1, x_2, \ldots, x_n$ and rearrange them in order, as $x_{(1)}, x_{(2)}, \ldots, x_{(n)}$, the values so derived are no longer independent nor identically distributed even though the original x's were so.

An expository account of many branches of the theory of order-statistics, with extensive tables, is given in the book edited by Sarhan and Greenberg (1962) and in the monograph by H. A. David (1970). The two large volumes of tables by Harter (1969) should also be consulted.

14.2 Notwithstanding lack of symmetry and independence, some remarkably simple results can be obtained. We have seen in **10.10** and **11.4** that, in samples of size n from any distribution function $F(x)$ with a continuous frequency function $f(x)$, the sampling distribution of $F(x_{(r)})$ is given by G_r, say, where

$$dG_r = \frac{\{F(x_{(r)})\}^{r-1}\{1-F(x_{(r)})\}^{n-r}\,dF(x_{(r)})}{B(r, n-r+1)} \tag{14.1}$$

is a Beta distribution. Likewise, the joint distribution of $F(x_{(r)})$ and $F(x_{(s)})$, $r<s$, is given (cf. **10.11**) by

$$dG_{r,s} = \frac{\{F(x_{(r)})\}^{r-1}\{F(x_{(s)})-F(x_{(r)})\}^{s-r-1}\{1-F(x_{(s)})\}^{n-s}\,dF(x_{(r)})\,dF(x_{(s)})}{B(r, s-r)B(s, n-s+1)}. \tag{14.2}$$

From (14.1), it follows at once that if μ_p' exists for the parent $F(x)$, it must also exist for $G_r(x)$; in the contrary case, the moment of $G_r(x)$ may or may not exist—Exercise 14.22 treats the Cauchy case.

In (14.2), $x_{(r)}$ and $x_{(s)}$ are always positively correlated, as is intuitively obvious—for a proof, see Bickel (1967) or Exercise 14.9.

The generalization to the joint distribution of several order-statistics is immediate. For instance, for $x_{(r)}, x_{(s)}, x_{(t)}, x_{(u)}$, $r<s<t<u$, we have, writing F_r for $F(x_{(r)})$, etc.,

$$dG_{r,s,t,u} = \frac{F_r^{r-1}(F_s-F_r)^{s-r-1}(F_t-F_s)^{t-s-1}(F_u-F_t)^{u-t-1}(1-F_u)^{n-u}\,dF_r\,dF_s\,dF_t\,dF_u}{B(r, s-r)B(s, t-s)B(t, u-t)B(u, n-u+1)}. \tag{14.3}$$

Distributions of the type (14.1) have been studied for various given forms of the parental distribution $F(x)$. When the parent is rectangularly distributed in the range 0 to 1, $F(x) = x$ and the distribution of $x_{(r)}$ is particularly simple, itself reducing to the Beta-distribution. A second type of parent leading to relatively simple results is the exponential $dF = \exp(-x/\sigma)\,dx/\sigma$. For others, and in particular the normal parent, the evaluation of exact results requires, in general, the quadrature of certain integrals.

Some important and interesting related results are given in Exercises 14.9 and 14.14.

The distribution function of an order-statistic

14.3 Even if $F(x)$ is discrete, the d.f. of $x_{(r)}$ is easily obtained, as H. A. David and Mishriky (1968) point out. We now suppose that $x_{(1)} \leqslant x_{(2)} \leqslant \ldots \leqslant x_{(n)}$, with equalities possible. The probability is $\binom{n}{j}\{F(x)\}^j\{1-F(x)\}^{n-j}$ that j of the n observations do not exceed a fixed value x. Thus

$$G_r(x) = \sum_{j=r}^{n}\binom{n}{j}\{F(x)\}^j\{1-F(x)\}^{n-j} \tag{14.4}$$

is the probability that at least r observations in the sample do not exceed x, i.e. that $x_{(r)} \leqslant x$. This is the d.f. of $x_{(r)}$. From (5.16), we have

$$G_r(x) = 1 - I_{1-F(x)}(n-r+1, r) = I_{F(x)}(r, n-r+1). \tag{14.5}$$

This result is completely general: we require differentiability of $F(x)$ only in order to obtain the f.f. of $x_{(r)}$, given at (14.1), from $G_r(x)$, as may be seen by writing out the Incomplete Beta Function as an integral.

For discrete variables taking non-negative integer values, Young (1970) gives recurrence relations for the frequency-generating function and the moments of the order-statistics. Vaughan and Venables (1972) generalize (14.1) to the case where the observations come from different d.f.'s.

Median and quartiles in the standardized normal case

14.4 Suppose that the sample size is odd, say $n = 2r+1$, so that the median is the $(r+1)$th order-statistic, $x_{(r+1)}$. Let us write

$$A_x = \frac{1}{\sqrt{(2\pi)}}\int_{-\infty}^{x} e^{-\frac{1}{2}t^2}\,dt.$$

Then the distribution of the median, from (14.1), is seen to be (writing A for A_x),

$$dG = n!\,A^r(1-A)^r\,dA/(r!)^2.$$

From symmetry, the expected value of the median is zero—cf. Exercise 14.7—and hence its variance is given by

$$\text{var}\, x_{(r+1)} = \int x^2\,dG = \frac{n!}{(r!)^2}\int_{-\infty}^{\infty} x^2 A^r(1-A)^r . \frac{1}{\sqrt{(2\pi)}} e^{-\frac{1}{2}x^2}\,dx. \tag{14.6}$$

We expand the term $(1-A)^r$ binomially to obtain

$$\text{var}\, x_{(r+1)} = \frac{n!}{(r!)^2}\int_{-\infty}^{\infty}\sum_{j=0}^{r}(-1)^j\binom{r}{j}A^{r+j}\frac{1}{\sqrt{(2\pi)}}x^2 e^{-\frac{1}{2}x^2}\,dx. \tag{14.7}$$

Each term may be integrated by parts, taking $xe^{-\frac{1}{2}x^2}$ for one part. We have

$$\frac{1}{\sqrt{(2\pi)}}\int_{-\infty}^{\infty} xA^{r+j}.xe^{-\frac{1}{2}x^2}\,dx = \left[\frac{-xA^{r+j}e^{-\frac{1}{2}x^2}}{\sqrt{(2\pi)}}\right]_{-\infty}^{\infty}$$
$$+\frac{1}{\sqrt{(2\pi)}}\int_{-\infty}^{\infty} e^{-\frac{1}{2}x^2}\left\{A^{r+j}+(r+j)A^{r+j-1}\frac{1}{\sqrt{(2\pi)}}xe^{-\frac{1}{2}x^2}\right\}dx. \quad (14.8)$$

The first term on the right vanishes. With a further integration by parts on the third term we find, for the integral on the left of (14.8),

$$\frac{1}{\sqrt{(2\pi)}}\int_{-\infty}^{\infty} e^{-\frac{1}{2}x^2}A^{r+j}\,dx+\frac{(r+j)(r+j-1)}{4\pi}\frac{1}{\sqrt{(2\pi)}}\int_{-\infty}^{\infty} A^{r+j-2}\exp(-\tfrac{3}{2}x^2)\,dx. \quad (14.9)$$

In (14.9), the first integral is simply $1/(r+j+1)$ and on substitution in (14.7) we find

$$\operatorname{var} x_{(r+1)} = \frac{n!}{(r!)^2}\sum_{j=0}^{r}(-1)^j\binom{r}{j}\frac{1}{r+j+1}$$
$$+\frac{n!}{(r!)^2 4\pi}\sum_{j=0}^{r}(-1)^j\binom{r}{j}(r+j)(r+j-1)\frac{1}{\sqrt{(2\pi)}}\int_{-\infty}^{\infty}A^{r+j-2}\exp(-\tfrac{3}{2}x^2)\,dx. \quad (14.10)$$

The first summation on the right of (14.10) yields unity, for

$$\sum_{j=0}^{r}(-1)^j\binom{r}{j}\frac{1}{r+j+1} = \int_0^1 \Sigma(-1)^j\binom{r}{j}t^{r+j}\,dt$$
$$= \int_0^1 t^r(1-t)^r\,dt = B(r+1,r+1). \quad (14.11)$$

Thus we have, from (14.10) and (14.11),

$$\operatorname{var} x_{(r+1)} = 1+\frac{n!}{4\pi(r!)^2}\Sigma(-1)^j\binom{r}{j}(r+j)(r+j-1)T_{r+j-2}, \quad (14.12)$$

where

$$T_{r+j-2} = \frac{1}{\sqrt{(2\pi)}}\int_{-\infty}^{\infty}A^{r+j-2}\exp(-\tfrac{3}{2}x^2)\,dx. \quad (14.13)$$

14.5 In this kind of way we can ascertain as many moments as we wish of any order-statistic. The results depend on integrals of type

$$\int_{-\infty}^{\infty}A^p\exp(-\tfrac{1}{2}qx^2)\,dx, \quad (14.14)$$

of which (14.13) is a particular case. These are sometimes known as Hojo's integrals, from the name of the author who first (1931) studied them in detail and tabulated them for a range of values of p and q. By an easy extension of the method, we may ascertain product-moments of two order-statistics, and hence the moments of such quantities as the interquartile range (the variate-distance between the upper and lower quartile) and the mid-range (half the sum of the smallest and greatest values). The quadratures involved are extensive and for ordinary purposes the methods of approximation given below are adequate. We may, however, quote a few of Hojo's exact results by way of illustration.

14.6 In normal samples the standard error of the mean is $\sigma/\sqrt{n}$. The standard error of the median is always larger than this, say $c_n\sigma/\sqrt{n}$ where $c_n \geqslant 1$ (and is only equal to unity for the trivial case $n = 2$). With a definition of the median for even $n = 2r$ as $\frac{1}{2}(x_{(r)}+x_{(r+1)})$ we have the following values for c_n :—

n	c_n	n	c_n
2	1·000	10	1·177
4	1·092	12	1·189
6	1·135	20	1·214
8	1·160	∞	1·253

This will be discussed further in **17.12**, Vol. 2.

Likewise, for the mid-range $\frac{1}{2}(x_{(1)} + x_{(n)})$ we have a standard error, say $d_n\sigma / \sqrt{n}$, but d_n tends to infinity with n, as we shall see below (Example 14.4). For increasing sample size the mid-range is an increasingly inferior estimator of the mean of the normal population. The following are some values of d_n :—

n	d_n	n	d_n
2	1·000	10	1·362
4	1·092	20	1·691
6	1·190	∞	∞

14.7 The interquartile distance, divided by twice 0·67449 (the distance between the quartiles in a normal population with unit variance), is sometimes used as an estimator of the parent standard deviation. It is interesting to compare the standard error of this quantity with the standard error of the sample s.d., namely $\sigma/\sqrt{(2n)}$. If the standard error is $e_n\sigma/\sqrt{(2n)}$, the following are some values of e_n :—

n	e_n	n	e_n
2	1·000	10	1·497
4	1·047	12	1·419
6	1·421	∞	1·649
8	1·313		

The values proceed somewhat irregularly owing to the arbitrary element in the definition of quartiles for sample sizes not of the form $4r+1$.

14.8 Hojo (1931, 1933) and K. Pearson (1931) give a number of further results of this character. For some more recent numerical work reference may be made to the book by Sarhan and Greenberg (1962). See also **14.21** below.

Pearson's expansion

14.9 In Chapter 6 we have considered the expansion of a distribution function $F(x)$ or a frequency function $f(x)$ as a power series in the variate x. We now consider an inverse expansion of x in terms of F, due to Karl Pearson. Let X_r be the parent value such that

$$F(X_r) = \frac{r}{n+1}. \tag{14.15}$$

We expand $x_{(r)}$ about X_r in a Taylor series

$$x_{(r)} = X_r + h_r X'_r + \frac{1}{2!} h_r^2 X''_r + \dots \tag{14.16}$$

where

$$\left.\begin{aligned} h_r &= F(x_{(r)}) - F(X_r) = F_r - r/(n+1), \\ X'_r &= \frac{dX_r}{dF} = \left.\frac{dx}{dF}\right|_{x=X_r}, \quad X''_r = \left.\frac{d^2x}{dF^2}\right|_{x=X_r} \end{aligned}\right\} \tag{14.17}$$

and so on.

From (14.16) we can express powers of x in series of powers of h; and the expectation of any power of h is easily derivable from (14.1). Provided, then, that our series converge in a suitable manner (or, more generally, give good approximations of an asymptotic kind) we may derive approximations to as many moments as we wish of any order-statistic.

Example 14.1

Consider again the distribution of the median $x_{(r+1)}$ in samples of $n = 2r+1$ observations from the standardized normal distribution.

We find, since $(r+1)/(n+1) = \frac{1}{2}, X_{r+1} = 0$,

$$X_r' = \left.\frac{dx}{dF}\right|_{x=0} = 1\Big/\left.\frac{dF}{dx}\right|_{x=0} = \sqrt{(2\pi)},$$

$$X_r'' = \left.\frac{d}{dF}\left(1\Big/\frac{dF}{dx}\right)\right|_{x=0} = \left.\frac{dx}{dF}\cdot\frac{d}{dx}\left(1\Big/\frac{dF}{dx}\right)\right|_{x=0} = \sqrt{(2\pi)}.0 = 0.$$

Similarly $X_r''' = (2\pi)^{3/2}$. Thus, from (14.16) and (14.4) with $h_{r+1} = A-\frac{1}{2}$, we have

$$x_{(r+1)} = (2\pi)^{\frac{1}{2}}(A-\tfrac{1}{2})+\tfrac{1}{6}(2\pi)^{3/2}(A-\tfrac{1}{2})^3+\ldots$$

and

$$x^2_{(r+1)} = 2\pi(A-\tfrac{1}{2})^2+\tfrac{1}{3}(2\pi)^2(A-\tfrac{1}{2})^4+\ldots$$

Substituting the expansion to this order for $x^2_{(r+1)}$ in (14.6), we find

$$\text{var}\, x_{(r+1)} = \frac{n!}{(r!)^2}\int_{-\infty}^{\infty} A^r(1-A)^r\,\{2\pi(A-\tfrac{1}{2})^2+\tfrac{1}{3}(2\pi)^2(A-\tfrac{1}{2})^4+\ldots\}\,dA,$$

which reduces to

$$\text{var}\, x_{(r+1)} = \frac{\pi}{2(n+2)}+\frac{\pi^2}{4(n+2)(n+4)}+o(n^{-2}). \tag{14.18}$$

This formula is accurate enough for ordinary purposes. For $n = 11$ it gives a variance of 0·137 agreeing to 3 d.p. with the known true value of 0·137,227.

Example 14.2

For the rectangular population

$$dF = dx, \qquad 0 \leqslant x \leqslant 1,$$

we find, for the median $x_{(r+1)}$ in samples of $n = 2r+1$ observations,

$$X' = 1, \qquad X^{(r)} = 0, \qquad r > 1.$$

The variance of the median is then the expectation of $(x-\frac{1}{2})^2$ in

$$dG = \frac{n!}{(r!)^2}x^r(1-x)^r\,dx, \qquad 0 \leqslant x \leqslant 1,$$

and this is obviously an exact result. We find

$$\text{var}\, x_{(r+1)} = 1/\{4(n+2)\}. \tag{14.19}$$

14.10 David and Johnson (1954) have pursued this subject systematically and give expansions for cumulants and product-cumulants of order-statistics up to and including the fourth order. They choose expansions in terms of $(n+2)^{-1}$ rather than n^{-1} because of the natural appearance of the former quantity in elementary cases—cf. equations (14.18) and (14.19). It is not clear how far these expansions have the required asymptotic properties, but they appear to work well in practice. Exercise 14.3 gives general results for the median, of which Examples 14.1 and 14.2 are special cases. See also Exercise 14.7.

Quantiles and extreme values: asymptotic distributions

14.11 In considering the asymptotic distributions of the order-statistics we may have two types of limiting process. In the first type we let r and n tend to infinity with the ratio r/n held constant: we are then considering the asymptotic distributions of *quantiles* of the sample. In the second type, r remains fixed as n tends to infinity: this leads to results of a radically different nature from those of the quantiles, and this branch of the subject is often designated as the theory of *extreme values*, the name deriving from the fact that the cases of particular interest are those in which r is close to 1 or to n.

14.12 We have already (**10.10–11** and Example 11.4) derived limiting results for quantiles. We saw there that the sample quantile corresponding to $x_{(r)}$ is asymptotically normal. Its mean is X_r (given by $F(X_r) = r/n$) and its variance is

$$p_r q_r/(n f_{r,n}^2), \tag{14.20}$$

where $p_r = r/n$, $q_r = 1-p_r$ and $f_{r,n}$ is the ordinate of the frequency function at X_r.

More rigorous proofs are given by Rényi (1953) and Walker (1968). Exercise 14.6 studies (14.20) as a function of p.

Similarly for two quantiles based on $x_{(r)}, x_{(s)}$ $(r < s)$ we have bivariate normality, variances given by (14.20) and covariance

$$p_r q_s/(n f_{r,n} f_{s,n}). \tag{14.21}$$

The result generalizes readily to more than two quantiles, the variances and covariances being of type (14.20) and (14.21), and the limiting distribution multivariate normal.

Chernoff *et al.* (1967) give general conditions under which functions of the order-statistics are asymptotically normal. See also Shorack (1969, 1972).

Asymptotic distributions of extreme values

14.13 We now turn to the asymptotic theory of extreme values, which we have not so far encountered. Gumbel (1958) treats the subject fully.

In (14.1), we make the transformation $y = nF(x_{(r)})$, obtaining as the distribution of y

$$dH_r(y) = \left(\frac{y}{n}\right)^{r-1}\left(1-\frac{y}{n}\right)^{n-r} d\left(\frac{y}{n}\right)\Big/ B(r, n-r+1), \tag{14.22}$$

so that y/n is a Beta variate of the first kind. As n tends to infinity, for any fixed r,

$$\lim_{n\to\infty} dH_r(y) = y^{r-1}e^{-y}\,dy/\Gamma(r), \tag{14.23}$$

a Gamma variate with parameter r. When $r = 1$, (14.23) becomes, for the transformed smallest value in the sample, an exponential distribution.

If we transform (14.1) by $y = n(1-F(x_{(r)}))$, we get a parallel result to (14.23), with $n-r$ written for $r-1$, and this also reduces to an exponential distribution for the case $r = n$.

These results apply to the transformed variable y. We now inquire directly into the distribution of the original variable $x_{(r)}$. If $F(x)$ is known explicitly, the transformation leading to (14.23) can, of course, be reversed to give the limiting distribution of $x_{(r)}$ itself.

14.14 In the case of the extreme values ($r = 1$ or n), direct progress can be made by the use of an argument due to Fisher and Tippett (1928). Consider the largest value, and suppose that $n = km$. The largest value $x_{(n)}$ may then be regarded as the largest of k largest values of samples of size m. If $x_{(n)}$ has a limiting distribution, it will be identical with that of the largest values in samples of size m, as m tends to infinity with k fixed, except possibly for location and scale factors due to the factor k. Thus, if $G(x)$ is the limiting distribution function, it satisfies the functional equation

$$G^k(x) = G(a_k x + b_k). \tag{14.24}$$

This therefore characterizes the possible forms of limiting distribution. To solve (14.24), we use a method due to Jenkinson (1955). Put

$$q(x) = -\log\{-\log G(x)\}, \tag{14.25}$$

so that on taking logarithms twice, (14.24) becomes

$$\log k = q(x) - q(a_k x + b_k). \tag{14.26}$$

We now expand $q(x)$ and $q(a_k x + b_k)$ in a Taylor series about the point x_{on} for which

$$q(x_{on}) = 0. \tag{14.27}$$

(14.26) becomes

$$\log k = -q(a_k x_{on} + b_k) + \sum_{r=1}^{\infty} \{q^{(r)}(x_{on}) - a_k^r q^{(r)}(a_k x_{on} + b_k)\} \frac{(x - x_{on})^r}{r!}. \tag{14.28}$$

This is an identity in x. Equating coefficients, we have

$$\log k = -q(a_k x_{on} + b_k) \tag{14.29}$$

and

$$q^{(r)}(x_{on}) = a_k^r q^{(r)}(a_k x_{on} + b_k), \qquad r \geqslant 1. \tag{14.30}$$

(14.30) gives

$$\frac{q^{(r)}(a_k x_{on} + b_k)}{\{q'(a_k x_{on} + b_k)\}^r} = \frac{q^{(r)}(x_{on})}{\{q'(x_{on})\}^r} = c_r, \tag{14.31}$$

where c_r is independent of k. Since a_k and b_k are constants depending on the value of k, which is arbitrary, we may rewrite (14.31) generally as

$$q^{(r)}(x) = c_r \{q'(x)\}^r. \tag{14.32}$$

Differentiating (14.32) with respect to x, we obtain

$$\begin{aligned} q^{(r+1)}(x) &= r c_r \{q'(x)\}^{r-1} . q''(x) \\ &= r c_r c_2 \{q'(x)\}^{r+1}, \end{aligned} \tag{14.33}$$

on using (14.32) with $r = 2$. Identifying (14.32) and (14.33), we have

$$c_{r+1} = r c_r c_2 = r!\, c_2^r, \qquad r \geqslant 1. \tag{14.34}$$

Putting (14.34) back into (14.32) gives

$$q^{(r)}(x) = (r-1)!\, c_2^{r-1} \{q'(x)\}^r, \qquad r \geqslant 1. \tag{14.35}$$

If (14.35) is used in the Taylor expansion of $q(x)$ about x_{on}, we obtain, using (14.27),

$$\begin{aligned} q(x) &= \sum_{r=1}^{\infty} c_2^{r-1} \{q'(x)\}^r (x - x_{on})^r / r \\ &= -\frac{1}{c_2} \log\{1 - c_2 q'(x_{on})(x - x_{on})\}. \end{aligned} \tag{14.36}$$

(14.25) and (14.36) give for the characteristic limiting form

$$G(x) = \exp[-\{1-c_2 q'(x_{on})(x-x_{on})\}^{1/c_2}]. \tag{14.37}$$

14.15 From **14.13** it follows that since $y = n(1-F(x_{(n)}))$ asymptotically has the distribution function $1-e^{-y}$, $G(x)$ may alternatively be written

$$G(x) = \exp[-n\{1-F(x)\}], \tag{14.38}$$

so that, from (14.25),

$$q(x) = -\log[n\{1-F(x)\}]. \tag{14.39}$$

At x_{on}, in virtue of (14.27) and (14.39), $q(x_{on}) = -\log[n\{1-F(x_{on})\}] = 0$, so that

$$n\{1-F(x_{on})\} = 1. \tag{14.40}$$

The left side of (14.40) is the expected number of values exceeding x_{on} in a sample of size n, so that x_{on} is that value of the variate which has expectation of being exceeded just once in such a sample.(*)

From (14.39) and (14.40)

$$q'(x) = F'(x)/\{1-F(x)\}, \tag{14.41}$$

$$q'(x_{on}) = nF'(x_{on}), \tag{14.42}$$

while, from (14.35),

$$c_2 = q''(x)/\{q'(x)\}^2 = \frac{d}{dx}\{-1/q'(x)\}$$

$$= -\frac{d}{dx}\left\{\frac{1-F(x)}{F'(x)}\right\}, \tag{14.43}$$

by (14.41). If

$$\lim_{x\to\infty} c_2 = 0, \tag{14.44}$$

the limiting form (14.37) becomes, on inserting (14.42) and (14.44),

$$G(x) = \exp[-\exp\{-nF'(x_{on})(x-x_{on})\}], \tag{14.45}$$

a limiting form put forward originally by Fisher and Tippett (1928), the explicit condition (14.44) being due to Von Mises (1936). For the exponential distribution

$$dF(x) = e^{-x}dx, \qquad 0 \leqslant x \leqslant \infty,$$

c_2 defined at (14.43) is zero identically in x, and by analogy distributions for which (14.44), and consequently (14.45), hold have been termed exponential-type.

Exercise 14.2 deduces (14.45) directly for the exponential and logistic distributions.

14.16 If $F(x)$ is not exponential-type, (14.45) is replaced by a different limiting form. From (14.43), we see that

$$c_2 = 1 + F''(x)\{1-F(x)\}/\{F'(x)\}^2. \tag{14.46}$$

If now

$$\lim_{x\to\infty} xF'(x)/\{1-F(x)\} = k > 0, \tag{14.47}$$

(14.46) and (14.47) give

$$\lim_{x\to\infty} c_2 = \lim[1 + xF''(x)/\{kF'(x)\}]. \tag{14.48}$$

(*) Gumbel (e.g., 1954) called x_{on} " the expected largest value " though, as he pointed out, this is confusing since it is not the expected value of the largest order-statistic.

On using L'Hôpital's rule, (14.47) becomes

$$\lim \{1+xF''(x)/F'(x)\} = -k, \tag{14.49}$$

and (14.48) and (14.49) imply that

$$\lim_{x\to\infty} c_2 = -1/k. \tag{14.50}$$

The use of (14.42), (14.47) and (14.50) in (14.37) gives us the limiting form

$$G(x) = \exp[-\{1+n(1-F(x_{on}))(x-x_{on})/x_{on}\}^{-k}],$$

which finally becomes, on using (14.40),

$$G(x) = \exp\left[-\left(\frac{x}{x_0}\right)^{-k}\right], \tag{14.51}$$

the second type of limiting form, due to Fréchet (1927) and to Fisher and Tippett (1928). The condition (14.47) is implied by

$$\lim_{x\to\infty} x^k \{1-F(x)\} = A>0, \tag{14.52}$$

as may be verified by applying L'Hôpital's rule to (14.52). Distributions satisfying (14.52) (and therefore (14.47)) have no moments of order k or more, and hence are termed Cauchy-type, after the Cauchy distribution of Example 3.3 for which $k = 1$. Exercise 14.2 deduces (14.51) directly in this case.

14.17 Finally, we consider an original distribution $F(x)$ with a finite terminal x_u where

$$F(x_u) = 1.$$

If, in addition

$$\left.\begin{array}{ll} F^{(r)}(x_u) = 0 & (r = 1,2,\ldots,k-1), \\ F^{(k)}(x_u) \neq 0, & \end{array}\right\} \tag{14.53}$$

we may expand the exponent of (14.38) in a Taylor series about x_u whose first non-vanishing term is that in $F^{(k)}(x_u)$. If $F^{(k+1)}(x)$ is bounded near x_u, the remainder term is negligible since $x-x_u$ is small, and we obtain

$$G(x) = \exp[-\{(-nF^{(k)}(x_u)/k!)^{1/k}(x-x_u)\}^k], \tag{14.54}$$

the third limiting form for extreme-value distributions given by Fisher and Tippett (1928). It is often known as the Weibull distribution, after W. Weibull who studied it in the 1950's in developing a theory of the strength of materials.

The reader should verify that the three limiting forms (14.45), (14.51) and (14.54) each satisfy the functional equation (14.24).

Gnedenko (1943) and Smirnov (1949) give full analyses of the three asymptotic distributions.

14.18 In **14.14–14.17** we have been discussing the possible limiting distributions of $x_{(n)}$. If we reverse the sign of the variable, $-x_{(1)}$ becomes $x_{(n)}$ and $G(x)$ becomes $1-G(x)$. Thus the distribution of the smallest value $x_{(1)}$ is simply

$$H(x) = 1-G(-x) \tag{14.55}$$

and the results for $x_{(n)}$ apply with only this modification.

Example 14.3

For the normal distribution

$$dF(x) = e^{-\frac{1}{2}x^2}\,dx/\sqrt{(2\pi)},$$

we have, on using (5.68) in (14.43), for large x,

$$c_2 = -\frac{d}{dx}\{(1-F(x))/F'(x)\} \sim -\frac{d}{dx}\left(\frac{1}{x}\right) = \frac{1}{x^2}$$

so that

$$\lim_{x\to\infty} c_2 = 0.$$

(14.44) is satisfied in this case, giving (14.45) as the limiting form of the distribution. To evaluate its exponent, we use the fact that at x_{on}, (14.40) holds, so that

$$nF'(x_{on})/x_{on} \sim n\{1-F(x_{on})\} = 1.$$

Thus

$$nF'(x_{on}) = x_{on},$$

and the exponent of (14.45) is simply $-x_{on}(x-x_{on})$. Taking logarithms,

$$\log n + \log F'(x_{on}) - \log x_{on} = 0,$$

i.e.

$$\log n - \tfrac{1}{2}\log(2\pi) - \tfrac{1}{2}x_{on}^2 = \log x_{on}.$$

As a first asymptotic approximation, this equation has the solution $x_{on} \sim (2\log n)^{\frac{1}{2}}$. For a better approximation, we may now solve the equation with $(2\log n)^{\frac{1}{2}}$ replacing x_{on} on its right-hand side. We find $x_{on} \sim (2\log n)^{\frac{1}{2}}\{1-\frac{1}{2}(\log 4\pi + \log\log n)/(2\log n)\}$.

McCord (1964) studies the asymptotic moments of the largest value for Cauchy-type, finite terminal, and some exponential-type distributions (excluding the normal and Gamma, but including the exponential itself and the logistic).

Bivariate and multivariate extremes

If a bivariate f.f. $f(x, y)$ has marginal f.f.'s $f_1(x)$, $f_2(y)$, the largest observations $x_{(n)}$ and $y_{(n)}$ will have a joint distribution function. Campbell and Tsokos (1973) show that if $E\left\{\frac{f(x,y)}{f_1(x)f_2(y)}\right\}$ is finite, the limiting form of this joint d.f. can only be

$$G(x, y) = G_1(x)\,G_2(y)\exp\{W(x, y)\},$$

where G_1, G_2 are the marginal limiting d.f.'s for $x_{(n)}$ and $y_{(n)}$. $W(x, y)$ reflects the dependence of $x_{(n)}$ and $y_{(n)}$, being zero when they are asymptotically independent. Remarkably, this is so for the bivariate normal distribution of Examples 3.17 and 15.1 (below), as well as for others—even though the initial variables x, y can be arbitrarily dependent, their extreme values tend to independence. See also Tiago de Oliveira (1962–3) and Posner *et al.* (1969). Galambos (1975) considers the joint distribution of extreme values (not necessarily the largest) for a multivariate initial distribution and obtains conditions for their independence.

Asymptotic distribution of mth extreme values

14.19 Reverting to (14.1), we now turn from the extreme values to a more general investigation of limiting distributions for other extreme order-statistics. We confine ourselves to the case of exponential-type distributions, which are the most important, and consider the mth largest order-statistic, $x_{(n-m+1)}$.

If we put $r = n-m+1$ in (14.1) and differentiate its logarithm, as in **10.10**, we obtain

$$(n-m)F'/F-(m-1)F'/(1-F)+F''/F' = 0 \qquad (14.56)$$

at the mode of $x_{(n-m+1)}$. From (14.41) and (14.43) it follows that if (14.44) holds

$$q''(x) = F''/(1-F)+(F')^2/(1-F)^2 \to 0$$

so that since $(1-F)\to 0$, F'' must also do so. Applying L'Hôpital's rule to the last term on the left of (14.56), we obtain

$$\lim F''/F' = \lim F'/\{-(1-F)\}, \tag{14.57}$$

so that (14.56) becomes

$$(n-m)F'/F - mF'/(1-F) = 0, \tag{14.58}$$

whence, at the mode $\tilde{x}$ of the distribution of $x_{(n-m+1)}$,

$$F(\tilde{x}) = 1-m/n. \tag{14.59}$$

Comparison of (14.59) with (14.40) shows that, in the exponential case, when $m = 1$ we have $x_{on} = \tilde{x}$, i.e. the mode of the distribution of $x_{(n)}$ coincides with the value which is on the average exceeded by one member.

Exercise 14.11 gives exact results for the mode in the normal case.

We now expand $F(x)$ in a Taylor series about $\tilde{x}$,

$$F(x) = F(\tilde{x})+(x-\tilde{x})F'(\tilde{x})+(x-\tilde{x})^2F''(\tilde{x})/2+\dots \tag{14.60}$$

Since, by (14.57) and (14.59)

$$F'' \sim -(F')^2/(1-F) = -n(F')^2/m, \tag{14.61}$$

we rewrite (14.60), using (14.59) and (14.61), as

$$F(x) = (1-m/n)+(x-\tilde{x})F'(\tilde{x})-n\{F'(\tilde{x})\}^2(x-\tilde{x})^2/(2m)+\dots \tag{14.62}$$

$$= 1-(m/n)\left[1-\frac{(x-\tilde{x})F'(\tilde{x})}{m/n}+\left\{\frac{(x-\tilde{x})F'(\tilde{x})}{m/n}\right\}^2\cdot\frac{1}{2!}+\dots\right]$$

$$\sim 1-(m/n)\exp\{-(x-\tilde{x})F'(\tilde{x})/(m/n)\} \tag{14.63}$$

approximately. Now (14.1) may be rewritten

$$dG_{n-m+1}(x) \propto \left(\frac{1-F}{F}\right)^{m-1} d(F^n). \tag{14.64}$$

If we write $-y_m$ for the exponent in (14.63) and substitute it into (14.64), we obtain

$$\left(\frac{1}{F}-1\right)^{m-1} = \left(\frac{1}{1-\frac{m}{n}e^{-y_m}}-1\right)^{m-1} \sim \left(\frac{m}{n}\right)^{m-1}\exp\{-(m-1)y_m\},$$

and

$$d(F^n) = n\left(1-\frac{m}{n}e^{-y_m}\right)^{n-1}\cdot\frac{m}{n}e^{-y_m}\,dy_m.$$

Thus, from (14.64), we have for large n,

$$dG_{n-m+1}(x) \propto \exp[-my_m - m\,e^{-y_m}]\,dy_m$$

and, on evaluating the constants by integration, we get

$$dG_{n-m+1}(x) = \frac{m^m}{(m-1)!}\exp[-my_m - m\,e^{-y_m}]\,dy_m. \tag{14.65}$$

In particular, when $m = 1$, we have

$$dG_n(x) = \exp[-y-e^{-y}]\,dy, \tag{14.66}$$

which is the frequency function corresponding to the distribution function (14.45).

14.20 The limiting form (14.65), due to Gumbel (1934), is far from normal for small m. From its c.f., which the reader is asked to find in Exercise 14.4, its c.g.f. is, with $\theta = it$,

$$\psi = \theta \log m + \log \{\Gamma(m-\theta)/\Gamma(m)\},$$

whence the mean is

$$\kappa_{1,m} = \left[\frac{\partial \psi}{\partial \theta}\right]_{\theta=0} = \log m + \left[\frac{\partial}{\partial \theta} \log \{\Gamma(m-\theta)/\Gamma(m)\}\right]_{\theta=0}$$
$$= \log m - \partial \log \Gamma(m)/\partial m,$$

which by a well-known property of the digamma function $\partial \log \Gamma(m)/\partial m = \Gamma'(m)/\Gamma(m)$ becomes

$$\kappa_{1,m} = \log m + \gamma - \sum_{r=1}^{m-1} r^{-1}, \tag{14.67}$$

where γ is Euler's constant 0·5772.... For the higher cumulants we have similarly

$$\kappa_{r,m} = \left[\frac{\partial^r \psi}{\partial \theta^r}\right]_{\theta=0} = (-1)^r \frac{\partial^r}{\partial m^r} \log \Gamma(m), \qquad r \geqslant 2, \tag{14.68}$$

so that the variance is

$$\kappa_{2,m} = \delta^2 \log \Gamma(m)/\partial m^2.$$

As $m \to \infty$, we see from (3.63) that $\partial \log \Gamma(m)/\partial m \sim \log m$ and $\delta^2 \log \Gamma(m)/\partial m^2 \sim m^{-1}$, so that the mean (14.67) tends to zero and the variance to m^{-1}. Similarly, the skewness (γ_1) tends to $m^{-\frac{1}{2}}$ and the kurtosis (γ_2) to $2m^{-1}$. For small m, some numerical values are given in the following table, derived from Gumbel's calculations :—

m	Mean	Variance	Skewness (γ_1)	Kurtosis (γ_2)
1	0·577	1·645	1·140	2·400
3	0·176	0·395	0·621	0·762
5	0·103	0·221	0·469	0·428
10	0·051	0·118	0·329	0·214

For fixed m, the limiting distributions are thus non-normal. They tend to normality as $m \to \infty$, for the distinction between extreme values and quantiles, made in **14.11**, is then lost. We thus see that the mean (14.67) tended to zero as $m \to \infty$ because the limiting normality drives the mean towards the mode which is the origin for y_m.

The asymptotic distributions of the extreme values themselves are obtained through the linear relation used in (14.63),

$$y_m = (n/m) F'(\tilde{x})(x - \tilde{x}). \tag{14.69}$$

The distribution of y corresponding to the mth smallest order-statistic, $x_{(m)}$, is similar by (14.55).

The c.f. and cumulants of (14.66) are given in more detail in Exercise 14.4.

J. J. Dronkers (1958) studied the asymptotic distributions of mth values in detail, deriving limiting forms for all three types of initial distribution. Transitional approximations are also given.

(14.65) is very slowly approached for the normal distribution. For $m = 1$, the approach may be accelerated by using $x_{(n)}^2$ instead of $x_{(n)}$ itself, for as the reader should verify, Exercise 14.4 and Example 14.3 imply that $E\{x_{(n)}^2\} \sim 2\log n$, $\text{var}(x_{(n)}^2) \sim 2\pi^2/3$, so that just as for $x_{(n)}$ in the exponential and logistic distributions in Exercise 14.2, $x_{(n)}^2$ in normal samples has mean $\propto \log n$ and constant variance asymptotically. Cf. Haldane and Jayakar (1963).

The order-statistics in the normal case

14.21 Many authors have considered the exact distributions of the order-statistics in samples of n observations from a normal population. The expected values of $x_{(r)}$ have been tabulated fairly extensively. See *Biometrika Tables for Statisticians*, Part 1, which gives the means of all normal order-statistics for $n = 2\ (1)\ 26\ (2)\ 50$ to at least two decimal places. Harter (1961) gives them to 5 d.p. for $n = 2(1)100(25)250(50)400$; his table is reproduced by F. N. David *et al.* (1968), in the *Biometrika Tables*, Vol. II, up to $n = 200$, and in Harter (1969), where all intermediate values of n with no prime factor exceeding 7 are also tabulated. Teichroew (1956) gives the means, and Sarhan and Greenberg (1956, 1958, 1962) give the variances and covariances, of all normal order-statistics to 10 decimal places for $n = 2(1)20$—the *Biometrika Tables*, Vol. II, reproduce the latter to 6 d.p. S. S. Gupta (1961) tabulates the 50, 75, 90 and 95 percentage points of the distributions of all normal order-statistics for $n = 1(1)10$, and for the extreme and central order-statistics for $n = 11(1)20$. Ruben (1954) gives recurrence relations for the moments of $x_{(r)}$ for sample size n in terms of those for lower sample sizes, tabulates the first ten moments for the largest order-statistic for $n = 1(1)50$, and from these calculates its first four central moments, and the measures of skewness and kurtosis to a minimum of 7 significant figures—his table is reproduced in Vol. II of the *Biometrika Tables*. For the largest order-statistic, Tippett (1925) obtained the results:—

n	Mean	Standard Deviation	β_1	β_2
2	0·564	0·826	0·019	3·062
5	1·163	0·669	0·092	3·202
10	1·539	0·587	0·168	3·331
100	2·508	0·429	0·429	3·765
500	3·037	0·370	0·570	4·003
1000	3·241	0·351	0·618	4·088
Asymptote (n = 1000)	3·872	0·345	1·300	5·400

The asymptotic values as $n \to \infty$ are obtainable from those in the table in **14.20** for $m = 1$ through (14.69), which here becomes, on using Example 14.3,

$$y = (2\log n)^{\frac{1}{2}}\{x-(2\log n)^{\frac{1}{2}}\}.$$

Thus $E(x) = (2\log n)^{-\frac{1}{2}}0{\cdot}577+(2\log n)^{\frac{1}{2}}$, $\text{var}\,x = 1{\cdot}645\,(2\log n)^{-1}$, and $\beta_1 = \gamma_1^2$ and $\beta_2 = \gamma_2+3$ are unchanged by the linear transformation. These asymptotic values are given for $n = 1000$ in the final row of the table above, and indicate that only the s.d. is close to its asymptotic value even for so large a sample. The values of β_1 and β_2 illustrate the point that, as n increases, the distribution of the extreme value diverges more and more from the normal form. It is tabulated for $n = 3(1)25(5)60,\ 100(100)1000$ to 7 d.p. in the *Biometrika Tables*, Vol. II, with percentage points in Vol. I for $n = 1(1)30$.

Govindarajulu (1963) reviews recurrence relations and identities among moments of normal order-statistics—cf. Exercises 14.5, 14.10 and 14.23 for general recurrence relations.

The general case

14.22 It is easy to show (by assuming the contrary and finding a contradiction) that in any sample $|x_{(r)}-\bar{x}| \leqslant (n-1)^{\frac{1}{2}}s$, where x, s are the observed mean and s.d.; in particular, this holds for $x_{(1)}$ and $x_{(n)}$. For the mean value of $x_{(n)}$, Hartley and David

(1954) have established an upper bound as follows. By the Cauchy–Schwarz inequality,

$$\int_0^1 n F^{n-1}\left(x+\frac{1}{a}\right) dF \leqslant \left[\left\{\int_0^1 (n F^{n-1})^2 dF\right\}\left\{\int_0^1 \left(x+\frac{1}{a}\right)^2 dF\right\}\right]^{\frac{1}{2}},$$

where F is any distribution function whatever, and the constant $a = (n-1)/(2n-1)^{\frac{1}{2}}$. If these integrals are all finite, this inequality becomes (on putting the mean of the distribution equal to zero and its variance to 1 without loss of generality)

$$\int_0^1 n F^{n-1} x\, dF + \frac{1}{a} \leqslant \left[\frac{n^2}{2n-1}\cdot\left(1+\frac{1}{a^2}\right)\right]^{\frac{1}{2}}.$$

The integral on the left is simply $E\{x_{(n)}\}$. Thus, on substituting for a, we have

$$E(x_{(n)}) \leqslant (n-1)/(2n-1)^{\frac{1}{2}} = a.$$

If the population is symmetrical about zero, we may use the fact that $F(x) = 1-F(-x)$ in the original inequality to obtain the sharper bound

$$E(x_{(n)}) \leqslant n\left[\left\{1-1\Big/\binom{2n-2}{n-1}\right\}\Big/\{2(2n-1)\}\right]^{\frac{1}{2}}. \tag{14.70}$$

This result is due to Moriguti (1951), who gives similar but more complicated results for upper and lower bounds of the variance and the coefficient of variation of $x_{(n)}$ in the symmetrical case. The obvious lower bound of zero for $E(x_{(n)})$ cannot in general be improved. It may be verified that it attains its general upper bound for the d.f.

$$F(x) = \left(\frac{ax+1}{n}\right)^{1/(n-1)}, \qquad -1/a \leqslant x \leqslant (n-1)/a. \tag{14.71}$$

For the smallest order-statistic, analogously, the same results hold with $E(x_{(n)})$ replaced by $|E(x_{(1)})|$. Hartley and David (1954) give the general result for any order-statistic

$$|E(x_{(m)})| \leqslant \left[\frac{B(2m-1, 2n-2m+1)}{\{B(m, n-m+1)\}^2} - 1\right]^{\frac{1}{2}}, \tag{14.72}$$

but this can only be attained by a distribution function if $m = 1$ or n, the special cases considered above.

C. Singh (1967) uses the Edgeworth expansion (6.42) with $\kappa_r = 0$, $r>5$ to investigate the distribution of $x_{(1)}$ and $x_{(n)}$ from moderately non-normal populations. Their d.f.s and mean values are more affected by skewness than kurtosis, especially as n increases, but their variances increase with kurtosis. Joshi (1969) gives bounds and approximations for any moment of any order-statistic, and C. Singh (1972) uses his (1967) results to find corrections for moderate non-normality to these moments.

The *Biometrika Tables*, Vol. II, give means of the order-statistics of the exponential distribution to 2 d.p. for $n = 1(1)60$, and of the Gamma distribution (6.20) with $\lambda = 0{\cdot}5$, $1{\cdot}5$, 2, $2{\cdot}5$, $3{\cdot}5$ to 3 d.p. for $n = 1(1)20$. Prescott (1974) gives 4 d.p. variances and covariances of the Gamma order-statistics with $\lambda = 2(1)5$ for $n = 2(1)10$.

Joint distribution of two order-statistics

14.23 If in (14.2) we make the transformation

$$y = nF(x_{(r)}) \qquad z = n\{1-F(x_{(s)})\}, \tag{14.73}$$

we obtain

$$dH_{r,s}(y,z) = \frac{\left(\frac{y}{n}\right)^{r-1}\left(1-\frac{y}{n}-\frac{z}{n}\right)^{s-r-1}\left(\frac{z}{n}\right)^{n-s} dy\,dz}{B(r,s-r)\,B(s,n-s+1).n^2}.$$

We keep r and $n-s$ fixed and small and let n tend to infinity. We find for the limit

$$dH = \frac{y^{r-1}e^{-y}\,dy}{\Gamma(r)} \cdot \frac{z^{n-s}e^{-z}\,dz}{\Gamma(n-s+1)}, \tag{14.74}$$

so that the order-statistics when transformed by (14.73) are asymptotically independently distributed Gamma variables. It follows from (14.74) that $x_{(r)}$ and $x_{(s)}$ themselves are also asymptotically independent. When $s = n-r+1$, (14.74) becomes

$$dH = \frac{y^{r-1}e^{-y}}{\Gamma(r)}\,dy \cdot \frac{z^{r-1}e^{-z}}{\Gamma(r)}\,dz, \tag{14.75}$$

the marginal distributions of the variables being identical if the order-statistics are symmetrically chosen. If $r = 1$, (14.75) is the product of two exponential distributions.

Pyke (1965) reviews the theory of the " spacings " $x_{(r)}-x_{(r-1)}$ between successive order-statistics. Exercise 14.1 gives their expectations, Exercise 14.14 gives the distribution of the transformed spacings, and Exercises 14.8 and 14.16 deal with the important case of the exponential parent distribution, which Exercises 14.17–18 use to confirm (14.75).

Ranges and mid-ranges

14.24 The range of a sample has been briefly mentioned in **2.17.** We now generalize the idea by defining the mth range as the difference R_m given by

$$R_m = x_{(n-m+1)} - x_{(m)}, \tag{14.76}$$

namely the distance between the mth values from either end of the sample. The range itself is the special case $m = 1$. We also define the mth mid-range analogously by

$$M_m = \tfrac{1}{2}(x_{(n-m+1)} + x_{(m)}). \tag{14.77}$$

The particular value M_1 is called the mid-range of the sample.

From (14.75) it is clear that the asymptotic distributions of ranges and mid-ranges are distributions of the differences and sums of transformed independent variates. We may, however, derive the exact distributions directly from (14.2) by putting $r = m = n-s+1$ and then transforming from (14.76) and (14.77). We obtain for the joint distribution of R_m and M_m,

$$\begin{aligned} dH_m \propto\; & \{F(M-\tfrac{1}{2}R)\}^{m-1}\{F(M+\tfrac{1}{2}R)-F(M-\tfrac{1}{2}R)\}^{n-2m} \\ & \times \{1-F(M+\tfrac{1}{2}R)\}^{m-1} f(M-\tfrac{1}{2}R) f(M+\tfrac{1}{2}R)\,dR\,dM, \end{aligned} \tag{14.78}$$

where the suffix m has been omitted on the right for convenience.

The marginal distributions of R_m and of M_m are obtained by integrating out the other variate from (14.78). A little care is necessary in determining the limits of integration, since these depend in each case on the range of the original distribution F. Equivalently, we may proceed direct from (14.2).

The exact distribution of the range

14.25 Putting $m = 1$ in (14.78), and integrating out M, we obtain for the frequency function of the range

$$dG(R) = dR\,.\,n(n-1)\int_{-\infty}^{\infty} \{F(x+R)-F(x)\}^{n-2} f(x+R)\,dF(x). \tag{14.79}$$

The distribution function is at once seen to be

$$G(R) = n\int_{-\infty}^{\infty} \{F(x+R)-F(x)\}^{n-1}\,dF(x). \tag{14.80}$$

For some particular cases, the distribution of the range is analytically obtainable. More generally, as in the normal case, quadrature is necessary.

For the mean value of the range, we have

$$\mathrm{E}(R) = \mathrm{E}(x_{(n)})-\mathrm{E}(x_{(1)}),$$

and using (14.1), this is

$$= \int_{-\infty}^{\infty} x\,n\,[\{F(x)\}^{n-1}-\{1-F(x)\}^{n-1}]f(x)\,dx. \tag{14.81}$$

Since $\{F(x)\}^{n-1}-\{1-F(x)\}^{n-1}$ is monotone non-decreasing, with value -1 as $x\to-\infty$ and $+1$ as $x\to+\infty$, (14.81) converges if and only if $\int_{-\infty}^{\infty} xf(x)\,dx$ does. Thus the mean range exists when the population mean exists. Integrating (14.81) by parts, we have

$$\mathrm{E}(R) = \left(x[\{F(x)\}^{n}+\{1-F(x)\}^{n}]\right)_{-\infty}^{\infty}-\int_{-\infty}^{\infty}[\{F(x)\}^{n}+\{1-F(x)\}^{n}]\,dx,$$

and since $F^{n}+(1-F)^{n} = 1$ at $x = \pm\infty$, we may write this as

$$\mathrm{E}(R) = \int_{-\infty}^{\infty}[1-\{F(x)\}^{n}-\{1-F(x)\}^{n}]\,dx. \tag{14.82}$$

Recurrence relations for E(R) like those in Exercise 14.19 are sometimes useful.

(14.82) is due to Tippett (1925); he also established the result, whose proof is simplified by Mardia (1965),

$$\mathrm{E}\{R-\mathrm{E}(R)\}^{m} = m(m-1)\int_{-\infty}^{\infty}\int_{-\infty}^{x_{(n)}}\{1-F_{n}^{\,n}-(1-F_{1})^{n}+(F_{n}-F_{1})^{n}\}$$
$$\times\{R-\mathrm{E}(R)\}^{m-2}\,dx_{(1)}dx_{(n)}-(m-1)\{-\mathrm{E}(R)\}^{m}$$

for $m\geqslant 2$. Here, F_{s} means $F\{x_{(s)}\}$ as before. As below (14.81) we may see from (14.2) with $r = 1$, $s = n$ that the mth moment of R exists if the population μ_{m}' exists.

The normal case has been studied by Tippett (1925) and E. S. Pearson (1926, 1932). Tippett found the first four moments of the distribution of the range, tabulated the mean values for values of n up to 1000 and gave a diagram for determining standard errors. His table of mean values to 5 d.p. is reproduced in the *Biometrika Tables*. The following values illustrate the general behaviour of the distribution:—

	Standard deviation	β_1 (approximate)	β_2 (approximate)
2	0·853	0·99	3·87
10	0·797	0·16	3·20
100	0·605	0·22	3·39
500	0·524	0·29	3·50
1000	0·497	0·31	3·54
Asymptote (n = 1000)	0·488	0·650	4·200

This table should be compared with that for $x_{(n)}$ in **14.21**. Since $x_{(1)}$ and $x_{(n)}$ are asymptotically independent by **14.23**, the cumulants of the distribution of R are asymptotically twice those of $x_{(n)}$ for any symmetrical parent distribution. Thus $\mathrm{E}(R) = 2\,\mathrm{E}(x_{(n)})$— this is of course an exact result, and the mean is omitted from our table because the entries would be exactly double those in the table in **14.21**—and the s.d. is increased by a factor $\sqrt{2}$; the tables show that even for $n = 100$, this approximation is little in error. Similarly, $\beta_1 = \kappa_3^2/\kappa_2^3$ is halved asymptotically, and for practical purposes at $n = 100$, and $\beta_2 - 3 = \kappa_4/\kappa_2^2$ is halved, so that β_2 here $= \frac{1}{2}(\beta_2+3)$ in **14.21**, again effectively at $n = 100$. Thus we may approximate the range in normal samples as the sum of two independent $x_{(n)}$ variables—see **14.29** below. As in **14.21**, we see that even at $n = 1000$, the distribution is still a long way from its asymptotic form, and that as n increases, it diverges more and more from the normal form.

E. S. Pearson and Hartley (1942) have tabulated to 4 d.p. the d.f. of R/σ in samples of size 2 to 20 from a normal population. Their results, too, are reproduced in the *Biometrika Tables*. Harter (1960) gives to 6 d.p. for $n = 2(1)20(2)40(10)100$ the values of R/σ for which the d.f. is equal to 0·0001, 0·0005, 0·001, 0·005, 0·01, 0·025, 0·05, 0·1 (0·1) 0·9, 0·95, 0·975, 0·99, 0·995, 0·999, 0.9995, 0·9999. He also gives the mean, variance, skewness and kurtosis to at least 8 significant figures for $n = 2(1)100$. These tables are reproduced in Harter (1969), which also contains an 8 d.p. table of the d.f. for the same range of n, an 8 d.p. table of the f.f. for $n = 2(1)16$ and many other related tables.

Subrahmaniam (1969) obtains the distribution and moments of $x_{(r)}$, and jointly of $x_{(r)}, x_{(s)}$, in samples from a population represented by an Edgeworth expansion. C. Singh (1967, 1970) similarly investigates the effect of moderate non-normality on the mean, variance and percentiles of the range. The variance is more sensitive than the mean is to parental skewness and kurtosis, but skewness always reduces both. High skewness and high kurtosis tend to have opposite, cancelling effects.

Gupta and Shah (1965) obtain the distribution of the range when the parent is logistic, and compare it with the normal case for $n = 2$ and 3.

Burr (1955) and Connor (1969) give the distribution of R in samples from a discrete population, respectively with and without replacement.

Exercise 14.24 gives the exact c.f. of R in the exponential case.

14.26 As in **14.22**, it may be seen by assuming the contrary and finding a contradiction that for any sample $R \leqslant (2n)^{\frac{1}{2}} s$. For the mean value of R, it follows from (14.70) that for symmetrical populations (with zero mean and unit variance)

$$0 \leqslant \mathrm{E}(R) \leqslant n\left[2\left\{1-1\Big/\binom{2n-2}{n-1}\right\}\Big/(2n-1)\right]^{\frac{1}{2}} \sim (n+\tfrac{1}{2})^{\frac{1}{2}},$$

a result due originally to Plackett (1947), who showed that it also holds for non-symmetrical populations. Hartley and David (1954) confirmed this and also derived upper and lower bounds for $\mathrm{E}(R)$ in samples from populations with finite range. The following table (extracted from Plackett (1947)) gives values for the upper bound, its asymptotic value and the exact values for the normal and rectangular populations.

n	Upper bound for E (R)	$(n+\frac{1}{2})^{\frac{1}{2}}$	Exact values	
			Normal	Rectangular
2	1·155	1·581	1·128	1·155
3	1·732	1·871	1·693	1·732
4	2·084	2·121	2·059	2·078
8	2·921	2·915	2·847	2·694
12	3·539	3·536	3·258	2·931

The asymptotic distribution of the range

14.27 We now turn to the limiting distribution of R as n becomes large. From (14.2) with $r = n-s+1 = 1$, we have, for the joint distribution of the smallest and largest observations,

$$dG_{1,n}(x_1, x_n) = n(n-1)(F_n - F_1)^{n-2}\, dF_1\, dF_n.$$

We now transform by the relations

$$u = 2n\{F_1(1-F_n)\}^{\frac{1}{2}}, \qquad v = \tfrac{1}{2}\log\{F_1/(1-F_n)\}, \tag{14.83}$$

the Jacobian of the transformation being $u/\{2n^2 f_1 f_n\}$. The joint distribution of u and v is then

$$dF(u,v) = \left(\frac{n-1}{2n}\right)u\left(1-\frac{u\cosh v}{n}\right)^{n-2} du\,dv. \tag{14.84}$$

As $n \to \infty$, (14.84) tends to

$$\lim_{n\to\infty} dF(u,v) = \tfrac{1}{2}u\exp(-u\cosh v)\,du\,dv. \tag{14.85}$$

The domain of variation in (14.84) is bounded by $u \geqslant 0$, $\cosh v \leqslant n/u$, so that in (14.85) $\cosh v$, and therefore v, can range to infinity. Hence we obtain from (14.85) the limiting marginal distribution of u:

$$\left.\begin{aligned} dF(u) &= u\,du\int_0^\infty \exp(-u\cosh v)\,dv \\ &= u\,du\int_1^\infty \exp(-ut)(t^2-1)^{-\frac{1}{2}}\,dt. \end{aligned}\right\} \tag{14.86}$$

Now u may be expressed as

$$u = 2n\{F(M-\tfrac{1}{2}R)[1-F(M+\tfrac{1}{2}R)]\}^{\frac{1}{2}}. \tag{14.87}$$

If now we assume that the mid-range M converges in probability to the known population mid-range μ, u will similarly behave like

$$u' = 2n\{F(\mu-\tfrac{1}{2}R)[1-F(\mu+\tfrac{1}{2}R)]\}^{\frac{1}{2}}. \tag{14.88}$$

If we may take $\mu = 0$ and if the population is symmetrical, (14.88) reduces to

$$u' = 2n\{F(-\tfrac{1}{2}R)\}, \tag{14.89}$$

so that u will tend to be simply the probability integral transform of the range, apart from constant multipliers.

The mean and variance of u are given in Exercise 14.20.

14.28 The value of these results, which are due to Elfving (1947), is restricted by the condition, imposed in order to reach (14.88), that the mid-range M converges to the population mid-range μ in probability. Even when the population is symmetrical (so that μ is the mean also) and exponential-type, this condition is not necessarily fulfilled, despite the fact that the symmetry implies $E(M) = \mu$ if it exists. To show this, we put $m = 1$ and $r = 2$ in (14.68), and find that the variance of the exponential-type extreme distribution (14.66) is $\pi^2/6$. By (14.45), the variate here is

$$y = nF'(x_{on})(x-x_{on}), \tag{14.90}$$

where x is written for $x_{(n)}$. Thus $\text{var}\,y = \pi^2/6$ and hence, for large samples,

$$\text{var}\,x = \pi^2/\{6[nF'(x_{on})]^2\}. \tag{14.91}$$

Since $x_{(n)}$ and $x_{(1)}$ are asymptotically independent, the mid-range, which is their mean, has variance

$$\text{var}\,M = \pi^2/\{12[nF'(x_{on})]^2\}. \tag{14.92}$$

Example 14.4

In the normal case with zero mean, we have seen (Example 14.3) that

$$nF'(x_{on}) \sim (2\log n)^{\frac{1}{2}}$$

so that, from (14.92), var $M = \pi^2/(24\log n)$. Thus, M converges to the population mid-range since $E(M) = 0$ here and var $M \to 0$ as $n \to \infty$.

Example 14.5

The double exponential (or Laplace) distribution

$$dF(x) = \tfrac{1}{2}\exp(-|x|)\,dx, \qquad -\infty \leqslant x \leqslant \infty,$$

has finite moments of all orders. Its distribution function is

$$F(x) = 1-\tfrac{1}{2}e^{-x}, \qquad x \geqslant 0. \qquad (14.93)$$

It is easily verified that (14.44) is satisfied. By (14.40) and (14.93), $x_{on} = \log(\frac{1}{2}n)$ and therefore

$$nF'(x_{on}) = \tfrac{1}{2}n\exp(-\log\tfrac{1}{2}n) = 1.$$

Thus for this case, from (14.92), var $M = \pi^2/12$ and since var M does not $\to 0$ as $n\to\infty$, M does not converge in probability to its expectation μ.

Exercises 14.12–13 give a number of results concerning the range and mid-range of the rectangular distribution.

The reduced range and mid-range

14.29 In view of the result of the last section, we return to the problem of **14.27**, and seek a more general solution of the asymptotic distribution problem for the range and mid-range, or some simple functions of them. We confine our attention to symmetrical exponential-type distributions.

If we equate (14.38) and (14.45), we find

$$1-F(x_{(n)}) = \frac{1}{n}\exp\{-nF'(x_{on})(x_{(n)}-x_{on})\},$$

and similarly, using (14.55) also,

$$F(x_{(1)}) = \frac{1}{n}\exp\{nF'(x_{on})(x_{(1)}+x_{on})\}.$$

Thus (14.83) becomes

$$\left.\begin{aligned} u &= 2\exp\{-\tfrac{1}{2}nF'(x_{on})(x_{(n)}-x_{(1)}-2x_{on})\}, \\ v &= \tfrac{1}{2}nF'(x_{on})(x_{(n)}+x_{(1)}). \end{aligned}\right\} \qquad (14.94)$$

We now define the *reduced range* R^* and the *reduced mid-range* M^* by

$$R^* = nF'(x_{on})(R-2x_{on}), \quad M^* = nF'(x_{on})M. \qquad (14.95)$$

(14.94) then becomes

$$u = 2\exp(-\tfrac{1}{2}R^*), \qquad (14.96)$$

$$v = M^*. \qquad (14.97)$$

If we apply (14.96) as a transformation of the distribution of u at (14.86), we obtain, for the distribution of R^*,

$$dF(R^*) = 2\exp(-R^*)\,dR^*.\int_0^\infty \exp\{-2\exp(-\tfrac{1}{2}R^*)\cosh v\}\,dv$$
$$= 2\exp(-R^*)\,K_0\{2\exp(-\tfrac{1}{2}R^*)\}\,dR^*, \qquad (14.98)$$

where K_0 is a Bessel function of zero order and imaginary argument. Gumbel (1949), who had earlier obtained (14.98) by a different method, has given tables of the frequency function and of the distribution function of R^*, and also tabulated its percentage points. Cox (1948), who also obtained this result, used the method of steepest descents to obtain a second approximation, which he compared with (14.98) and Elfving's result at (14.86). The latter is the most accurate in the normal case, and the steepest descents approximation, as might be expected, is also better than (14.98). But as will be seen in the next paragraph, even (14.98) gives a good approximation for small n.

The range proper, R, is a linear function of R^*, by (14.95). As in **14.28**, we know that, asymptotically,

$$\left.\begin{aligned} \mathrm{E}(R^*) &= \mathrm{E}(y_{(n)})-\mathrm{E}(y_{(1)}) = 2\gamma, \\ \mathrm{var}(R^*) &= \mathrm{var}(y_{(n)})+\mathrm{var}(y_{(1)}) = \pi^2/3. \end{aligned}\right\} \qquad (14.99)$$

Using (14.95), we have from (14.99), asymptotically,

$$\mathrm{E}(R) = 2\gamma/\{nF'(x_{on})\}+2x_{on}, \quad \mathrm{var}(R) = \pi^2/[3\{nF'(x_{on})\}^2]. \qquad (14.100)$$

By using this and the exact normal parent values of $\mathrm{E}(R)$ and $\mathrm{var}(R)$ (tabulated by Tippett (1925) and E. S. Pearson (1926)) for small n, Gumbel (1947) estimated x_{on} and $F'(x_{on})$ and investigated the closeness of the approximation to the distribution of R obtained from (14.98) by comparing the results it gave with those obtained from the exact tables given by Pearson and Hartley (1942). He found that the distribution function of R was closely approximated by (14.98) for n as low as 6. Thus in the normal case, at least, (14.98) solves the problem of the distribution of the range.

14.30 Turning now to the reduced mid-range M^*, we see from (14.97) that its asymptotic distribution, obtained by integrating u out of (14.85), is

$$dG(M^*) = \tfrac{1}{2}dM^*\int_0^\infty u\exp(-u\cosh M^*)\,du. \qquad (14.101)$$

The substitution $z = u\cosh M^*$ reduces this to

$$dG(M^*) = dM^*/(2\cosh^2 M^*) = 2\exp(-2M^*)\,dM^*/\{1+\exp(-2M^*)\}^2, \qquad (14.102)$$

so that $2M^*$ has the logistic distribution of Exercise 4.21. (14.95) immediately gives the distribution of the mid-range proper, M, from (14.102). It is symmetrical, with zero mean, so that M is, for symmetrical exponential-type populations, an unbiased estimator of the population mid-range (mean), which is zero. But Example 14.5 shows that the even moments of M are heavily dependent on the form of the population.

Exercise 14.21 indicates another proof of (14.102).

Tables for extreme-value distributions

14.31 *Probability Tables for the Analysis of Extreme-Value Data* (National Bureau of Standards, Applied Mathematics Series, 22, Washington, 1953) gives :—

(1) The distribution function and frequency function of the exponential-type extreme-value distribution (14.45) and (14.66) to 7 decimal places over the interval $-3{\cdot}0$ to $17{\cdot}0$, including $-2{\cdot}40\,(0{\cdot}05)\,0\,(0{\cdot}1)\,4{\cdot}0$, with auxiliary inverse tables giving the extreme value to 5 d.p. for values of its distribution function $0\,(0{\cdot}0001)\,0{\cdot}005\,(0{\cdot}001)\,0{\cdot}988\,(0{\cdot}0001)\,0{\cdot}9994\,(0{\cdot}000{,}01)\,1$, and giving the ordinate to 5 d.p. for distribution function values $0\,(0{\cdot}0001)\,0{\cdot}010\,(0{\cdot}001)\,0{\cdot}999$.

(2) Percentage points for the distributions of mth values (14.65) for $m = 1\,(1)\,15\,(5)\,50$. The 12 percentage points range between 0·005 and 0·995.

(3) The distribution and frequency functions of the reduced range for exponential-type distributions (14.98), for values of the reduced range $1{\cdot}0\,(0{\cdot}05)\,11{\cdot}0\,(0{\cdot}5)\,20{\cdot}0$, to 7 d.p., with inverse tables giving the reduced range to 3 or 4 d.p. for selected values of its distribution function.

14.32 When we come in Chapters 19 and 32 (Volume 2) to consider problems of estimation and testing hypotheses, we shall have to turn to some additional sampling properties of the order-statistics.

EXERCISES

14.1 Show, using (14.4), that in samples of size n from a continuous distribution function $F(x)$,

$$E\{x_{(r+1)}-x_{(r)}\} = \binom{n}{r}\int_{-\infty}^{\infty}\{F(x)\}^r\{1-F(x)\}^{n-r}\,dx$$

Hence establish (14.82) for the mean of the range. Verify that when $f(x)$ is the uniform distribution on $(0,1)$, $E\{x_{(r+1)}-x_{(r)}\} = (n+1)^{-1}$, and that if $f(x) = e^{-x}$, $x \geqslant 0$, it is $(n-r)^{-1}$.

(The integral form is due to K. Pearson in 1902 —see E. S. Pearson (1926) and Sillitto (1951).)

14.2 For the exponential d.f. $F(x) = 1-e^{-x}$, $x \geqslant 0$ and also for the logistic d.f. $F(x) = (1+e^{-x})^{-1}$, show directly that the limiting d.f. of $z = x_{(n)}-\log n$ is $G(z) = \exp(-e^{-z})$ and verify that this is equivalent to (14.45) in each case.

Similarly, for the Cauchy d.f. $F(x) = 0{\cdot}5+\pi^{-1}\arctan x$, show directly that the limiting d.f. of $z = \pi x_{(n)}/n$ is $G(z) = \exp(-z^{-1})$, equivalent to (14.51).

14.3 In the notation of **14.9** show that for the median $x_{(r+1)}$ in samples of $n = 2r+1$ observations, to order $(n+2)^{-2}$,

$$E(x_{(r+1)}) = X_{r+1}+\frac{1}{8(n+2)}X''_{r+1}+\frac{1}{128(n+2)^2}X^{(iv)}_{r+1},$$

$$\text{var}\, x_{(r+1)} = \frac{1}{4(n+2)}(X'_{r+1})^2+\frac{1}{32(n+2)}\{2X'_{r+1}X'''_{r+1}+(X''_{r+1})^2\},$$

$$\gamma_1(x_{(r+1)}) = \frac{3}{2(n+2)^{\frac{1}{2}}}\cdot\frac{X''_{r+1}}{X'_{r+1}}, \qquad \gamma_2(x_{(r+1)}) = \frac{1}{n+2}\left\{\frac{X'''_{r+1}}{X'_{r+1}}+\frac{3(X''_{r+1})^2}{(X'_{r+1})^2}-6\right\}.$$

(David and Johnson, 1954)

14.4 Show that the distribution (14.65) has c.f. $\phi(t) = m^{it}\Gamma(m-it)/\Gamma(m)$ and hence that when $m = 1$, (14.66) has cumulants

$$\kappa_1 = \gamma \text{ (Euler's constant, } = 0{\cdot}5772); \quad \kappa_2 = \Sigma n^{-2} = \pi^2/6 = 1{\cdot}645;$$
$$\kappa_3 = 2\Sigma n^{-3} = 2{\cdot}404; \quad \kappa_4 = 6\Sigma n^{-4} = \pi^4/15 = 6{\cdot}494.$$

Hence derive $\qquad \beta_1 = 1{\cdot}299, \ \beta_2 = 5{\cdot}4.$

Show further that (14.66) has mode zero and median $-\log\log 2 = 0{\cdot}3665$, so that (2.13) is approximately satisfied.

14.5 Show that if $G_{r,n}$ refers to the distribution (14.1) for the rth order-statistic in a sample of n,

$$(n+1)\,dG_{r,n} = r\,dG_{r+1,n+1}+(n-r+1)\,dG_{r,n+1}$$

and hence that the c.f. and moments of $x_{(r)}$ obey a similar recurrence relation. Similarly, if $G_{r,s,n}$ refers to (14.2), show that

$$(n+1)\,dG_{r,s,n} = r\,dG_{r+1,s+1,n+1}+(s-r)\,dG_{r,s+1,n+1}+(n-s+1)\,dG_{r,s,n+1}.$$

14.6 Rewriting (14.20) as $V_p = p(1-p)/(nf_p^2)$, $f_p > 0$, show that if f'_p denotes $\left(\frac{\partial f}{\partial x}\right)_p$,

$$\frac{\partial \log V_p}{\partial p} = \frac{1-2p}{p(1-p)}-2\frac{f'_p}{f_p^2}, \quad \text{and} \quad \frac{\partial^2 \log V_p}{\partial p^2} = -\left(\frac{1}{p^2}+\frac{1}{(1-p)^2}\right)+4\left(\frac{f'_p}{f_p^2}\right)^2-2\frac{f''_p}{f_p^3}.$$

Hence show that V_p has a stationary value at the sample median ($p = 0{\cdot}5$) if and only if f_p has a stationary value there; and that V_p is a local maximum there if $f'_{0\cdot5} = 0$, $f''_{0\cdot5} > -4f^3_{0\cdot5}$ and a local minimum if $f'_{0\cdot5} = 0$, $f''_{0\cdot5} < -4f^3_{0\cdot5}$. Show also that if f_p has a stationary value at $p_0 \neq 0{\cdot}5$, V_p has no stationary value at p_0.

(Cf. Sen, 1961)

14.7 Show from (14.1) and Exercise 1.20 that if $f(x)$ is symmetrical about a value ξ, so is the distribution of the sample median $x_{(r+1)}$ in samples of $n = 2r+1$ observations from $f(x)$. If $f(x)$ has an absolute maximum at ξ, show that

$$\text{var}\, x_{(r+1)} \geqslant [4\,\{f(\xi)\}^2.(n+2)]^{-1},$$

and verify that the equality holds for the rectangular distribution. Hence show, using Example 11.4, that the asymptotic value for the variance effectively provides a lower bound for the small-sample variance.

(Cf. Chu, 1955)

14.8 Show that in samples of n independent observations from $dF = e^{-x}\,dx$, $0 \leqslant x \leqslant \infty$, the conditional distribution of the n observations, given that their sum is X, is $(n-1)!\; X^{-(n-1)}$, and that when $X = 1$ this is precisely the distribution of the spacings $x_{(r)} - x_{(r-1)}$ ($r = 1, 2, \ldots n$, $x_{(0)} \equiv 0$, $x_{(n)} \equiv 1$) between the order-statistics in a sample of size $(n-1)$ from the uniform distribution on (0, 1).

14.9 Using (14.1) and (14.2), show that the conditional distribution of $x_{(s)}$ given the value of $x_{(r)}$ is exactly the distribution (14.1) of $x_{(s-r)}$ in samples of size $(n-r)$ from the same distribution truncated at $x_{(r)}$, i.e. with d.f.

$$\{F(x) - F(x_{(r)})\}/\{1 - F(x_{(r)})\} \quad \text{for } x \geqslant x_{(r)}.$$

Hence show that $\text{E}(x_{(s)} \mid x_{(r)})$ is a monotone increasing function of the value of $x_{(r)}$. Use this to show that $\text{cov}(x_{(s)}, x_{(r)}) \geqslant 0$.

14.10 In (14.1), expand $\{1 - F(x_{(r)})\}^p$ in binomial series and show that for $0 < p \leqslant n - r_1$

$$dG_{r,n} = \frac{1}{B(r, n-r+1)} \sum_{q=0}^{p} (-1)^q \binom{p}{q} B(r+q, n-r-p+1)\, dG_{r+q,\, n-p+q}.$$

Similarly, expanding $\{F(x_{(r)})\}^p$, show that for $0 < p \leqslant r-1$,

$$dG_{r,n} = \frac{1}{B(r, n-r+1)} \sum_{q=0}^{p} (-1)^q \binom{p}{q} B(r-p, n-r+q+1)\, dG_{r-p,\, n-p+q}.$$

Hence derive corresponding recurrence relations for the c.f. and moments of order-statistics. Derive the first result of Exercise 14.5 as a special case.

14.11 Show from (14.1) that the modal value of the distribution of $x_{(r)}$ in samples of size n from $f(x) = (2\pi)^{-\frac{1}{2}} \exp(-\frac{1}{2}x^2)$ is the root of

$$(r-1)\, f(x)\, \{1 - F(x)\} - (n-r)\, f(x)\, F(x) = xF(x)\{1 - F(x)\},$$

reducing when $r = n$ to $(n-1)\, f(x) = xF(x)$.

Use this and (5.68) to verify the asymptotic exponential-type result (14.59) in the normal case.

(S. S. Gupta (1961) tabulated the mode of $x_{(n)}$ for $n = 1\,(1)\,25\,(5)\,50\,(10)\,100$.)

14.12 Show that the joint distribution of the sample range and mid-range in samples from a distribution rectangular on the interval $(-\frac{1}{2}, \frac{1}{2})$ is

$$dF(R, M) = n(n-1)\, R^{n-2}\, dR\, dM, \quad 0 \leqslant R \leqslant 1 - 2\,|M| \leqslant 1,$$

that the marginal distribution of M is

$$dG(M) = n(1 - 2\,|M|)^{n-1}\, dM, \quad |M| \leqslant \tfrac{1}{2},$$

with moments
$$\left.\begin{aligned}\mu_{2k-1} &= 0\\ \mu_{2k} &= \left\{2^{2k}\binom{2k+n}{2k}\right\}^{-1}\end{aligned}\right\} k = 1,2,\ldots$$

and that if $t = nM$, $\lim_{n\to\infty} dP(t) = \exp(-2\,|\,t\,|)\,dt.$

Show further that the marginal distribution of R is
$$dH(R) = n(n-1)R^{n-2}(1-R)\,dR, \quad 0 \leqslant R \leqslant 1,$$
with moments
$$\mu'_k = n(n-1)/\{(n+k)(n+k-1)\}.$$
In particular
$$\mu'_1 = (n-1)/(n+1)$$
$$\mu_2 = 2(n-1)/\{(n+1)^2(n+2)\}.$$

Show that the conditional distribution of M given R is rectangular on the interval $(-\frac{1}{2}(1-R), \frac{1}{2}(1-R))$ and that the limiting f.f. of $u = n(1-R)$ is $u\exp(-u)$.

14.13 Continuing Exercise 14.12, show that the distribution of the ratio M/R is given by
$$dF\left(\frac{M}{R}\right) = (n-1)\left\{1+2\left|\frac{M}{R}\right|\right\}^{-n} d\left(\frac{M}{R}\right),$$
with moments
$$\left.\begin{aligned}\mu_{2k-1} &= 0,\\ \mu_{2k} &= 1\Big/\left\{2^{2k}\binom{n-2}{2k}\right\}\end{aligned}\right\} \quad k = 1,2,\ldots$$

Show that when $n = 2$, $2M/R$ is identical with "Student's" t statistic, defined at (11.42). Compare the distribution of t in this rectangular case with the normal form (11.44) with $n = 2$, which is the Cauchy distribution. (Cf. Carlton, 1946)

14.14 From (14.2), show that the joint distribution of $y = F(x_{(s)}) - F(x_{(r)})$ and $z = F(x_{(r)})$ is
$$dH_{y,z} = \frac{z^{r-1}y^{s-r-1}(1-y-z)^{n-s}\,dy\,dz}{B(r, s-r)\,B(s, n-s+1)}, \quad 0 \leqslant y+z \leqslant 1,$$
and hence that y is distributed in the first Beta form (6.14) with parameters $p = s-r$, $q = n-s+r+1$, depending only on $(s-r)$. In particular, when $s = r+1$, $p = 1$ and $q = n$ for any r, so that the transformed spacings are identically distributed. Show from (14.1) that $F(x_{(1)})$ and $1-F(x_{(n)})$ also have this distribution.

Show further that $w = y/z$ is distributed in the second kind of Beta distribution (6.17) with parameters $p = s-r$, $q = r$.

14.15 Show that in a sample of size $n \geqslant 2$ from a standardized normal distribution, $(x_{(j)} - \bar{x})$ is independent of $\bar{x}$. Hence show that
$$\begin{aligned}\kappa_r(x_{(j)} - \bar{x}) &= \kappa_r(x_{(j)}), \quad r > 2\\ &= \kappa_2(x_{(j)}) - n^{-1}, \quad r = 2,\end{aligned}$$
and that $\sum_{k=1}^{n} \text{cov}(x_{(j)}, x_{(k)}) = 1$ for any j. (Cf. McKay, 1935)

14.16 If $z_1, \ldots, z_n$ are independently exponentially distributed, so that
$$dF(z_i) = \exp(-z_i)\,dz_i, \quad 0 \leqslant z_i \leqslant \infty,$$
show that
$$y_1 = nz_{(1)},$$
$$y_r = (n+1-r)\{z_{(r)} - z_{(r-1)}\}, \quad r = 2, 3, \ldots, n$$
are also independently exponentially distributed.

(P. V. Sukhatme, 1937. These are the only n linear functions which are independent of each other—see Tanis (1964).)

14.17 By making the transformation $z_i = -\log F\{x_i\}$ for a random sample $x_1, \ldots, x_n$ from any continuous $F(x)$, use Exercise 14.16 to show that

$$y_n = -n \log F\{x_{(n)}\}$$

$$y_r = -r \log \frac{F\{x_{(r)}\}}{F\{x_{(r+1)}\}}, \quad r = 1, 2, \ldots, n-1$$

are independently exponentially distributed, and hence that

$$[F\{x_{(n)}\}]^n, \left[\frac{F\{x_{(r)}\}}{F\{x_{(r+1)}\}}\right]^r, \quad r = 1, 2, \ldots, n-1$$

are independently uniformly distributed on (0, 1). (Rényi, 1953)

14.18 In Exercise 14.16, show that $z_{(r)} = \sum_{s=1}^{r} \frac{y_s}{n+1-s}$, and that as $n \to \infty$ with r fixed, $nz_{(r)} \sim \sum_{s=1}^{r} y_s$ is a Gamma variable with parameter r. Hence show that in Exercise 14.17, $-n \log F\{x_{(n-r+1)}\}$ has this limiting distribution and that this implies the limiting distributions obtained for y and z in **14.23.**

(Rényi, 1953)

14.19 From (14.82), show that if n is odd, and R_n is the range in samples of size n,

$$2\mathrm{E}(R_n) = \sum_{s=2}^{n-1} (-1)^s \binom{n}{s} \mathrm{E}(R_s) \quad \text{and} \quad (n+1)\,\mathrm{E}(R_n) = \sum_{s=2}^{n-1} (-1)^s \binom{n+1}{s} \mathrm{E}(R_s),$$

and hence that also $$(n-1)\,\mathrm{E}(R_n) = \sum_{s=2}^{n-1} (-1)^s \binom{n}{s-1} \mathrm{E}(R_s).$$

In particular, $2\mathrm{E}(R_3) = 3\mathrm{E}(R_2)$.

14.20 Show that u in (14.86) has mean $\frac{1}{2}\pi$ and variance $(4-\frac{1}{4}\pi^2)$. (Elfving, 1947)

14.21 Use the c.f.'s of Exercises 4.21 and 14.4 to establish the result (14.102) for the limiting distribution of the reduced mid-range.

14.22 Show from (14.1) that for the Cauchy distribution $dF(x) = dx/\{\pi(1+x^2)\}$, the pth moment of $x_{(r)}$ exists if and only if $p < \min(r, n-r+1)$, i.e., if at least p order-statistics lie on each side of $x_{(r)}$. (Example 3.3 with $m = 1$ is here the special case $n = r = 1$.)

14.23 Writing (14.5) as $G_{r,n}(x) = I_{F(x)}(r, n-r+1)$, show by inverting the order of summation and integration that

$$\sum_{r=1}^{n} r^{-1} G_{r,n}(x) \equiv \sum_{r=1}^{n} r^{-1} G_{1,r}(x)$$

and

$$\sum_{r=1}^{n} (n-r+1)^{-1} G_{r,n}(x) \equiv \sum_{r=1}^{n} r^{-1} G_{r,r}(x),$$

with corresponding relations between the c.f.'s, moments and f.f.'s in the continuous case.

(Joshi, 1973)

14.24 In Exercise 14.18, show that $z_{(r)}$ in a sample of size n is distributed exactly as $z_{(r-p)}$ in a sample of size $n-p$ from the exponential distribution truncated at $z_{(p)}$—cf. Exercise 14.9.

Show further that the range $R = z_{(n)} - z_{(1)}$ has exact c.f. $\mathrm{E}(e^{itR}) = \Gamma(n)\Gamma(1-it)/\Gamma(n-it)$, with $\mathrm{E}(R) = \sum_{s=2}^{n} (n+1-s)^{-1}$ and $\mathrm{var}\, R = \sum_{s=2}^{n} (n+1-s)^{-2}$, and that R itself is exponentially distributed when $n = 2$—cf. Exercise 11.21.

CHAPTER 15

THE MULTIVARIATE NORMAL DISTRIBUTION AND QUADRATIC FORMS

15.1 The univariate normal distribution

$$dF = (2\pi)^{-\frac{1}{2}}\exp\left\{-\tfrac{1}{2}\left(\frac{x-\mu}{\sigma}\right)^2\right\}\frac{dx}{\sigma}, \qquad -\infty \leqslant x \leqslant \infty,$$

has been discussed in many contexts in earlier chapters. Apart from location and scale parameters, μ and σ, the exponent of the frequency function is simply $-\frac{1}{2}x^2$. We now seek to generalize the normal distribution to the joint distribution of two or more variables, and it is natural to look for this generalization in a frequency function which has as its exponent a quadratic form in the variables, p in number, i.e.

$$dF \propto \exp\left\{-\tfrac{1}{2}\sum_{j=1}^{p}\sum_{k=1}^{p} a_{jk}\left(\frac{x_j-\mu_j}{\sigma_j}\right)\left(\frac{x_k-\mu_k}{\sigma_k}\right)\right\}\prod_{j=1}^{p}\frac{dx_j}{\sigma_j}, \qquad -\infty \leqslant x_j \leqslant \infty. \tag{15.1}$$

Here, as in the univariate case, the μ_j and σ_j are simply location and scale parameters which may be removed by a standardizing transformation.

In this chapter, and generally when dealing with multivariate problems, we shall find it convenient to use vector and matrix notation for economy and clarity in representation. A bold-face capital letter will represent a matrix, a bold-face small letter a column vector, and a prime will denote transposition, so that a row vector will be represented by a bold-face primed small letter. With this notation, (15.1) becomes

$$dF \propto \exp\{-\tfrac{1}{2}(\mathbf{x}-\boldsymbol{\mu})'\mathbf{A}(\mathbf{x}-\boldsymbol{\mu})\}\prod dx_j, \tag{15.2}$$

where $\mathbf{A}$ is a symmetric $(p \times p)$ matrix.

Characteristic function and moments

15.2 The condition that (15.2) be a properly defined frequency function is simply that its exponent is non-negative definite. For in such a case (and only in such a case) we can find a real linear transformation which transforms the exponent into a negative sum of squares, and the integral of dF will converge. In particular there is an *orthogonal* transformation

$$\mathbf{x} = \mathbf{B}\mathbf{y} \tag{15.3}$$

which transforms the exponent of (15.2) into a sum of squares, the coefficients in this sum being the latent roots of $\mathbf{A}$. Confining ourselves to the non-singular case, where the rank of $\mathbf{A}$ is p,(*) and measuring from the vector $\boldsymbol{\mu}$, we now apply this transformation to obtain the characteristic function of the distribution, and incidentally to evaluate the proportionality constant which ensures that its integral has the value unity.

In our notation, the characteristic function is (cf. (4.27))

(*) If $\mathbf{A}$ has rank $r < p$, the distribution degenerates into r dimensions, i.e. one or more variates are redundant.

$$\phi(\mathbf{t}) \propto \int_{-\infty}^{\infty} \cdots \int_{-\infty}^{\infty} \exp(i\mathbf{t}'\mathbf{x} - \tfrac{1}{2}\mathbf{x}'\mathbf{A}\mathbf{x}) \Pi\, dx \tag{15.4}$$

which, on applying (15.3), becomes

$$\phi(\mathbf{t}) \propto \int_{-\infty}^{\infty} \cdots \int_{-\infty}^{\infty} \exp(i\mathbf{t}'\mathbf{B}\mathbf{y} - \tfrac{1}{2}\mathbf{y}'\mathbf{C}\mathbf{y}) \prod_j dy_j, \tag{15.5}$$

the Jacobian being unity since $\mathbf{B}$ is orthogonal. We have written

$$\mathbf{C} = \mathbf{B}'\mathbf{A}\mathbf{B} \tag{15.6}$$

in (15.5). Since $\mathbf{C}$ is a diagonal matrix, the right side of (15.5) factorizes into p single integrals of type

$$I_j = \int_{-\infty}^{\infty} \exp(iu_j b_j y_j - \tfrac{1}{2}c_{jj} y_j^2)\, dy_j,$$

where u_j is linear in the dummy variables t_j. Then I_j is obtained directly from the characteristic function of the univariate normal distribution (Example 3.11) as

$$I_j = \left(\frac{2\pi}{c_{jj}}\right)^{\frac{1}{2}} \exp\left(-\tfrac{1}{2}\frac{u_j^2 b_j^2}{c_{jj}}\right). \tag{15.7}$$

Applying (15.7) to each term in (15.5) gives

$$\phi(\mathbf{t}) \propto (2\pi)^{\frac{1}{2}p} (\prod_j c_{jj})^{-\frac{1}{2}} \exp\left(-\tfrac{1}{2}\Sigma \frac{u_j^2 b_j^2}{c_{jj}}\right). \tag{15.8}$$

Since $\mathbf{C}$ is diagonal, its determinant is

$$|C| = \Pi\, c_{jj}$$

and that of its inverse matrix $\mathbf{C}^{-1}$ is

$$|C^{-1}| = |C|^{-1} = (\prod_j c_{jj})^{-1}. \tag{15.9}$$

Using (15.9) in (15.8) and rewriting its exponent in matrix form, we have

$$\phi(\mathbf{t}) \propto (2\pi)^{\frac{1}{2}p} |C^{-1}|^{\frac{1}{2}} \exp(-\tfrac{1}{2}\mathbf{t}'\mathbf{B}\mathbf{C}^{-1}\mathbf{B}'\mathbf{t}). \tag{15.10}$$

Since, by (15.6),

$$\mathbf{B}\mathbf{C}^{-1}\mathbf{B}' = \mathbf{B}(\mathbf{B}'\mathbf{A}\mathbf{B})^{-1}\mathbf{B}' = \mathbf{B}\mathbf{B}^{-1}\mathbf{A}^{-1}(\mathbf{B}')^{-1}\mathbf{B}' = \mathbf{A}^{-1}, \tag{15.11}$$

we have, since $\mathbf{B}$ is orthogonal, $\mathbf{B}\mathbf{B}' = \mathbf{I}$ and

$$|A^{-1}| = |BC^{-1}B'| = |B||C^{-1}||B'| = |C^{-1}|. \tag{15.12}$$

We may substitute (15.11) and (15.12) into (15.10) to give

$$\phi(\mathbf{t}) \propto (2\pi)^{\frac{1}{2}p} |A^{-1}|^{\frac{1}{2}} \exp(-\tfrac{1}{2}\mathbf{t}'\mathbf{A}^{-1}\mathbf{t}). \tag{15.13}$$

If we put $\mathbf{t} = \mathbf{0}$ in (15.13), we obtain immediately for the proportionality constant of the multivariate normal distribution

$$(2\pi)^{-\frac{1}{2}p} |A^{-1}|^{-\frac{1}{2}} = (2\pi)^{-\frac{1}{2}p} |A|^{\frac{1}{2}}.$$

Thus (15.2) becomes

$$dF = (2\pi)^{-\frac{1}{2}p} |A|^{\frac{1}{2}} \exp\{-\tfrac{1}{2}(\mathbf{x}-\boldsymbol{\mu})'\mathbf{A}(\mathbf{x}-\boldsymbol{\mu})\} \Pi\, dx \tag{15.14}$$

with characteristic function

$$\phi(\mathbf{t}) = \exp(-\tfrac{1}{2}\mathbf{t}'\mathbf{A}^{-1}\mathbf{t})\exp(i\mathbf{t}'\boldsymbol{\mu}), \tag{15.15}$$

the second factor on the right of (15.15) arising if the location vector $\boldsymbol{\mu}$ is non-null, as may be seen from the effect on (15.5) of putting $(\mathbf{x}-\boldsymbol{\mu})$ for $\mathbf{x}$ in the second term of the exponent of (15.4) and in (15.3).

15.3 From the characteristic function (15.15), the moments of the distribution, or, more conveniently, its cumulants, are easily obtainable. The cumulant-generating function is

$$\psi(\mathbf{t}) = -\tfrac{1}{2}\mathbf{t}'\mathbf{A}^{-1}\mathbf{t}+i\mathbf{t}'\boldsymbol{\mu}. \tag{15.16}$$

It is clear from the form of (15.16), which has no terms of degree above 2, that all cumulants of order greater than 2 are equal to zero, thus generalizing the univariate property of the normal distribution. The mean of the jth variate is

$$\left[\frac{\partial\psi(\mathbf{t})}{\partial(it_j)}\right]_{\mathbf{t}=\mathbf{0}} = \mu_j,$$

so that the location vector $\boldsymbol{\mu}$ is the vector of means of the variables. The variance of the jth variate is

$$\left[\frac{\partial^2\psi(\mathbf{t})}{\partial(it_j)^2}\right]_{\mathbf{t}=\mathbf{0}} = a_{jj}^{-1}, \tag{15.17}$$

the element in the jth row and jth column of $\mathbf{A}^{-1}$. The covariance of the jth and kth variates is

$$\left[\frac{\partial^2\psi(\mathbf{t})}{\partial(it_j)\,\partial(it_k)}\right]_{\mathbf{t}=\mathbf{0}} = a_{jk}^{-1}. \tag{15.18}$$

(15.17) and (15.18) show that $\mathbf{A}^{-1}$ is the matrix of variances and covariances of the variates, which we shall call their *dispersion matrix*. We may now make our notation more suggestive by writing $\mathbf{V}$ for $\mathbf{A}^{-1}$, and obtain, as our final expression for the p-variate normal distribution in terms of its first- and second-order moments,

$$dF = (2\pi)^{-\frac{1}{2}p}|V^{-1}|^{\frac{1}{2}}\exp\{-\tfrac{1}{2}(\mathbf{x}-\boldsymbol{\mu})'\mathbf{V}^{-1}(\mathbf{x}-\boldsymbol{\mu})\}\Pi\,dx \tag{15.19}$$

with characteristic function

$$\phi(\mathbf{t}) = \exp(-\tfrac{1}{2}\mathbf{t}'\mathbf{V}\mathbf{t})\exp(i\mathbf{t}'\boldsymbol{\mu}). \tag{15.20}$$

If and only if $\mathbf{V}$ is a diagonal matrix, the variates are independent.

The dispersion matrix $\mathbf{V}$ of any (not necessarily multinormal) variates is non-negative definite since the quadratic form in $u_1, \ldots, u_p$,

$$\mathbf{u}'\mathbf{V}\mathbf{u} = E\left\{\sum_{j=1}^{p}(x_j-\mu_j)u_j\right\}^2 \geqslant 0.$$

In the non-singular case, $\mathbf{V}$ is strictly positive definite.

Example 15.1 The bivariate normal distribution

We have already encountered the bivariate specializations of (15.19) and (15.20) in Example 3.17 and elsewhere. In this case the matrix of variances and covariances is

$$\mathbf{V} = \begin{pmatrix}\sigma_1^2 & \rho\sigma_1\sigma_2\\ \rho\sigma_1\sigma_2 & \sigma_2^2\end{pmatrix},$$

where ρ, the correlation, coefficient is defined by $\mu_{11}/\sigma_1\sigma_2$. Hence

$$\mathbf{V}^{-1} = \begin{pmatrix} \dfrac{1}{\sigma_1^2(1-\rho^2)} & \dfrac{-\rho}{\sigma_1\sigma_2(1-\rho^2)} \\ \dfrac{-\rho}{\sigma_1\sigma_2(1-\rho^2)} & \dfrac{1}{\sigma_2^2(1-\rho^2)} \end{pmatrix},$$

and the frequency function with zero means is, from (15.19),

$$dF = \frac{dx_1\,dx_2}{2\pi\sigma_1\sigma_2(1-\rho^2)^{\frac{1}{2}}}\exp\left\{-\frac{1}{2(1-\rho^2)}\left(\frac{x_1^2}{\sigma_1^2}-\frac{2\rho x_1 x_2}{\sigma_1\sigma_2}+\frac{x_2^2}{\sigma_2^2}\right)\right\},$$

with characteristic function, from (15.20),

$$\phi(t_1,t_2) = \exp\{-\tfrac{1}{2}(\sigma_1^2 t_1^2+2\rho\sigma_1\sigma_2 t_1 t_2+\sigma_2^2 t_2^2)\}.$$

Linear functions of normal variates

15.4 We have seen (Example 11.2) that a linear function of independent univariate normal variates is itself normally distributed. We may now obtain a much more general result, to the effect that a set of variates, each of which is a linear function of a set of multinormal variates (i.e. variates which have a multivariate normal distribution), itself has a multinormal distribution. (15.20) is the characteristic function of a p-variate non-singular normal distribution. It may be written (putting $\boldsymbol{\mu} = \mathbf{0}$)

$$E\{\exp(i\mathbf{t}'\mathbf{x})\} = \exp(-\tfrac{1}{2}\mathbf{t}'\mathbf{V}\mathbf{t}). \tag{15.21}$$

Suppose that we write

$$\mathbf{t}' = \mathbf{s}'\mathbf{A}, \tag{15.22}$$

where $\mathbf{s}'$ is a vector of q dummy variables, just as $\mathbf{t}'$ is a vector of p such variables, and $\mathbf{A}$ is any $(q\times p)$ matrix. (15.22) substituted into (15.21) gives

$$E\{\exp(i\mathbf{s}'\mathbf{A}\mathbf{x})\} = \exp(-\tfrac{1}{2}\mathbf{s}'\mathbf{A}\mathbf{V}\mathbf{A}'\mathbf{s}). \tag{15.23}$$

(15.23) shows that the set of q linear functions of the p x-variates defined by $\mathbf{y} = \mathbf{A}\mathbf{x}$ has a q-variate multinormal c.f. with dispersion matrix $\mathbf{AVA}'$. If $\mathbf{A}$ is a non-singular matrix, the multivariate form of the Inversion Theorem (**4.17**) gives us a multinormal frequency function (15.19) for $\mathbf{y}$, with $\mathbf{AVA}'$ replacing $\mathbf{V}$.

The converse is also true : if (15.23) holds for every $\mathbf{A}$, (15.22) leads us back to (15.21) and the multinormal distribution of $\mathbf{x}$. In particular, putting $q = 1$ shows that the multinormal distribution may be characterized by the property that every linear function of its variates has a univariate normal distribution.

If $\mathbf{A}$ is singular, $(\mathbf{AVA}')^{-1}$ is not defined, but (15.23) is nevertheless the c.f. of a degenerate multinormal distribution. In particular, this is so if $q>p$. Looking at the matter from this standpoint, it is evident that any degenerate multinormal distribution may be regarded, through its c.f., as having arisen by a singular transformation from a non-degenerate one. Such singularity causes no difficulty in practice, since a distribution can be made non-singular by discarding redundant variates.

An important corollary of the result of this section, obtained by making $\mathbf{A}$ an identity matrix augmented by zeros, with $q\leqslant p-1$, is that the (marginal) distribution of any subset of the original variates is multinormal, with variances and covariances unchanged, although the inverse of their dispersion matrix, and hence the coefficients in the exponent of their (marginal) frequency function, will of course be changed.

Exercise 15.1 discusses the conditional distributions.

The multivariate normal integral

15.5 Apart from the location and scale parameters, which may be removed by standardization, the frequency function (15.19) contains $\frac{1}{2}p(p-1)$ parameters, namely the covariances between the p variates. It is thus clear that tabulation of its distribution function is a formidable undertaking for $p>2$. Bivariate tables are given in *Tables of the Bivariate Normal Distribution and Related Functions* (National Bureau of Standards, Applied Mathematics Series, 50, Washington, 1959) for $\pm\rho = 0(0{\cdot}05)\,0{\cdot}95\,(0{\cdot}01)1$ and variates in the range $0\,(0{\cdot}1)\,4$, to 6 or 7 d.p. Zelen and Severo (1960) give charts for reading the bivariate normal integral value with an error of 1 per cent. or less.

D. B. Owen (1956, 1957) gives tables for computing the bivariate integral over regions bounded by straight lines and certain other regions. G. P. Steck (1958b) gives tables for computing the trivariate integral. S. S. Gupta (1963a, b) surveys the literature for $p \geqslant 2$ and gives a bibliography. For any p, Dutt (1973) gives rapidly convergent formulae by using integral transforms in tetrachoric series like (15.33) below.

15.6 A special problem which frequently arises in statistical theory is the evaluation of the multivariate normal integral over that part of the range of variation for which all p variates exceed their means, i.e., if we measure from their means, for which they are positive. Although this is a considerable simplification of the general problem, it still presents analytical difficulties for $p>3$, and a number of special devices have been used to overcome these in particular cases.

We deal first with the two- and three-variate cases, the only ones for which analytical solutions of the problem are known.

In the two-variate case, we write the frequency function as the Fourier transform (4.31) of its c.f. (15.20) and evaluate its integral over the positive quadrant as

$$\mathrm{P}_2^0 = \int_0^\infty \int_0^\infty \left\{ \frac{1}{(2\pi)^2} \int_{-\infty}^\infty \int_{-\infty}^\infty \exp(-i\mathbf{t}'\mathbf{x} - \tfrac{1}{2}\mathbf{t}'\mathbf{V}\mathbf{t})\, dt_1\, dt_2 \right\} dx_1\, dx_2, \tag{15.24}$$

where the vectors and matrix are now of order two. Reversing the order of integration in (15.24), we integrate out x_1 and x_2 and obtain

$$4\pi^2 \mathrm{P}_2^0 = \int_{-\infty}^\infty \int_{-\infty}^\infty \exp(-\tfrac{1}{2}\mathbf{t}'\mathbf{V}\mathbf{t}) \frac{dt_1}{it_1} \frac{dt_2}{tt_2}. \tag{15.25}$$

The coefficient of $(-t_1 t_2)$ in the exponent of the integrand is the covariance of the two variates, which is (see Example 15.1) $\rho\sigma_1\sigma_2$. Differentiating (15.25) with respect to ρ, we obtain

$$4\pi^2 \frac{\partial \mathrm{P}_2^0}{\sigma_1 \sigma_2 \partial \rho} = \int_{-\infty}^\infty \int_{-\infty}^\infty \exp(-\tfrac{1}{2}\mathbf{t}'\mathbf{V}\mathbf{t})\, dt_1\, dt_2. \tag{15.26}$$

The integrand on the right of (15.26) is a bivariate normal form in t_1 and t_2 apart from the constant term. The integral in (15.26) therefore equals the reciprocal of that constant, and (15.26) becomes

$$\frac{\partial \mathrm{P}_2^0}{\sigma_1 \sigma_2 \partial \rho} = \frac{1}{4\pi^2} \cdot \frac{2\pi}{|V|^{\frac{1}{2}}} = \frac{1}{2\pi |V|^{\frac{1}{2}}} = \{2\pi\sigma_1\sigma_2(1-\rho^2)^{\frac{1}{2}}\}^{-1}, \tag{15.27}$$

from Example 15.1. On integration, (15.27) gives $\mathrm{P}_2^0 = \arcsin\rho/(2\pi)$ + a constant, and the constant is seen to be $\frac{1}{4}$ by putting $\rho = 0$. Thus

$$\mathrm{P}_2^0 = \tfrac{1}{4} + \arcsin\rho/(2\pi). \tag{15.28}$$

(15.28) is usually called Sheppard's (1898) theorem on median dichotomy, although it was given earlier by Stieltjes. By symmetry, the same result holds for the negative quadrant, while the two "mixed" quadrants have content $(\frac{1}{2}-P_2^0)$ each. As is to be expected, P_2^0 does not depend on the variances, which cannot affect the proportion of the total frequency in a quadrant. P_2^0 is tabulated by National Bureau of Standards (1959)—cf. reference in **15.5**—for $\rho = 0(0{\cdot}01)1$.

15.7 The corresponding three-variate result is derived from this fairly easily. We have seen (Exercise 7.1) that the probability of occurrence of at least one of a set of p events is

$$\Sigma P_1 - \Sigma P_2 + \Sigma P_3 - \ldots + (-1)^{p-1} P_p, \tag{15.29}$$

the summations being of the probabilities of occurrence of the events singly, in pairs, and so on to the last term, which contains the probability that all p events occur jointly.

If these p events are the positiveness of p multinormal variates, the probability that at least one is positive is the complement of the probability that none is positive, and, by symmetry, also the complement of the probability that none is negative, which is what we seek. Using (15.29) and writing the zero superscript as before, we therefore have

$$1-P_p^0 = \Sigma P_1^0 - \Sigma P_2^0 + \Sigma P_3^0 \ldots + (-1)^{p-1} P_p^0. \tag{15.30}$$

If p is even, the last term on the right has a negative sign, and (15.30) is an identity; but if p is odd, it becomes

$$P_p^0 = \tfrac{1}{2}\{1 - \Sigma P_1^0 + \Sigma P_2^0 \ldots + (-1)^{p-1} \Sigma P_{p-1}^0\}. \tag{15.31}$$

(15.31) enables us to obtain P_p^0 for odd p from lower-order probabilities.

With $p = 3$, (15.31) gives

$$\begin{aligned} P_3^0 &= \tfrac{1}{2}(1 - \Sigma P_1^0 + \Sigma P_2^0) = \tfrac{1}{2}\{1 - 3.\tfrac{1}{2} + \Sigma(\tfrac{1}{4} + \text{arc } \sin\rho)/(2\pi)\} \\ &= \tfrac{1}{8} + \frac{1}{4\pi}(\text{arc } \sin\rho_{12} + \text{arc } \sin\rho_{13} + \text{arc } \sin\rho_{23}), \end{aligned} \tag{15.32}$$

a very simple generalization of Sheppard's formula.

15.8 No such simple result is available for $p > 3$. Abrahamson (1964) gives tables for calculating P_4^0—some special covariance structures are treated by Cheng (1969). Childs (1967) reduces P_4^0 to the evaluation of three single integrals and also obtains results for P_6^0. For general p, Kendall (1941) gave a power series expansion in the covariance parameters. We may assume all variances to be unity, without losing generality. The Fourier inversion of the characteristic function (cf. (15.24)) can then be expanded as a power series in the correlations ρ_{ij}. Thus

$$\begin{aligned} P_p^0 &= \int_0^\infty \cdots \int_0^\infty \left\{\frac{1}{(2\pi)^p} \int_{-\infty}^\infty \cdots \int_{-\infty}^\infty \exp(-i\mathbf{t}'\mathbf{x} - \tfrac{1}{2}\mathbf{t}'\mathbf{V}\mathbf{t})\, dt_1 \ldots dt_p\right\} dx_1 \ldots dx_p \\ &= \int_0^\infty \cdots \int_0^\infty \left\{\frac{1}{(2\pi)^p} \int_{-\infty}^\infty \cdots \int_{-\infty}^\infty \exp(-i\mathbf{t}'\mathbf{x} - \tfrac{1}{2}\mathbf{t}'\mathbf{t}).\Sigma\left[(-1)^{n_{\cdot\cdot}} \frac{\rho_{12}^{n_{12}} \rho_{13}^{n_{13}} \cdots}{n_{12}!\, n_{13}! \ldots}\right]\right. \\ &\qquad \left. \times\, t_1^{n_{1\cdot}}\, t_2^{n_{2\cdot}} \ldots t_p^{n_{p\cdot}}\, dt_1 \ldots dt_p\right\} dx_1 \ldots dx_p, \end{aligned} \tag{15.33}$$

where the summation is over all possible sets of the ρ_{ij} taken over all non-negative values of the n_{ij}; $n_{i.} = \sum_j (n_{ij}+n_{ji})$ and $n_{..} = \sum_i n_{i.}$.

Interchanging summation and integration, we may re-write (15.33) as

$$P_p^0 = \Sigma\left[(-1)^{n_{..}}\frac{\rho_{12}^{n_{12}}\rho_{13}^{n_{13}}\cdots}{n_{12}!\,n_{13}!\cdots}\right]\prod_{j=1}^{p} G_{n_{j.}}, \tag{15.34}$$

where
$$G_{n_{j.}} = \frac{1}{2\pi}\int_0^\infty\int_{-\infty}^\infty t_j^{n_{j.}}\exp(-it_j x_j - \tfrac{1}{2}t_j^2)\,dt_j\,dx_j. \tag{15.35}$$

If $n_{j.} = 0$, $G_{n_{j.}} = \frac{1}{2}$ from (15.35). If $n_{j.}$ is an integer, we integrate for x_j as at (15.24) to obtain

$$G_{n_{j.}} = \frac{1}{2\pi i}\int_{-\infty}^\infty t_j^{n_{j.}-1}e^{-\frac{1}{2}t_j^2}dt_j,$$

and by the formulae for the moments of the univariate normal distribution (5.60), we have

$$G_{n_{j.}}\begin{cases} = 0 & \text{if } n_{j.} \text{ is even} \\ = \dfrac{(2m)!}{\sqrt{(2\pi)}\,i\,2^m\,m!} & \text{if } n_{j.} \text{ is odd} = 2m+1. \end{cases}$$

Thus finally

$$G_{n_{j.}} = \begin{cases} \frac{1}{2} & \text{if } n_{j.} = 0 \\ 0 & \text{if } n_{j.} = 2m \quad (m = 1,2\ldots) \\ \dfrac{(2m)!}{\sqrt{(2\pi)}\,i\,2^m\,m!} & \text{if } n_{j.} = 2m+1 \quad (m = 0,1,2\ldots). \end{cases} \tag{15.36}$$

Thus partitions vanish from (15.34) which contain any t_i an even number of times.

(15.34) and (15.36) formally solve the problem. But although the series (15.34) for P_p^0 always converges, it does so only slowly if the ρ_{ij} are not small. Moreover, the terms in (15.34) are not all positive. In the four-variate case (cf. Moran (1948)), (15.34) becomes

$$P_4^0 = \frac{1}{16}+\frac{1}{8\pi}\sum_{i,j}\rho_{ij}+\frac{1}{4\pi^2}\sum_{ijkl}\rho_{ij}\rho_{kl}+\ldots$$

but the coefficient of $\rho_{12}\,\rho_{13}\,\rho_{14}$ is negative, being $-1/(4\pi^2)$.

15.9 In the special case when all the correlations are equal, a remarkable simplification is possible.

Let $x_0, x_1, \ldots, x_p$ be a set of $p+1$ independent normal variates with zero means and unit variances. Consider the derived set of variables

$$y_i = x_i - bx_0, \qquad i = 1,\ldots p, \qquad b>0. \tag{15.37}$$

These will be multinormally distributed with zero means, variances $(1+b^2)$ and co-variances b^2. The correlation between any pair of y's is thus

$$\rho = b^2/(1+b^2).$$

This ranges from zero when b is zero to 1 when $b\to\infty$. The y_i can thus be taken to represent the general multinormal distribution with equal (positive) correlations. The value of P_p^0 is immediately deducible from the construction of the distribution; it is simply

$$P_p^0 = \int_{-\infty}^{\infty} (2\pi)^{-\frac{1}{2}} e^{-\frac{1}{2}x^2} \left\{ \int_{bx}^{\infty} (2\pi)^{-\frac{1}{2}} e^{-\frac{1}{2}t^2} dt \right\}^p dx, \tag{15.38}$$

the probability that p independent normal variates exceed bx, integrated over a standard normal distribution. This is an integral of the Hojo type (cf. **14.5**) and can be simply calculated with little error by replacing the integration by a summation. Evidently, the argument leading to the result is basically unaffected if some y_i are defined by $y_i = x_i + bx_0$ so that some of the y_i have negative correlations with the others, of magnitude $-\rho$. In this case, of course, (15.38) must be suitably modified.

Ruben (1954) has tabulated P_p^0 in the equal-correlations case for $\rho = 1/j$, $p = 1, 2, \ldots, 51-j$, where $j = 2, 3, \ldots, 12$.

The case of equal negative correlations is covered by the generalization of (15.37) given in Exercise 15.11, which deals with the order-statistics for any ρ. Gupta *et al.* (1973) tabulate the upper percentiles of $y_{(p)}$ for 17 positive values of ρ and $p = 1(1)10(2)50$.

G. P. Steck (1962) reviews the equal-correlations case and gives new results, especially for $p = 4$. See also Bacon (1963), S. S. Gupta (1963a), Dutt (1973) and Bechhofer and Tamhane (1974).

Exercise 15.12 generalizes (15.38) to the case $\rho_{ij} = a_i a_j$.

Example 15.2

To illustrate the approximation of the outer integral in (15.38) by a sum, we consider the case $p = 2$ when $b = 1$ in (15.37), i.e. $\rho = \frac{1}{2}$. Replacing the integral by the sum of the values at $x = \pm 3, \pm 2, \pm 1, 0$, we find from Appendix Tables 1 and 2 the following results:—

x	(1) $= \int_x^{\infty} (2\pi)^{-\frac{1}{2}} e^{-\frac{1}{2}t^2} dt$	(2) $= \{(1)\}^2$	(3) = Sums of (2)	(4) $= (2\pi)^{-\frac{1}{2}} e^{-\frac{1}{2}x^2}$	(3) × (4)
−3 +3	0·99865 0·00135	0·99730 0·00000	0·99730	0·00443	0·00442
−2 +2	0·97725 0·02275	0·95502 0·00052	0·95554	0·05399	0·05159
−1 +1	0·84134 0·15866	0·70785 0·02517	0·73302	0·24197	0·17737
0	0·50000	0·25000	0·25000	0·39894	0·09973
					0·33311

The value obtained is 0·3331 to 4 d.p. The exact value, given by (15.28), is $\frac{1}{3}$. Greater accuracy could have been obtained by using a sum of more than seven values. For example, if we add in a further pair of terms for $x = \pm 4$, they contribute 0·00013 to the sum, giving a total of 0·33324. To 5 d.p., no further gain in accuracy results from adding terms for $x = \pm 5$, but accuracy as great as is desired could be obtained by spacing the values at narrower intervals, say 0·5 instead of 1.

Moran (1956) gives a bound for the relative error in the sum compared to the integral. It is, in our notation,

$$2\exp\left\{-\frac{\pi^2}{h^2(1+pb^2)}\right\},$$

where h is the interval of summation. In our example, this is

$$2\exp\left(-\frac{\pi^2}{3h^2}\right).$$

With $h = 1$, this gives a bound considerably in excess of the error obtained above with nine values. With smaller h, the bound decreases sharply and also seems to be closer to the actual error.

Quadratic forms in normal variates

15.10 Suppose that $\mathbf{x}$ is distributed with the multinormal density

$$dF \propto \exp(-\tfrac{1}{2}\mathbf{x}'\mathbf{A}\mathbf{x}) \prod_{j=1}^{p} dx_j,$$

where $\mathbf{A}$ is not necessarily non-singular. Consider the distribution of the quadratic form $\mathbf{x}'\mathbf{A}\mathbf{x}$ itself. Its characteristic function is, writing $\theta = it$,

$$\phi(t) \propto \int_{-\infty}^{\infty} \cdots \int_{-\infty}^{\infty} \exp\{-\tfrac{1}{2}\mathbf{x}'\mathbf{A}\mathbf{x}(1-2\theta)\} \prod_j dx_j. \tag{15.39}$$

The transformation (15.3) orthogonalizes the quadratic form as in **15.2** and (15.39) becomes

$$\phi(t) \propto \int_{-\infty}^{\infty} \cdots \int_{-\infty}^{\infty} \exp\{-\tfrac{1}{2}\mathbf{y}'\mathbf{C}\mathbf{y}(1-2\theta)\} \prod_j dy_j. \tag{15.40}$$

The number of non-zero diagonal elements in $\mathbf{C}$ is precisely the number of non-zero latent roots of $\mathbf{A}$, say r. If we integrate out the $(p-r)$ variables corresponding to the zero latent roots, the exponent of (15.40) is unaffected. To the remaining r-fold integral, we apply the transformation

$$z_j = y_j c_{jj}(1-2\theta)^{\frac{1}{2}}$$

to obtain

$$\phi(t) \propto (1-2\theta)^{-\frac{1}{2}r} \int_{-\infty}^{\infty} \cdots \int_{-\infty}^{\infty} \exp(-\tfrac{1}{2}\mathbf{z}'\mathbf{z}) \prod_j dz_j. \tag{15.41}$$

The integral in (15.41) is independent of θ. Thus

$$\phi(t) = (1-2\theta)^{-\frac{1}{2}r}, \tag{15.42}$$

the elimination of constants being verified by the fact that $\phi(0) = 1$. By the Inversion Theorem and Example 3.6, (15.42) shows that the quadratic form in the exponent of a multinormal distribution has a distribution known as the χ^2 distribution with r degrees of freedom, defined by (11.8) with $n = r$ (see also Chapter 16), where r is the rank (of the quadratic form) of the distribution. This generalizes the result of (11.8), which deals effectively with the case where $\mathbf{A}$ is diagonal and non-singular.

Example 15.3 The χ^2 goodness-of-fit distribution

Given a sample of n observations from a multinomial distribution (**5.30**) with probabilities p_i $(i = 1, \ldots, k)$, it follows from the multivariate form of the Central Limit theorem that the observed numbers x_i in the k groups will tend, with increasing n, to have a multivariate normal distribution, with means np_i and dispersion matrix given (as at (5.80)) by

$$\mathbf{V} = n\begin{pmatrix} p_1(1-p_1), & -p_1p_2, & -p_1p_3, & \ldots, & -p_1p_k \\ -p_2p_1, & p_2(1-p_2), & -p_2p_3, & \ldots, & -p_2p_k \\ \cdot & \cdot & \cdot & \cdots & \cdot \\ \cdot & \cdot & \cdot & \cdots & \cdot \\ \cdot & \cdot & \cdot & \cdots & \cdot \\ -p_kp_1, & -p_kp_2, & \cdots & \ldots, & p_k(1-p_k) \end{pmatrix}$$

Because of the linear restriction $\sum_{j=1}^{k} x_j = n$, the multinormal distribution is singular, of rank $(k-1)$, so $|\mathbf{V}| = 0$ and we cannot invert $\mathbf{V}$. However, we may omit one of the x_i (say x_k), as it is redundant, the remaining $(k-1)$ x_i being multinormally distributed with means np_i and dispersion matrix $\mathbf{V}^*$ which is simply $\mathbf{V}$ with its last row and column eliminated. It is easily verified by a direct matrix multiplication that

$$(\mathbf{V}^*)^{-1} = \frac{1}{n}\begin{pmatrix} \frac{1}{p_1}+\frac{1}{p_k}, & \frac{1}{p_k}, & \frac{1}{p_k}, & \cdots & \frac{1}{p_k} \\ \frac{1}{p_k}, & \frac{1}{p_2}+\frac{1}{p_k}, & \frac{1}{p_k}, & \cdots & \frac{1}{p_k} \\ \cdot & \cdot & \cdot & \cdots & \cdot \\ \cdot & \cdot & \cdot & \cdots & \cdot \\ \cdot & \cdot & \cdot & \cdots & \cdot \\ \frac{1}{p_k}, & \frac{1}{p_k}, & \frac{1}{p_k}, & \cdots & \frac{1}{p_{k-1}}+\frac{1}{p_k} \end{pmatrix}$$

The exponent of the multinormal distribution thus has the quadratic form

$$\begin{aligned}(\mathbf{x}-\boldsymbol{\mu})'(\mathbf{V}^*)^{-1}(\mathbf{x}-\boldsymbol{\mu}) &= \sum_{i=1}^{k-1}\frac{(x_i-np_i)^2}{np_i}+\frac{1}{np_k}\sum_{i=1}^{k-1}\sum_{j=1}^{k-1}(x_i-np_i)(x_j-np_j) \\ &= \sum_{i=1}^{k-1}\frac{(x_i-np_i)^2}{np_i}+\frac{1}{np_k}\left\{\sum_{i=1}^{k-1}(x_i-np_i)\right\}^2 \\ &= \sum_{i=1}^{k}\frac{(x_i-np_i)^2}{np_i}.\end{aligned}$$

By the result of **15.10**, this variate is distributed like χ^2 with $(k-1)$ degrees of freedom. This result, due to Karl Pearson (1900), is the basis of the χ^2 goodness-of-fit test which we shall discuss in Chapter 30, Volume 2.

15.11 We now suppose that $\mathbf{x}$ is a vector of p independent standardized normal variates, and consider the distribution of the general quadratic form

$$Q = \mathbf{x}'\mathbf{A}\mathbf{x},$$

where the real matrix $\mathbf{A}$ may be assumed symmetric without loss of generality. By the orthogonal transformation (15.3), $\mathbf{x}$ is transformed to $\mathbf{y}$, a vector of independent standardized normal variates by Example 11.2, and using (15.6)

$$Q = \mathbf{y}'\mathbf{C}\mathbf{y} = \sum_{i=1}^{p} a_i y_i^2, \tag{15.43}$$

where the a_i are latent roots of $\mathbf{A}$—only the r ($\leqslant p$) non-zero a_i contribute to (15.43), where r is the rank of $\mathbf{A}$, so that the distribution of Q depends only upon them. The c.g.f. of $a_i y_i^2$, a constant multiple of a χ^2 distribution with 1 d.fr., is (cf. **15.10** and **16.3** below)

$$\psi_i(t) = -\tfrac{1}{2}\log(1-2\theta a_i), \tag{15.44}$$

where $\theta = it$. By the additive property of c.g.f.'s, that of Q is therefore

$$\psi(t) = -\tfrac{1}{2}\sum_{i=1}^{p}\log(1-2\theta a_i). \tag{15.45}$$

Expanding the logarithm in series near $t = 0$, it follows that the cumulants of Q are

$$\kappa_s = 2^{s-1}(s-1)!\sum_{i=1}^{p} a_i^s. \tag{15.46}$$

Now $\sum_{i=1}^{p} a_i^s$ is the trace (i.e. the sum of elements in the leading diagonal) of the sth power of $\mathbf{A}$, and we denote this by $\operatorname{tr}\mathbf{A}^s$. Thus (15.46) is

$$\kappa_s = 2^{s-1}(s-1)!\operatorname{tr}\mathbf{A}^s \tag{15.47}$$

and the c.g.f. (15.45) may therefore be written in the form

$$\psi(t) = \sum_{s=1}^{\infty} 2^{s-1}\operatorname{tr}(\theta\mathbf{A})^s/s. \tag{15.48}$$

The c.f. is, using (15.45),

$$\begin{aligned}\phi(t) = \mathrm{E}\{\exp(\mathbf{x}'\theta\mathbf{A}\mathbf{x})\} &= \prod_{i=1}^{p}(1-2\theta a_i)^{-\frac{1}{2}} \\ &= |\mathbf{I}-2\theta\mathbf{A}|^{-\frac{1}{2}},\end{aligned} \tag{15.49}$$

since $1-2\theta a_i$ is a latent root of $\mathbf{I}-2\theta\mathbf{A}$ and the determinant of a matrix is the product of its latent roots.

If (15.49) is to be of the form (15.42) of the c.f. of a χ^2 distribution with ν degrees of freedom, we must therefore have, identically in θ,

$$\prod_{i=1}^{p}(1-2\theta a_i) = (1-2\theta)^{\nu}.$$

Thus ν of the a_i must be equal to 1, $\nu \leqslant p$, and the other $p-\nu$ equal to 0, whence ν must equal r, the rank of $\mathbf{A}$, and (15.46) becomes $\kappa_s = 2^s(s-1)!\,r$.

From (15.6), $\mathbf{A} = \mathbf{BCB}'$, so $\mathbf{A}^2 = \mathbf{BCB}'\mathbf{BCB}' = \mathbf{BC}^2\mathbf{B}' = \mathbf{BCB}'$ since $\mathbf{C}$ is a diagonal matrix containing only elements 0 and 1. Thus $\mathbf{A} = \mathbf{A}^2$. Conversely, $\mathbf{A} = \mathbf{A}^2$ implies that r of the $a_i = 1$ and the others 0. A matrix $\mathbf{A} = \mathbf{A}^2$ is called *idempotent*. We have established that Q is distributed as χ^2 with r d.fr. if and only if $\mathbf{A}$ is idempotent of rank r.

Series representations of the distribution of Q for any positive definite $\mathbf{A}$ are unified by Kotz *et al.* (1967a, b), who use the more general assumptions of **15.21** below. See also Gideon and Gurland (1976).

Tables of the distribution of Q were given by Solomon (1960) for $r = 2$ and 3 and by Johnson and Kotz (1967a, b, 1968) for $r = 4$ and 5. Sankaran (1959) and Jensen and Solomon (1972) used the method of **16.7** below, as generalized by Haldane (1938), to find the fractional power h of Q that is most closely normally distributed, and obtained $h = 1-2\,\text{tr}\,\mathbf{A}\,\text{tr}\,\mathbf{A}^3/\{3(\text{tr}\,\mathbf{A}^2)^2\}$, reducing to $h = \frac{1}{3}$ when $\mathbf{A}$ is idempotent and distributed as χ^2, agreeing with **16.7**.

If the a_i are equal in pairs, (15.49) becomes $\phi(t) = \prod_{i=1}^{\frac{1}{2}p}(1-2\theta a_i)^{-1}$, which may be expanded in partial fractions as $\phi(t) = \sum_{i=1}^{\frac{1}{2}p} c_j(1-2\theta a_i)^{-1}$, whence the Inversion Theorem for c.f.'s gives $\text{Prob}\{Q > d\} = \sum_{i=1}^{\frac{1}{2}p} c_j \text{Prob}\{z_i > d/a_i\}$, where the z_i are independent χ^2 variates with 2 d.f., so that $\frac{1}{2}z_i$ are independent exponential variates. A more general result of this kind is given in Exercise 36.14, Vol. 3.

15.12 If we now turn to the joint distribution of two quadratic forms

$$Q_1 = \mathbf{x}'\mathbf{A}\mathbf{x}, \qquad Q_2 = \mathbf{x}'\mathbf{B}\mathbf{x},$$

we see at once that its c.f. $\phi(t_1, t_2) = E\{\exp[\mathbf{x}'(\theta_1\mathbf{A}+\theta_2\mathbf{B})\mathbf{x}]\}$ is simply (15.49) with $\theta\mathbf{A}$ replaced by $(\theta_1\mathbf{A}+\theta_2\mathbf{B})$. Thus we have

$$\phi(t_1, t_2) = |\mathbf{I}-2\theta_1\mathbf{A}-2\theta_2\mathbf{B}|^{-\frac{1}{2}} \tag{15.50}$$

and for the c.g.f. similarly, from (15.48),

$$\psi(t_1, t_2) = \sum_{s=1}^{\infty} 2^{s-1}\,\text{tr}\,(\theta_1\mathbf{A}+\theta_2\mathbf{B})^s/s. \tag{15.51}$$

In particular, (15.51) gives with $s = 2, 4$ respectively the product-cumulants

$$\left.\begin{aligned} \kappa_{11} &= \text{tr}\,(\mathbf{AB}+\mathbf{BA}) = 2\,\text{tr}\,\mathbf{AB}, \\ \kappa_{22} &= 8\,\text{tr}\,\{\mathbf{A}^2\mathbf{B}^2+(\mathbf{AB})^2+\mathbf{AB}^2\mathbf{A}+\mathbf{BA}^2\mathbf{B}+(\mathbf{BA})^2+\mathbf{B}^2\mathbf{A}^2\} \\ &= 16\,\text{tr}\,\{2\mathbf{A}^2\mathbf{B}^2+(\mathbf{AB})^2\}. \end{aligned}\right\} \tag{15.52}$$

From **4.16–17**, Q_1 and Q_2 will be independently distributed if and only if their joint c.f. factorizes into their individual c.f.'s. From (15.49–50), this condition is

$$|\mathbf{I}-2\theta_1\mathbf{A}-2\theta_2\mathbf{B}| = |\mathbf{I}-2\theta_1\mathbf{A}|\,|\mathbf{I}-2\theta_2\mathbf{B}|, \tag{15.53}$$

a result due to Cochran (1934).

15.13 The right-hand side of (15.53) is equal to

$$|\mathbf{I}-2\theta_1\mathbf{A}-2\theta_2\mathbf{B}+4\theta_1\theta_2\mathbf{AB}| \tag{15.54}$$

and this is equal to the left-hand side of (15.53) if

$$\mathbf{AB} = \mathbf{0}. \tag{15.55}$$

The converse is also true, as we shall see below. This is A. T. Craig's (1943) theorem: Q_1 and Q_2 are independent if and only if (15.55) holds.

Before we prove the converse, we observe that (15.55) implies

$$\text{tr}\,(\mathbf{A}^2\mathbf{B}^2) = \text{tr}\,(\mathbf{A}.\mathbf{AB}.\mathbf{B}) = 0, \tag{15.56}$$

and it is easy to see that (15.56) also implies (15.55), for we always have

$$\text{tr}\,(\mathbf{A}^2\mathbf{B}^2) = \text{tr}\,(\mathbf{BA}.\mathbf{AB}) = \text{tr}\,\{(\mathbf{AB})'\mathbf{AB}\} \geqslant 0,$$

since $\text{tr}\{(\mathbf{AB})'\mathbf{AB}\}$ is the sum of the squares of the elements of **AB**. If this sum is zero, all the elements of **AB** must also be zero. Hence (15.56) is equivalent to (15.55), and is itself a necessary and sufficient condition for the independence of Q_1 and Q_2, due to Lancaster (1954).

Now the converse of Craig's theorem follows easily, for if Q_1 and Q_2 are independent, their product-cumulants are zero by Example 12.7, and in particular (15.52) gives

$$\kappa_{22} = 16\,\text{tr}\{2\mathbf{A}^2\mathbf{B}^2+(\mathbf{AB})^2\} = 0.$$

Thus we may write

$$0 = 2\,\text{tr}\{2\mathbf{A}^2\mathbf{B}^2+(\mathbf{AB})^2\} = \text{tr}(\mathbf{AB}+\mathbf{BA})^2+2\,\text{tr}\{(\mathbf{AB})'\mathbf{AB}\}.$$

Since **A** and **B** are symmetric, so is $(\mathbf{AB}+\mathbf{BA})$, and its square has non-negative real latent roots and trace; and we have seen above that the latter holds for $(\mathbf{AB})'\mathbf{AB}$. Thus both traces on the right must be zero and (15.56) must hold, a proof essentially given by Ogasawara and Takahashi (1951).

15.14 Further, since **A** and **B** are symmetric, we may write $\mathbf{A} = \mathbf{TT}'$, $\mathbf{B} = \mathbf{UU}'$, and $\mathbf{AB} = \mathbf{TT}'\mathbf{UU}'$. Hence

$$\begin{aligned}\text{tr}(\mathbf{AB}) &= \text{tr}(\mathbf{TT}'\mathbf{U}.\mathbf{U}') = \text{tr}(\mathbf{U}'\mathbf{T}.\mathbf{T}'\mathbf{U}) \\ &= \text{tr}\{(\mathbf{T}'\mathbf{U})'(\mathbf{T}'\mathbf{U})\}.\end{aligned}$$

The leading diagonal of $(\mathbf{T}'\mathbf{U})'(\mathbf{T}'\mathbf{U})$ contains only the squares of all the elements of $(\mathbf{T}'\mathbf{U})$, but these are not necessarily all real. But if **A** and **B** are *non-negative* forms, the elements are all real, and hence

$$\text{tr}(\mathbf{AB}) = \text{tr}\{(\mathbf{T}'\mathbf{U})'(\mathbf{T}'\mathbf{U})\} = 0 \qquad (15.57)$$

implies $\mathbf{T}'\mathbf{U} = \mathbf{0}$ and hence $\mathbf{AB} = \mathbf{T}.\mathbf{T}'\mathbf{U}.\mathbf{U}' = \mathbf{0}$.

Thus for non-negative quadratic forms, (15.57) implies (15.55) and independence, a result due to Matérn (1949). From (15.52), we therefore see that for non-negative quadratic forms, $\kappa_{11} = 0$ is equivalent to independence.

In particular, **A** and **B** are non-negative if Q_1 and Q_2 are χ^2 variates. In this case, the equivalence of (15.57) and (15.55) may be seen directly since **A**, **B** are idempotent and $\mathbf{AB} = \mathbf{A}^2\mathbf{B}^2$, so that (15.57) implies (15.56) and hence (15.55) and independence.

Example 15.4 The independence of sample mean and variance in normal samples

We have seen (Examples 11.3, 11.7) that in samples of size n from a univariate normal distribution, which without loss of generality we may take to have zero mean and unit variance, the sample mean $\bar{x}$ is itself normally distributed with mean 0 and variance $1/n$, and the sample sum of squares $\sum_{i=1}^{n}(x_i-\bar{x})^2$ is distributed like χ^2 with $(n-1)$ degrees of freedom. It follows that $n\bar{x}^2$ is also distributed like χ^2 with 1 degree of freedom. The quadratic forms $n\bar{x}^2$ and $\sum_{i=1}^{n}(x_i-\bar{x})^2$ have matrices

$$\mathbf{A} = \frac{1}{n}\mathbf{U}, \qquad \mathbf{B} = \mathbf{I}-\frac{1}{n}\mathbf{U},$$

where $\mathbf{U}$ is the matrix with unit for each of its elements. Thus

$$\text{tr}(\mathbf{AB}) = \text{tr}(\mathbf{A}-\mathbf{A}^2) = 0$$

by the idempotency of χ^2 variates. Hence, from (15.57), $n\bar{x}^2$ and $\sum_{i=1}^{n}(x_i-\bar{x})^2$ are independently distributed, as we have already seen in Example 11.3, where we obtained the result in a slightly different form.

15.15 As a consequence of Craig's condition (15.55), we may obtain a necessary and sufficient condition for the independence of a quadratic form $Q = \mathbf{x}'\mathbf{A}\mathbf{x}$ and a linear form $L = \mathbf{b}'\mathbf{x}$. For Q and L are independent if and only if Q and L^2 are independent, i.e. if $\mathbf{x}'\mathbf{A}\mathbf{x}$ and $(\mathbf{b}'\mathbf{x})'.\mathbf{b}'\mathbf{x}$ are. From (15.55), this will be so if and only if

$$\mathbf{A}.\mathbf{b}\mathbf{b}' = \mathbf{0}. \tag{15.58}$$

From (15.58) it follows that

$$\mathbf{A}\mathbf{b}\mathbf{b}'\mathbf{b} = \mathbf{0}, \tag{15.59}$$

and as $\mathbf{b}'\mathbf{b}$ is a non-zero scalar, we may drop it from (15.59) to give

$$\mathbf{A}\mathbf{b} = \mathbf{0}. \tag{15.60}$$

This only establishes the necessity of (15.60) for (15.58). Its sufficiency follows immediately.

Example 15.5

We may apply (15.60) to obtain the result of Example 15.4. Here the quadratic form is, using $\mathbf{1}$ to denote a vector of units,

$$\sum_{i=1}^{n}(x_i-\bar{x})^2 = \mathbf{x}'(\mathbf{I}-\frac{1}{n}\mathbf{U})\mathbf{x},$$

and the linear form is $\bar{x} = (1/n)\mathbf{1}'\mathbf{x}$. The product of these matrices is

$$\left(\mathbf{I}-\frac{1}{n}\mathbf{U}\right).\frac{1}{n}\mathbf{1} = \frac{1}{n}\mathbf{1}-\frac{1}{n^2}.n\mathbf{1} = \mathbf{0},$$

so (15.60) is obeyed and $\sum_{i=1}^{n}(x_i-\bar{x})^2$ and $\bar{x}$ are independent.

The decomposition of quadratic forms in independent normal variates

15.16 We now prove a remarkable result due to Cochran (1934), whose fundamental theorem has been amplified by James (1952) and Lancaster (1954). With $\mathbf{x}$ a vector of p independent standardized normal variates as before, suppose that the sum of squares $Q = \mathbf{x}'\mathbf{x}$ is decomposed into k quadratic forms $Q_i = \mathbf{x}'\mathbf{A}_i\mathbf{x}$ with ranks r_i, i.e. that

$$\sum_{i=1}^{k}\mathbf{x}'\mathbf{A}_i\mathbf{x} = \mathbf{x}'\mathbf{I}\mathbf{x}. \tag{15.61}$$

Then any one of the following three conditions implies the other two :—

(a) The ranks r_i of the Q_i add to that of Q.
(b) Each of the Q_i is distributed like χ^2.
(c) Each Q_i is independent of every other.

15.17 First, we deduce (b) from (a) and conversely.

Select an arbitrary Q_i, say Q_1. If we make an orthogonal transformation (15.3) which diagonalizes $\mathbf{A}_1$, we obtain

$$\mathbf{y}'\mathbf{B}'\mathbf{A}_1\mathbf{B}\mathbf{y}+\mathbf{y}'\mathbf{B}'(\mathbf{I}-\mathbf{A}_1)\mathbf{B}\mathbf{y} = \mathbf{y}'\mathbf{I}\mathbf{y}, \tag{15.62}$$

the identity matrix, and the ranks on the left, remaining invariant under orthogonal transformation. Since the first and last quadratic forms in (15.62) are diagonal, so must the second be. Moreover, since $(p-r_1)$ of the leading diagonal elements of $\mathbf{B}'\mathbf{A}_1\mathbf{B}$ are zero, the corresponding elements of $\mathbf{B}'(\mathbf{I}-\mathbf{A}_1)\mathbf{B}$ are unity since they must add to those of $\mathbf{I}$. Now if condition (a) holds, the rank of $\mathbf{B}'(\mathbf{I}-\mathbf{A}_1)\mathbf{B}$ is $(p-r_1)$, the other elements of its leading diagonal are zero, and the corresponding elements of $\mathbf{B}'\mathbf{A}_1\mathbf{B}$ are unity. Hence, from **15.11**, Q_1 is a χ^2 variable, $\mathbf{A}_1$ being idempotent. The same result holds for the other $\mathbf{A}_i$. Thus we have established (b) from (a).

Further, from (15.61)

$$\mathbf{I} = \sum_i \mathbf{A}_i. \tag{15.63}$$

If (b) holds, $\mathbf{A}_i$ has r_i latent roots of unity and $(p-r_i)$ of zero, and from (15.63) it follows, on taking traces of both sides, that

$$p = \sum_{i=1}^{k} r_i,$$

which establishes (a) from (b).

15.18 Now we deduce (c) from (b) and conversely.

Since the $\mathbf{A}_i$ are idempotent if (b) holds, (15.63) squared gives

$$\mathbf{I} = \sum_i \mathbf{A}_i + \sum_{i\neq j} \mathbf{A}_i\mathbf{A}_j$$

or

$$\sum_{i\neq j} \mathbf{A}_i\mathbf{A}_j = \mathbf{0}. \tag{15.64}$$

Taking traces on both sides of (15.64),

$$\sum_{i\neq j} \text{tr}(\mathbf{A}_i\mathbf{A}_j) = 0. \tag{15.65}$$

But since the $\mathbf{A}_i$ are idempotent,

$$\text{tr}(\mathbf{A}_i\mathbf{A}_j) = \text{tr}(\mathbf{A}_i^2\mathbf{A}_j^2) \geqslant 0, \qquad (i \neq j), \tag{15.66}$$

as we saw below (15.56). Thus, as there, (15.65–6) imply

$$\text{tr}(\mathbf{A}_i^2\mathbf{A}_j^2) = 0, \qquad (\text{all } i \neq j), \tag{15.67}$$

which implies $\mathbf{A}_i\mathbf{A}_j = \mathbf{0}$ (all $i \neq j$), and the independence of Q_i from Q_j. Thus (c) is established from (b).

Conversely, taking powers of (15.63) gives, if $\mathbf{A}_i\mathbf{A}_j = \mathbf{0}$ for all $i \neq j$,

$$\sum_i \mathbf{A}_i^s = \mathbf{I} \qquad (\text{all } s), \tag{15.68}$$

and taking traces of both sides of (15.68) gives

$$\text{tr}\sum \mathbf{A}_i^s = p \qquad (\text{all } s). \tag{15.69}$$

(15.69) can only hold if every non-zero latent root of each $\mathbf{A}_i$ is unity, i.e. if each Q_i is distributed like χ^2. This establishes (b) from (c).

15.19 We have thus seen that (a) implies (b) and conversely, and also that (b) implies (c) and conversely. This establishes the sufficiency of any one of the conditions for the other two.

15.20 Generalizing (15.61), suppose now that Q on its right-hand side may be any quadratic form $\mathbf{x}'\mathbf{A}\mathbf{x}$ where $\mathbf{A}$ is idempotent of rank $r \leqslant p$. If we define $P = \mathbf{x}'(\mathbf{I}-\mathbf{A})\mathbf{x}$ we have $Q+P = \mathbf{x}'\mathbf{x}$ and write this as

$$\sum_{i=1}^{k} \mathbf{x}'\mathbf{A}_i\mathbf{x}+\mathbf{x}'(\mathbf{I}-\mathbf{A})\mathbf{x} = \mathbf{x}'\mathbf{I}\mathbf{x},$$

which is precisely of the original form of (15.61), so that **15.16–19** apply here. Now $(\mathbf{I}-\mathbf{A})$ is idempotent since $\mathbf{A}$ is, so by **15.16** P and Q are independent and P has rank $(p-r)$. Also, if each $\mathbf{A}_i\mathbf{A}_j = \mathbf{0}$, we see as at (15.69) that $\text{tr} \sum_{i=1}^{k} \mathbf{A}_i^s = \text{tr}\,\mathbf{A}^s = r$, all s, so each $\mathbf{A}_i$ is idempotent and $\mathbf{A}_i(\mathbf{I}-\mathbf{A}) = \mathbf{A}_i-\mathbf{A}_i^2 = \mathbf{0}$ also. Thus any one of the conditions (a) to (c) of **15.16** which holds for the k Q_i also holds for the Q_i and P together and hence implies the other two conditions for the Q_i and P, and thus for the Q_i alone. Cochran's theorem is therefore unchanged.

Exercise 15.16 shows that for $k=2$, idempotency of $\mathbf{A}_1$ and *non-negativity* of $\mathbf{A}_2$ suffice for Cochran's theorem to hold.

It is not true generally that if two variates are distributed like χ^2 and their sum is also distributed like χ^2 with degrees of freedom equal to the sum of their individual degrees of freedom, then the variates are independent. Example 7.6 above is a counter-example due to James (1952): the restriction to quadratic forms in standard normal variates is crucial here.

Example 15.6

Reverting to the distribution of mean and variance in normal samples, discussed in Example 15.4, we see that the decomposition of the sum of squares

$$\sum_{i=1}^{n} (x_i-\bar{x})^2+n\bar{x}^2 = \sum_{i=1}^{n} x_i^2$$

obeys the conditions for the application of Cochran's theorem. Condition (b) of **15.16** was established in Examples 11.3 and 11.7, and immediately implies the independence of the two sums on the left. It is easily verified that their ranks are $(n-1)$ and 1 respectively (agreeing with the degrees of freedom in their χ^2 distributions), and this satisfaction of condition (a) is enough to imply both their χ^2 distributions and their independence.

15.21 From **15.11** onwards, we have supposed $\mathbf{x}$ to be a vector of independent standardized normal variates, but no essential difference is made to our results if we take $\mathbf{x}$ to be multinormally and non-singularly distributed with mean $\mathbf{0}$ and dispersion matrix $\mathbf{V}$. For since $\mathbf{V}$ is positive definite, we may write

$$\mathbf{V} = \mathbf{T}\mathbf{T}' \tag{15.70}$$

where $\mathbf{T}$ has real elements, and the transformation $\mathbf{x} = \mathbf{T}\mathbf{y}$ transforms the exponent of the distribution of $\mathbf{x}$ from $\mathbf{x}'\mathbf{V}^{-1}\mathbf{x}$ to

$$\mathbf{y}'\mathbf{T}'\mathbf{V}^{-1}\mathbf{T}\mathbf{y} = \mathbf{y}'\mathbf{T}'(\mathbf{T}')^{-1}\mathbf{T}^{-1}\mathbf{T}\mathbf{y} = \mathbf{y}'\mathbf{y},$$

using (15.70), and we may treat the vector of independent variates $\mathbf{y}$ as before. Thus the quadratic forms $\mathbf{x}'\mathbf{A}\mathbf{x}$, $\mathbf{x}'\mathbf{B}\mathbf{x}$ become $\mathbf{y}'.\mathbf{T}'\mathbf{A}\mathbf{T}.\mathbf{y}$ and $\mathbf{y}'.\mathbf{T}'\mathbf{B}\mathbf{T}.\mathbf{y}$ respectively, and Craig's condition (15.55) becomes, using (15.70),

$$\mathbf{0} = \mathbf{T}'\mathbf{A}\mathbf{T}.\mathbf{T}'\mathbf{B}\mathbf{T} = \mathbf{T}'.\mathbf{A}\mathbf{V}\mathbf{B}.\mathbf{T}$$

or

$$\mathbf{A}\mathbf{V}\mathbf{B} = \mathbf{0}, \tag{15.71}$$

a generalization due to Aitken (1950), while (15.60) generalizes as in Exercise 15.13. Similarly, Lancaster's condition (15.56) becomes

$$0 = \text{tr}\,\{(\mathbf{T}'\mathbf{A}\mathbf{T})^2(\mathbf{T}'\mathbf{B}\mathbf{T})^2\} = \text{tr}\,\{(\mathbf{A}\mathbf{V})^2(\mathbf{B}\mathbf{V})^2\}, \tag{15.72}$$

and Matérn's condition (15.57) for non-negative forms becomes

$$0 = \text{tr}\,\{(\mathbf{T}'\mathbf{A}\mathbf{T})(\mathbf{T}'\mathbf{B}\mathbf{T})\} = \text{tr}(\mathbf{A}\mathbf{V}\mathbf{B}\mathbf{V}). \tag{15.73}$$

Similarly, Cochran's theorem holds with obvious modifications (Exercise 15.17).

Exercises 15.14 and 15.18 generalize even further to the singular case.

Subrahmaniam (1966) investigates the effect of non-normality on the distributions of quadratic forms.

There are theorems, analogous to those concerning quadratic forms in normal variables, for linear forms in order-statistics from an exponential distribution—see Tanis (1964).

Characterizations of the normal distribution

15.22 Many, although not all, of the properties of the normal distribution that we have discussed are characterizing properties, shared by no other distribution. Thus, we saw in **15.4** that the normality of all linear functions of the variates is a characterizing property; on the other hand, Exercise 11.23 showed that the result of Example 11.21 is not unique to the normal distribution. We now discuss some of the other important characterizations of normality.

15.23 We have seen in Example 12.7 that the independence of sample mean and variance characterizes the univariate normal distribution among distributions with all cumulants finite. This result may be generalized to the multivariate case: the independence of the sample mean vector and the elements of the sample dispersion matrix characterizes the multinormal distribution among all multivariate distributions with finite dispersion matrices. It will be noticed that the existence of higher-order cumulants is not necessary to the result, which is due to Lukacs (1942). We first give his proof for the univariate case, and sketch the generalization.

We suppose a population to have finite mean μ and variance σ^2, and characteristic function (c.f.) $\phi(t)$. The joint c.f. of the sample mean $\bar{x}$ and variance s^2 in samples of size n is

$$\phi_{12}(t_1, t_2) = \int \cdots \int \exp(it_1\bar{x}+it_2 s^2)\,dF_n. \tag{15.74}$$

A necessary and sufficient condition for the independence of $\bar{x}$ and s^2 is that (15.74) factorizes, i.e. that

$$\phi_{12}(t_1, t_2) = \phi_1(t_1)\phi_2(t_2). \tag{15.75}$$

Differentiating (15.75) with respect to t_2 we obtain

$$\left[\frac{\partial \phi_{12}}{\partial t_2}\right]_{t_2=0} = \phi_1(t_1)\left[\frac{\partial \phi_2}{\partial t_2}\right]_{t_2=0}. \tag{15.76}$$

Now in (15.76), we may use the relation connecting the c.f. of the sample mean with that of the parent distribution

$$\phi_1(t_1) = \left\{\phi\left(\frac{t_1}{n}\right)\right\}^n \tag{15.77}$$

Further
$$\left[\frac{\partial \phi_2}{\partial t_2}\right]_{t_2=0} = i\mathrm{E}(s^2) = i.\frac{(n-1)}{n}\sigma^2. \tag{15.78}$$

Finally, from (15.74),

$$\frac{\partial \phi_{12}}{\partial t_2} = i\int\cdots\int s^2 \exp(it_1\bar{x}+it_2 s^2)\,dF_n, \tag{15.79}$$

and since

$$s^2 = \frac{1}{n}\sum_i x_i^2 - \bar{x}^2 = \frac{1}{n^2}\left\{(n-1)\sum_i x_i^2 - \sum_{i\neq j}\sum x_i x_j\right\}, \tag{15.80}$$

(15.79) and (15.80) give

$$\begin{aligned}\left[\frac{\partial \phi_{12}}{\partial t_2}\right]_{t_2=0} &= \frac{i}{n^2}\int\cdots\int\left\{(n-1)\sum_i x_i^2 - \sum_{i\neq j}\sum x_i x_j\right\}\exp\left(it_1\sum_i x_i/n\right)dF_n \\ &= \frac{i(n-1)}{n}\left\{\int x^2\exp(it_1x/n)\,dF_1.\left[\phi\left(\frac{t_1}{n}\right)\right]^{n-1}\right. \\ &\qquad \left.-\left[\int x\exp(it_1x/n)\,dF_1\right]^2\left[\phi\left(\frac{t_1}{n}\right)\right]^{n-2}\right\} \\ &= \frac{i(n-1)}{n}\left[\phi\left(\frac{t_1}{n}\right)\right]^{n-2}\cdot\left\{-\phi\left(\frac{t_1}{n}\right)\cdot\frac{\partial^2\phi\left(\frac{t_1}{n}\right)}{\partial\left(\frac{t_1}{n}\right)^2}+\left[\frac{\partial\phi\left(\frac{t_1}{n}\right)}{\partial\left(\frac{t_1}{n}\right)}\right]^2\right\}.\end{aligned} \tag{15.81}$$

Writing t for $\frac{t_1}{n}$, we put (15.81), (15.77) and (15.78) into (15.76) and obtain

$$-\phi(t)\phi''(t)+\{\phi'(t)\}^2 = \{\phi(t)\}^2\sigma^2. \tag{15.82}$$

(15.82) may be rewritten

$$\frac{d}{dt}\left\{\frac{d}{dt}\log\phi(t)\right\} = -\sigma^2, \tag{15.83}$$

so that integration of (15.83) gives

$$\phi'(t)/\phi(t) = -\sigma^2 t + c. \tag{15.84}$$

From the initial conditions $\phi(0) = 1$, $\phi'(0) = i\mu$, we find in (15.84)

$$c = i\mu,$$

and (15.84) becomes
$$\frac{d}{dt}\log\phi(t) = i\mu - \sigma^2 t. \tag{15.85}$$

Integrating (15.85), we have

$$\log\phi(t) = i\mu t - \tfrac{1}{2}\sigma^2 t^2,$$

the constant of integration being zero since $\log\phi(0) = 0$. Thus finally

$$\phi(t) = \exp(i\mu t - \tfrac{1}{2}\sigma^2 t^2), \tag{15.86}$$

which is the c.f. unique to the normal distribution. (15.86) thus establishes the independence of sample mean and variance as a characterization of the normal distribution alone among distributions with finite variance.

Kawata and Sakamoto (1949) relax even the restriction of requiring a finite variance.

15.24 The generalization of the result of **15.23** to the multivariate case is straightforward. If the distribution of the sample mean vector is independent of $s_{l,m}$, the sample covariance of x_l and x_m, we obtain, by the argument leading to (15.82),

$$-\phi''_{l,m}/\phi+\phi'_l\phi'_m/\phi^2 = \sigma_{lm}, \qquad (15.87)$$

where ϕ is the c.f. of the parent distribution and suffixes denote the t-variable with respect to which differentiation has been carried out. If (15.87) holds for $l, m = 1, 2, \ldots, p$, we have a system of partial differential equations which may be written in matrix form, corresponding to (15.83),

$$\left(\frac{\partial}{\partial t_l}\frac{\partial}{\partial t_m}\log\phi\right) = -\mathbf{V}, \qquad (15.88)$$

so that on applying the initial conditions, (15.88) yields after two integrations

$$\phi(\mathbf{t}) = \exp(i\mathbf{t}'\boldsymbol{\mu}-\tfrac{1}{2}\mathbf{t}'\mathbf{V}\mathbf{t}),$$

the c.f. of the multinormal distribution.

15.25 In our discussion of the geometrical method of deriving sampling distributions in Chapter 11, we saw that the probability density of a sample of independent observations from a standardized univariate normal distribution is constant on the sphere Σx^2 = constant, since the density is a function of Σx^2 only. The question arises whether this property of radial symmetry is possessed by any other distribution. The following argument, due to Bartlett (1934), shows that it is not.

If the density may be written

$$L = \prod_{i=1}^{n} f(x_i) \propto g(\Sigma x^2), \qquad (15.89)$$

we must have

$$\frac{\partial L}{\partial x_i} = \frac{\partial \log L}{\partial x_i} = 0$$

for

$$\sum_i x_i^2 = \text{constant}.$$

Using Lagrange's method with a multiplier λ, we therefore have

$$\frac{\partial \log f(x_i)}{\partial x_i}+\lambda x_i = 0 \quad \text{(all } i\text{)}. \qquad (15.90)$$

Integration yields

$$\log f(x_i) = -\tfrac{1}{2}\lambda x_i^2+k_i$$

or

$$f(x_i) \propto \exp(-\tfrac{1}{2}\lambda x_i^2). \qquad (15.91)$$

(15.89) thus characterizes the normal distribution.

15.26 The argument generalizes at once to the multivariate case. Instead of radial symmetry with respect to a single vector of length $(\Sigma x^2)^{\frac{1}{2}}$, we now consider a group of p vectors. If the distribution has a density which is a function only of the squared lengths of these vectors (the sums of squares) and the angles between them (the correlation coefficients—cf. **16.24**), it must be multinormal. For if we define $\mathbf{x}_i$ as the $(p \times 1)$ vector containing the ith set of observations on the p variates,(*) and $\mathbf{X}'$ as the $(p \times n)$ matrix formed by these vectors, the sums of squares and products are the elements of $\mathbf{X}'\mathbf{X}$, and we have

$$L = \prod_i f(\mathbf{x}_i) \propto g(\mathbf{X}'\mathbf{X}). \tag{15.92}$$

For $\mathbf{X}'\mathbf{X}$ constant, we now introduce a $(p \times p)$ matrix $\mathbf{\Lambda}$ of Lagrange multipliers, and differentiate logarithmically to obtain

$$\frac{\partial \log f(\mathbf{x}_i)}{\partial \mathbf{x}_i} + \mathbf{\Lambda}\mathbf{x}_i = \mathbf{0} \qquad \text{(all } i\text{)},$$

whence

$$f(\mathbf{x}_i) \propto \exp(-\tfrac{1}{2}\mathbf{x}_i'\mathbf{\Lambda}\mathbf{x}_i),$$

so that (15.92) characterizes the multinormal distribution.

15.27 Finally, we prove a characterization of the normal distribution which illuminates the position of central importance which it holds in statistical theory. We have seen (Examples 11.2 and 11.3) that an orthogonal transformation from a set of independent standardized normal variates yields a set of linear functions of the original variates which are also independent standardized normal variates. We now prove that if $\mathbf{x}$ is a vector of (not necessarily identical) independent standardized variates with finite cumulants, and the non-trivial transformation

$$\mathbf{x} = \mathbf{C}\mathbf{y} \tag{15.93}$$

gives a vector $\mathbf{y}$ of independent standardized variates, then each of the x_i is normal (and hence so is each y_i) and the transformation is orthogonal. The result is due to Lancaster (1954).

The orthogonality of $\mathbf{C}$ follows at once, for since the variates are standardized

$$\sum_j c_{ij}^2 = 1, \tag{15.94}$$

while since they are independent

$$\sum_j c_{ij}\,c_{kj} = 0 \qquad (i \neq k). \tag{15.95}$$

(15.94) and (15.95) are the conditions for orthogonality. Thus $\mathbf{C}' = \mathbf{C}^{-1}$ and

$$\mathbf{y} = \mathbf{C}'\mathbf{x}. \tag{15.96}$$

Since the y_i are independent, (15.93) gives for the sth cumulant of x_i, κ_{si},

$$\kappa_{si} = \sum_k c_{ik}^s \lambda_{sk} \qquad \text{(all } s\text{)}, \tag{15.97}$$

where λ_{sk} is the sth cumulant of y_k. Conversely, from (15.96),

$$\lambda_{si} = \sum_j (c_{ij}')^s \kappa_{sj} \qquad \text{(all } s\text{)}. \tag{15.98}$$

(*) These new vectors are, of course, distinct from the $(n \times 1)$ vectors of observations on single variates just discussed.

Combination of (15.97) and (15.98) gives

$$\kappa_{si} = \sum \kappa_{sj} \sum_k c_{ik}^s c_{jk}^s \qquad \text{(all } s\text{)}. \tag{15.99}$$

For $s = 1, 2$, (15.99) reduces to $\kappa_{1i} = 0$, $\kappa_{2i} = 1$, as it must for standardized variates. Now

$$\sum_j \left| \sum_k c_{ik}^s c_{jk}^s \right| \leqslant \sum_j \sum_k |c_{ik}^s| \, |c_{jk}^s|, \tag{15.100}$$

and using (15.94), (15.100) becomes, for $s \geqslant 3$,

$$\sum_j \left| \sum_k c_{ik}^s c_{jk}^s \right| < 1. \tag{15.101}$$

Taking absolute values in (15.99), we have

$$|\kappa_{si}| \leqslant \sum_j |\kappa_{sj}| \left| \sum_k c_{ik}^s c_{jk}^s \right|. \tag{15.102}$$

(15.102) has on the right-hand side a weighted sum of the κ_{sj}, the weights adding to less than 1 by (15.101). If for fixed s we take that value of i for which κ_{si} is largest, (15.102) cannot hold unless that $\kappa_{si} = 0$, i.e. all $\kappa_{si} = 0$. This holds for all s. Thus

$$\kappa_{si} = 0 \qquad \text{(all } s \geqslant 3\text{)} \tag{15.103}$$

and each x_i is normal. The normality of the y_i follows at once.

15.28 Lukacs (1956) reviews characterization methods and results generally. See also Patil and Seshadri (1964). Lancaster (1960) reviews the characterizations of normality by independence properties, and shows in particular that the latter imply the existence of all moments, which therefore need not be assumed. A general exposition of characterizations is given in the book by Kagan *et al.* (1973).

Exercise 15.22 characterizes Gamma distributions.

EXERCISES

15.1 From a multivariate normal distribution, a conditional distribution is found by fixing the values of certain of the variates. Show that this distribution is itself multivariate normal, with a mean vector which is a function of the eliminated variates and with the same elements in the inverse variance matrix as the variates possessed in the original distribution.

15.2 A trivariate distribution, not necessarily normal, has correlations $\rho_{12}, \rho_{13}, \rho_{23}$. Show that $1+2\rho_{12}\rho_{13}\rho_{23} \geqslant \rho_{12}^2+\rho_{13}^2+\rho_{23}^2$.

15.3 In a p-variate distribution all the correlations ρ are equal. Show that this is possible if, and only if, $\rho \geqslant -1/(p-1)$.

15.4 Show that the frequency function of the multivariate normal distribution satisfies the system of equations

$$\frac{\partial f}{\partial \rho_{ij}} = \frac{\partial^2 f}{\partial x_i \, \partial x_j}$$

where ρ_{ij} is the correlation coefficient of x_i and x_j.

15.5 By considering the bivariate normal distribution with independent variates, show that

$$\int_{-x}^{x} (2\pi)^{-\frac{1}{2}} \exp\left(-\tfrac{1}{2}t^2\right) dt \leqslant \{1-\exp\left(-2x^2/\pi\right)\}^{\frac{1}{2}}.$$

(The error in taking the equality to hold is always less than $\frac{3}{4}$ per cent.)

(Pólya, 1945; J. D. Williams, 1946)

15.6 In a standardized bivariate normal distribution with covariance ρ, show that

$$\mathrm{E}\{|x_1||x_2|\} = 4\int_0^{\infty}\int_0^{\infty} x_1 x_2 \, dF = 4I, \text{ say.}$$

Show as in **15.6** that

$$4\pi^2 I = \int_{-\infty}^{\infty}\int_{-\infty}^{\infty} e^{-\frac{1}{2}\mathbf{t}'\mathbf{V}\mathbf{t}} \frac{dt_1\, dt_2}{(it_1)^2 (it_2)^2}$$

and hence that $\delta^2 I/\delta\rho^2 = \delta P_2^0/\delta\rho$, so that

$$I = \{\rho \arcsin \rho + (1-\rho^2)^{\frac{1}{2}}\}/(2\pi).$$

Use this result to establish the exact variance (10.40) of the mean deviation in normal samples.

15.7 Shots are fired at a vertical circular target of unit radius. The distribution of horizontal and vertical deviations from the centre of the target is bivariate normal, with zero means, equal variances v and correlation ρ. Show that the probability of hitting the target is

$$\frac{(1-\rho^2)^{\frac{1}{2}}}{2\pi}\int_0^{2\pi}\left[1-\exp\left\{-\frac{1+\rho\cos 2\theta}{2v(1-\rho^2)}\right\}\right]\frac{d\theta}{1+\rho\cos 2\theta},$$

reducing to $1-\exp\{-1/(2v)\}$ when $\rho = 0$.

15.8 The standardized variates x and y are non-singularly distributed in a bivariate normal form. Show that $x^2+2axy+y^2$ cannot be independent of $x^2+2bxy+y^2$ unless a and b are both unity in absolute value and of opposite sign.

15.9 From **15.9**, show that for a standardized p-variate multinormal distribution with correlations all equal to $\frac{1}{2}$, the probability that all the variates exceed their expectation is $1/(p+1)$.

(Cf. Foster and Stuart, 1954)

15.10 In the previous exercise, if, instead of the correlations being $\frac{1}{2}$, the dispersion matrix is given by

$$\mathbf{V} = \begin{pmatrix} 1 & -\frac{1}{2} & 0 & 0 & . & . & . & . & 0 \\ -\frac{1}{2} & 1 & -\frac{1}{2} & 0 & . & . & . & . & 0 \\ 0 & -\frac{1}{2} & 1 & -\frac{1}{2} & . & . & . & . & 0 \\ 0 & 0 & -\frac{1}{2} & 1 & . & . & . & . & 0 \\ . & . & . & . & . & . & . & . & . \\ . & . & . & . & . & . & . & . & 0 \\ 0 & . & . & . & . & . & . & 1 & -\frac{1}{2} \\ 0 & 0 & . & . & . & . & 0 & -\frac{1}{2} & 1 \end{pmatrix}$$

show that the probability becomes $1/(p+1)!$.

15.11 In **15.9**, let x_0 have correlation coefficient λ with each of the other p x_i, which remain mutually independent, and replace (15.37) by $y_i = ax_0+bx_i$. Show that if the variances of the y_i are equated to unity, they are multinormally distributed with equal correlations $\rho = 1-b^2$. By putting $\lambda = 0$ for $\rho \geqslant 0$ and $\lambda = -a/b$ for $\rho<0$, show that $y_i = |\rho|^{\frac{1}{2}}x_0+(1-\rho)^{\frac{1}{2}}x_i$ and that if $y_{(i)}$, $x_{(i)}$ denote the order-statistics of the p y's and of the p x's, the c.f. of $y_{(i)}$ is related to that of $x_{(i)}$ by

$$\phi_{y(i)}(t) = \exp(-\tfrac{1}{2}\rho t^2)\,\phi_{x(i)}(t(1-\rho)^{\frac{1}{2}}),$$

so that in particular we have

$$\begin{aligned} E(y_{(i)}) &= (1-\rho)^{\frac{1}{2}}E(x_{(i)}), \\ \text{var}(y_{(i)}) &= \rho+(1-\rho)\,\text{var}(x_{(i)}), \end{aligned}$$

connecting the order-statistics of p equally correlated multinormal variates with univariate normal order-statistics in samples of size p, discussed in **14.21.**

(Similar results are available for the product-moments of pairs of order-statistics—cf. Owen and Steck (1962). For $p = 2$, Clark (1961) gave the first four moments of the larger of bivariate normal variates with arbitrary means and variances.)

15.12 By considering the variates $y_i = x_i-b_ix_0$, where the x_i are independent standardized normal variates with frequency function f, generalize (15.38) by showing that for a multinormal distribution with correlations $\rho_{ij} = a_i a_j\ (i \neq j)$ the probability that all the variables are positive is

$$\int_{-\infty}^{\infty}\left\{\prod_i \int_{b_i x}^{\infty} f(t)\,dt\right\} f(x)\,dx.$$

(Stuart, 1958)

15.13 If $\mathbf{x}$ has a multinormal distribution with mean $\mathbf{0}$ and non-singular dispersion matrix $\mathbf{V}$, show that a necessary and sufficient condition for the independence of a quadratic form $\mathbf{x'Ax}$ and a linear form $\mathbf{b'x}$ is that $\mathbf{AVb} = \mathbf{0}$.

(Aitken, 1950)

15.14 If $\mathbf{x'Ax}$ and $\mathbf{x'Bx}$ are quadratic forms, and $\mathbf{a'x}$, $\mathbf{b'x}$ are linear forms, in multinormal variables $\mathbf{x}$ whose dispersion matrix $\mathbf{V}$ is not necessarily non-singular, show by using c.f.'s that the independence conditions in (15.71) and Exercise 15.13 are replaced by $\mathbf{VAVBV} = \mathbf{0}$ and by $\mathbf{VAVb} = \mathbf{0}$ respectively, and that the linear forms are independent if and only if $\mathbf{a'Vb} = 0$.

(Good, 1963, 1966)

15.15 If in **15.12** Q_1 and Q_2 are distributed like χ^2 with n and $m\,(<n)$ degrees of freedom, show that a necessary and sufficient condition that $Q_1 = Q_2+Q_3$, where Q_3 is also a χ^2 variate, is that their correlation coefficient takes the value $(m/n)^{\frac{1}{2}}$. (Lancaster, 1954)

15.16 Show that if $\mathbf{x}'\mathbf{A}\mathbf{x} = \mathbf{x}'\mathbf{A}_1\mathbf{x}+\mathbf{x}'\mathbf{A}_2\mathbf{x}$, where $\mathbf{A}$ and $\mathbf{A}_1$ are idempotent and $\mathbf{A}_2$ is non-negative, then $\mathbf{A}_2$ is idempotent, so that we obtain a decomposition as in **15.16.**

(Hogg and Craig, 1958)

15.17 If $\mathbf{x}$ is multinormal with mean $\mathbf{0}$ and non-singular dispersion matrix $\mathbf{V}$, show as in **15.11** that the c.f. of $\mathbf{x}'\mathbf{A}\mathbf{x}$ is (15.42) if and only if $\mathbf{AV}$ is idempotent of rank r. If $\sum_{i=1}^{k} \mathbf{x}'\mathbf{A}_i\mathbf{x} = \mathbf{x}'\mathbf{A}\mathbf{x}$, where $\mathbf{AV}$ is idempotent, show that any of the conditions (a), (b), (c) of **15.16** implies the other two, where (b) here requires the idempotency of each $\mathbf{A}_i\mathbf{V}$ and (c) requires that $\mathbf{A}_i\mathbf{V}\mathbf{A}_j = \mathbf{0}$, all $i \neq j$.

15.18 If $\mathbf{V}$ may be singular in Exercise 15.17, show that necessary and sufficient conditions that the c.f. of $\mathbf{x}'\mathbf{A}\mathbf{x}$ is (15.42) are :

(a) $\mathbf{VAVAV} = \mathbf{VAV}$ and $\text{rank}(\mathbf{VAV}) = \text{tr}(\mathbf{AV}) = r$;
or (b) $\mathbf{AVAV} = \mathbf{AV}$, if and only if $\text{rank}(\mathbf{AV}) = \text{tr}(\mathbf{AV}) = r$;
or (c) $\mathbf{AVAV} = \mathbf{AV}$, if and only if $\text{rank}(\mathbf{AV}) = \text{rank}(\mathbf{VAV}) = r$.

(Cf. Styan, 1970)

15.19 $x_1, x_2, \ldots, x_n$ are distributed in a multivariate normal form, the covariance of x_i and x_j being c_{ij}. Show that their covariance, conditional upon $l_1x_1+l_2x_2+\ldots+l_nx_n =$ constant, is

$$c_{ij}-\frac{(\sum_k l_k c_{ik})(\sum_k l_k c_{jk})}{\sum_k\sum_m l_k l_m c_{km}}.$$

15.20 x and y have a standardized bivariate normal distribution. Whenever the signs of x and y differ, the sign of x is changed. Show that when this is done, the two variates are marginally normally distributed as before, but that their joint distribution is not bivariate normal. Generalize to show that, for a set of n variates, it is possible for any subset of $p<n$ variates to be jointly normally distributed without the n variates being so distributed.

(Exercise 16.22 gives another instance.)

15.21 Show that the joint distribution of k independent Poisson variables, conditional upon the value of their sum, is a multinomial distribution. Hence establish the asymptotic χ^2 distribution of Example 15.3.

15.22 If $x_1, \ldots, x_k$ are independent variables having (not necessarily identical) Gamma-distributions (6.20), show by using moment-generating functions that any scale-free function of them, $h(x_1, \ldots, x_k)$, is distributed independently of $S = \sum_{i=1}^{k} x_i$. In particular this holds for $h(x_1,\ldots,x_k) = \sum_{i=1}^{n} a_i x_i \Big/ \sum_{i=1}^{n} x_i = R$. Hence show that for a vector $\mathbf{x}$ of p independent standardized normal variates, $\mathbf{x}'\mathbf{A}\mathbf{x}/\mathbf{x}'\mathbf{x}$ is independent of $\mathbf{x}'\mathbf{x}$.

(Pitman (1937b). Laha (1954) gives a converse: if the x_i are independently *and identically* distributed with finite variance, the independence of R and S implies that each x_i has a Gamma distribution.)

15.23 As in **15.6**, write the d.f. of the bivariate normal distribution as

$$F_\rho(u,v) = \int_{-\infty}^{u}\int_{-\infty}^{v}\left\{(2\pi)^{-2}\int_{-\infty}^{\infty}\int_{-\infty}^{\infty}\exp(-i\mathbf{t}'\mathbf{x}-\tfrac{1}{2}\mathbf{t}'\mathbf{V}\mathbf{t})\,dt_1\,dt_2\right\}dx_1\,dx_2$$

and show that for any fixed (u,v), $\frac{\partial}{\partial\rho}F_\rho(u,v)>0$. Hence show that if a bivariate normal distribution is doubly dichotomized at any point (u, v), with frequencies $\begin{array}{c|c} a & b \\ \hline c & d \end{array}$ in the four resulting quadrants, the function

$$R_{u,v.\rho} \equiv bc - ad$$

is a monotone increasing function of ρ, always having the same sign as ρ.

(The result is originally due to G. U. Yule, who called any bivariate distribution *isotropic* if R does not change sign as a function of u and v.)

15.24 Using Exercise 6.24, show that the bivariate distributions defined by (6.83) are isotropic. Show also that the distributions in Exercise 1.22 are isotropic.

CHAPTER 16

DISTRIBUTIONS ASSOCIATED WITH THE NORMAL

16.1 In the course of our investigations in earlier chapters, we have encountered several distributions, related to the normal distribution, which play important parts in the theory of statistics precisely because they are the forms taken by the sampling distributions of various statistics in samples from normal populations. The special position which the normal distribution holds, mainly by virtue of the Central Limit theorem in one or other of its forms, is reflected in the positions of central importance occupied by these related distributions. Whenever some statistic (under regularity conditions) has a sampling distribution which is asymptotically normal, other statistics exist which have the corresponding related distributions. We have already encountered an instance of this in discussing the χ^2 goodness-of-fit test in Example 15.3.

In this chapter, the three important distributions related to the univariate normal, namely the χ^2 distribution, " Student's " t-distribution, and Fisher's variance ratio (F) distribution, will in turn be examined, and the salient properties of each described. Later in the chapter, sampling distributions arising from the bivariate normal distribution will be discussed.

The χ^2 distribution

16.2 We have already seen (cf. **11.2**, Example 11.6) that the sum z of the squares of n independent standardized normal variates is distributed with frequency function

$$dF = \frac{1}{2^{\frac{1}{2}n}\Gamma(\frac{1}{2}n)}\exp(-\tfrac{1}{2}z)z^{\frac{1}{2}n-1}\,dz, \qquad 0\leqslant z\leqslant\infty, \qquad (16.1)$$

and that if we impose p independent linear constraints upon these variates, i.e. consider the distribution of z conditional upon $p\,(<n)$ linear relations among the normal variates being satisfied, the effect of the constraints is simply to replace n in (16.1) by $(n-p)$. Further, we have also seen (Example 11.7) that if the sum of squares is taken about the sample mean instead of the population mean, the effect on the distribution of z is to impose one linear constraint, replacing n in (16.1) by $(n-1)$.

The distribution (16.1) is called the χ^2 distribution with n degrees of freedom—see **16.10** for an explanation. If rewritten, with $z = \chi^2$ and $n = \nu$, as

$$dF = \frac{1}{\Gamma(\frac{1}{2}\nu)}\exp(-\tfrac{1}{2}\chi^2)(\tfrac{1}{2}\chi^2)^{\frac{1}{2}\nu-1}\,d(\tfrac{1}{2}\chi^2), \qquad \nu>0\,;\ 0\leqslant\chi^2\leqslant\infty, \qquad (16.2)$$

it is at once evident that $\frac{1}{2}\chi^2$ is distributed in the Gamma distribution with parameter $\frac{1}{2}\nu$. (16.2) is a Pearson Type III distribution. Although we have so far considered distributions of the type (16.2) with ν a positive integer, we have taken the opportunity here to widen the definition of the distribution to all cases for which ν is positive. For such cases, the distribution function always converges, so that the distribution is always properly defined.

By differentiation of (16.2), we find that for $\nu > 2$, the χ^2 distribution rises from the origin to a unique mode at $\nu-2$. For $\nu = 2$, the distribution is J-shaped, with maximum ordinate at zero, while for $0 < \nu < 2$ the distribution is J-shaped and has infinite ordinate at the origin.

Properties of the χ^2 distribution

16.3 The characteristic function of χ^2 is obtained at once from (16.1) (see Example 3.6) as

$$\phi(t) = \frac{1}{(1-2it)^{\frac{1}{2}\nu}} \tag{16.3}$$

whence, for the cumulants, we have

$$\kappa_r = \nu 2^{r-1}(r-1)! \tag{16.4}$$

and for moments about the mean

$$\left.\begin{aligned} \mu_2 &= 2\nu \\ \mu_3 &= 8\nu \\ \mu_4 &= 48\nu+12\nu^2 \\ \mu_5 &= 32\nu(5\nu+12) \\ \mu_6 &= 40\nu(3\nu^2+52\nu+96). \end{aligned}\right\} \tag{16.5}$$

Since κ_r is linear in ν, μ_r, which can contain only $[\frac{1}{2}r]$ powers of μ_2, must be of degree $[\frac{1}{2}r]$ in ν, i.e. $\frac{1}{2}r$ if r is even and $\frac{1}{2}(r-1)$ if r is odd.

Exercise 16.3 gives a recurrence relation between the central moments.

As ν tends to infinity the χ^2 distribution tends to normality, for in standard measure we have

$$\phi(t) = e^{-\frac{\nu it}{\sqrt{(2\nu)}}}\left(1-\frac{2it}{\sqrt{(2\nu)}}\right)^{-\frac{1}{2}\nu}$$

and

$$\log\phi(t) = \frac{-\nu it}{\sqrt{(2\nu)}}-\frac{\nu}{2}\left\{\frac{-2it}{\sqrt{(2\nu)}}-\frac{1}{2}\left(\frac{2it}{\sqrt{(2\nu)}}\right)^2+o(\nu^{-1})\right\} \to -\tfrac{1}{2}t^2.$$

The tendency is, however, rather slow, and there are better approximations, as we shall see in a moment.

16.4 The distribution function of (16.2) is an incomplete Gamma-function. We have

$$F(\zeta) = \int_0^\zeta \frac{1}{2^{\frac{1}{2}\nu}\Gamma(\frac{1}{2}\nu)}\exp(-\tfrac{1}{2}z)z^{\frac{1}{2}\nu-1}\,dz = \Gamma_{\frac{1}{2}\zeta}(\tfrac{1}{2}\nu)/\Gamma(\tfrac{1}{2}\nu),$$

or, in the notation of Pearson's tables,

$$= I\left(\frac{\frac{1}{2}\zeta}{\sqrt{(\frac{1}{2}\nu)}},\ \tfrac{1}{2}\nu-1\right).$$

Series expansions of the d.f. are given in Exercise 16.7, which also gives the complementary relation with the Poisson d.f. implicit in (5.23).

Tables of the χ^2 distribution

(a) E. S. Pearson and Hartley (1950) give the distribution function of χ^2 to 5 d.p. for $\nu = 1(1)20(2)70$ and $\chi^2 = 0{\cdot}001(0{\cdot}001)0{\cdot}01(0{\cdot}01)0{\cdot}1(0{\cdot}1)2{\cdot}0(0{\cdot}2)10(0{\cdot}5)20(1)40(2)120$. This table is fully reproduced in the *Biometrika Tables.*

(b) Thompson (1941) gives a table of the percentage points of the distribution, i.e. the values of χ^2 corresponding to the values $(1-P)$ of the distribution function, for $P = 0{\cdot}995$, $0{\cdot}990$, $0{\cdot}975$, $0{\cdot}950$, $0{\cdot}900$, $0{\cdot}750$, $0{\cdot}500$, $0{\cdot}250$, $0{\cdot}100$, $0{\cdot}050$, $0{\cdot}025$, $0{\cdot}010$, $0{\cdot}005$ and $\nu = 1\,(1)\,30\,(10)\,100$. This table, too, is reproduced in the *Biometrika Tables*, with the addition of entries for $P = 0{\cdot}001$.

(c) Fisher and Yates (1953) give percentage points for $P = 0{\cdot}99$, $0{\cdot}98$, $0{\cdot}95$, $0{\cdot}90\,(0{\cdot}10)\,0{\cdot}10$, $0{\cdot}05$, $0{\cdot}02$, $0{\cdot}01$, $0{\cdot}001$ and $\nu = 1\,(1)\,30$. A selection from this table is given as Appendix Table 3.

(d) A table by Yule, reproduced in Appendix Table 4, gives P for $\nu = 1$, $\chi^2 = 0\,(0{\cdot}01)\,1\,(0{\cdot}1)\,10$.

(e) Harter (1964a) gives a 9 d.p. table of the d.f. and 23 percentage points to 6 significant figures for $\nu = 1(1)150(2)330$. For $\nu = 1(1)100$, the latter appear in Harter (1964b). The *Biometrika Tables*, Vol. II, extend the percentage points to the range $\nu = 0{\cdot}1\,(0{\cdot}1)\,3{\cdot}0\,(0{\cdot}2)\,10{\cdot}0\,(1)\,100$.

(f) Khamis and Rudert (1965) give a 10 d.p. table of the d.f. for $\nu = 0{\cdot}1(0{\cdot}1)\,20\,(0{\cdot}2)\,40\,(0{\cdot}5)\,140$ at the values $0{\cdot}0001\,(0{\cdot}0001)\,0{\cdot}001\,(0{\cdot}001)\,0{\cdot}01\,(0{\cdot}01)\,1\,(0{\cdot}05)\,6\,(0{\cdot}1)\,16\,(0{\cdot}5)\,66\,(1)\,166\,(2)\,250$.

Square-root and cube-root transformations of χ^2

16.5 Two approximations to the distribution of χ^2 are in common use, each of which selects an appropriate function of the variate which is normally distributed to a closer approximation than is χ^2 itself. They are:

(a) Fisher's result that $\sqrt{(2\chi^2)}$ is approximately normally distributed with mean $\sqrt{(2\nu-1)}$ and unit variance;

(b) Wilson and Hilferty's (1931) result that $(\chi^2/\nu)^{1/3}$ is approximately normally distributed with mean $1-2/(9\nu)$ and variance $2/(9\nu)$.

The second of these is the more accurate approximation, but involves more computation in applications.

16.6 The relative speed of approach to normality of χ^2 and $\sqrt{(2\chi^2)}$ may be compared as follows:—

For χ^2 we have, from (16.5), the skewness and kurtosis coefficients

$$\gamma_1 = \sqrt{\beta_1} = (8/\nu)^{\frac{1}{2}}, \qquad \gamma_2 = \beta_2 - 3 = 12/\nu. \tag{16.6}$$

For Fisher's square-root transformation, we need the moments of χ, for which we have

$$\begin{aligned} \mu_r' &= \frac{1}{2^{\frac{1}{2}(\nu-2)}\,\Gamma(\frac{1}{2}\nu)} \int_0^\infty e^{-\frac{1}{2}\chi^2} \chi^{\nu+r-1}\,d\chi \\ &= \frac{2^{\frac{1}{2}r}\,\Gamma\{\frac{1}{2}(\nu+r)\}}{\Gamma(\frac{1}{2}\nu)}. \end{aligned} \tag{16.7}$$

Thus

$$\mu_1' = \sqrt{2}\,\frac{\Gamma\{\frac{1}{2}(\nu+1)\}}{\Gamma(\frac{1}{2}\nu)}.$$

Using Stirling's expansion

$$\log \Gamma(x+1) \sim \tfrac{1}{2}\log(2\pi) + (x+\tfrac{1}{2})\log x - x + \frac{1}{12x} - \frac{1}{360x^3} + o\left(\frac{1}{x^3}\right),$$

we find after substitution and reduction

$$\mu_1' = \sqrt{\nu}\left(1-\frac{1}{4\nu}+\frac{1}{32\nu^2}+\frac{5}{128\nu^3}+\ldots\right),$$

whence
$$\mu_1'^2 = \nu\left(1-\frac{1}{2\nu}+\frac{1}{8\nu^2}+\frac{1}{16\nu^3}+\ldots\right).$$

Also
$$\mu_2' = \nu, \qquad \mu_3' = (\nu+1)\mu_1', \qquad \mu_4' = (\nu+2)\nu,$$

whence we find for moments about the mean

$$\mu_2 = \frac{1}{2}-\frac{1}{8\nu}+o\left(\frac{1}{\nu}\right), \qquad \mu_3 = \frac{1}{4\sqrt{\nu}}+o\left(\frac{1}{\sqrt{\nu}}\right), \qquad \mu_4 = \frac{3}{4}-\frac{3}{8\nu}+o\left(\frac{1}{\nu}\right)$$

while for the moments of $\sqrt{(2\chi^2)}$ we have

$$\mu_2 = 1-\frac{1}{4\nu}+o\left(\frac{1}{\nu}\right)$$

$$\left.\begin{aligned}\gamma_1 &= \frac{1}{\sqrt{(2\nu)}}+o\left(\frac{1}{\sqrt{\nu}}\right)\\ \gamma_2 &= \frac{3}{4\nu^2}+o\left(\frac{1}{\nu^2}\right).\end{aligned}\right\} \tag{16.8}$$

A comparison of (16.8) with (16.6) shows that $\sqrt{(2\chi^2)}$ tends to normality with considerably greater rapidity than χ^2. Moreover, the expression for μ_1' of χ is equal to $\sqrt{(\nu-\frac{1}{2})}$ to order $\nu^{-3/2}$ and hence $\sqrt{(2\chi^2)}$ is distributed about mean $\sqrt{(2\nu-1)}$ to that order, with variance which is unity to order ν^{-1}.

The *Biometrika Tables* give μ_1', $\mu_2^{1/2}$, β_1 and β_2 of $\chi/\nu^{1/2}$ to at least 4 d.p. for $\nu = 1\,(1)\,20\,(5)\,50\,(10)\,100$. Hodges and Lehmann (1967) study the moments in detail.

16.7 For the Wilson–Hilferty cube-root transformation, consider the distribution of χ^2 about its mean value ν. Let us find the distribution of $\left(\frac{\chi^2}{\nu}\right)^h = y$, say, h as yet being undetermined. Write $\xi = \chi^2-\nu$. Then

$$\begin{aligned} y &= \left(1+\frac{\xi}{\nu}\right)^h \\ &= 1+\frac{h\xi}{\nu}+\frac{h(h-1)}{2!}\frac{\xi^2}{\nu^2}+\ldots \end{aligned} \tag{16.9}$$

Taking mean values and using the results of (16.5), we find, after some reduction,

$$\begin{aligned}\mu_1'(y) &= 1+\frac{h(h-1)}{2!}\frac{\mu_2(\chi^2)}{\nu^2}+\ldots \\ &= 1+\frac{h(h-1)}{1}\cdot\frac{1}{\nu}+\frac{h(h-1)(h-2)(3h-1)}{6\nu^2} \\ &\quad +\frac{h^2(h-1)^2(h-2)(h-3)}{6\nu^3}+O(\nu^{-4}).\end{aligned} \tag{16.10}$$

If in (16.9) we put rh for h, we obtain $\mu_r'(y)$. We then find in particular, using (3.9),

$$\mu_3(y) = \frac{4(3h-1)}{\nu^2} + O(\nu^{-3}). \quad (16.11)$$

Thus if we take $h = \frac{1}{3}$, μ_3 will be zero to order ν^{-2}, and the distribution will presumably be brought closer to symmetry and normality. We then find, with $h = \frac{1}{3}$,

$$\mu_1'(y) = 1 - \frac{2}{9\nu} + \frac{80}{3^7\nu^3} + O(\nu^{-4}),$$

$$\mu_2'(y) = 1 - \frac{2}{9\nu} + \frac{4}{3^4\nu^2} + \frac{56}{3^7\nu^3} + O(\nu^{-4}),$$

$$\mu_3'(y) = 1,$$

$$\mu_4'(y) = 1 + \frac{4}{9\nu} - \frac{4}{3^3\nu^2} + \frac{80}{3^7\nu^3} + O(\nu^{-4}),$$

or, transferring to the mean,

$$\left.\begin{aligned} \mu_2(y) &= \frac{2}{9\nu} - \frac{104}{3^7\nu^3} + O(\nu^{-4}), \\ \mu_3(y) &= \frac{32}{3^6\nu^3} + O(\nu^{-4}), \\ \mu_4(y) &= \frac{4}{3^3\nu^2} - \frac{16}{3^6\nu^3} + O(\nu^{-4}). \end{aligned}\right\} \quad (16.12)$$

We now find

$$\left.\begin{aligned} \gamma_1 &= \frac{2^{7/2}}{3^3\nu^{3/2}} + o(\nu^{-3/2}) \\ \gamma_2 &= -\frac{4}{9\nu} + o(\nu^{-1}). \end{aligned}\right\} \quad (16.13)$$

Comparison with (16.6) and (16.8) shows that $\left(\frac{\chi^2}{\nu}\right)^{1/3}$ tends to symmetry, as measured by γ_1, more rapidly than either χ^2 or $\sqrt{(2\chi^2)}$. To order ν^{-2} the variance is, from (16.12), equal to $\frac{2}{9\nu}$ and the mean $1-\frac{2}{9\nu}$. The result may also be expressed by saying that

$$\left\{\left(\frac{\chi^2}{\nu}\right)^{1/3} + \frac{2}{9\nu} - 1\right\}\left(\frac{9\nu}{2}\right)^{1/2} \quad (16.14)$$

is distributed normally about zero mean with unit variance.

Exercise 37.10 (Vol. 3, 3rd edn) gives Haldane's (1938) generalization of the method of this section to a class of variates. In general, $h = 1 - \kappa_1\kappa_3/(3\kappa_2^2)$, reducing to $\frac{1}{3}$ in the χ^2 case on substitution from (16.4).

16.8 The following table, quoted from Garwood (1936), shows some comparisons of the two approximations with the exact values. In each case there have been worked out the values of $\frac{1}{2}\chi_0^2$ corresponding to $P = \int_{\frac{1}{2}\chi_0^2}^{\infty} dF(\frac{1}{2}\chi^2) = 0{\cdot}01, 0{\cdot}05, 0{\cdot}95$ and $0{\cdot}99$. The exact values are denoted by m_T, those given by Fisher's approximation by m_F, and those by Wilson and Hilferty's approximation by m_W.

Table 16.1—Comparison of approximations to the χ^2 d.f. (Garwood (1936))

	ν	m_T	m_F	$m_T - m_F$	m_W	$m_T - m_W$
$P = 0{\cdot}99$	40	11·082	10·764	0·318	11·070	0·012
	60	18·742	18·414	0·328	18·732	0·010
	80	26·770	26·436	0·334	26·761	0·009
	100	35·032	34·694	0·338	35·025	0·007
$P = 0{\cdot}95$	40	13·255	13·116	0·139	13·254	0·001
	60	21·594	21·455	0·139	21·594	0·000
	80	30·196	30·056	0·140	30·196	0·000
	100	38·965	38·825	0·140	38·965	0·000
$P = 0{\cdot}05$	42	29·062	28·919	0·143	29·060	0·002
	62	40·691	40·548	0·143	40·689	0·002
	82	52·069	51·926	0·143	52·068	0·001
	102	63·287	63·144	0·143	63·286	0·001
$P = 0{\cdot}01$	42	33·103	32·700	0·403	33·113	−0·010
	62	45·401	45·003	0·398	45·409	−0·008
	82	57·347	56·953	0·394	57·355	−0·008
	102	69·067	68·676	0·391	69·074	−0·007

The m_W approximation is evidently very good and the m_F approximation is fair. These are comparisons of approximate values of the variate for fixed P. Mathur (1961) has compared the three normal approximations that we have discussed in respect of the maximum error in the d.f. committed over the range of the variate—cf. Exercise 16.24.

Slutsky (1950) tabulates the χ^2 probabilities using the approximate normality of $\{\sqrt{(2\chi^2)}-\sqrt{(2\nu)}\}$ for $\sqrt{(2/\nu)} = 0$ to $0{\cdot}25$, i.e. from $\nu = \infty$ down to $\nu = 32$, to 5 d.p.

16.9 Another approximation may be obtained by the Cornish–Fisher method of **6.25** and **6.26**, and in fact our Example 6.4 was virtually based on the χ^2 distribution. The Gamma variate there with parameter λ corresponds to $\frac{1}{2}\chi^2$ with 2λ degrees of freedom.

What appears to be the best available normal approximation is given by Peizer and Pratt (1968) and Pratt (1968).

"Student's" t-distribution

16.10 In Example 11.8, we have seen that, in samples from a normal population, the ratio of the difference between the sample and population means to its estimated standard error, $t = (\bar{x}-\mu)\sqrt{\{n(n-1)\}}/\sqrt{\Sigma(x-\bar{x})^2} = (\bar{x}-\mu)\sqrt{(n/s^2)}$, where $s^2 = \Sigma(x-\bar{x})^2/(n-1)$, has a sampling distribution given by

$$dF = dt\Big/\left\{\nu^{\frac{1}{2}}B\left(\tfrac{1}{2},\tfrac{1}{2}\nu\right)\left(1+\frac{t^2}{\nu}\right)^{\frac{1}{2}(\nu+1)}\right\}, \qquad \nu>0\,;\ -\infty\leqslant t\leqslant\infty. \tag{16.15}$$

As before we write ν for the *degrees of freedom*, here equal to $n-1$. The origin of this term requires a few words of explanation.

It has been seen (Example 11.7) that the variance of a normal sample is distributed like the sum of $(n-1)$ squares of independent variates, and generally, that if there are k linear relations connecting the original variates, the sum of squares of the originals is distributed as the sum of $n-k$ independent normal standardized variates. Each linear relation reduces the dimension of the variation, as it were, by unity. It is thus natural to speak of the number of degrees of freedom, ν, of a function such as χ^2, meaning thereby that it is distributed as the sum of squares of ν independent standardized normal variates. The expression only has this natural meaning where normal variation is concerned.

It so happens that the quantity t depends on a quantity ν which is convenient for tabulating its distribution function and is also the number of degrees of freedom of the statistic s^2 entering into the denominator of t. ν may thus, by an extension of the term, be called the number of degrees of freedom of t.

(16.15) is a distribution of the Pearson Type VII. It is evidently symmetrical about the origin and unimodal, and extends to infinity in both directions. It is now universally known as "Student's" t-distribution, the name "Student" having been the pseudonym of its discoverer, W. S. Gosset. The case $\nu = 1$ is the Cauchy distribution $\{\pi(1+t^2)\}^{-1}$.

It is clear, from the fact that the estimated standard error of a sample mean converges to the true standard error in large samples, that the distribution (16.15) must tend with large ν to the standardized normal distribution, for the mean of a normal sample is exactly normally distributed (Example 11.12). Compare Example 4.8.

Properties of "Student's" t-distribution

16.11 The characteristic function of the distribution has already been examined in Example 3.13, where for our present notation we replace m by $\frac{1}{2}(\nu+1)$. The moments μ_r of the distribution exist only for $r<\nu$, and are then equal to zero by symmetry for odd-order moments, while for even moments (Example 3.3)

$$\mu_{2r} = \nu^r \frac{\Gamma(r+\frac{1}{2})\,\Gamma(\frac{1}{2}\nu-r)}{\Gamma(\frac{1}{2})\,\Gamma(\frac{1}{2}\nu)}, \qquad 2r<\nu. \tag{16.16}$$

Thus the skewness and kurtosis coefficients are

$$\gamma_1 = 0, \quad \gamma_2 = 6/(\nu-4), \quad \nu>4.$$

The distribution function of t is

$$F(\tau) = \int_{-\infty}^{\tau} dF$$

whence, using the symmetry of (16.15),

$$2F-1 = \frac{2}{B\left(\frac{\nu}{2}, \frac{1}{2}\right)} \int_0^{\tau} \frac{1}{\left(1+\frac{t^2}{\nu}\right)^{\frac{1}{2}(\nu+1)}} \frac{dt}{\sqrt{\nu}},$$

and by putting $\xi = \left(1+\frac{t^2}{\nu}\right)^{-1}$ this is seen to be

$$2F-1 = \frac{1}{B\left(\frac{\nu}{2},\frac{1}{2}\right)}\int_{\xi(\tau)}^{1}\xi^{\frac{1}{2}\nu-1}(1-\xi)^{-\frac{1}{2}}\,d\xi = 1-I_{\xi}\left(\frac{\nu}{2},\frac{1}{2}\right),$$

whence
$$F = 1-\tfrac{1}{2}I_{\xi}\left(\frac{\nu}{2},\frac{1}{2}\right). \tag{16.17}$$

The values of the argument for which I has the values 0·50, 0·25, 0·10, 0·05, 0·025, 0·01, 0·005 and $\nu = 1(1)30$, 40, 60, 120, ∞, have been tabled to five significant figures by C. M. Thompson *et al.* (1941) and can hence be used to derive the values of t for the corresponding percentage points. These tables are reproduced in the *Biometrika Tables.*

In the Cauchy case $\nu = 1$, $F(x) = \frac{1}{2}+\frac{1}{\pi}\arctan x$.

16.12 Except for special purposes, however, the use of the B-function is unnecessary, since the distribution function of t itself and tables based thereon are available. We have

$$-\log\left(1+\frac{t^2}{\nu}\right) = -\frac{t^2}{\nu}+\frac{t^4}{2\nu^2}-\ldots+\frac{(-t^2)^j}{j\nu^j}+\ldots$$

and hence

$$-\tfrac{1}{2}(\nu+1)\log\left(1+\frac{t^2}{\nu}\right) = -\tfrac{1}{2}t^2+\ldots-\frac{j(-t^2)^{j+1}+(j+1)(-t^2)^j}{2j(j+1)\nu^j}+\ldots \tag{16.18}$$

Further, from the expansion for $\log\Gamma(1+x)$ we find

$$\log\left\{\frac{\Gamma\{\frac{1}{2}(\nu+1)\}}{\Gamma(\frac{1}{2}\nu)}\sqrt{\frac{2}{\nu}}\right\} = -\frac{1}{4\nu}+\frac{1}{24\nu^3}-\frac{1}{20\nu^5}+\ldots \tag{16.19}$$

Now as ν tends to infinity, t tends to the standardized normal form, as we have already seen. Writing

$$y = \frac{1}{\sqrt{(2\pi)}}\exp(-\tfrac{1}{2}t^2), \tag{16.20}$$

we use (16.18), (16.19) and (16.20) to find, for the logarithm of the ordinate of (16.15), in descending powers of ν,

$$\log y+\frac{1}{4\nu}(t^4-2t^2-1)-\frac{1}{12\nu^2}(2t^6-3t^4)+\frac{1}{24\nu^3}(3t^8-4t^6+1)$$
$$-\frac{1}{40\nu^4}(4t^{10}-5t^8)+\frac{1}{60\nu^5}(5t^{12}-6t^{10}-3)-\ldots \tag{16.21}$$

Taking the exponential of (16.21) and integrating from t to ∞, we find

$$1-F = \int_t^{\infty}y\,dt+y\left\{\frac{1}{4\nu}t(t^2+1)+\frac{1}{96\nu^2}(3t^6-7t^4-5t^2-3)t+\frac{1}{384\nu^3}(t^{10}-11t^8+14t^6+6t^4\right.$$
$$-3t^2-15)t+\frac{1}{92160\nu^4}(15t^{14}-375t^{12}+2225t^{10}-2141t^8-939t^6-213t^4+915t^2$$
$$\left.+945)t+\ldots\right\}. \tag{16.22}$$

(16.22) is the expression, due to Fisher, which was used by "Student" himself in calculating the distribution function of t. For values of $\nu > 20$ the first four terms of (16.22) give F with a maximum error of 0·000,005.

Exercise 16.10 gives a finite series expansion for the d.f. for even ν.

Tables of Student's *t*-distribution

16.13

(a) The *Biometrika Tables* give the d.f. for $t = 0(0{\cdot}1)8$ and $\nu = 1(1)24; 30; 40; 60; 120$ and ∞.

(b) Fisher and Yates' (1953) table of percentage points of the distribution, i.e. the values of t corresponding to the values of the distribution function F, satisfying $2(1-F) = 0{\cdot}9(0{\cdot}1)0{\cdot}1, 0{\cdot}05, 0{\cdot}02, 0{\cdot}01, 0{\cdot}001$ for $\nu = 1(1)30; 40; 60; 120$ and ∞. This table is given as Appendix Table 5.

(c) Thompson's tables of percentage points discussed in **16.11** above.

(d) Baldwin's (1946) percentage points table for $2(1-F) = 0{\cdot}05, 0{\cdot}01$ with $\nu = 1(1)30(2)100$.

(e) Federighi (1959) gives percentage points for 20 values of $(1-F)$ between 0·25 and 0·000,000,1 and $\nu = 1(1)30(5)60(10)100; 200; 500; 1000; 2000; 10{,}000$ and ∞.

(f) Smirnov (1961) gives the d.f. and f.f. to 6 d.p. at $t = 0(0{\cdot}1)3{\cdot}00(0{\cdot}02)4{\cdot}50(0{\cdot}05)6{\cdot}50$ for $\nu = 1(1)12$, and at $t = 0(0{\cdot}01)2{\cdot}50(0{\cdot}02)3{\cdot}50(0{\cdot}05)6{\cdot}50$ for $\nu = 13(1)25$; and also the d.f. at $t = 0(0{\cdot}01)2{\cdot}50(0{\cdot}02)3{\cdot}50(0{\cdot}05)5{\cdot}00$ for $\nu = 25(1)35$. There are auxiliary tables for larger values of t and ν. Tables of percentage points of the d.f. are given for $2(1-F) = 0{\cdot}4, 0{\cdot}25, 0{\cdot}10, 0{\cdot}05, 0{\cdot}01, 0{\cdot}005, 0{\cdot}0025, 0{\cdot}001, 0{\cdot}0005$ and $\nu = 1(1)30(10)100, 120, 150(50)500$, etc.

(g) Lempers and Louter (1971) give the percentage points of t for $F = k/16$, $k = 1(1)15$ and $\nu = 1(1)30, 40, 60, 120, \infty$.

16.14 Before leaving the t-distribution, we should mention that in many applications it is the distribution of t^2, rather than that of t, which is of interest. There is, of course, no trouble in passing from t to t^2 in (16.15). It turns out, however, that the distribution of t^2 is merely a special case of a more general distribution, Fisher's F or z, and it is to this distribution, the last of the three basic sampling distributions related to the normal, that we now turn.

Fisher's *F*- and *z*-distributions

16.15 In discussing the distribution of a ratio, we saw, in effect (Example 11.20), that in independent samples from two normal populations with equal variances the sampling distribution of the sample variance ratio, which we shall now call

$$F = \frac{\sum_{i=1}^{n_1} (x_{1i} - \bar{x}_1)^2/(n_1 - 1)}{\sum_{j=1}^{n_2} (x_{2j} - \bar{x}_2)^2/(n_2 - 1)}, \tag{16.23}$$

is given by $$dG = \frac{\nu_1^{\frac{1}{2}\nu_1}\nu_2^{\frac{1}{2}\nu_2}F^{\frac{1}{2}(\nu_1-2)}\,dF}{B(\frac{1}{2}\nu_1,\frac{1}{2}\nu_2)(\nu_1 F+\nu_2)^{\frac{1}{2}(\nu_1+\nu_2)}}, \qquad \nu_1,\nu_2>0\,;\;\; 0\leqslant F\leqslant\infty, \tag{16.24}$$
where we have written $\nu_1 = n_1-1$, $\nu_2 = n_2-1$, for the "degrees of freedom" as before. If we put $x = (\nu_1/\nu_2)F$ in (16.24), we reduce it to a Beta distribution of the second kind as at (6.17), with parameters $p = \frac{1}{2}\nu_1$, $q = \frac{1}{2}\nu_2$.

On comparing (16.24) with (16.15), it is evident that if we put $\nu_1 = 1$, $\nu_2 = \nu$, (16.24) becomes identical with the distribution of t^2 obtainable immediately from (16.15). Further, it will be seen that in Example 11.20 the only facts made use of in deriving the equivalent of (16.24) were that $\sum\limits_i(x_{1i}-\bar{x}_1)^2/\sum\limits_j(x_{2j}-\bar{x}_2)^2$ be a ratio of two independently distributed χ^2 variates with ν_1 and ν_2 degrees of freedom for numerator and denominator respectively. It follows that (16.24) holds for the distribution of any such ratio. For example, reverting to (16.23) above, suppose that the normal populations we are sampling have different variances, say σ_1^2 and σ_2^2. It then follows that $\frac{\sigma_2^2}{\sigma_1^2}F$, where F is as at (16.23), has the distribution (16.24). For, in each sample, it is the quantity
$$\sum_i(x_{pi}-\bar{x}_p)^2/\sigma_p^2, \qquad p = 1,2,$$
which is distributed like χ^2, since the standard deviation is the scale factor.

When $\nu_1 = \nu_2$, another relation between (16.24) and (16.15) is given in Exercise 16.2.

16.16 The discussion of Example 11.20 was largely concerned with the simple transformation of (16.23)
$$z = \tfrac{1}{2}\log F, \tag{16.25}$$
whose distribution, given at (11.82) or immediately obtained from (16.24), is
$$dH = \frac{2\nu_1^{\frac{1}{2}\nu_1}\nu_2^{\frac{1}{2}\nu_2}}{B(\frac{1}{2}\nu_1,\frac{1}{2}\nu_2)}\cdot\frac{\exp(\nu_1 z)\,dz}{\{\nu_1\exp(2z)+\nu_2\}^{\frac{1}{2}(\nu_1+\nu_2)}}, \qquad \nu_1,\nu_2>0\,;\;\; -\infty\leqslant z\leqslant\infty. \tag{16.26}$$
It was in this form that the distribution was first obtained by R. A. Fisher. Modern practice is almost entirely in favour of using the simpler statistic F, although z was tabulated earlier, and the original mathematical investigations of the distribution were carried out in terms of z.

In the special case $\nu_1 = \nu_2 = 2$, the distribution of $2z$ is the logistic of Exercise 4.21, as may be seen either from (16.26) or from its c.f. (16.29) below.

Properties of the F- and z-distributions

16.17 The general shape of the F-distribution (16.24) is indicated by the fact that, while it always ranges from zero to infinity, it is J-shaped if $\nu_1\leqslant 2$, while if $\nu_1>2$ it is a unimodal skew distribution with modal value at
$$\hat{F} = \frac{\nu_1-2}{\nu_1}\,\frac{\nu_2}{\nu_2+2}. \tag{16.27}$$
It is easily verified directly from (16.24) that the mean and variance of F are
$$\left.\begin{aligned}\mu_1' &= \nu_2/(\nu_2-2), \qquad \nu_2>2,\\ \mu_2 &= 2\nu_2^2(\nu_1+\nu_2-2)/\{\nu_1(\nu_2-2)^2(\nu_2-4)\}, \qquad \nu_2>4,\end{aligned}\right\} \tag{16.28}$$
the conditions on ν_2 being required for the existence of these moments.

Comparison of (16.27) and (16.28) shows that for $\nu_1, \nu_2 > 2$ the F-distribution always has its modal value below $F = 1$ and its mean value above $F = 1$, confirming that the distribution is always " positively " skew, as is otherwise obvious. That $E(F) > 1$ irrespective of parent normality is a consequence of Exercise 9.15.

Higher moments of (16.24) are given in Exercise 16.1.

The z-distribution (16.26) is symmetric about zero only when $\nu_1 = \nu_2$, but always has a unique mode at zero.

The c.f. of z is most easily obtained from the F-distribution as $E(e^{\theta \frac{1}{2} \log F}) = E(F^{\frac{1}{2}\theta})$ where $\theta = it$. This is formally the $(\frac{1}{2}\theta)$th moment of the F-distribution, and Exercise 16.1 shows this to be

$$\phi(t) = \left(\frac{\nu_2}{\nu_1}\right)^{\frac{1}{2}\theta} \frac{\Gamma\{\frac{1}{2}(\nu_1+\theta)\}\Gamma\{\frac{1}{2}(\nu_2-\theta)\}}{\Gamma(\frac{1}{2}\nu_1)\Gamma(\frac{1}{2}\nu_2)}. \tag{16.29}$$

All the moments of the z-distribution are finite, as may be seen from (16.26). We shall only obtain their asymptotic forms here. Using the Stirling expansion

$$\log \Gamma(1+x) = \tfrac{1}{2}\log 2\pi + (x+\tfrac{1}{2})\log x - x + \frac{1}{12x} - \ldots,$$

we find from (16.29), for the c.g.f. of z,

$$\log \phi(t) = \frac{\theta}{2}\left(\frac{1}{\nu_2} - \frac{1}{\nu_1}\right) + \frac{\theta^2}{4}\left(\frac{1}{\nu_1} + \frac{1}{\nu_2}\right) + o\left(\frac{1}{\nu}\right)\ldots. \tag{16.30}$$

For large ν_1 and ν_2, z is thus distributed approximately normally with moments

$$\mu_1' = \tfrac{1}{2}\left(\frac{1}{\nu_2} - \frac{1}{\nu_1}\right), \qquad \mu_2 = \tfrac{1}{2}\left(\frac{1}{\nu_1} + \frac{1}{\nu_2}\right). \tag{16.31}$$

The asymptotic cumulants of z are given in more detail at (16.34) below. For the exact cumulants see Exercise 32.12, Vol. 2.

16.18 The distribution function of F (or z) may be obtained from tables of the Incomplete Beta-function by making the transformation

$$\xi = (1 + \nu_1 F/\nu_2)^{-1}$$

in (16.24). We may then, as in **16.11** above, use Incomplete Beta Function tables, the d.f. of F being given by $1 - I_\xi\left(\frac{\nu_2}{2}, \frac{\nu_1}{2}\right)$, of which (16.17) is a special case—the factor $\frac{1}{2}$ in (16.17) disappears if we work in terms of t^2 instead of t. However, special tables of the distribution exist which make these procedures unnecessary in the usual applications.

Tables of F and z

16.19 Because there are two " degrees of freedom " entries to any table of F, a full tabulation of the distribution would require a table of triple entry, and it is therefore more convenient to use inverse tables of the quantiles of the distribution. These have been provided by :

(a) Fisher and Yates (1953) who give the percentage points of F (and z) corresponding to the values $(1-P)$ of the distribution function for $P = 0{\cdot}20$, $0{\cdot}10$, $0{\cdot}05$, $0{\cdot}01$ and $0{\cdot}001$; $\nu_1 = 1(1)6$; 8, 12, 24, ∞ and $\nu_2 = 1(1)30$; 40, 60, 120, ∞. The upper 5% and 1% points tables are reproduced as Appendix Tables 6–9.

(b) The *Biometrika Tables* give the percentage points of F to 2 d.p. for $P = 0{\cdot}25$, $0{\cdot}10$, $0{\cdot}05$, $0{\cdot}025$, $0{\cdot}01$, $0{\cdot}005$ and $0{\cdot}001$; $\nu_1 = 1(1)10$; 12, 15, 20, 24, 30, 40, 60, 120, ∞ and $\nu_2 = 1(1)30$; 40, 60, 120, ∞; Vol. II of the *Biometrika Tables* give 5 significant figure tables, including also $P = 0{\cdot}5$ and $0{\cdot}0025$, with an auxiliary table for fractional d.fr.

(c) Harter (1964a) gives 23 percentage points to 7 significant figures for *even* $\nu_1, \nu_2 \leqslant 80$.

(d) Vogler (1964) gives 17 percentage points to 6 significant figures for $\nu_1 = 1(1)10$, 12, 15, 20, 24, 30, 40, 60, 120 and $\nu_2 = 1(0{\cdot}1)2$, $2{\cdot}2$, $2{\cdot}5(0{\cdot}5)5$, $6(1)10$, 12, 15, 20, 24, 30, 40, 60, 120, ∞.

16.20 As is obvious from the definition of F as a ratio, we may interchange its numerator and denominator and obtain a distribution of exactly the same form, but with ν_1 and ν_2 interchanged. This may be verified by making the transformation $F = 1/y$ in (16.24). In terms of the distribution function of F, say $G_{\nu_1,\nu_2}(F)$, this implies

$$G_{\nu_1,\nu_2}(F) = 1-G_{\nu_2,\nu_1}\left(\frac{1}{F}\right). \tag{16.32}$$

Since there is a one-one correspondence between F and z, and $\log\frac{1}{F} = -\log F$, this is equivalent to

$$H_{\nu_1,\nu_2}(z) = 1-H_{\nu_2,\nu_1}(-z). \tag{16.33}$$

The relations (16.32) and (16.33) make it unnecessary to tabulate the distributions at both extremes. In fact, if we are interested in the distribution function for fractional values of F (i.e. negative values of z) we have only to invert the ratio (i.e. change the sign of z) and use these relations to obtain the value of the distribution function which we are seeking.

16.21 Various approximations have been given for the case when ν_1 and ν_2 are not large enough to justify the use of the normal approximation of **16.18.**

(a) (Cornish and Fisher, 1937.) The method is that of **6.25** and depends on the expansion of the distribution in a Gram-Charlier series. From the successive derivatives of $\log\Gamma(1+x)$ we can find those of $\log\phi(t)$, and hence ascertain the cumulants of z. Writing $r_1 = 1/\nu_1$ and $r_2 = 1/\nu_2$, we find the leading terms in the cumulants

$$\left.\begin{aligned}
\kappa_1 &= -\tfrac{1}{2}(r_1-r_2)-\tfrac{1}{6}(r_1^2-r_2^2), & \kappa_4 &= r_1^3+r_2^3+3(r_1^4+r_2^4),\\
\kappa_2 &= \tfrac{1}{2}(r_1+r_2)+\tfrac{1}{2}(r_1^2+r_2^2)+\tfrac{1}{3}(r_1^3+r_2^3), & \kappa_5 &= -3(r_1^4-r_2^4),\\
\kappa_3 &= -\tfrac{1}{2}(r_1^2-r_2^2)-(r_1^3-r_2^3), & \kappa_6 &= 12(r_1^5+r_2^5).
\end{aligned}\right\} \tag{16.34}$$

Hence, putting $\sigma = r_1+r_2$ and $\delta = r_1-r_2$, we find for the l's of **6.25** ($m = 0$, variance $= \frac{1}{2}\sigma$) :—

$$l_1 = -\sqrt{\frac{2}{\sigma}}\left(\tfrac{1}{2}\delta+\tfrac{1}{6}\delta\sigma\right), \qquad l_2 = \tfrac{1}{2}\left(\sigma+\frac{\delta^2}{\sigma}\right)+\tfrac{1}{6}(\sigma^2+3\delta^2),$$

and so on. After some reduction we find, for the value of z corresponding to a probability α (which in turn corresponds to a normal deviate ξ),—

$$z = \xi\sqrt{\frac{\sigma}{2}} - \tfrac{1}{6}\delta(\xi^2+2) + \sqrt{\frac{\sigma}{2}}\left\{\frac{\sigma}{24}(\xi^3+3\xi) + \frac{1}{72}\frac{\delta^2}{\sigma}(\xi^3+11\xi)\right\}$$
$$-\frac{\delta\sigma}{120}(\xi^4+9\xi^2+8) + \frac{\delta^3}{3240\sigma}(3\xi^4+7\xi^2-16) + \sqrt{\frac{\sigma}{2}}\left\{\frac{\sigma^2}{1920}(\xi^5+20\xi^3+15\xi)\right.$$
$$\left.+\frac{\delta^4}{2880}(\xi^5+44\xi^3+183\xi) + \frac{\delta^4}{155520\sigma^2}(9\xi^5-284\xi^3-1513\xi)\right\}. \tag{16.35}$$

(b) (Fisher, extended by Cochran (1940).) Writing n indifferently for ν_1 and ν_2, we have, from (16.35), to order $n^{-3/2}$:—

$$z = \xi\sqrt{\frac{\sigma}{2}} - \tfrac{1}{6}\delta(\xi^2+2) + \sqrt{\frac{\sigma}{2}}\left\{\frac{\sigma}{24}(\xi^3+3\xi) + \frac{1}{72}\frac{\delta^2}{\sigma}(\xi^3+11\xi)\right\}.$$

Put $h = 2/\sigma$. Then

$$z = \frac{\xi}{\sqrt{h}} - \tfrac{1}{6}\delta(\xi^2+2) + \frac{1}{\sqrt{h}}\left\{\frac{\xi^3+3\xi}{12h} + \frac{\xi^3+11\xi}{144}h\delta^2\right\}. \tag{16.36}$$

Now
$$\frac{\xi}{\sqrt{(h-\lambda)}} = \frac{\xi}{\sqrt{h}} + \frac{\lambda\xi}{2h\sqrt{h}} + o(h^{-3/2}).$$

Hence, if we put

$$z = \frac{\xi}{\sqrt{(h-\lambda)}} - \tfrac{1}{6}\delta(\xi^2+2), \tag{16.37}$$

the difference of (16.37) from (16.36) is

$$\frac{(\xi^3+11\xi)\,\delta^2\sqrt{h}}{144},$$

provided that we take $\lambda = \dfrac{\xi^2+3}{6}$.

The difference is small in virtue of the large denominator and the factor $\delta^2 = \left(\dfrac{1}{\nu_1} - \dfrac{1}{\nu_2}\right)^2$ which is small if ν_1 and ν_2 are not too different. Thus we may take z as approximately given by (16.37). The values of λ for various values of the d.f. $(1-P)$ are

$100P$ %	40%	30%	20%	10%	5%	1%	0·1%
λ	0·51	0·55	0·62	0·77	0·95	1·40	2·09

For the commoner values of P the form taken by (16.37) is

$$20 \text{ per cent.}: \frac{0{\cdot}8416}{\sqrt{(h-\lambda)}} - 0{\cdot}4514\delta \tag{16.38}$$

$$5 \text{ per cent.}: \frac{1{\cdot}6449}{\sqrt{(h-\lambda)}} - 0{\cdot}7843\delta \tag{16.39}$$

$$1 \text{ per cent.}: \frac{2{\cdot}3263}{\sqrt{(h-\lambda)}} - 1{\cdot}235\delta \tag{16.40}$$

$$0{\cdot}1 \text{ per cent.}: \frac{3{\cdot}0902}{\sqrt{(h-\lambda)}} - 1{\cdot}925\delta. \tag{16.41}$$

The accuracy of the approximation for $\nu_1 = 24$, $\nu_2 = 60$ may be judged from the following comparison :—

100 P %	Value of z from (16.37)	Exact value
20	0·1337	0·1338
1	0·3748	0·3746
0·1	0·4966	0·4955

(c) (Paulson, 1942.) The Wilson–Hilferty approximation to χ^2 of **16.7** indicates that $\left(\frac{\chi^2}{\nu}\right)^{1/3}$ is distributed normally about mean $1-\frac{2}{9\nu}$ with variance $\frac{2}{9\nu}$. Let $p^2 = \frac{s_1^2}{s_2^2}$ be the ratio of two independent quantities distributed as χ^2/ν with ν_1 and ν_2 degrees of freedom. Further, in virtue of Geary's theorem (Exercise 11.11) the ratio $\frac{m_1 - m_2 p}{(\sigma_1^2 + \sigma_2^2 p^2)^{\frac{1}{2}}}$ is normally distributed in standard measure. We may thus regard

$$u = \frac{\left(1-\frac{2}{9\nu_2}\right)\left(\frac{s_1}{s_2}\right)^{\frac{2}{3}} - \left(1-\frac{2}{9\nu_1}\right)}{\left\{\frac{2}{9\nu_2}\left(\frac{s_1}{s_2}\right)^{\frac{4}{3}} + \frac{2}{9\nu_1}\right\}^{\frac{1}{2}}} \tag{16.42}$$

as approximately normally distributed in standard measure. The approximation seems remarkably good. For instance, the following shows the exact and approximate values of p^2 for $\nu_1 = 6$, $\nu_2 = 12$.

100 P %	$\left(\frac{s_1}{s_2}\right)^2 = p^2$, from (16.42)	Exact value
20	1·72	1·72
5	3·00	3·00
1	4·85	4·82
0·1	8·58	8·38

An even better normal approximation is given by Peizer and Pratt (1968) and Pratt (1968).

Wise (1960) shows that $(-\log \xi)^{1/3}$ is approximately normal, ξ being the Beta variable obtained from F by the transformation in **16.18.**

16.22 We have now completed our study of the three fundamental sampling distributions arising from the univariate normal. Before proceeding to the development of the sampling theory of the bivariate normal distribution, it will be as well here to gather together explicitly the various relations existing between these three derived distributions and the parent normal distribution. In brief, these are as follows :—

(1) The sum of squares of n standardized normal variates about population (or sample) mean has a χ^2 distribution with $\nu = n$ (or $(n-1)$) degrees of freedom.

(2) The χ^2 distribution itself is approximately of the normal form for increasing degrees of freedom.

(3) The ratio of the square of a normal variate with zero mean to an independent estimator of its variance σ^2 (distributed like $\chi^2\sigma^2/\nu$) has "Student's" t^2-distribution, and its square root a "Student's" t-distribution.

(4) The "Student" t-distribution tends to the standardized normal form as its degrees of freedom increase.

(5) The ratio of two independent variates, each distributed like χ^2/ν, has a Fisher's F-distribution, and half its natural logarithm has the z-distribution. The t^2-distribution is the special case of the F-distribution where the numerator variate has only one degree of freedom—see (3) and (1) above.

(6) As degrees of freedom increase for both numerator and denominator, the F-distribution tends to normality.

(7) If denominator degrees of freedom, ν_2, alone tend to infinity, it is easily seen that $\nu_1 F$ tends to a χ^2 distribution with ν_1 degrees of freedom, for then we effectively return from (5) to (1) above.

Use of these relationships between the distributions often makes it possible to use one set of tables for many purposes.

The bivariate normal distribution

16.23 We have seen (Example 15.1) that the normal distribution in two variates may be written

$$dF = \frac{dx\,dy}{2\pi\sigma_1\sigma_2(1-\rho^2)^{\frac{1}{2}}}\exp\left\{-\frac{1}{2(1-\rho^2)}\left(\frac{x^2}{\sigma_1^2}-\frac{2\rho xy}{\sigma_1\sigma_2}+\frac{y^2}{\sigma_2^2}\right)\right\}, \tag{16.43}$$

where σ_1^2, σ_2^2 are the variances of the variates, μ_{11} their covariance and $\rho = \mu_{11}/(\sigma_1\sigma_2)$ is called the *correlation coefficient* between x and y in the distribution. In (16.43), we have measured each variate from its mean, a simplification which affects none of the results which we shall discuss.

By rewriting the exponent in (16.43) alternatively as

$$-\frac{1}{2(1-\rho^2)}\left\{\left(\frac{x}{\sigma_1}-\rho\frac{y}{\sigma_2}\right)^2+\frac{y^2}{\sigma_2^2}(1-\rho^2)\right\} \tag{16.44}$$

and

$$-\frac{1}{2(1-\rho^2)}\left\{\left(\frac{y}{\sigma_2}-\rho\frac{x}{\sigma_1}\right)^2+\frac{x^2}{\sigma_1^2}(1-\rho^2)\right\}, \tag{16.45}$$

we see from (16.44) that the conditional distribution of x, given any fixed value of y, is univariate normal with mean and variance

$$\left.\begin{aligned} \mathrm{E}(x\,|\,y) &= \rho y\sigma_1/\sigma_2, \\ \mathrm{var}(x\,|\,y) &= \sigma_1^2(1-\rho^2). \end{aligned}\right\} \tag{16.46}$$

From (16.45), analogous results follow for y, given x, which are (16.46) with x, y and suffixes *1*, *2* interchanged.

It follows from (16.46) that the conditional mean of either variate, given the value of the other, is a linear function of that value, and that its conditional variance is constant, independent of that value. We may express this by saying that the ***regression*** of x on y (and of y on x) is linear and homoscedastic.(*)

(*) The terms "homoscedastic" for "of constant variance", and its opposite "heteroscedastic", are due to K. Pearson.

In Chapters 26–28 (Vol. 2) we shall be discussing the theory of correlation and regression in detail. In the remaining sections of this chapter, we shall be concerned to derive formally the sampling distributions of certain statistics in samples from a bivariate normal population.

16.24 The joint probability of n sample values $(x_1, y_1) \ldots (x_n, y_n)$ from a bivariate normal population with means μ_1 and μ_2 is

$$dF = \frac{1}{(2\pi)^n \sigma_1^n \sigma_2^n (1-\rho^2)^{\frac{1}{2}n}} \exp\left[-\frac{1}{2(1-\rho^2)}\left\{\Sigma\left(\frac{x-\mu_1}{\sigma_1}\right)^2 - 2\rho\Sigma\frac{(x-\mu_1)(y-\mu_2)}{\sigma_1\sigma_2} + \Sigma\left(\frac{y-\mu_2}{\sigma_2}\right)^2\right\}\right] dx_1 dy_1 \ldots dx_n dy_n. \quad (16.47)$$

The exponent in (16.47) may be expressed solely in terms of the five parameters of the distribution $(\mu_1, \mu_2, \sigma_1^2, \sigma_2^2, \rho)$ and the corresponding sample statistics

$$\left.\begin{aligned} \bar{x} &= \frac{1}{n}\Sigma x, \bar{y} = \frac{1}{n}\Sigma y, \\ s_1^2 &= \frac{1}{n}\Sigma(x-\bar{x})^2, s_2^2 = \frac{1}{n}\Sigma(y-\bar{y})^2, \\ r &= \frac{1}{n}\Sigma(x-\bar{x})(y-\bar{y})/(s_1 s_2). \end{aligned}\right\} \quad (16.48)$$

For example, leaving aside the factor $-\frac{1}{2(1-\rho^2)}$, the first term in the exponent of (16.47) is

$$\Sigma\left(\frac{x-\mu_1}{\sigma_1}\right)^2 = \Sigma\left\{\frac{(x-\bar{x})+(\bar{x}-\mu_1)}{\sigma_1}\right\}^2 = \frac{n}{\sigma^2}\{s_1^2+(\bar{x}-\mu)^2\}, \quad (16.49)$$

the cross-product term vanishing since $\Sigma(x-\bar{x}) = 0$. The other two terms in the exponent of (16.47) can similarly be resolved into forms like (16.49), so we obtain for the complete exponent

$$n\left\{\left(\frac{\bar{x}-\mu_1}{\sigma_1}\right)^2 - 2\rho\left(\frac{\bar{x}-\mu_1}{\sigma_1}\right)\left(\frac{\bar{y}-\mu_2}{\sigma_2}\right) + \left(\frac{\bar{y}-\mu_2}{\sigma_2}\right)^2\right\} + n\left(\frac{s_1^2}{\sigma_1^2} - \frac{2\rho r s_1 s_2}{\sigma_1\sigma_2} + \frac{s_2^2}{\sigma_2^2}\right). \quad (16.50)$$

We proceed to find the joint sampling distribution of the five statistics defined at (16.48). To do this, we must transform (16.47) suitably. With the aid of (16.50) we can express the whole of the non-differential part of (16.47) in terms of the five statistics and the five parameters. We now consider the differential element.

Generalizing the geometrical approach of Chapter 11, we consider a sample space of n dimensions for x, and another such space for y, superimposed upon the first. The sample point varies in each of the two spaces, but not independently so. In fact, if P represents the point $(x_1, \ldots, x_n)$ in the x-space and Q the point $(y_1, \ldots, y_n)$ in the y-space, and if O_1, O_2 are the points $(\bar{x}, \ldots, \bar{x})$, $(\bar{y}, \ldots, \bar{y})$, then for any given r we have

$$r = \frac{\Sigma(x-\bar{x})(y-\bar{y})}{n s_1 s_2} = \frac{\Sigma(x-\bar{x})(y-\bar{y})}{\{\Sigma(x-\bar{x})^2\Sigma(y-\bar{y})^2\}^{\frac{1}{2}}}$$

and thus r is the cosine of the angle, say θ, between O_1P and O_2Q, so that if P and r are fixed, Q varies on the cone in the y-space obtained by rotating O_2Q such that the angle made with O_1P is constant.

The element in the x-space is proportional to $s_1^{n-2}\,ds_1\,d\bar{x}$, as was seen in Example 11.7. For given $\bar{y}$ and s_2 the point Q likewise varies on the $(n-2)$-dimensional surface of the hypersphere of radius $s_2\sqrt{n}$, centre $\bar{y}$ and $(n-1)$ dimensions, but another dimension is lost by keeping θ constant, so that Q lies on a surface sphere of $(n-3)$ dimensions. This $(n-3)$-dimensional surface has radius $s_2\sqrt{n}\sin\theta = s_2\sqrt{n}(1-r^2)^{\frac{1}{2}}$ and "width" $s_2\sqrt{n}\,d\theta = \dfrac{s_2\sqrt{n}\,dr}{(1-r^2)^{\frac{1}{2}}}$ and thus its content is proportional to

$$\{s_2\sqrt{n}(1-r^2)^{\frac{1}{2}}\}^{n-3}\frac{s_2\sqrt{n}\,dr}{(1-r^2)^{\frac{1}{2}}},$$

that is, to $s_2^{n-2}(1-r^2)^{\frac{1}{2}(n-4)}$.

Thus the volume element may be written

$$\begin{aligned} dv &\propto s_1^{n-2}\,ds_1\,d\bar{x}\,.\,s_2^{n-2}(1-r^2)^{\frac{1}{2}(n-4)}\,ds_2\,d\bar{y}\,dr \\ &\propto s_1^{n-2}s_2^{n-2}\,ds_1\,ds_2(1-r^2)^{\frac{1}{2}(n-4)}\,dr\,d\bar{x}\,d\bar{y}. \end{aligned} \tag{16.51}$$

From (16.47), (16.50) and (16.51), the joint frequency element of the five variables is proportional to

$$\exp\left(-\frac{n}{2(1-\rho^2)}\left[\left\{\frac{(\bar{x}-\mu_1)^2}{\sigma_1^2}-2\rho\frac{(\bar{x}-\mu_1)(\bar{y}-\mu_2)}{\sigma_1\sigma_2}+\frac{(\bar{y}-\mu_2)^2}{\sigma_2^2}\right\}\right.\right.$$
$$\left.\left.+\left\{\frac{s_1^2}{\sigma_1^2}-2\rho r\frac{s_1 s_2}{\sigma_1\sigma_2}+\frac{s_2^2}{\sigma_2^2}\right\}\right]\right)dv. \tag{16.52}$$

This fundamental result is due to R. A. Fisher (1915).

16.25 One important property of (16.52) may be remarked. The distribution may be factorized into two parts, one containing only $\bar{x}$ and $\bar{y}$ and the other only s_1, s_2 and r, namely

$$dF \propto \exp\left[-\frac{n}{2(1-\rho^2)}\left\{\frac{(\bar{x}-\mu_1)^2}{\sigma_1^2}-2\rho\frac{(\bar{x}-\mu_1)(\bar{y}-\mu_2)}{\sigma_1\sigma_2}+\frac{(\bar{y}-\mu_2)^2}{\sigma_2^2}\right\}\right]d\bar{x}\,d\bar{y} \tag{16.53}$$

and

$$dF \propto \exp\left[-\frac{n}{2(1-\rho^2)}\left\{\frac{s_1^2}{\sigma_1^2}-\frac{2\rho r s_1 s_2}{\sigma_1\sigma_2}+\frac{s_2^2}{\sigma_2^2}\right\}\right]s_1^{n-2}s_2^{n-2}(1-r^2)^{\frac{1}{2}(n-4)}\,ds_1\,ds_2\,dr. \tag{16.54}$$

Thus we see that in normal samples the distribution of means is entirely independent of that of the variances and covariance. This is a characteristic property of multivariate normality—cf. **15.24.**

Before leaving (16.53), we may also note that the means are themselves distributed exactly in the bivariate normal form, with $\text{E}(\bar{x}) = \mu_1$, $\text{E}(\bar{y}) = \mu_2$, $\text{var}\,\bar{x} = \dfrac{\sigma_1^2}{n}$, $\text{var}\,\bar{y} = \dfrac{\sigma_2^2}{n}$ (all of which results are already familiar), and

$$\text{cov}(\bar{x},\bar{y}) = \frac{\sigma_1\sigma_2\rho}{n}, \tag{16.55}$$

so that the correlation between $\bar{x}$ and $\bar{y}$ is ρ, the correlation in the parent population. We indicated a more general result in Exercise 13.2.

16.26 We may now use (16.54) to obtain the distribution of the sample correlation coefficient by integrating with respect to s_1 and s_2 from 0 to ∞. Let us first of all evaluate the constant to be attached to it from the consideration that $\int dF = 1$.

Make the variate-transformation

$$\left.\begin{aligned} a &= \frac{s_1^2}{\sigma_1^2}\cdot\frac{n}{2(1-\rho^2)} \\ b &= \frac{rs_1s_2}{\sigma_1\sigma_2}\cdot\frac{n}{2(1-\rho^2)} \\ c &= \frac{s_2^2}{\sigma_2^2}\cdot\frac{n}{2(1-\rho^2)}. \end{aligned}\right\} \tag{16.56}$$

We have for the reciprocal of the Jacobian of the transformation

$$\frac{\partial(a,b,c)}{\partial(s_1,r,s_2)} = \begin{vmatrix} \frac{2s_1}{\sigma_1^2}\cdot\frac{n}{2(1-\rho^2)} & 0 & 0 \\ \frac{rs_2}{\sigma_1\sigma_2}\cdot\frac{n}{2(1-\rho^2)} & \frac{s_1s_2}{\sigma_1\sigma_2}\cdot\frac{n}{2(1-\rho^2)} & \frac{rs_1}{\sigma_1\sigma_2}\cdot\frac{n}{2(1-\rho^2)} \\ 0 & 0 & \frac{2s_2}{\sigma_2^2}\cdot\frac{n}{2(1-\rho^2)} \end{vmatrix}$$

$$= \frac{s_1^2 s_2^2 n^3}{2\sigma_1^3\sigma_2^3(1-\rho^2)^3} = \frac{2acn}{\sigma_1\sigma_2(1-\rho^2)}$$

and also the relation

$$r^2 = \frac{b^2}{ac}.$$

The integral of (16.54) then becomes

$$\int \exp[-a+2\rho b-c].\left\{\frac{2a\sigma_1^2(1-\rho^2)}{n}\right\}^{\frac{1}{2}(n-2)}\left\{\frac{2c\sigma_2^2(1-\rho^2)}{n}\right\}^{\frac{1}{2}(n-2)}\left(1-\frac{b^2}{ac}\right)^{\frac{1}{2}(n-4)} \times\frac{\sigma_1\sigma_2(1-\rho^2)}{2acn}\,da\,db\,dc$$

$$= \frac{2^{n-3}\sigma_1^{n-1}\sigma_2^{n-1}(1-\rho^2)^{n-1}}{n^{n-1}}\int \exp[-a+2\rho b-c](ac-b^2)^{\frac{1}{2}(n-4)}\,da\,db\,dc \tag{16.57}$$

where the limits of a and c are 0 to ∞ and those of b are $\pm\sqrt{ac}$. This integral may be evaluated in terms of the Γ-function. Putting $\xi = a-\frac{b^2}{c}$ we find

$$\int \exp(-\xi)\exp\left(+2\rho b-c-\frac{b^2}{c}\right)\xi^{\frac{1}{2}(n-4)}c^{\frac{1}{2}(n-4)}\,d\xi\,db\,dc,$$

$$0\leqslant\xi\leqslant\infty,\quad -\infty\leqslant b\leqslant\infty,\quad 0\leqslant c\leqslant\infty,$$

$$= \Gamma\{\tfrac{1}{2}(n-2)\}\int \exp\left[-\frac{1}{c}\{(b-\rho c)^2+(1-\rho^2)c^2\}\right]c^{\frac{1}{2}(n-4)}\,db\,dc$$

$$= \Gamma\{\tfrac{1}{2}(n-2)\}\sqrt{\pi}\int \exp\{-(1-\rho^2)c\}c^{\frac{1}{2}(n-3)}\,dc$$

$$= \sqrt{\pi}\frac{\Gamma\{\tfrac{1}{2}(n-2)\}\Gamma\{\tfrac{1}{2}(n-1)\}}{(1-\rho^2)^{\frac{1}{2}(n-1)}}$$

$$= \frac{\pi\Gamma(n-2)}{2^{n-3}(1-\rho^2)^{\frac{1}{2}(n-1)}}, \tag{16.58}$$

since $\Gamma(x)\Gamma(x+\frac{1}{2}) = \pi^{\frac{1}{2}}\Gamma(2x)/2^{2x-1}$. The reciprocal of the product of the constants in (16.57) and (16.58) is the constant that we require for (16.54). Thus the joint distribution of s_1, s_2 and r is

$$dF = \frac{n^{n-1}}{\pi\sigma_1^{n-1}\sigma_2^{n-1}(1-\rho^2)^{\frac{1}{2}(n-1)}\Gamma(n-2)}\exp\left[-\frac{n}{2(1-\rho^2)}\left\{\frac{s_1^2}{\sigma_1^2}-\frac{2\rho r s_1 s_2}{\sigma_1\sigma_2}+\frac{s_2^2}{\sigma_2^2}\right\}\right]$$
$$\times s_1^{n-2}s_2^{n-2}(1-r^2)^{\frac{1}{2}(n-4)}\,ds_1\,ds_2\,dr. \tag{16.59}$$

16.27 Now put

$$\zeta = \frac{s_1 s_2}{\sigma_1\sigma_2}, \qquad \beta = \log\frac{\sigma_2 s_1}{\sigma_1 s_2}, \qquad r = r.$$

We find, for the reciprocal of the Jacobian of the transformation,

$$\frac{\partial(\zeta,\beta,r)}{\partial(s_1,s_2,r)} = \begin{vmatrix} \frac{s_2}{\sigma_1\sigma_2} & \frac{s_1}{\sigma_1\sigma_2} & 0 \\ \frac{1}{s_1} & -\frac{1}{s_2} & 0 \\ 0 & 0 & 1 \end{vmatrix} = -\frac{2}{\sigma_1\sigma_2}.$$

The exponent in (16.59) becomes

$$\exp\left[-\frac{n}{2(1-\rho^2)}\{\zeta e^{\beta}-2\rho r\zeta+\zeta e^{-\beta}\}\right]$$

and after a little reduction the distribution becomes

$$dF = \frac{n^{n-1}}{2\pi(1-\rho^2)^{\frac{1}{2}(n-1)}\Gamma(n-2)}\exp\left[-\frac{n}{(1-\rho^2)}\zeta(\cosh\beta-\rho r)\right]\zeta^{n-2}\,d\zeta\,d\beta\,(1-r^2)^{\frac{1}{2}(n-4)}\,dr.$$

On integration with respect to ζ over its range $(0, \infty)$, we have

$$dF = \frac{(1-\rho^2)^{\frac{1}{2}(n-1)}}{2\pi\Gamma(n-2)}\frac{(1-r^2)^{\frac{1}{2}(n-4)}\Gamma(n-1)}{(\cosh\beta-\rho r)^{n-1}}\,d\beta\,dr. \tag{16.60}$$

(16.60) is an even function of β, so we integrate out β from 0 to ∞, dropping the factor 2 from the denominator to compensate.

Putting $-\rho r = \cos\theta$ we have, since

$$\int_0^{\infty}\frac{d\beta}{\cosh\beta+\cos\theta} = \frac{\theta}{\sin\theta},$$

$$dF = \frac{(1-\rho^2)^{\frac{1}{2}(n-1)}}{\pi\Gamma(n-2)}(1-r^2)^{\frac{1}{2}(n-4)}\frac{d^{n-2}}{d(-\cos\theta)^{n-2}}\left(\frac{\theta}{\sin\theta}\right)dr$$

$$= \frac{(1-\rho^2)^{\frac{1}{2}(n-1)}}{\pi\Gamma(n-2)}(1-r^2)^{\frac{1}{2}(n-4)}\frac{d^{n-2}}{d(r\rho)^{n-2}}\left\{\frac{\arccos(-\rho r)}{\sqrt{(1-\rho^2 r^2)}}\right\}dr. \tag{16.61}$$

These results are due to Fisher (1915).

16.28 In the particular case $\rho = 0$, (16.61) reduces, as can also be seen directly from the factorization of (16.59) into three independent parts, to

$$dF = \frac{1}{B\{\frac{1}{2}, \frac{1}{2}(n-2)\}}(1-r^2)^{\frac{1}{2}(n-4)}\,dr, \tag{16.62}$$

a form surmised by "Student" (1908b). Its distribution function may be obtained from tables of the Incomplete Beta-function, or more directly by putting

$$t = \{(n-2)r^2/(1-r^2)\}^{\frac{1}{2}}, \tag{16.63}$$

which reduces (16.62) to a "Student's" t-distribution (16.15) with $\nu = n-2$ degrees of freedom.

16.29 The general distribution (16.61) has been studied in some detail. We will here indicate only the more important features of the results.

First, as to the shape of the frequency curves. When $n = 2$ the distribution of (16.61) becomes nugatory because of the factor $\Gamma(n-2)$. This is understandable because, for samples of two, r must be either $+1$ or -1. In such a case, then, we have a discontinuous distribution which we may regard as an extreme case of a U-shaped distribution.

When $n = 3$ we find for the ordinate of the f.f., say y,

$$y = y_0\left\{\frac{1}{\sin^2\theta} - \frac{\theta\cos\theta}{\sin^3\theta}\right\}\frac{1}{(1-r^2)^{\frac{1}{2}}},$$

again a U-shaped distribution. For $n = 4$,

$$y = \frac{y_0}{\sin^3\theta}(\theta - 3\cot\theta + 3\,\theta\cot^2\theta).$$

If $\rho = 0$ this reduces to the rectangular form $y = \frac{1}{2}$. In other cases the curve is J-shaped.

For $n > 4$ the frequency curves are unimodal and increasingly skew as $|\rho|$ increases, as follows from the facts that the mode moves with ρ and that r, being the cosine of an angle, satisfies $r^2 \leqslant 1$. For any ρ, as $n \to \infty$, the distribution of r tends to normality, though slowly. Some interesting photographs of models of these curves are given in the *Co-operative Study* (1917).

16.30 We may express the frequency function (16.61) as a hypergeometric function, as Hotelling (1953) pointed out. Consider the integral

$$I_{n-1} = \int_0^\infty \frac{d\beta}{(\cosh\beta - \rho r)^{n-1}}. \tag{16.64}$$

The substitution $\cosh\beta = (1-\rho r z)/(1-z)$ transforms (16.64) to

$$I_{n-1} = \frac{1}{\sqrt{2}(1-\rho r)^{n-3/2}}\int_0^1 z^{-\frac{1}{2}}(1-z)^{n-2}\{1-\tfrac{1}{2}(1+\rho r)z\}^{-\frac{1}{2}}dz,$$

and if we expand the factor in braces under the integral sign into a uniformly convergent series, and integrate the series term-by-term, we obtain

$$I_{n-1} = \frac{B(\frac{1}{2}, n-1)}{\sqrt{2}(1-\rho r)^{n-3/2}}F(\tfrac{1}{2}, \tfrac{1}{2}, n-\tfrac{1}{2}, \tfrac{1}{2}(1+\rho r)) \tag{16.65}$$

in the usual hypergeometric notation, so that (16.60), (16.64) and (16.65) yield the alternative expression for (16.61)

$$dF = \frac{(n-2)\,dr}{(n-1)\sqrt{2}B(\frac{1}{2},n-\frac{1}{2})}(1-\rho^2)^{\frac{1}{2}(n-1)}(1-r^2)^{\frac{1}{2}(n-4)}(1-\rho r)^{3/2-n}F(\tfrac{1}{2},\tfrac{1}{2},n-\tfrac{1}{2},\tfrac{1}{2}(1+\rho r)). \tag{16.66}$$

(16.66) converges rapidly even for small n. For large n, the first term is often enough. The error caused by stopping at any stage is considerably less than $2/(1-\rho r)$ times the last term used.

A recurrence relation and a differential equation for the f.f. of r are given in Exercises 16.14–15.

16.31 Term-by-term integration of the uniformly convergent series (16.66) gives us an expression for the distribution function of r. For the powers of $(1+\rho r)$ can be expressed as powers of $\{2-(1-\rho r)\}$ and expanded binomially as a power series in $(1-\rho r)$. For $r>\rho$, Hotelling (1953) obtained the result

$$1-F(r) = \frac{(n-2)}{(n-1)\sqrt{2}B(\frac{1}{2},n-\frac{1}{2})}\left\{M_0+\frac{2M_0-M_1}{4(2n-1)}+\frac{9(4M_0-4M_1+M_2)}{32(2n-1)(2n+1)}+\dots\right\} \tag{16.67}$$

where $M_k = \int_r^1 (1-\rho^2)^{\frac{1}{2}(n-1)}(1-x^2)^{\frac{1}{2}(n-4)}(1-\rho x)^{3/2+k-n}\,dx, \qquad k = 0,1,2\dots,$

and the asymptotic formula, for $r>\rho$,

$$1-F(r) \sim \frac{(n-2)}{(n-1)^2\sqrt{2}B(\frac{1}{2},n-\frac{1}{2})}\cdot\frac{(1-\rho^2)^{\frac{1}{2}(n-1)}(1-r^2)^{\frac{1}{2}(n-2)}}{(r-\rho)(1-\rho r)^{n-5/2}}\{1+O(n^{-1})\}. \tag{16.68}$$

The ordinates and distribution function of the correlation coefficient have been tabulated by F. N. David (1938) for values of $n = 3\,(1)\,25$, 50, 100, 200 and 400; for $\rho = 0{\cdot}0\,(0{\cdot}1)\,0{\cdot}9$; and for $r = -1{\cdot}00\,(0{\cdot}05)+1{\cdot}00$, with finer intervals in places—Boomsma (1975) corrects errors in the d.f. table for $n = 100$, $\rho = 0{\cdot}4$. Lee (1972) gives 5% and 1% points of $|r|$ for $|\rho| = 0\,(0{\cdot}1)\,0{\cdot}9$ and $(n-1)^{\frac{1}{2}} = 60/\nu$ with $\nu = 1\,(1)\,6\,(2)\,20$.

16.32 The moments of the distribution are also expressible in terms of hypergeometric functions. Returning to (16.59), let us write

$$\alpha_1^2 = \frac{\sigma_1^2(1-\rho^2)}{n}, \quad \alpha_2^2 = \frac{\sigma_2^2(1-\rho^2)}{n}.$$

After a little rearrangement (16.59) becomes

$$dF = \frac{(1-\rho^2)^{\frac{1}{2}(n-1)}}{\pi\Gamma(n-2)}\exp\left\{-\frac{1}{2}\frac{s_1^2}{\alpha_1^2}-\frac{1}{2}\frac{s_2^2}{\alpha_2^2}+\rho r\frac{s_1 s_2}{\alpha_1\alpha_2}\right\}\left(\frac{s_1}{\alpha_1}\right)^{n-2}\left(\frac{s_2}{\alpha_2}\right)^{n-2}(1-r^2)^{\frac{1}{2}(n-4)}\left(\frac{ds_1}{\alpha_1}\right)\left(\frac{ds_2}{\alpha_2}\right)dr. \tag{16.69}$$

Putting $u_1 = \frac{s_1}{\alpha_1}$, $u_2 = \frac{s_2}{\alpha_2}$ and expanding the term in $\exp\left(\rho r \frac{s_1 s_2}{\alpha_1 \alpha_2}\right)$ in (16.69), we have

$$dF = \frac{(1-\rho^2)^{\frac{1}{2}(n-1)}}{\pi \Gamma(n-2)} \exp\left(-\tfrac{1}{2}u_1^2 - \tfrac{1}{2}u_2^2\right) u_1^{n-2} u_2^{n-2} (1-r^2)^{\frac{1}{2}(n-4)} \times \sum_{j=0}^{\infty} \frac{(\rho r u_1 u_2)^j}{j!} du_1\, du_2\, dr. \quad (16.70)$$

Integrating u_2 from 0 to ∞ in (16.70) we find for the distribution of u_1 and r

$$dF = \frac{(1-\rho^2)^{\frac{1}{2}(n-1)}}{\pi \Gamma(n-2)} \exp\left(-\tfrac{1}{2}u_1^2\right) u_1^{n-2} (1-r^2)^{\frac{1}{2}(n-4)} \times \Sigma \frac{(\rho r u_1)^j}{j!} \Gamma\left(\frac{n-1+j}{2}\right) \cdot 2^{\frac{1}{2}(n+j-3)} du_1\, dr. \quad (16.71)$$

Multiplying the right-hand side of (16.71) by r and integrating from -1 to $+1$ we find, for the mean of r,

$$\frac{(1-\rho^2)^{\frac{1}{2}(n-1)}}{\pi \Gamma(n-2)} \exp\left(-\tfrac{1}{2}u_1^2\right) u_1^{n-2} \times \sum_{j=0}^{\infty} \frac{(\rho u_1)^{2j+1}}{(2j+1)!} \Gamma\left(\frac{n+2j}{2}\right) B\left(\frac{n-2}{2}, \frac{2j+3}{2}\right) . 2^{\frac{1}{2}(n+2j-2)} du_1$$

and finally, integrating with respect to u_1, we obtain

$$\mu_1'(r) = \frac{(1-\rho^2)^{\frac{1}{2}(n-1)}}{\pi \Gamma(n-2)} \sum_{=0}^{\infty} \frac{\rho^{2j+1}}{(2j+1)!} \Gamma\left(\frac{n+2j}{2}\right) B\left(\frac{n-2}{2}, \frac{2j+3}{2}\right) \Gamma\left(\frac{n+2j}{2}\right) 2^{n+2j-2}. \quad (16.72)$$

Substituting for the Beta-function in terms of Gamma-functions and remembering that $\Gamma(x)\Gamma(x+\frac{1}{2}) = \frac{\pi^{\frac{1}{2}}}{2^{2x-1}}\Gamma(2x)$, we find

$$\mu_1'(r) = \frac{\rho(1-\rho^2)^{\frac{1}{2}(n-1)} \Gamma^2(\frac{1}{2}n)}{\Gamma\{\frac{1}{2}(n-1)\}\Gamma\{\frac{1}{2}(n+1)\}} \left\{1 + \frac{\frac{1}{2}n . \frac{1}{2}n}{\frac{1}{2}(n+1)} \frac{\rho^2}{1!} + \frac{\frac{1}{2}n(\frac{1}{2}n+1) . \frac{1}{2}n(\frac{1}{2}n+1)}{\frac{1}{2}(n+1) . \frac{1}{2}(n+3)} \frac{\rho^4}{2!} + \dots\right\}$$

$$= \frac{\rho(1-\rho^2)^{\frac{1}{2}(n-1)} \Gamma^2(\frac{1}{2}n)}{\Gamma\{\frac{1}{2}(n-1)\}\Gamma\{\frac{1}{2}(n+1)\}} F(\tfrac{1}{2}n, \tfrac{1}{2}n, \tfrac{1}{2}(n+1), \rho^2)$$

and since $F(\alpha, \beta, \gamma, x) = (1-x)^{\gamma-\alpha-\beta} F(\gamma-\alpha, \gamma-\beta, \gamma, x)$

$$\mu_1'(r) = \frac{\rho \Gamma^2(\frac{1}{2}n)}{\Gamma\{\frac{1}{2}(n-1)\}\Gamma\{\frac{1}{2}(n+1)\}} F(\tfrac{1}{2}, \tfrac{1}{2}, \tfrac{1}{2}(n+1), \rho^2)$$

$$= \rho\left\{1 - \frac{(1-\rho^2)}{2n} + O\left(\frac{1}{n^2}\right)\right\}. \quad (16.73)$$

Similarly (cf. Hotelling (1953)) we find

$$\mu_2(r) = \frac{(1-\rho^2)^2}{n-1}\left(1 + \frac{11\rho^2}{2n}\right) + O\left(\frac{1}{n^3}\right), \quad (16.74)$$

and

$$\gamma_1 = \frac{-6\rho}{n^{\frac{1}{2}}} + o(n^{-\frac{1}{2}}), \qquad \gamma_2 = \frac{6(12\rho^2-1)}{n} + o(n^{-1}).$$

Thus the skewness increases with $|\rho|$ and declines only as $n^{-\frac{1}{2}}$.

B. K. Ghosh (1966) gives the first four moments of r as power series in $(n+6)^{-1}$ to 4 or more terms. Exercise 16.17 shows that when $\rho = 0$, (16.73–4) agree with the exact results for *any* continuous independent variates.

Fisher's transformation of r

16.33 Fisher (1921*b*) found a remarkable transformation of r which tends to normality very much faster than r, with a variance almost independent of ρ. Putting

$$r = \tanh z, \quad z = \tfrac{1}{2}\log\frac{1+r}{1-r}, \quad \rho = \tanh\zeta, \quad \zeta = \tfrac{1}{2}\log\frac{1+\rho}{1-\rho}, \tag{16.75}$$

(which is easily done using a special table of the transformation in the *Biometrika Tables*) we may expand the frequency function of r in powers of $z-\zeta$, $= x$ say, and inverse powers of n. Fisher gives the following expansion:

$$\begin{aligned} f = \frac{n-2}{\sqrt{\{2\pi(n-1)\}}} e^{-\frac{1}{2}(n-1)x^2} \Bigg\{ & 1 + \tfrac{1}{2}\rho x + \left(\frac{2+\rho^2}{8(n-1)} + \frac{4-\rho^2}{8}x^2 + \frac{n-1}{12}x^4\right) \\ & + \rho x\left(\frac{4-\rho^2}{16(n-1)} + \frac{4+3\rho^2}{48}x^2 + \frac{n-1}{24}x^4\right) + \left(\frac{4+12\rho^2+9\rho^4}{128(n-1)^2} + \frac{8-2\rho^2+3\rho^4}{64(n-1)}x^2\right. \\ & \left. + \frac{8+4\rho^2-5\rho^4}{128}x^4 + \frac{28-15\rho^2}{1440}x^6(n-1)\right) + \frac{(n-1)^2}{288}x^8 + \dots \Bigg\}. \end{aligned} \tag{16.76}$$

Taking moments about $x = 0$ we find, on transferring to the mean,(*)

$$\left.\begin{aligned} \mu_1' &= \frac{\rho}{2(n-1)}\left\{1 + \frac{5+\rho^2}{4(n-1)} + \frac{11+2\rho^2+3\rho^4}{8(n-1)^2} + \dots\right\} \\ \mu_2 &= \frac{1}{n-1}\left\{1 + \frac{4-\rho^2}{2(n-1)} + \frac{22-6\rho^2-3\rho^4}{6(n-1)^2} + \dots\right\} \\ \mu_3 &= \frac{\rho^3}{(n-1)^3} + \dots \\ \mu_4 &= \frac{1}{(n-1)^2}\left\{3 + \frac{14-3\rho^2}{n-1} + \frac{184-48\rho^2-21\rho^4}{4(n-1)^2} + \dots\right\} \end{aligned}\right\} \tag{16.77}$$

whence

$$\left.\begin{aligned} \gamma_1 &= \frac{\rho^3}{(n-1)^{3/2}} + \dots \\ \gamma_2 &= \frac{2}{n-1} + \frac{4+2\rho^2-3\rho^4}{(n-1)^2} + \dots \end{aligned}\right\} \tag{16.78}$$

Thus the variance of $z-\zeta$ is almost independent of ρ, while γ_1 declines as $n^{-\frac{3}{2}}$, and we may take $z-\zeta$ to be approximately normally distributed with mean and variance as given by (16.77). As a slightly rougher approximation we may take

$$\mu_1'(z-\zeta) = \frac{\rho}{2(n-1)}, \qquad \operatorname{var}(z-\zeta) = \frac{1}{n-1} + \frac{4-\rho^2}{2(n-1)^2}, \tag{16.79}$$

which is approximately equal for small ρ to

$$\frac{1}{n-1} + \frac{2}{(n-1)^2} = \frac{1}{n-3} \text{ approximately.} \tag{16.80}$$

When n is moderate we may take a still rougher approximation by assuming $z-\zeta$ to be

(*) Equations (16.77), as given in Fisher's original paper, contained some errors which were corrected by Gayen (1951). The main result, equation (16.80), remains unaffected.

normally distributed about zero mean with variance $\frac{1}{n-3}$. Some comparisons of the various approximations are given in the introduction to F. N. David's (1938) tables, and it appears that for $n > 50$ the forms (16.79) and (16.80) are adequate. The approximation given by (16.77) appears to hold satisfactorily for values of n as low as 11. See also Gurland (1968) and Gurland and Milton (1970).

When $\rho = 0$, the exact distribution of z is given in Exercise 16.21.

Hotelling (1953) has investigated the possibility of improving upon the z transformation. He found (cf. Exercises 16.18–19) that

$$z^* = z-(3z+r)/(4n) \tag{16.81}$$

has moments

$$\mu_1' = \left(\zeta - \frac{3\zeta+\rho}{4n}\right) + \frac{\rho}{2(n-1)} + \frac{3\rho}{8(n-1)^2} + O(n^{-3}), \qquad \mu_2 = \frac{1}{n-1} + O(n^{-3}) \tag{16.82}$$

and is more closely normally distributed. A further improvement z^{**} is also discussed.

Ruben (1966) used the identity given in Exercise 16.6 and the device of **12.19** to obtain an approximation to the distribution of r which is much more accurate than that obtained from z, and about as accurate as z^* at (16.81). Gurland (1968) gives a finite sum for the d.f. of $r^2/(1-r^2)$ when n is even, infinite sums for odd or even n, and an approximation to the f.f. See also Gurland and Milton (1970).

Samiuddin (1970) shows that if (16.63) is generalized to

$$t = (r-\rho)\{(n-2)/[(1-r^2)(1-\rho^2)]\}^{\frac{1}{2}},$$

t remains approximately distributed in "Student's" form with $(n-2)$ d.fr. for $\rho \neq 0$ and $n \geq 8$. See also Kraemer (1973).

16.34 Just as r is defined as the ratio of sample covariance to the product of sample standard deviations, and ρ as the corresponding population quantity, we now define (see (16.48))

$$b_1 = \frac{1}{n}\Sigma(x-\bar{x})(y-\bar{y})/s_2^2, \qquad b_2 = \frac{1}{n}\Sigma(x-\bar{x})(y-\bar{y})/s_1^2, \tag{16.83}$$

the population analogues being $\beta_1 = \mu_{11}/\sigma_2^2$ and $\beta_2 = \mu_{11}/\sigma_1^2$ respectively. b_1 and b_2 are called the sample *regression coefficients*, of x on y and of y on x respectively. β_1 and β_2 are the corresponding population regression coefficients.

Referring to (16.46), we see that

$$\mathrm{E}(x\,|\,y) = \beta_1 y, \qquad \mathrm{E}(y\,|\,x) = \beta_2 x.$$

We proceed to consider the sampling distribution of b_2. That for b_1 is immediately obtainable from this by permutation of suffixes.

Distribution of regression coefficients in normal samples

16.35 Turning again to equation (16.59) we have, substituting $b_2 = \frac{rs_2}{s_1}$, the joint frequency function of s_1, s_2 and b_2,

$$dF \propto \exp\left[-\frac{n}{2(1-\rho^2)}\left\{\frac{s_1^2}{\sigma_1^2} - \frac{2\rho s_1^2 b_2}{\sigma_1\sigma_2} + \frac{s_2^2}{\sigma_2^2}\right\}\right] s_1^{n-1} s_2^{n-3}\left(1-\frac{s_1^2 b_2^2}{s_2^2}\right)^{\frac{1}{2}(n-4)} ds_1\, ds_2\, db_2. \tag{16.84}$$

Integration of (16.84) with respect to s_2 gives, for the distribution of s_1 and b_2,

$$dF \propto \exp\left[-\frac{n s_1^2}{2(1-\rho^2)}\left\{\frac{1}{\sigma_1^2}-\frac{2\rho b_2}{\sigma_1\sigma_2}+\frac{b_2^2}{\sigma_2^2}\right\}\right] s_1^{n-1}\,ds_1\,db_2. \tag{16.85}$$

A further integration of (16.85) with respect to s_1 gives for the distribution of b_2

$$dF \propto \frac{db_2}{\left(1-\frac{2\rho\sigma_1}{\sigma_2}b_2+\frac{\sigma_1^2}{\sigma_2^2}b_2^2\right)^{\frac{1}{2}n}}$$

or, on evaluation of the constant,

$$dF = \frac{\Gamma(\frac{1}{2}n)\,\sigma_2^{n-1}(1-\rho^2)^{\frac{1}{2}(n-1)}}{\sqrt{\pi}\,\Gamma\left(\frac{n-1}{2}\right)\sigma_1^{n-1}}\,\frac{db_2}{\left\{\frac{\sigma_2^2}{\sigma_1^2}(1-\rho^2)+\left(b_2-\frac{\rho\sigma_2}{\sigma_1}\right)^2\right\}^{\frac{1}{2}n}}. \tag{16.86}$$

It follows that if we write

$$g = (b_2-\beta_2)\sigma_1/\{\sigma_2(1-\rho^2)^{\frac{1}{2}}\}, \tag{16.87}$$

where $\beta_2 = \frac{\rho\sigma_2}{\sigma_1}$, the population regression coefficient, then $g(n-1)^{\frac{1}{2}}$ has " Student's " distribution (16.15) with $\nu = n-1$. Thus b_2 is distributed symmetrically about β_2. It tends to normality fairly rapidly, and the use of the standard error for regressions is therefore valid for lower values of n than in the case of the correlation coefficient. For small samples, however, (16.87) is useless since it depends on the unknown quantities σ_1, σ_2 and ρ.

16.36 To avoid this difficulty, we substitute s_1, s_2 and r for σ_1, σ_2 and ρ in (16.87), obtaining $h = (b_2-\beta_2)s_1/\{s_2(1-r^2)^{\frac{1}{2}}\}$. To find its distribution, we return to that of the quantities a, b, c of equation (16.56), namely, from (16.57),

$$dF \propto \exp[-a+2\rho b-c]\,(ac-b^2)^{\frac{1}{2}(n-4)}\,da\,db\,dc. \tag{16.88}$$

Now

$$h = \frac{b-\rho a}{\sqrt{(ac-b^2)}},$$

and on transforming from c to h in (16.88) we have, after a little reduction,

$$dF \propto \frac{\exp\left[-a\left(1+\frac{\rho^2}{h^2}\right)\right]da\,dh}{ah^{n-1}}\,.\exp\left[\frac{h^2+1}{h^2}\left(2\rho b-\frac{b^2}{a}\right)\right](b-\rho a)^{n-2}\,db. \tag{16.89}$$

Integration of the second factor on the right of (16.89) will be found to give a result proportional to

$$\exp\left\{\frac{a\rho^2(h^2+1)}{h^2}\right\}\cdot\left(\frac{ah^2}{h^2+1}\right)^{\frac{1}{2}(n-1)} \tag{16.90}$$

and hence for the distribution of a and h we find, from (16.89) and (16.90),

$$dF \propto a^{\frac{1}{2}(n-3)}\exp\{a(\rho^2-1)\}\,da\,.\frac{dh}{(1+h^2)^{\frac{1}{2}(n-1)}}. \tag{16.91}$$

Hence the distributions of a and h are independent, and for that of h we have

$$dF \propto \frac{dh}{(1+h^2)^{\frac{1}{2}(n-1)}}. \tag{16.92}$$

This distribution does not contain any of the parent parameters. It follows that

$$t = h\sqrt{(n-2)} = \frac{(b_2-\beta_2)s_1\sqrt{(n-2)}}{s_2(1-r^2)^{\frac{1}{2}}} \tag{16.93}$$

is distributed in " Student's " form (16.15) with $\nu = n-2$ degrees of freedom. Exercise 16.6 proves our results differently.

Only one degree of freedom has been lost in passing from g of (16.87) to h here, since s_2 is the only statistic not already used in b_2. When $\beta_2 = 0$, (16.93) coincides with (16.63), since then $\rho = 0$ also.

16.37 In Volumes 2 and 3, we shall be dealing with the applications of the distributions of r and b which have been formally derived in the preceding sections, and also with the generalization of our results to multivariate normal distributions with more than two variates.

EXERCISES

16.1 Show that the rth moment of the variance-ratio distribution at (16.24) is

$$\mu'_r = \left(\frac{\nu_2}{\nu_1}\right)^r \frac{\Gamma(\frac{1}{2}\nu_1+r)\Gamma(\frac{1}{2}\nu_2-r)}{\Gamma(\frac{1}{2}\nu_1)\Gamma(\frac{1}{2}\nu_2)}$$

if $2r < \nu_2$, and does not exist if $2r \geqslant \nu_2$. Hence show that

$$\beta_1 = \mu_3^2/\mu_2^3 = \frac{8(\nu_2-4)(2\nu_1+\nu_2-2)^2}{\nu_1(\nu_2-6)^2(\nu_1+\nu_2-2)}$$

and that

$$\beta_2 = \mu_4/\mu_2^2 = \frac{3}{\nu_2-8}\{\nu_2-4+\tfrac{1}{2}(\nu_2-6)\beta_1\}.$$

16.2 In the variance-ratio distribution (16.24), show that when $\nu_1 = \nu_2 = \nu$, $\frac{1}{2}\nu^{\frac{1}{2}}(F^{\frac{1}{2}}-F^{-\frac{1}{2}})$ is distributed in " Student's " t-distribution (16.15).

(Cf. Cacoullos, 1965)

16.3 Show that, for the χ^2 distribution with ν degrees of freedom, the moments about the mean obey the relation

$$\mu_{j+1} = 2j(\mu_j + \nu\mu_{j-1}), \quad j \geqslant 1,$$

and hence verify (16.5).

16.4 n variates $x_1, \ldots, x_n$ are independently distributed in the rectangular form $dF = dx$, $0 \leqslant x \leqslant 1$. Show that $-2\log(x_1 x_2 \ldots x_n)$ is distributed as χ^2 with $2n$ degrees of freedom.

16.5 X_i $(i = 1, \ldots, n)$ are a given set of values with zero mean and ε_i $(i = 1, \ldots, n)$ a set of values of a normal random variable with zero mean and variance σ^2. Show that if

$$y_i = \beta_0 + \beta_1 X_i + \varepsilon_i,$$

and the constants β_0, β_1 are estimated as b_0, b_1 where $b_0 = \bar{y}$, $b_1 = \Sigma(y_i-\bar{y})X_i/\Sigma X_i^2$,

then if

$$y'_i = y_i - (b_0 + b_1 X_i),$$

we have

$$\Sigma(y_i-\bar{y})^2 = \Sigma(y'_i)^2 + \frac{\{\Sigma X_i(y_i-\bar{y})\}^2}{\Sigma X_i^2}.$$

Hence show that $\Sigma(y'_i)^2$ is distributed as $\sigma^2\chi^2$ with $(n-2)$ degrees of freedom independently of $b_1^2\Sigma X_i^2$ which is distributed as $\sigma^2\chi^2$ with one degree of freedom.

If x and y are distributed normally and independently, deduce that $(n-2)r^2/(1-r^2)$ is distributed as " Student's " t^2 with $(n-2)$ degrees of freedom.

16.6 In the notation of **16.23–36**, define

$$v^2 = ns_1^2/\sigma_1^2, \qquad u^2 = \frac{\sigma_1^2(b_2-\beta_2)^2}{\sigma_2^2(1-\rho^2)}v^2, \qquad w^2 = \frac{ns_2^2(1-r^2)}{\sigma_2^2(1-\rho^2)},$$

$$\tilde{r} = r/(1-r^2)^{\frac{1}{2}}, \quad \tilde{\rho} = \rho/(1-\rho^2)^{\frac{1}{2}}.$$

Show that $\tilde{r} \equiv (u+\tilde{\rho}v)/w$ for any bivariate population. For the bivariate normal case, show that u^2, v^2 and w^2 are independent χ^2 variables, with respectively 1, $(n-1)$ and $(n-2)$ degrees of freedom. Derive the results of **16.35–6**.

(Ruben, 1966)

16.7 Denoting the χ^2 f.f. (16.1) by $f_\nu(z)$ and its d.f. by $F_\nu(z)$, and writing $G(x)$ for the d.f. of a standardized normal distribution, show that for positive integers s,

$$F_{\nu+2s}(z) = F_\nu(z) - 2\sum_{r=1}^{s} f_{\nu+2r}(z),$$

and hence

$$F_{2s-1}(z) = 2G(z^{\frac{1}{2}}) - 1 - 2\sum_{r=2}^{s} f_{2r-1}(z),$$

$$F_{2s}(z) = 1 - 2\sum_{r=1}^{s} f_{2r}(z),$$

so that $F_{2s}(z)$ is the complement to unity of the sum of the first s terms of a Poisson distribution with parameter $\frac{1}{2}z$.

Letting $s \to \infty$, show that for any value $z \geqslant 0$

$$\sum_{r=2}^{\infty} f_{2r-1}(z) = G(z^{\frac{1}{2}}) - \tfrac{1}{2},$$

$$\sum_{r=1}^{\infty} f_{2r}(z) = \tfrac{1}{2}, \quad \text{and hence that}$$

$$\sum_{\nu=2}^{\infty} f_\nu(z) = G(z^{\frac{1}{2}}).$$

16.8 If x, y are standardized bivariate normal variates with correlation parameter ρ, show that the c.f. of their product is

$$\phi(t) = \{(1-\rho it)^2 + t^2\}^{-\frac{1}{2}}$$

and hence that $E(xy) = \rho$, $\text{var}(xy) = 1+\rho^2$.

Verify var (xy) from Exercise 10.23. Use Exercise 4.3 to show that when $\rho = 0$ we may obtain $g(z)$ in Exercise 11.21. Show that the case $\rho = 1$ reduces to Exercise 4.18.

16.9 Show directly from (16.15) that "Student's" t tends to normality as ν tends to infinity.

16.10 By the substitution $q = (1+t^2/\nu)^{-1}$ show that the distribution function $F(t)$ of "Student's" t is, for even ν, given by

$$2F(t) - 1 = (1-q)^{\frac{1}{2}}\left\{1 + \sum_{s=1}^{\frac{1}{2}(\nu-2)} \frac{q^s}{2^{2s-1}.s.B(s,s)}\right\},$$

the expression in braces on the right-hand side being the cumulative sum of terms of the binomial expansion $(1-q)^{-\frac{1}{2}}$.

(Fisher, 1935)

16.11 x, y, z are multinormally distributed with all population correlations zero, so that in a sample of n observations, each of the three sample correlations is distributed in the form (16.62). Show that each of these correlations is distributed independently of each of the others, but that the three correlations are not completely independent of each other.

16.12 Verify (16.29), the characteristic function of the z-distribution, and by expanding $\log \phi(t)$ verify the cumulants at (16.34). Deduce the asymptotic normality at (16.31).

16.13 Using the methods of Chapter 10, show that in large samples from a bivariate normal population the variance of a sample regression coefficient is approximately

$$\text{var}\left(\frac{m_{11}}{m_{20}}\right) = \frac{\sigma_2^2}{\sigma_1^2}\frac{(1-\rho^2)}{n}.$$

16.14 By differentiation of (16.64) with respect to (ρr) show that

$$\{1-(\rho r)^2\}I_2 = 1+\rho r I_1$$

and hence that

$$n\{1-(\rho r)^2\}I_{n+1}-(2n-1)\rho r\, I_n-(n-1)\,I_{n-1} = 0.$$

Hence show that the frequency function, f_n, of r based on a sample of size n satisfies the recurrence formula

$$(n-1)\{1-(\rho r)^2\}f_{n+2} = (2n-1)\,\rho r\,(1-\rho^2)^{\frac{1}{2}}(1-r^2)^{\frac{1}{2}}f_{n+1}+(n-1)^2(1-\rho^2)(1-r^2)f_n/(n-2).$$

(*Co-operative Study*, 1917; Hotelling, 1953)

16.15 Verify that the frequency function of r satisfies the relation

$$r\frac{\partial f_n}{\partial r}+\frac{(n-4)\,r^2 f_n}{1-r^2} = \rho\frac{\partial f_n}{\partial \rho}+\frac{(n-1)\rho^2 f_n}{1-\rho^2}.$$

(Hotelling, 1953)

16.16 From (16.60), show that the distribution of $v = \dfrac{s_1}{\sigma_1}\Big/\dfrac{s_2}{\sigma_2}$ and r is given by

$$dF \propto \frac{v^{n-2}(1-r^2)^{\frac{1}{2}(n-4)}}{(1-2\rho r v+v^2)^{n-1}}\,dr\,dv.$$

Integrating for r from -1 to $+1$ by putting

$$r = \frac{u(\lambda+\mu)-(1-u)(\lambda-\mu)}{u(\lambda+\mu)+(1-u)(\lambda-\mu)},\quad \lambda = 1+v^2,\ \mu = 2\rho v,$$

show that the distribution of v is

$$dF = \frac{2(1-\rho^2)^{\frac{1}{2}(n-1)}}{B\left(\frac{n-1}{2},\frac{n-1}{2}\right)}\frac{v^{n-2}}{(1+v^2)^{n-1}}\left\{1-\frac{4\rho^2 v^2}{(1+v^2)^2}\right\}^{-\frac{1}{2}n}dv,$$

and that when $\rho = 0$, the distribution of $F = v^2$ is (16.24) with $\nu_1 = \nu_2 = n-1$.

(S. S. Bose, 1935; Finney, 1938)

16.17 If x and y are independent continuous variates, show that in samples of size n from their joint distribution the sample correlation coefficient r has zero mean and variance $(n-1)^{-1}$.

(Pitman, 1937a)

16.18 If t is a statistic which in large samples tends to a parent parameter θ, and the variance of t is some function of θ, say $f(\theta)/n+o(n^{-1})$, show using (10.14) that the variate

$$z = \int^t \{f(\theta)\}^{-\frac{1}{2}}\,d\theta$$

has variance $1/n+o(n^{-1})$.

Hence, from the large-sample result obtained from (16.74), $\text{var}\, r = (1-\rho^2)^2/(n-1)$, obtain the transformation $z = \tanh^{-1} r$, with variance $(n-1)^{-1}+o(n^{-1})$.

16.19 Noting that, from (16.77), the variance of z is

$$\frac{1}{n-1}\left(1+\frac{4-\rho^2}{2n}\right)+o(n^{-2}),$$

show that z^*, defined by

$$z^* \propto \int^r \frac{d\rho}{(1-\rho^2)\left(1+\frac{4-\rho^2}{2n}\right)^{\frac{1}{2}}} = z-\frac{3z+r}{4n}+o(n^{-1}),$$

has variance

$$\text{var}\, z^* = \frac{1}{n-1}+o(n^{-2}).$$

(Hotelling, 1953; Kendall, 1953)

16.20 In (16.54), transform from (s_1, s_2, r) to the variables

$$u = \frac{2s_1 s_2 r}{s_1^2+s_2^2}, \quad v = s_1^2+s_2^2, \quad w = s_2^2,$$

and then from (u, v, w) to (u, v, t), where $t = (2w-v)/\{v(1-u^2)^{\frac{1}{2}}\}$. Integrate with respect to v and transform the resulting distribution of u and t to that of u and z, where $t = (1-z)/(1+z)$. Finally, integrate with respect to z to obtain the distribution of u:

$$p(u) = \frac{(n-2)2^{n-3}}{\pi} B(\tfrac{1}{2}n, \tfrac{1}{2}n-1)\left(\frac{1-\rho^2}{\sigma_1^2\sigma_2^2}\right)^{\frac{1}{2}(n-1)} \frac{a}{(a^2-b^2)^{\frac{1}{2}n}}(1-u^2)^{\frac{1}{2}(n-3)},$$

where $a = \frac{1}{2}\left(\frac{1}{\sigma_1^2}+\frac{1}{\sigma_2^2}\right)-\frac{\rho u}{\sigma_1\sigma_2}$ and $b = \frac{1}{2}\left(\frac{1}{\sigma_2^2}-\frac{1}{\sigma_1^2}\right)(1-u^2)^{\frac{1}{2}}$.

Show that $|u| \leqslant |r|$ and that if $\sigma_1 = \sigma_2$ and $\rho = 0$, $p(u)$ is exactly (16.62) with $(n+1)$ written for n.

(DeLury, 1938; Mehta and Gurland, 1969)

16.21 Using Exercise 16.2 on (16.63), show that when $\rho = 0$ the exact distribution of Fisher's z defined at (16.75) is the variance-ratio z-distribution (16.26) with $\nu_1 = \nu_2 = n-2$. From its c.f. (16.29), use the relation for positive integers p

$$\frac{\Gamma''(p)}{\Gamma(p)}-\left\{\frac{\Gamma'(p)}{\Gamma(p)}\right\}^2 = \frac{\pi^2}{6}-\sum_{r=1}^{p-1}\frac{1}{r^2}$$

to show that for even $n \geqslant 6$ the exact variance of z is

$$\mu_2 = \pi^2/12-\frac{1}{2}\sum_{r=1}^{\frac{1}{2}n-2}\frac{1}{r^2},$$

which for $n = 6$ equals 0·32 against the value 0·33 obtained from (16.80).

16.22 If u is a Gamma variable with parameter p, and v is independently a Beta variable distributed as at (6.14) with parameters $(q, p-q)$, show that $w = uv$ and $y = u(1-v)$ are independent Gamma variables with parameters q and $p-q$. (If q or $p-q = \frac{1}{2}$, $(2w)^{\frac{1}{2}}$ or $(2y)^{\frac{1}{2}}$ is normal.) Show that the result of Exercise 11.26 is a special case with $p = 1$, $q = \frac{1}{2}$.

Use this result in (16.59) to show that in bivariate normal samples when $\rho = 0$, the distribution of $z_i = rs_i/\sigma_i$ is exactly normal with mean zero and variance $\frac{1}{n}$ $(i = 1, 2)$, but that z_1 and z_2 are not bivariate normally distributed.

(Stuart (1962); the product of other independent variables can have a Gamma or a normal distribution—cf. Kotlarski (1965) and Exercise 11.9. See also Aitchison (1963).)

16.23 In Exercise 16.22, show that conversely if the distributions of u and w are as stated, that of v follows if it is independent of u.

Let $x_1, x_2, \ldots, x_n$ be a random sample from a normal distribution with variance σ^2, $\bar{x} = \sum_{i=1}^{n} x_i/n$, $d_i^2 = (x_i-\bar{x})^2$ and $s^2 = \sum_{i=1}^{n} d_i^2/(n-1)$. Show that $w_i = d_i^2 n/\{(n-1)2\sigma^2\}$ and $u = (n-1)s^2/(2\sigma^2)$ are Gamma variables with parameters $\frac{1}{2}$ and $\frac{1}{2}(n-1)$ respectively, and that $v_i = w_i/u$ is distributed independently of u. Hence show that for any $i = 1, 2, \ldots, n$,

$$v_i = \frac{n}{(n-1)^2}\frac{(x_i-\bar{x})^2}{s^2}$$

is distributed in the Beta form (6.14) with parameters $(\frac{1}{2}, \frac{1}{2}(n-2))$, so that $t = \{(n-2)v_i/(1-v_i)\}^{\frac{1}{2}}$ is distributed in Student's form with $\nu = n-2$.

16.24 Using the Gram–Charlier method of Exercise 6.11, show that for the Wilson–Hilferty normal approximation to the χ^2 distribution, with standardized cumulants (16.13), the maximum absolute error in the d.f. resulting from the approximation satisfies

$$|e(x)| < 0{\cdot}001, \quad n \geqslant 15,$$
$$|e(x)| < 0{\cdot}007, \quad n \geqslant 3.$$

Similarly, use (16.8) to show that for Fisher's normal approximation, we must have $n \geqslant 23$ for $|e(x)| < 0{\cdot}01$, while from (16.6), the approximation of χ^2 itself requires $n \geqslant 354$ for $|e(x)| < 0{\cdot}01$.

(Mathur, 1961)

16.25 Given σ^2, x is normally distributed with mean zero and variance σ^2, and ν/σ^2 is distributed in the χ^2 form with ν d.f. Show that the distribution of x is Student's t with ν d.f. (Exercise 5.25 ends with the special case $\nu = 1$, as may be seen from Exercise 11.25.)

16.26 If t_m and t_n are independent Student's t-variables, distributed as at (16.15) with m, n d.fr. respectively, show, using **11.8**, that their difference $d_{m,n} = t_m - t_n$ has the same distribution f as their sum, and that f is symmetrical about its unique mode zero and also symmetric in (m, n).

Writing $t_j = y_j/(u_j/j)^{\frac{1}{2}}$, where the y_j are standardized normal and the u_j are χ^2 variables with j d.fr., show that $x = u_m/(u_m+u_n)$ has the Beta distribution (6.14) with parameters $(\frac{1}{2}m, \frac{1}{2}n)$ and that the conditional distribution of $d_{m,n}$ given x is a $\left\{\dfrac{m(1-x)+nx}{2(m+n)x(1-x)}\right\}^{\frac{1}{2}} t_{m+n}$ variable, so that the distribution of $d_{m,n}$ may be represented as the product of two independent random variables

$$t_{m+n} \times \left\{\frac{(1+F)(m+nF)}{2(m+n)F}\right\}^{\frac{1}{2}},$$

where F has the variance ratio distribution (16.24) with n and m d.fr.

(Ruben, 1960; B. K. Ghosh (1975) gives tables when $m = n$ and a good approximation for the general case.)

16.27 Generalizing Exercise 16.25, show that if ν/σ^2 is distributed as there in the joint distribution of the mean $\bar{x}$ and variance s^2 in samples from $N(0, \sigma^2)$ given in Examples 11.3 and 11.7, the resulting conditional distribution of $\bar{x}$ given the value of s^2 is such that

$$\left\{\frac{n(n-1+\nu)}{ns^2+\nu}\right\}^{\frac{1}{2}} \bar{x}$$

has a Student's t-distribution with $(n-1+\nu)$ d.fr. (Exercise 16.25 is the case $n = 1$.)

APPENDIX TABLES

1 The frequency function of the normal distribution

2 The distribution function of the normal distribution

3 Quantiles of the d.f. of χ^2

4a The distribution function of χ^2 for one degree of freedom, $0 \leqslant \chi^2 \leqslant 1$

4b The distribution function of χ^2 for one degree of freedom, $1 \leqslant \chi^2 \leqslant 10$

5 Quantiles of the d.f. of t

6 5 per cent. points of z

7 5 per cent. points of F

8 1 per cent. points of z

9 1 per cent. points of F

10 Augmented symmetric functions in terms of power-sums and vice versa

11 Multiple k-statistics in terms of augmented symmetric functions and vice versa

Appendix Table 1 Frequency function of the normal distribution $y = \frac{1}{\sqrt{(2\pi)}} e^{-\frac{1}{2}x^2}$ with first and second differences

x	y	$\Delta^1(-)$	Δ^2	x	y	$\Delta^1(-)$	Δ^2
0·0	0·39894	199	−392	2·5	0·01753	395	+79
0·1	0·39695	591	−374	2·6	0·01358	316	+66
0·2	0·39104	965	−347	2·7	0·01042	250	+53
0·3	0·38139	1312	−308	2·8	0·00792	197	+45
0·4	0·36827	1620	−265	2·9	0·00595	152	+36
0·5	0·35207	1885	−212	3·0	0·00443	116	+27
0·6	0·33322	2097	−159	3·1	0·00327	89	+23
0·7	0·31225	2256	−104	3·2	0·00238	66	+17
0·8	0·28969	2360	− 52	3·3	0·00172	49	+13
0·9	0·26609	2412	0	3·4	0·00123	36	+10
1·0	0·24197	2412	+ 46	3·5	0·00087	26	+ 7
1·1	0·21785	2366	+ 84	3·6	0·00061	19	+ 6
1·2	0·19419	2282	+118	3·7	0·00042	13	+ 4
1·3	0·17137	2164	+143	3·8	0·00029	9	+ 2
1·4	0·14973	2021	+161	3·9	0·00020	7	+ 3
1·5	0·12952	1860	+173	4·0	0·00013	4	—
1·6	0·11092	1687	+177	4·1	0·00009	3	—
1·7	0·09405	1510	+177	4·2	0·00006	2	—
1·8	0·07895	1333	+170	4·3	0·00004	2	—
1·9	0·06562	1163	+162	4·4	0·00002	—	—
2·0	0·05399	1001	+150	4·5	0·00002	—	—
2·1	0·04398	851	+137	4·6	0·00001	—	—
2·2	0·03547	714	+120	4·7	0·00001	—	—
2·3	0·02833	594	+108	4·8	0·00000	—	—
2·4	0·02239	486	+ 91				

Appendix Table 2 Distribution function of the normal distribution

The table shows the area under the curve $y = (2\pi)^{-\frac{1}{2}}e^{-\frac{1}{2}x^2}$ lying to the left of specified deviates x; e.g. the area corresponding to a deviate 1·86 (= 1·5 + 0·36) is 0·9686.

Deviate	0·0 +	0·5 +	1·0 +	1·5 +	2·0 +	2·5 +	3·0 +	3·5 +
0·00	5000	6915	8413	9332	9772	9^2379	9^2865	9^377
0·01	5040	6950	8438	9345	9778	9^2396	9^2869	9^378
0·02	5080	6985	8461	9357	9783	9^2413	9^2874	9^378
0·03	5120	7019	8485	9370	9788	9^2430	9^2878	9^379
0·04	5160	7054	8508	9382	9793	9^2446	9^2882	9^380
0·05	5199	7088	8531	9394	9798	9^2461	9^2886	9^381
0·06	5239	7123	8554	9406	9803	9^2477	9^2889	9^381
0·07	5279	7157	8577	9418	9808	9^2492	9^2893	9^382
0·08	5319	7190	8599	9429	9812	9^2506	9^2897	9^383
0·09	5359	7224	8621	9441	9817	9^2520	9^2900	9^383
0·10	5398	7257	8643	9452	9821	9^2534	9^303	9^384
0·11	5438	7291	8665	9463	9826	9^2547	9^306	9^385
0·12	5478	7324	8686	9474	9830	9^2560	9^310	9^385
0·13	5517	7357	8708	9484	9834	9^2573	9^313	9^386
0·14	5557	7389	8729	9495	9838	9^2585	9^316	9^386
0·15	5596	7422	8749	9505	9842	9^2598	9^318	9^387
0·16	5636	7454	8770	9515	9846	9^2609	9^321	9^387
0·17	5675	7486	8790	9525	9850	9^2621	9^324	9^388
0·18	5714	7517	8810	9535	9854	9^2632	9^326	9^388
0·19	5753	7549	8830	9545	9857	9^2643	9^329	9^389
0·20	5793	7580	8849	9554	9861	9^2653	9^331	9^389
0·21	5832	7611	8869	9564	9864	9^2664	9^334	9^390
0·22	5871	7642	8888	9573	9868	9^2674	9^336	9^390
0·23	5910	7673	8907	9582	9871	9^2683	9^338	9^404
0·24	5948	7704	8925	9591	9875	9^2693	9^340	9^408
0·25	5987	7738	8944	9599	9878	9^2702	9^342	9^412
0·26	6026	7764	8962	9608	9881	9^2711	9^344	9^415
0·27	6064	7794	8980	9616	9884	9^2720	9^346	9^418
0·28	6103	7823	8997	9625	9887	9^2728	9^348	9^422
0·29	6141	7852	9015	9633	9890	9^2736	9^350	9^425
0·30	6179	7881	9032	9641	9893	9^2744	9^352	9^428
0·31	6217	7910	9049	9649	9896	9^2752	9^353	9^431
0·32	6255	7939	9066	9656	9898	9^2760	9^355	9^433
0·33	6293	7967	9082	9664	9901	9^2767	9^357	9^436
0·34	6331	7995	9099	9671	9904	9^2774	9^358	9^439
0·35	6368	8023	9115	9678	9906	9^2781	9^360	9^441
0·36	6406	8051	9131	9686	9909	9^2788	9^361	9^443
0·37	6443	8078	9147	9693	9911	9^2795	9^362	9^446
0·38	6480	8106	9162	9699	9913	9^2801	9^364	9^448
0·39	6517	8133	9177	9706	9916	9^2807	9^365	9^450
0·40	6554	8159	9192	9713	9918	9^2813	9^366	9^452
0·41	6591	8186	9207	9719	9920	9^2819	9^368	9^454
0·42	6628	8212	9222	9726	9922	9^2825	9^369	9^456
0·43	6664	8238	9236	9732	9925	9^2831	9^370	9^458
0·44	6700	8264	9251	9738	9927	9^2836	9^371	9^459
0·45	6736	8289	9265	9744	9929	9^2841	9^372	9^461
0·46	6772	8315	9279	9750	9931	9^2846	9^373	9^463
0·47	6808	8340	9292	9756	9932	9^2851	9^374	9^464
0·48	6844	8365	9306	9761	9934	9^2856	9^375	9^466
0·49	6879	8389	9319	9767	9936	9^2861	9^376	9^467

Note—Decimal points in the body of the table are omitted. Repeated 9's are indicated by powers, e.g. 9^371 stands for 0·99971.

Appendix Table 3 Quantiles of the d.f. of χ^2

(Reproduced from Table III of Sir Ronald Fisher's *Statistical Methods for Research Workers*, Oliver and Boyd Ltd., Edinburgh, by kind permission of the author and publishers)

$P = 1 - F$	0·99	0·98	0·95	0·90	0·80	0·70	0·50	0·30	0·20	0·10	0·05	0·02	0·01
$\nu =$ 1	0·0^{3}157	0·0^{3}628	0·0^{2}393	0·0158	0·0642	0·148	0·455	1·074	1·642	2·706	3·841	5·412	6·635
2	0·0201	0·0404	0·103	0·211	0·446	0·713	1·386	2·408	3·219	4·605	5·991	7·824	9·210
3	0·115	0·185	0·352	0·584	1·005	1·424	2·366	3·665	4·642	6·251	7·815	9·837	11·345
4	0·297	0·429	0·711	1·064	1·649	2·195	3·357	4·878	5·989	7·779	9·488	11·668	13·277
5	0·554	0·752	1·145	1·160	2·343	3·000	4·351	6·064	7·289	9·236	11·070	13·388	15·086
6	0·872	1·134	1·635	2·204	3·070	3·828	5·348	7·231	8·558	10·645	12·592	15·033	16·812
7	1·239	1·564	2·167	2·833	3·822	4·671	6·346	8·383	9·803	12·017	14·067	16·622	18·475
8	1·646	2·032	2·733	3·490	4·594	5·527	7·344	9·524	11·030	13·362	15·507	18·168	20·090
9	2·088	2·532	3·325	4·168	5·380	6·393	8·343	10·656	12·242	14·684	16·919	19·679	21·666
10	2·358	3·059	3·940	4·865	6·179	7·267	9·342	11·781	13·442	15·987	18·307	21·161	23·209
11	3·053	3·609	4·575	5·578	6·989	8·148	10·341	12·899	14·631	17·275	19·675	22·618	24·725
12	3·571	4·178	5·226	6·304	7·807	9·034	11·340	14·011	15·821	18·549	21·026	24·054	26·217
13	4·107	4·765	5·892	7·042	8·634	9·926	12·340	15·119	16·985	19·812	22·362	25·472	27·688
14	4·660	5·368	6·571	7·790	9·467	10·821	13·339	16·222	18·151	21·064	23·685	26·873	29·141
15	5·229	5·985	7·261	8·547	10·307	11·721	14·339	17·322	19·311	22·307	24·996	28·259	30·578
16	5·812	6·614	7·962	9·312	11·152	12·624	15·338	18·418	20·465	23·542	26·296	29·633	32·000
17	6·408	7·255	8·672	10·085	12·002	13·531	16·338	19·511	21·615	24·769	27·587	30·995	33·409
18	7·015	7·906	9·390	10·865	12·857	14·440	17·338	20·601	22·760	25·989	28·869	32·346	34·805
19	7·633	8·567	10·117	11·651	13·716	15·352	18·338	21·689	23·900	27·204	30·144	33·687	36·191
20	8·260	9·237	10·851	12·443	14·578	16·266	19·337	22·775	25·038	28·412	31·410	35·020	37·566
21	8·897	9·915	11·591	13·240	15·445	17·182	20·337	23·858	26·171	29·615	32·671	36·343	38·932
22	9·542	10·600	12·338	14·041	16·314	18·101	21·337	24·939	27·301	30·813	33·924	37·659	40·289
23	10·196	11·293	13·091	14·848	17·187	19·021	22·337	26·018	28·429	32·007	35·172	38·968	41·638
24	10·856	11·992	13·848	15·659	18·062	19·943	23·337	27·096	29·553	33·196	36·415	40·270	42·980
25	11·524	12·697	14·611	16·473	18·940	20·867	24·337	28·172	30·675	34·382	37·652	41·566	44·314
26	12·198	13·409	15·379	17·292	19·820	21·792	25·336	29·246	31·795	35·563	38·885	42·856	45·642
27	12·879	14·125	16·151	18·114	20·703	22·719	26·336	30·319	32·912	36·741	40·113	44·140	46·963
28	13·565	14·847	16·928	18·939	21·588	23·647	27·336	31·391	34·027	37·916	41·337	45·419	48·278
29	14·256	15·574	17·708	19·768	22·475	24·577	28·336	32·461	35·139	39·087	42·557	46·693	49·588
30	14·953	16·306	18·493	20·599	23·364	25·508	29·336	33·530	36·250	40·256	43·773	47·962	50·892

Note—For values of ν greater than 30 the quantity $\sqrt{(2\chi^2)}$ may be taken to be distributed normally about mean $\sqrt{(2\nu - 1)}$ with unit variance.

Appendix Table 4a Distribution function of χ^2 for one degree of freedom for values of $\chi^2 = 0$ to $\chi^2 = 1$ by steps of 0·01

χ^2	$P = 1-F$	Δ	χ^2	$P = 1-F$	Δ
0	1·00000	7966	0·50	0·47950	436
0·01	0·92034	3280	0·51	0·47514	430
0·02	0·88754	2505	0·52	0·47084	423
0·03	0·86249	2101	0·53	0·46661	418
0·04	0·84148	1842	0·54	0·46243	411
0·05	0·82306	1656	0·55	0·45832	406
0·06	0·80650	1516	0·56	0·45426	400
0·07	0·79134	1404	0·57	0·45026	395
0·08	0·77730	1312	0·58	0·44631	389
0·09	0·76418	1235	0·59	0·44242	384
0·10	0·75183	1169	0·60	0·43858	379
0·11	0·74014	1111	0·61	0·43479	374
0·12	0·72903	1060	0·62	0·43105	369
0·13	0·71843	1015	0·63	0·42736	365
0·14	0·70828	974	0·64	0·42371	360
0·15	0·69854	938	0·65	0·42011	355
0·16	0·68916	905	0·66	0·41656	351
0·17	0·68011	874	0·67	0·41305	346
0·18	0·67137	845	0·68	0·40959	343
0·19	0·66292	820	0·69	0·40616	338
0·20	0·65472	795	0·70	0·40278	334
0·21	0·64677	773	0·71	0·39944	330
0·22	0·63904	752	0·72	0·39614	326
0·23	0·63152	731	0·73	0·39288	322
0·24	0·62421	713	0·74	0·38966	318
0·25	0·61708	696	0·75	0·38648	315
0·26	0·61012	679	0·76	0·38333	311
0·27	0·60333	663	0·77	0·38022	308
0·28	0·59670	648	0·78	0·37714	304
0·29	0·59022	634	0·79	0·37410	301
0·30	0·58388	620	0·80	0·37109	297
0·31	0·57768	607	0·81	0·36812	294
0·32	0·57161	595	0·82	0·36518	291
0·33	0·56566	583	0·83	0·36227	287
0·34	0·55983	572	0·84	0·35940	285
0·35	0·55411	560	0·85	0·35655	281
0·36	0·54851	551	0·86	0·35374	278
0·37	0·54300	540	0·87	0·35096	276
0·38	0·53760	530	0·88	0·34820	272
0·39	0·53230	521	0·89	0·34548	270
0·40	0·52709	512	0·90	0·34278	267
0·41	0·52197	503	0·91	0·34011	264
0·42	0·51694	495	0·92	0·33747	261
0·43	0·51199	487	0·93	0·33486	258
0·44	0·50712	479	0·94	0·33228	256
0·45	0·50233	471	0·95	0·32972	253
0·46	0·49762	463	0·96	0·32719	251
0·47	0·49299	457	0·97	0·32468	248
0·48	0·48842	449	0·98	0·32220	246
0·49	0·48393	443	0·99	0·31974	243
0·50	0·47950	436	1·00	0·31731	241

Appendix Table 4b Distribution function of χ^2 for one degree of freedom for values of χ^2 from 1 to 10 by steps of 0·1

χ^2	$P = 1-F$	Δ	χ^2	$P = 1-F$	Δ
1·0	0·31731	2304	5·5	0·01902	106
1·1	0·29427	2095	5·6	0·01796	99
1·2	0·27332	1911	5·7	0·01697	94
1·3	0·25421	1749	5·8	0·01603	89
1·4	0·23672	1605	5·9	0·01514	83
1·5	0·22067	1477	6·0	0·01431	79
1·6	0·20590	1361	6·1	0·01352	74
1·7	0·19229	1258	6·2	0·01278	71
1·8	0·17971	1163	6·3	0·01207	66
1·9	0·16808	1078	6·4	0·01141	62
2·0	0·15730	1000	6·5	0·01079	59
2·1	0·14730	929	6·6	0·01020	56
2·2	0·13801	864	6·7	0·00964	52
2·3	0·12937	803	6·8	0·00912	50
2·4	0·12134	749	6·9	0·00862	47
2·5	0·11385	699	7·0	0·00815	44
2·6	0·10686	651	7·1	0·00771	42
2·7	0·10035	609	7·2	0·00729	39
2·8	0·09426	568	7·3	0·00690	38
2·9	0·08858	532	7·4	0·00652	35
3·0	0·08326	497	7·5	0·00617	33
3·1	0·07829	465	7·6	0·00584	32
3·2	0·07364	436	7·7	0·00552	30
3·3	0·06928	408	7·8	0·00522	28
3·4	0·06520	383	7·9	0·00494	26
3·5	0·06137	359	8·0	0·00468	25
3·6	0·05778	337	8·1	0·00443	24
3·7	0·05441	316	8·2	0·00419	23
3·8	0·05125	296	8·3	0·00396	21
3·9	0·04829	279	8·4	0·00375	20
4·0	0·04550	262	8·5	0·00355	19
4·1	0·04288	246	8·6	0·00336	18
4·2	0·04042	231	8·7	0·00318	17
4·3	0·03811	217	8·8	0·00301	16
4·4	0·03594	205	8·9	0·00285	15
4·5	0·03389	192	9·0	0·00270	14
4·6	0·03197	181	9·1	0·00256	14
4·7	0·03016	170	9·2	0·00242	13
4·8	0·02846	160	9·3	0·00229	12
4·9	0·02686	151	9·4	0·00217	12
5·0	0·02535	142	9·5	0·00205	10
5·1	0·02393	134	9·6	0·00195	11
5·2	0·02259	126	9·7	0·00184	10
5·3	0·02133	119	9·8	0·00174	9
5·4	0·02014	112	9·9	0·00165	8
5·5	0·01902	106	10·0	0·00157	8

Appendix Table 5 Quantiles of the d.f. of t

(Reproduced from Sir Ronald Fisher and Dr F. Yates: *Statistical Tables for Biological, Medical and Agricultural Research*, Oliver and Boyd Ltd., Edinburgh, by kind permission of the authors and publishers)

$P=2(1-F)$	0·9	0·8	0·7	0·6	0·5	0·4	0·3	0·2	0·1	0·05	0·02	0·01	0·001
$\nu=$ 1	0·158	0·325	0·510	0·727	1·000	1·376	1·963	3·078	6·314	12·706	31·821	63·657	636·619
2	0·142	0·289	0·445	0·617	0·816	1·061	1·386	1·886	2·920	4·303	6·965	9·925	31·598
3	0·137	0·277	0·424	0·584	0·765	0·978	1·250	1·638	2·353	3·182	4·541	5·841	12·924
4	0·134	0·271	0·414	0·569	0·741	0·941	1·190	1·533	2·132	2·776	3·747	4·604	8·610
5	0·132	0·267	0·408	0·559	0·727	0·920	1·156	1·476	2·015	2·571	3·365	4·032	6·869
6	0·131	0·265	0·404	0·553	0·718	0·906	1·134	1·440	1·943	2·447	3·143	3·707	5·959
7	0·130	0·263	0·402	0·549	0·711	0·896	1·119	1·415	1·895	2·365	2·998	3·499	5·408
8	0·130	0·262	0·399	0·546	0·706	0·889	1·108	1·397	1·860	2·306	2·896	3·355	5·041
9	0·129	0·261	0·398	0·543	0·703	0·883	1·100	1·383	1·833	2·262	2·821	3·250	4·781
10	0·129	0·260	0·397	0·542	0·700	0·879	1·093	1·372	1·812	2·228	2·764	3·169	4·587
11	0·129	0·260	0·396	0·540	0·697	0·876	1·088	1·363	1·796	2·201	2·718	3·106	4·437
12	0·128	0·259	0·395	0·539	0·695	0·873	1·083	1·356	1·782	2·179	2·681	3·055	4·318
13	0·128	0·259	0·394	0·538	0·694	0·870	1·079	1·350	1·771	2·160	2·650	3·012	4·221
14	0·128	0·258	0·393	0·537	0·692	0·868	1·076	1·345	1·761	2·145	2·624	2·977	4·140
15	0·128	0·258	0·393	0·536	0·691	0·866	1·074	1·341	1·753	2·131	2·602	2·947	4·073
16	0·128	0·258	0·392	0·535	0·690	0·865	1·071	1·337	1·746	2·120	2·583	2·921	4·015
17	0·128	0·257	0·392	0·534	0·689	0·863	1·069	1·333	1·740	2·110	2·567	2·898	3·965
18	0·127	0·257	0·392	0·534	0·688	0·862	1·067	1·330	1·734	2·101	2·552	2·878	3·922
19	0·127	0·257	0·391	0·533	0·688	0·861	1·066	1·328	1·729	2·093	2·539	2·861	3·883
20	0·127	0·257	0·391	0·533	0·687	0·860	1·064	1·325	1·725	2·086	2·528	2·845	3·850
21	0·127	0·257	0·391	0·532	0·686	0·859	1·063	1·323	1·721	2·080	2·518	2·831	3·819
22	0·127	0·256	0·390	0·532	0·686	0·858	1·061	1·321	1·717	2·074	2·508	2·819	3·792
23	0·127	0·256	0·390	0·532	0·685	0·858	1·060	1·319	1·714	2·069	2·500	2·807	3·767
24	0·127	0·256	0·390	0·531	0·685	0·857	1·059	1·318	1·711	2·064	2·492	2·797	3·745
25	0·127	0·256	0·390	0·531	0·684	0·856	1·058	1·316	1·708	2·060	2·485	2·787	3·725
26	0·127	0·256	0·390	0·531	0·684	0·856	1·058	1·315	1·706	2·056	2·479	2·779	3·707
27	0·127	0·256	0·389	0·531	0·684	0·855	1·057	1·314	1·703	2·052	2·473	2·771	3·690
28	0·127	0·256	0·389	0·530	0·683	0·855	1·056	1·313	1·701	2·048	2·467	2·763	3·674
29	0·127	0·256	0·389	0·530	0·683	0·854	1·055	1·311	1·699	2·045	2·462	2·756	3·659
30	0·127	0·256	0·389	0·530	0·683	0·854	1·055	1·310	1·697	2·042	2·457	2·750	3·646
40	0·126	0·255	0·388	0·529	0·681	0·851	1·050	1·303	1·684	2·021	2·423	2·704	3·551
60	0·126	0·254	0·387	0·527	0·679	0·848	1·046	1·296	1·671	2·000	2·390	2·660	3·460
120	0·126	0·254	0·386	0·526	0·677	0·845	1·041	1·289	1·658	1·980	2·358	2·617	3·373
∞	0·126	0·253	0·385	0·524	0·674	0·842	1·036	1·282	1·645	1·960	2·326	2·576	3·291

Appendix Table 6 5 per cent. points of the distribution of z (values at which the d.f. = 0·95)

(Reprinted from Table VI of Sir Ronald Fisher's *Statistical Methods for Research Workers*, Oliver and Boyd Ltd., Edinburgh, by kind permission of the author and publishers)

Values of ν_2	Values of ν_1: 1	2	3	4	5	6	8	12	24	∞
1	2·5421	2·6479	2·6870	2·7071	2·7194	2·7276	2·7380	2·7484	2·7588	2·7693
2	1·4592	1·4722	1·4765	1·4787	1·4800	1·4808	1·4819	1·4830	1·4840	1·4851
3	1·1577	1·1284	1·1137	1·1051	1·0994	1·0953	1·0899	1·0842	1·0781	1·0716
4	1·0212	0·9690	0·9429	0·9272	0·9168	0·9093	0·8993	0·8885	0·8767	0·8639
5	0·9441	0·8777	0·8441	0·8236	0·8097	0·7997	0·7862	0·7714	0·7550	0·7368
6	0·8948	0·8188	0·7798	0·7558	0·7394	0·7274	0·7112	0·6931	0·6729	0·6499
7	0·8606	0·7777	0·7347	0·7080	0·6896	0·6761	0·6576	0·6369	0·6134	0·5862
8	0·8355	0·7475	0·7014	0·6725	0·6525	0·6378	0·6175	0·5945	0·5682	0·5371
9	0·8163	0·7242	0·6757	0·6450	0·6238	0·6080	0·5862	0·5613	0·5324	0·4979
10	0·8012	0·7058	0·6553	0·6232	0·6009	0·5843	0·5611	0·5346	0·5035	0·4657
11	0·7889	0·6909	0·6387	0·6055	0·5822	0·5648	0·5406	0·5126	0·4795	0·4387
12	0·7788	0·6786	0·6250	0·5907	0·5666	0·5487	0·5234	0·4941	0·4592	0·4156
13	0·7703	0·6682	0·6134	0·5783	0·5535	0·5350	0·5089	0·4785	0·4419	0·3957
14	0·7630	0·6594	0·6036	0·5677	0·5423	0·5233	0·4964	0·4649	0·4269	0·3782
15	0·7568	0·6518	0·5950	0·5585	0·5326	0·5131	0·4855	0·4532	0·4138	0·3628
16	0·7514	0·6451	0·5876	0·5505	0·5241	0·5042	0·4760	0·4428	0·4022	0·3490
17	0·7466	0·6393	0·5811	0·5434	0·5166	0·4964	0·4676	0·4337	0·3919	0·3366
18	0·7424	0·6341	0·5753	0·5371	0·5099	0·4894	0·4602	0·4255	0·3827	0·3253
19	0·7386	0·6295	0·5701	0·5315	0·5040	0·4832	0·4535	0·4182	0·3743	0·3151
20	0·7352	0·6254	0·5654	0·5265	0·4986	0·4776	0·4474	0·4116	0·3668	0·3057
21	0·7322	0·6216	0·5612	0·5219	0·4938	0·4725	0·4420	0·4055	0·3599	0·2971
22	0·7294	0·6182	0·5574	0·5178	0·4894	0·4679	0·4370	0·4001	0·3536	0·2892
23	0·7269	0·6151	0·5540	0·5140	0·4854	0·4636	0·4325	0·3950	0·3478	0·2818
24	0·7246	0·6123	0·5508	0·5106	0·4817	0·4598	0·4283	0·3904	0·3425	0·2749
25	0·7225	0·6097	0·5478	0·5074	0·4783	0·4562	0·4244	0·3862	0·3376	0·2685
26	0·7205	0·6073	0·5451	0·5045	0·4752	0·4529	0·4209	0·3823	0·3330	0·2625
27	0·7187	0·6051	0·5427	0·5017	0·4723	0·4499	0·4176	0·3786	0·3287	0·2569
28	0·7171	0·6030	0·5403	0·4992	0·4696	0·4471	0·4146	0·3752	0·3248	0·2516
29	0·7155	0·6011	0·5382	0·4969	0·4671	0·4444	0·4117	0·3720	0·3211	0·2466
30	0·7141	0·5994	0·5362	0·4947	0·4648	0·4420	0·4090	0·3691	0·3176	0·2419
60	0·6933	0·5738	0·5073	0·4632	0·4311	0·4064	0·3702	0·3255	0·2654	0·1644
∞	0·6729	0·5486	0·4787	0·4319	0·3974	0·3706	0·3309	0·2804	0·2085	0

Appendix Table 7 5 per cent. points of the variance ratio *F* (values at which the d.f. = 0·95)

(Reproduced from Sir Ronald Fisher and Dr F. Yates: *Statistical Tables for Biological, Medical and Agricultural Research*, Oliver and Boyd Ltd., Edinburgh, by kind permission of the authors and publishers)

ν_2 \ ν_1	1	2	3	4	5	6	8	12	24	∞
1	161·40	199·50	215·70	224·60	230·20	234·00	238·90	243·90	249·00	254·30
2	18·51	19·00	19·16	19·25	19·30	19·33	19·37	19·41	19·45	19·50
3	10·13	9·55	9·28	9·12	9·01	8·94	8·84	8·74	8·64	8·53
4	7·71	6·94	6·59	6·39	6·26	6·16	6·04	5·91	5·77	5·63
5	6·61	5·79	5·41	5·19	5·05	4·95	4·82	4·68	4·53	4·36
6	5·99	5·14	4·76	4·53	4·39	4·28	4·15	4·00	3·84	3·67
7	5·59	4·74	4·35	4·12	3·97	3·87	3·73	3·57	3·41	3·23
8	5·32	4·46	4·07	3·84	3·69	3·58	3·44	3·28	3·12	2·93
9	5·12	4·26	3·86	3·63	3·48	3·37	3·23	3·07	2·90	2·71
10	4·96	4·10	3·71	3·48	3·33	3·22	3·07	2·91	2·74	2·54
11	4·84	3·98	3·59	3·36	3·20	3·09	2·95	2·79	2·61	2·40
12	4·75	3·88	3·49	3·26	3·11	3·00	2·85	2·69	2·50	2·30
13	4·67	3·80	3·41	3·18	3·02	2·92	2·77	2·60	2·42	2·21
14	4·60	3·74	3·34	3·11	2·96	2·85	2·70	2·53	2·35	2·13
15	4·54	3·68	3·29	3·06	2·90	2·79	2·64	2·48	2·29	2·07
16	4·49	3·63	3·24	3·01	2·85	2·74	2·59	2·42	2·24	2·01
17	4·45	3·59	3·20	2·96	2·81	2·70	2·55	2·38	2·19	1·96
18	4·41	3·55	3·16	2·93	2·77	2·66	2·51	2·34	2·15	1·92
19	4·38	3·52	3·13	2·90	2·74	2·63	2·48	2·31	2·11	1·88
20	4·35	3·49	3·10	2·87	2·71	2·60	2·45	2·28	2·08	1·84
21	4·32	3·47	3·07	2·84	2·68	2·57	2·42	2·25	2·05	1·81
22	4·30	3·44	3·05	2·82	2·66	2·55	2·40	2·23	2·03	1·78
23	4·28	3·42	3·03	2·80	2·64	2·53	2·38	2·20	2·00	1·76
24	4·26	3·40	3·01	2·78	2·62	2·51	2·36	2·18	1·98	1·73
25	4·24	3·38	2·99	2·76	2·60	2·49	2·34	2·16	1·96	1·71
26	4·22	3·37	2·98	2·74	2·59	2·47	2·32	2·15	1·95	1·69
27	4·21	3·35	2·96	2·73	2·57	2·46	2·30	2·13	1·93	1·67
28	4·20	3·34	2·95	2·71	2·56	2·44	2·29	2·12	1·91	1·65
29	4·18	3·33	2·93	2·70	2·54	2·43	2·28	2·10	1·90	1·64
30	4·17	3·32	2·92	2·69	2·53	2·42	2·27	2·09	1·89	1·62
40	4·08	3·23	2·84	2·61	2·45	2·34	2·18	2·00	1·79	1·51
60	4·00	3·15	2·76	2·52	2·37	2·25	2·10	1·92	1·70	1·39
120	3·92	3·07	2·68	2·45	2·29	2·17	2·02	1·83	1·61	1·25
∞	3·84	2·99	2·60	2·37	2·21	2·09	1·94	1·75	1·52	1·00

Lower 5 per cent. points are found by interchange of ν_1 and ν_2, i.e. ν_1 must always correspond to the greater mean square.

Appendix Table 8 1 per cent. points of the distribution of z (values at which the d.f. = 0·99)

(Reprinted from Table VI of Sir Ronald Fisher's *Statistical Methods for Research Workers*, Oliver and Boyd Ltd., Edinburgh, by kind permission of the author and publishers)

Values of ν_2	Values of ν_1: 1	2	3	4	5	6	8	12	24	∞
1	4·1535	4·2585	4·2974	4·3175	4·3297	4·3379	4·3482	4·3585	4·3689	4·3794
2	2·2950	2·2976	2·2984	2·2988	2·2991	2·2992	2·2994	2·2997	2·2999	2·3001
3	1·7649	1·7140	1·6915	1·6786	1·6703	1·6645	1·6569	1·6489	1·6404	1·6314
4	1·5270	1·4452	1·4075	1·3856	1·3711	1·3609	1·3473	1·3327	1·3170	1·3000
5	1·3943	1·2929	1·2449	1·2164	1·1974	1·1838	1·1656	1·1457	1·1239	1·0997
6	1·3103	1·1955	1·1401	1·1068	1·0843	1·0680	1·0460	1·0218	0·9948	0·9643
7	1·2526	1·1281	1·0672	1·0300	1·0048	0·9864	0·9614	0·9335	0·9020	0·8658
8	1·2106	1·0787	1·0135	0·9734	0·9459	0·9259	0·8983	0·8673	0·8319	0·7904
9	1·1786	1·0411	0·9724	0·9299	0·9006	0·8791	0·8494	0·8157	0·7769	0·7305
10	1·1535	1·0114	0·9399	0·8954	0·8646	0·8419	0·8104	0·7744	0·7324	0·6816
11	1·1333	0·9874	0·9136	0·8674	0·8354	0·8116	0·7785	0·7405	0·6958	0·6408
12	1·1166	0·9677	0·8919	0·8443	0·8111	0·7864	0·7520	0·7122	0·6649	0·6061
13	1·1027	0·9511	0·8737	0·8248	0·7907	0·7652	0·7295	0·6882	0·6386	0·5761
14	1·0909	0·9370	0·8581	0·8082	0·7732	0·7471	0·7103	0·6675	0·6159	0·5500
15	1·0807	0·9249	0·8448	0·7939	0·7582	0·7314	0·6937	0·6496	0·5961	0·5269
16	1·0719	0·9144	0·8331	0·7814	0·7450	0·7177	0·6791	0·6339	0·5786	0·5064
17	1·0641	0·9051	0·8229	0·7705	0·7335	0·7057	0·6663	0·6199	0·5630	0·4879
18	1·0572	0·8970	0·8138	0·7607	0·7232	0·6950	0·6549	0·6075	0·5491	0·4712
19	1·0511	0·8897	0·8057	0·7521	0·7140	0·6854	0·6447	0·5964	0·5366	0·4560
20	1·0457	0·8831	0·7985	0·7443	0·7058	0·6768	0·6355	0·5864	0·5253	0·4421
21	1·0408	0·8772	0·7920	0·7372	0·6984	0·6690	0·6272	0·5773	0·5150	0·4294
22	1·0363	0·8719	0·7860	0·7309	0·6916	0·6620	0·6196	0·5691	0·5056	0·4176
23	1·0322	0·8670	0·7806	0·7251	0·6855	0·6555	0·6127	0·5615	0·4969	0·4068
24	1·0285	0·8626	0·7757	0·7197	0·6799	0·6496	0·6064	0·5545	0·4890	0·3967
25	1·0251	0·8585	0·7712	0·7148	0·6747	0·6442	0·6006	0·5481	0·4816	0·3872
26	1·0220	0·8548	0·7670	0·7103	0·6699	0·6392	0·5952	0·5422	0·4748	0·3784
27	1·0191	0·8513	0·7631	0·7062	0·6655	0·6346	0·5902	0·5367	0·4685	0·3701
28	1·0164	0·8481	0·7595	0·7023	0·6614	0·6303	0·5856	0·5316	0·4626	0·3624
29	1·0139	0·8451	0·7562	0·6987	0·6576	0·6263	0·5813	0·5269	0·4570	0·3550
30	1·0116	0·8423	0·7531	0·6954	0·6540	0·6226	0·5773	0·5224	0·4519	0·3481
60	0·9784	0·8025	0·7086	0·6472	0·6028	0·5687	0·5189	0·4574	0·3746	0·2352
∞	0·9462	0·7636	0·6651	0·5999	0·5522	0·5152	0·4604	0·3908	0·2913	0

Appendix Table 9 1 per cent. points of the variance ratio F
(values at which the d.f. = 0·99)

(Reproduced from Sir Ronald Fisher and Dr F. Yates: *Statistical Tables for Biological, Medical and Agricultural Research*, Oliver and Boyd Ltd., Edinburgh, by kind permission of the authors and publishers)

ν_2 \ ν_1	1	2	3	4	5	6	8	12	24	∞
1	4052	4999	5403	5625	5764	5859	5981	6106	6234	6366
2	98·49	99·00	99·17	99·25	99·30	99·33	99·36	99·42	99·46	99·50
3	34·12	30·81	29·46	28·71	28·24	27·91	27·49	27·05	26·60	26·12
4	21·20	18·00	16·69	15·98	15·52	15·21	14·80	14·37	13·93	13·46
5	16·26	13·27	12·06	11·39	10·97	10·67	10·27	9·89	9·47	9·02
6	13·74	10·92	9·78	9·15	8·75	8·47	8·10	7·72	7·31	6·88
7	12·25	9·55	8·45	7·85	7·46	7·19	6·84	6·47	6·07	5·65
8	11·26	8·65	7·59	7·01	6·63	6·37	6·03	5·67	5·28	4·86
9	10·56	8·02	6·99	6·42	6·06	5·80	5·47	5·11	4·73	4·31
10	10·04	7·56	6·55	5·99	5·64	5·39	5·06	4·71	4·33	3·91
11	9·65	7·20	6·22	5·67	5·32	5·07	4·74	4·40	4·02	3·60
12	9·33	6·93	5·95	5·41	5·06	4·82	4·50	4·16	3·78	3·36
13	9·07	6·70	5·74	5·20	4·86	4·62	4·30	3·96	3·59	3·16
14	8·86	6·51	5·56	5·03	4·69	4·46	4·14	3·80	3·43	3·00
15	8·68	6·36	5·42	4·89	4·56	4·32	4·00	3·67	3·29	2·87
16	8·53	6·23	5·29	4·77	4·44	4·20	3·89	3·55	3·18	2·75
17	8·40	6·11	5·18	4·67	4·34	4·10	3·79	3·45	3·08	2·65
18	8·28	6·01	5·09	4·58	4·25	4·01	3·71	3·37	3·00	2·57
19	8·18	5·93	5·01	4·50	4·17	3·94	3·63	3·30	2·92	2·49
20	8·10	5·85	4·94	4·43	4·10	3·87	3·56	3·23	2·86	2·42
21	8·02	5·78	4·87	4·37	4·04	3·81	3·51	3·17	2·80	2·36
22	7·94	5·72	4·82	4·31	3·99	3·76	3·45	3·12	2·75	2·31
23	7·88	5·66	4·76	4·26	3·94	3·71	3·41	3·07	2·70	2·26
24	7·82	5·61	4·72	4·22	3·90	3·67	3·36	3·03	2·66	2·21
25	7·77	5·57	4·68	4·18	3·86	3·63	3·32	2·99	2·62	2·17
26	7·72	5·53	4·64	4·14	3·82	3·59	3·29	2·96	2·58	2·13
27	7·68	5·49	4·60	4·11	3·78	3·56	3·26	2·93	2·55	2·10
28	7·64	5·45	4·57	4·07	3·75	3·53	3·23	2·90	2·52	2·06
29	7·60	5·42	4·54	4·04	3·73	3·50	3·20	2·87	2·49	2·03
30	7·56	5·39	4·51	4·02	3·70	3·47	3·17	2·84	2·47	2·01
40	7·31	5·18	4·31	3·83	3·51	3·29	2·99	2·66	2·29	1·80
60	7·08	4·98	4·13	3·65	3·34	3·12	2·82	2·50	2·12	1·60
120	6·85	4·79	3·95	3·48	3·17	2·96	2·66	2·34	1·95	1·38
∞	6·64	4·60	3·78	3·32	3·02	2·80	2·51	2·18	1·79	1·00

Lower 1 per cent. points are found by interchange of ν_1 and ν_2, i.e. ν_1 must always correspond to the greater mean square.

Appendix Table 10 Augmented symmetric functions in terms of power-sums and vice versa

(Reproduced from F. N. David and Kendall (1949) by kind permission of Prof. David and the editors of *Biometrika*)

weight 1 $(1) = [1]$

weight 2

	$[2]$	$[1^2]$
(2)	1	-1
$(1)^2$	1	1

weight 3

	$[3]$	$[21]$	$[1^3]$
(3)	1	-1	2
$(2)(1)$	1	1	-3
$(1)^3$	1	3	1

weight 4

	$[4]$	$[31]$	$[2^2]$	$[21^2]$	$[1^4]$
(4)	1	-1	-1	2	-6
$(3)(1)$	1	1	·	-2	8
$(2)^2$	1	·	1	-1	3
$(2)(1)^2$	1	2	1	1	-6
$(1)^4$	1	4	3	6	1

weight 5

	$[5]$	$[41]$	$[32]$	$[31^2]$	$[2^21]$	$[21^3]$	$[1^5]$
(5)	1	-1	-1	2	2	-6	24
$(4)(1)$	1	1	·	-2	-1	6	-30
$(3)(2)$	1	·	1	-1	-2	5	-20
$(3)(1)^2$	1	2	1	1	·	-3	20
$(2)^2(1)$	1	1	2	·	1	-3	15
$(2)(1)^3$	1	3	4	3	3	1	-10
$(1)^5$	1	5	10	10	15	10	1

weight 6

	$[6]$	$[51]$	$[42]$	$[41^2]$	$[3^2]$	$[321]$	$[31^3]$	$[2^3]$	$[2^21^2]$	$[21^4]$	$[1^6]$
(6)	1	-1	-1	2	-1	2	-6	2	-6	24	-120
$(5)(1)$	1	1	·	-2	·	-1	6	·	4	-24	144
$(4)(2)$	1	·	1	-1	·	-1	3	-3	5	-18	90
$(4)(1)^2$	1	2	1	1	·	·	-3	·	-1	12	-90
$(3)^2$	1	·	·	·	1	-1	2	·	2	-8	40
$(3)(2)(1)$	1	1	1	·	1	1	-3	·	-4	20	-120
$(3)(1)^3$	1	3	3	3	1	3	1	·	·	-4	40
$(2)^3$	1	·	3	·	·	·	·	1	-1	3	-15
$(2)^2(1)^2$	1	2	3	1	2	4	·	1	1	-6	45
$(2)(1)^4$	1	4	7	6	4	16	4	3	6	1	-15
$(1)^6$	1	6	15	15	10	60	20	15	45	15	1

To express the [ ] functions in terms of ( ), read downwards up to and including the main diagonal, e.g. $[41^2] = 2(6) - 2(5)(1) - (4)(2) + (4)(1)^2$. To express the ( ) functions in terms of [ ], read across up to and including the main diagonal, e.g. $(4)(1)^2 = [6] + 2[51] + [42] + [41^2]$.

Appendix Table 11 Multiple *k*-statistics in terms of augmented symmetric functions and vice versa

(Reproduced from Wishart (1952) by kind permission of the late author and the editors of *Biometrika*)

1st order $l_1 = [1]/n$

2nd order

	l_{11}	l_2
$[1^2]/n^{[2]}$	1	−1
$[2]/n$	1	1

3rd order

	l_{111}	l_{21}	l_3
$[1^3]/n^{[3]}$	1	−1	2
$[21]/n^{[2]}$	1	1	−3
$[3]/n$	1	3	1

4th order

	l_{1111}	l_{211}	l_{22}	l_{31}	l_4
$[1^4]/n^{[4]}$	1	−1	1	2	−6
$[21^2]/n^{[3]}$	1	1	−2	−3	12
$[2^2]/n^{[2]}$	1	2	1	·	−3
$[31]/n^{[2]}$	1	3	·	1	−4
$[4]/n$	1	6	3	4	1

5th order

	l_{11111}	l_{2111}	l_{221}	l_{311}	l_{32}	l_{41}	l_5
$[1^5]/n^{[5]}$	1	−1	1	2	−2	−6	24
$[21^3]/n^{[4]}$	1	1	−2	−3	5	12	−60
$[2^21]/n^{[3]}$	1	2	1	·	−3	−3	30
$[31^2]/n^{[3]}$	1	3	·	1	−1	−4	20
$[32]/n^{[2]}$	1	4	3	1	1	·	−10
$[41]/n^{[2]}$	1	6	3	4	·	1	−5
$[5]/n$	1	10	15	10	10	5	1

6th order

	l_{111111}	l_{21111}	l_{2211}	l_{3111}	l_{222}	l_{321}	l_{411}	l_{33}	l_{42}	l_{51}	l_6
$[1^6]/n^{[6]}$	1	−1	1	2	−1	−2	−6	4	6	24	−120
$[21^4]/n^{[5]}$	1	1	−2	−3	3	5	12	−12	−18	−60	360
$[2^21^2]/n^{[4]}$	1	2	1	·	−3	−3	−3	9	15	30	−270
$[31^3]/n^{[4]}$	1	3	·	1	·	−1	−4	4	4	20	−120
$[2^3]/n^{[3]}$	1	3	3	·	1	·	·	·	−3	·	30
$[321]/n^{[3]}$	1	4	3	1	·	1	·	−6	−4	−10	120
$[41^2]/n^{[3]}$	1	6	3	4	·	·	1	·	−1	−5	30
$[3^2]/n^{[2]}$	1	6	9	2	·	6	·	1	·	·	−10
$[42]/n^{[2]}$	1	7	9	4	3	4	1	·	1	·	−15
$[51]/n^{[2]}$	1	10	15	10	·	10	5	·	·	1	−6
$[6]/n$	1	15	45	20	15	60	15	10	15	6	1

To express l's in terms of symmetrics read downwards as far as and including the main diagonal, e.g. $l_{222} = -[1^6]/n^{[6]} + 3\,[21^4]/n^{[5]} - 3\,[2^21^2]/n^{[4]} + [2^3]/n^{[3]}$.

To express symmetrics in terms of l's read horizontally up to and including the diagonal, e.g. $[2^21^2]/n^{[4]} = l_{111111} + 2l_{21111} + l_{2211}$.

REFERENCES

Note. References to R. A. Fisher, J. Neyman, K. Pearson, " Student " and G. U. Yule that are marked with an asterisk are reproduced in the following collections:

R. A. Fisher, *Contributions to Mathematical Statistics*, Wiley, New York, 1950.
A Selection of Early Statistical Papers of J. Neyman, Cambridge Univ. Press, 1967.
Karl Pearson's Early Statistical Papers, Cambridge Univ. Press, London, 1948.
" *Student's* " *Collected Papers*, Biometrika Office, University College London, 1942.
Statistical Papers of George Udny Yule, Griffin, London, 1971.

ABDEL ATY, S. H. (1954). Tables of generalised k-statistics. *Biometrika*, **41,** 253.

ABRAHAMSON, I. G. (1964). Orthant probabilities for the quadrivariate normal distribution. *Ann. Math. Statist.*, **35,** 1685.

AITCHISON, J. (1963). Inverse distributions and independent gamma-distributed products of random variables. *Biometrika*, **50,** 505.

AITCHISON, J. and BROWN, J. A. C. (1957). *The lognormal distribution with special reference to its uses in economics.* Cambridge Univ. Press.

AITKEN, A. C. (1950). Statistical independence of quadratic forms in normal variates. *Biometrika*, **37,** 93.

ALI, M. M. (1974). Stochastic ordering and kurtosis measure. *J. Amer. Statist. Ass.*, **69,** 543.

AMOS, D. E. (1963). Additional percentage points for the incomplete beta distribution. *Biometrika*, **50,** 449.

ANSCOMBE, F. J. (1950). Sampling theory of the negative-binomial and logarithmic series distributions. *Biometrika*, **37,** 358.

BACON, R. H. (1963). Approximations to multivariate normal orthant probabilities. *Ann. Math. Statist.*, **34,** 191.

BAIN, L. J. and ENGELHARDT, M. (1975). A two-moment chi-square approximation for the statistic log $(\bar{x}/\tilde{x})$. *J. Amer. Statist. Ass.*, **70,** 948.

BALDWIN, E. M. (1946). Percentage points of the t distribution. *Biometrika*, **33,** 362.

BARTLETT, M. S. (1934). The vector representation of a sample. *Proc. Camb. Phil. Soc.*, **30,** 327.

BARTON, D. E. (1957). The modality of Neyman's contagious distribution of Type A. *Trab. Estadist.*, **8,** 13.

BARTON, D. E. and DENNIS, K. E. (1952). The conditions under which Gram–Charlier and Edgeworth curves are positive definite and unimodal. *Biometrika*, **39,** 425.

BECHHOFER, R. E. and TAMHANE, A. C. (1974). An iterated integral representation for a multivariate normal integral having block covariance structure. *Biometrika*, **61,** 615.

BELZ, M. H. (1947). Note on the Liapounoff inequality for absolute moments. *Ann. Math. Statist.*, **18,** 604.

BERNDT, G. D. (1957). The regions of unimodality and positivity in the abbreviated Edgeworth and Gram–Charlier series. *J. Amer. Statist. Ass.*, **52,** 253.

BICKEL, P. J. (1967). Some contributions to the theory of order statistics. *Proc. 5th Berkeley Symp. Math. Statist. and Prob.*, **1,** 575.

BOHRNSTEDT, G. W. and GOLDBERGER, A. S. (1969). On the exact covariance of products of random variables. *J. Amer. Statist. Ass.*, **64,** 1439.

BOOMSMA, A. (1975). An error in F. N. David's tables of the correlation coefficient. *Biometrika*, **62,** 711.

BOSE, S. S. (1935). On the distribution of the ratio of variances of two samples drawn from a given normal bivariate correlated population. *Sankhyā*, **2,** 65.

Box, G. E. P. and Muller, M. E. (1958). A note on the generation of random normal deviates. *Ann. Math. Statist.*, **29**, 610.

Brown, G. W. and Tukey, J. W. (1946). Some distributions of sample means. *Ann. Math. Statist.*, **17**, 1.

Buckle, N., Kraft, C. and van Eeden, C. (1969). An approximation to the Wilcoxon–Mann–Whitney distribution. *J. Amer. Statist. Ass.*, **64**, 591.

Bukač, J. (1972). Fitting S_B curves using symmetrical percentage points. *Biometrika*, **59**, 688.

Burr, I. W. (1942). Cumulative frequency functions. *Ann. Math. Statist.*, **13**, 215.

Burr, I. W. (1955). Calculation of exact sampling distribution of ranges from a discrete population. *Ann. Math. Statist.*, **26**, 530.

Burr, I. W. and Cislak, P. J. (1968). On a general system of distributions. I. Its curve-shape characteristics. II. The sample median. *J. Amer. Statist. Ass.*, **63**, 627.

Cacoullos, T. (1965). A relation between t and F-distributions. *J. Amer. Statist. Ass.*, **60**, 528.

Campbell, J. W. and Tsokos, C. P. (1973). The asymptotic distribution of maxima in bivariate samples. *J. Amer. Statist. Ass.*, **68**, 734.

Carleman, T. (1925). *Les fonctions quasi-analytiques*. Gauthier-Villars, Paris.

Carlton, G. A. (1946). Estimating the parameters of a rectangular distribution. *Ann. Math. Statist.*, **17**, 355.

Chakrabarti, M. C. (1948). On the ratio of mean deviation to standard deviation. *Calcutta Statist. Ass. Bull.*, **1**, 187.

Chambers, J. M. (1967). On methods of asymptotic approximation for multivariate distributions. *Biometrika*, **54**, 367.

Charlier, C. V. L. (1931). *Applications [de la théorie des probabilités] à l'astronomie.* (Part of the *Traité* edited by Borel.) Gauthier-Villars, Paris.

Cheng, M. C. (1969). The orthant probabilities of four Gaussian variates. *Ann. Math. Statist.*, **40**, 152.

Chernoff, H., Gastwirth, J. L. and Johns, M. V., Jr. (1967). Asymptotic distribution of linear combinations of functions of order statistics with applications to estimation. *Ann. Math. Statist.*, **38**, 52.

Childs, D. R. (1967). Reduction of the multivariate normal integral to characteristic form. *Biometrika*, **54**, 293.

Chu, J. T. (1955). The "inefficiency" of the sample median for many familiar symmetric distributions. *Biometrika*, **42**, 520.

Churchill, E. (1946). Information given by odd moments. *Ann. Math. Statist.*, **17**, 244.

Clark, C. E. (1961). The greatest of a finite set of random variables. *Operations Research*, **9**, 145.

Cochran, W. G. (1934). The distribution of quadratic forms in a normal system, with applications to the analysis of covariance. *Proc. Camb. Phil. Soc.*, **30**, 178.

Cochran, W. G. (1940). Note on an approximative formula for significance levels of z. *Ann. Math. Statist.*, **11**, 93.

Connor, R. J. (1969). The sampling distribution of the range from discrete uniform finite populations and a range test for homogeneity. *J. Amer. Statist. Ass.*, **64**, 1443.

Cook, M. B. (1951). Bivariate k-statistics and cumulants of their joint sampling distribution. *Biometrika*, **38**, 179.

Co-operative Study (1917). On the distribution of the correlation coefficient in small samples. *Biometrika*, **11**, 328.

Cornish, E. A., and Fisher, R. A. (1937).* Moments and cumulants in the specification of distributions. *Rev. Int. Statist. Inst.*, **5**, 307.

Cox, D. R. (1948). Asymptotic distribution of the range. *Biometrika*, **35**, 310.

Craig, A. T. (1943). Note on the independence of certain quadratic forms. *Ann. Math. Statist.*, **14**, 195.

Craig, C. C. (1936a). A new exposition and chart for the Pearson system of frequency curves. *Ann. Math. Statist.*, **7**, 16.

CRAIG, C. C. (1936b). Sheppard's corrections for a discrete variable. *Ann. Math. Statist.*, **7**, 55.

CRAMÉR, H. (1926). On some classes of series used in mathematical statistics. *Skandinaviske Matematikercongres*, Copenhagen.

CRAMÉR, H. (1928). On the composition of elementary errors. *Skand. Aktuartidskr.*, **11**, 13 and 141.

CRAMÉR, H. (1937). *Random Variables and Probability Distributions.* Cambridge Univ. Press.

CRAMÉR, H., and WOLD, H. (1936). Some theorems on distribution functions. *J. Lond. Math. Soc.*, **11**, 290.

D'AGOSTINO, R. and PEARSON, E. S. (1973). Tests for departure from normality. Empirical results for the distributions of b_2 and $\sqrt{b_1}$. *Biometrika*, **60**, 613.

DANIELS, H. E. (1954). Saddlepoint approximations in statistics. *Ann. Math. Statist.*, **25**, 631.

DAS, S. C. (1956). The numerical evaluation of a class of integral. II. *Proc. Camb. Phil. Soc.*, **52**, 442.

DAVID, F. N. (1938). *Tables of the Correlation Coefficient.* Cambridge Univ. Press.

DAVID, F. N. (1949a). Note on the application of Fisher's k-statistics. *Biometrika*, **36**, 383.

DAVID, F. N. (1949b). Moments of the z and F distributions. *Biometrika*, **36**, 394.

DAVID, F. N., and JOHNSON, N. L. (1951). The effect of non-normality on the power function of the F-test. *Biometrika*, **38**, 43.

DAVID, F. N., and JOHNSON, N. L. (1954). Statistical treatment of censored data. Part I. Fundamental formulae. *Biometrika*, **41**, 225.

DAVID, F. N., and KENDALL, M. G. (1949, 1951, 1953, 1955). Tables of symmetric functions. *Biometrika*, **36**, 431; **38**, 435; **40**, 427; **42**, 223.

DAVID, F. N., BARTON, D. E., GANESHALINGAM, S., HARTER, H. L., KIM, P. J., MERRINGTON, M. and WALLEY, D. (1968). *Normal Centroids, Medians and Scores for Ordinal Data.* Cambridge University Press.

DAVID, H. A. (1968). Gini's mean difference rediscovered. *Biometrika*, **55**, 573.

DAVID, H. A. (1970). *Order Statistics.* Wiley, New York.

DAVID, H. A. and MISHRIKY, R. S. (1968). Order statistics for discrete populations and for grouped samples. *J. Amer. Statist. Ass.*, **63**, 1390.

DAVIES, O. L. (1933–4). On asymptotic formulae for the hypergeometric series, I and II. *Biometrika*, **25**, 295 and **26**, 59.

DE LURY, D. B. (1938). Note on correlations. *Ann. Math. Statist.*, **9**, 149.

DEMING, W. E. and GLASSER, G. J. (1959). On the problem of matching lists by samples. *J. Amer. Statist. Ass.*, **54**, 403.

DRAPER, N. R. and TIERNEY, D. E. (1972). Regions of positive and unimodal series expansion of the Edgeworth and Gram–Charlier approximations. *Biometrika*, **59**, 463.

DRONKERS, J. J. (1958). Approximate formulae for the statistical distributions of extreme values. *Biometrika*, **45**, 447.

DURBIN, J. and WATSON, G. S. (1951). Testing for serial correlation in least squares regression. II. *Biometrika*, **38**, 159.

DUTT, J. E. (1973). A representation of multivariate normal probability integrals by integral transforms. *Biometrika*, **60**, 637.

DWYER, P. S. (1964). Properties of polykays of deviates. *Ann. Math. Statist.*, **35**, 1167.

DYWER, P. S. and TRACY, D. S. (1964). A combinatorial method for products of two polykays with some general formulae. *Ann. Math. Statist.*, **35**, 1174.

EDGEWORTH, F. Y. (1904). The Law of Error. *Trans. Camb. Phil. Soc.*, **20**, 36 and 113 (with an Appendix not printed in the *T.C.P.S.* but issued with reprints).

ELDERTON, W. P. (1938). Correzioni dei momenti quando la curva è simmetrica. *Giorn. Ist. Ital. Attuari*, **9**, 145.

ELDERTON, W. P. and JOHNSON, N. L. (1969). *Systems of Frequency Curves.* Cambridge University Press, London.

ELFVING, G. (1947). Asymptotical distribution of range in samples from a normal population. *Biometrika,* **34,** 111.

FAMA, E. F. and ROLL, R. (1968). Some properties of symmetric stable distributions. *J. Amer. Statist. Ass.,* **63,** 817.

FEDERIGHI, E. T. (1959). Extended tables of the percentage points of Student's t-distribution. *J. Amer. Statist. Ass.,* **54,** 683.

FELLER, W. (1950). *An Introduction to Probability Theory and its Applications.* Vol. 1. John Wiley and Sons, New York. Chapman and Hall, London.

FINNEY, D. J. (1938). The distribution of the ratio of estimates of the two variances in a sample from a normal bivariate population. *Biometrika,* **30,** 190.

FINUCAN, H. M. (1964). A note on kurtosis. *J. R. Statist. Soc.,* **B, 26,** 111.

FISHER, R. A. (1915). Frequency-distribution of the values of the correlation coefficient in samples from an indefinitely large population. *Biometrika,* **10,** 507.

FISHER, R. A. (1920).* A mathematical examination of the methods of determining the accuracy of an observation by the mean error and by the mean square error. *Month. Not. R. Astr. Soc.,* **80,** 758.

FISHER, R. A. (1921a).* On the mathematical foundations of theoretical statistics. *Phil. Trans. Roy. Soc.,* **A, 222,** 309.

FISHER, R. A. (1921b). On the probable error of a coefficient of correlation deduced from a small sample. *Metron,* **1,** No. 4, 1.

FISHER, R. A. (1929).* Moments and product moments of sampling distributions. *Proc. Lond. Math. Soc.,* (2), **30,** 199.

FISHER, R. A. (1930)*. The moments of the distribution for normal samples of measures of departure from normality. *Proc. Roy. Soc.,* **A, 130,** 16.

FISHER, R. A. (1935).* The mathematical distributions used in the common tests of significance. *Econometrica,* **3,** 353.

FISHER, R. A. and CORNISH, E. A. (1960). The percentile points of distributions having known cumulants. *Technometrics,* **2,** 209.

FISHER, R. A., and TIPPETT, L. H. C. (1928).* Limiting forms of the frequency-distribution of the largest or smallest member of a sample. *Proc. Camb. Phil. Soc.,* **24,** 180.

FISHER, R. A., and WISHART, J. (1931). The derivation of the pattern formulae of two-way partitions from those of simpler patterns. *Proc. Lond. Math. Soc.,* **33,** 195.

FISHER, R. A., and YATES, F. (1963). *Statistical Tables for Biological, Agricultural and Medical Research.* 6th edition. Oliver and Boyd, Edinburgh.

FISHER, R. A., CORBET, A. S., and WILLIAMS, C. B. (1943).* The relation between the number of species and the number of individuals. *J. Animal Ecology,* **12,** 42.

FOSTER, F. G., and STUART, A. (1954). Distribution-free tests in time-series based on the breaking of records. *J. R. Statist. Soc.,* **B, 16,** 1.

FRAME, J. S. (1945). Mean deviation of the binomial distribution. *Amer. Math. Monthly,* **52,** 377.

FRÉCHET, M. (1927). Sur la loi de probabilité de l'écart maximum. *Annales de la Soc. Polonaise. de Math.,* **6,** 92.

FRÉCHET, M. (1937). *Recherches théoriques modernes.* (Part of the *Traité* edited by Borel.) Gauthier-Villars, Paris.

FRÉCHET, M. (1951). Sur les tableaux de corrélation dont les marges sont données. *Ann. Univ. Lyon,* **A, (3), 14,** 53.

FRISCH, R. (1925). Recurrence formulae for moments of the point binomial. *Biometrika,* **17,** 165.

FRISCH, R. (1926). Sur les semi-invariants et moments employés dans l'étude des distributions statistiques. Oslo, *Skrifter af det Norske Videnskaps Academie, II, Hist.-Filos. Klasse,* No. 3.

GALAMBOS, J. (1975). Order statistics of samples from multivariate distributions. *J. Amer. Statist. Ass.,* **70,** 674.

GARWOOD, F. (1936). Fiducial limits for the Poisson distribution. *Biometrika*, **28**, 437.
GASTWIRTH, J. L. (1974). Large sample theory of some measures of income inequality. *Econometrica*, **42**, 191.
GAYEN, A. K. (1951). The frequency distribution of the product-moment correlation coefficient in random samples of any size drawn from non-normal universes. *Biometrika*, **38**, 219.
GEARY, R. C. (1930). The frequency distribution of the quotient of two normal variables. *J. R. Statist. Soc.*, **93**, 442.
GEARY, R. C. (1936). The distribution of "Student's" ratio for non-normal samples. *Supp. J. R. Statist. Soc.*, **3**, 178.
GEARY, R. C. (1944). Extension of a theorem by Harald Cramér on the frequency distribution of the quotient of two variables. *J. R. Statist. Soc.*, **107**, 56.
GEBHARDT, F. (1969). Some numerical comparisons of several approximations to the binomial distribution. *J. Amer. Statist. Ass.*, **64**, 1638.
GHOSH, B. K. (1966). Asymptotic expansions for the moments of the distribution of correlation coefficient. *Biometrika*, **53**, 258.
GHOSH, B. K. (1975). On the distribution of the difference of two t-variables. *J. Amer. Statist. Ass.*, **70**, 463.
GIDEON, R. A. and GURLAND, J. (1976). Series expansions for quadratic forms in normal variables. *J. Amer. Statist. Ass.*, **71**, 227.
GINI, C. (1912). Variabilità e Mutabilità, contributo allo studio delle distribuzioni e relazioni statistiche. *Studi Economico-Giuridici dell' Univ. di Cagliari*, **3**, part 2, 1–158.
GLASER, R. E. (1976). The ratio of the geometric mean to the arithmetic mean for a random sample from a Gamma distribution. *J. Amer. Statist. Ass.*, **71**, 480.
GNEDENKO, B. V. (1943). Sur la distribution limite du terme maximum d'une série aléatoire. *Ann. Math.*, **(2)**, **44**, 423.
GODWIN, H. J. (1945). On the distribution of the estimate of mean deviation obtained from samples from a normal population. *Biometrika*, **33**, 254.
GODWIN, H. J. (1948). A further note on the mean deviation. *Biometrika*, **35**, 304.
GODWIN, H. J. (1955). Generalisations of Tchebycheff's inequality. *J. Amer. Statist. Ass.*, **50**, 923.
GOOD, I. J. (1963, 1966). On the independence of quadratic expressions. *J. R. Statist. Soc.*, **B**, **25**, 377; corrigenda: **28**, 584.
GOODMAN, L. A. (1952). On the analysis of samples from k lists. *Ann. Math. Statist.*, **23**, 632.
GOODMAN, L. A. (1960). On the exact variance of products. *J. Amer. Statist. Ass.*, **55**, 708.
GOODMAN, L. A. (1962). The variance of the product of K random variables. *J. Amer. Statist. Ass.*, **57**, 54.
GOVINDARAJULU, Z. (1963). On moments of order statistics and quasi-ranges from normal populations. *Ann. Math. Statist.*, **34**, 633.
GRAM, H. L., COBERLY, W. A. and LEWIS, T. O. (1975). On Edgeworth expansions with unknown cumulants. *Ann. Statist.*, **3**, 741.
GRAY, H. L. and ODELL, P. L. (1966). On sums and products of rectangular variates. *Biometrika*, **53**, 615.
GREENWOOD, M., and YULE, G. U. (1920).* An inquiry into the nature of frequency-distributions of multiple happenings, etc. *J. R. Statist. Soc.*, **83**, 255.
GULDBERG, S. (1935). Recurrence formulae for the semi-invariants of some discontinuous frequency distributions of n variables. *Skand. Aktuartidskr.*, **18**, 270.
GUMBEL, E. J. (1934). Les valeurs extrêmes des distributions statistiques. *Ann. Inst. H. Poincaré*, **5**, 115.
GUMBEL, E. J. (1947). The distribution of the range. *Ann. Math. Statist*, **18**, 384.
GUMBEL, E. J. (1949). Probability tables for the range. *Biometrika*, **36**, 142.
GUMBEL, E. J. (1954). *Statistical Theory of Extreme Values and some Practical Applications.* National Bureau of Standards, A.M.S. 33.
GUMBEL, E. J. (1958). *Statistics of extremes.* Columbia Univ. Press, N.Y.
GUMBEL, E. J. (1960). Bivariate exponential distributions. *J. Amer. Statist. Ass.*, **55**, 698.

GUPTA, S. S. (1961). Percentage points and modes of order statistics from the normal distribution. *Ann. Math. Statist.*, **32**, 888.

GUPTA, S. S. (1963a). Probability integrals of multivariate normal and multivariate *t*. *Ann. Math. Statist.*, **34**, 792.

GUPTA, S. S. (1963b). Bibliography on the multivariate normal integrals and related topics. *Ann. Math. Statist.*, **34**, 829.

GUPTA, S. S. and SHAH, B. K. (1965). Exact moments and percentage points of the order statistics and the distribution of the range from the logistic distribution. *Ann. Math. Statist.*, **36**, 907.

GUPTA, S. S., NAGEL, K. and PANCHAPAKESAN, S. (1973). On the order statistics from equally correlated normal random variables. *Biometrika*, **60**, 403.

GURLAND, J. (1968). A relatively simple form of the distribution of the multiple correlation coefficient. *J. R. Statist. Soc.*, **B**, **30**, 276.

GURLAND, J. and MILTON, R. (1970). Further consideration of the distribution of the multiple correlation coefficient. *J. R. Statist. Soc.*, **B**, **32**, 381.

HALDANE, J. B. S. (1938). The approximate normalization of a class of frequency distributions. *Biometrika*, **29**, 392.

HALDANE, J. B. S. (1939). Cumulants and moments of the binomial distribution. *Biometrika*, **31**, 392.

HALDANE, J. B. S. (1942). Mode and median of a nearly normal distribution with given cumulants. *Biometrika*, **32**, 294.

HALDANE, J. B. S. (1948). Note on the median of a multivariate distribution. *Biometrika*, **35**, 414.

HALDANE, J. B. S. and JAYAKAR, S. D. (1963). The distribution of extremal and nearly extremal values in samples from a normal distribution. *Biometrika*, **50**, 89.

HALL, P. (1927). The distribution of means for samples of size *N* drawn from a population in which the variate takes values between 0 and 1, all such values being equally probable. *Biometrika*, **19**, 240.

HAMBURGER, H. (1920, 1921). Über eine Erweiterung des Stieltjesschen Moment-problems. *Math. Ann.*, **81**, 235; **82**, 120 and 168.

HARTER, H. L. (1960). Tables of range and studentized range. *Ann. Math. Statist.*, **31**, 1122.

HARTER, H. L. (1961). Expected values of normal order statistics. *Biometrika*, **48**, 151, 476.

HARTER, H. L. (1964a). *New tables of the incomplete gamma-function ratio and of percentage points of the chi-square and beta distributions.* Aerospace Research Laboratories, Office of Aerospace Research, U.S.A., Report 64–123, Wright-Patterson A.F.B., Ohio.

HARTER, H. L. (1964b). A new table of percentage points of the chi-square distribution. *Biometrika*, **51**, 231.

HARTER, H. L. (1969). *Order Statistics and their Use in Testing and Estimation.* Volume 1, *Tests Based on Range and Studentized Range of Samples from a Normal Population.* Volume 2, *Estimates Based on Order Statistics of Samples from Various Populations.* Aerospace Research Laboratories, U.S. Air Force; U.S. Government Printing Office, Washington.

HARTLEY, H. O. (1945). Note on the calculation of the distribution of the estimate of mean deviation in normal samples. Tables of the probability integral of the mean deviation in normal samples. *Biometrika*, **33**, 257.

HARTLEY, H. O., and DAVID, H. A. (1954). Universal bounds for mean range and extreme observation. *Ann. Math. Statist.*, **25**, 85.

HASTINGS, C. Jr. (1955). *Approximations for digital computers*, Princeton U.P. and Oxford U.P.

HATKE, M. A. (1949). A certain cumulative probability function. *Ann. Math. Statist.*, **20**, 461.

HELMERT, F. R. (1876). Die Genauigkeit der Formel von Peters zur Berechnung des wahrscheinlichen Beobachtungsfehlers direkter Beobachtungen gleicher Genauigkeit. *Astronomische Nachrichten*, **88**, No. 2096.

HEYDE, C. C. (1963). On a property of the lognormal distribution. *J. R. Statist. Soc.*, **B**, **25**, 392.

HILL, G. W. and DAVIS, A. W. (1968). Generalised asymptotic expansions of Cornish–Fisher type. *Ann. Math. Statist.*, **39**, 1264.

HINKLEY, D. V. (1969). On the ratio of two correlated normal random variables. *Biometrika*, **56**, 635 ; correction, **57**, 683.

HODGES, J. L., Jr, and LEHMANN, E. L. (1967). Moments of chi and power of *t*. *Proc. 5th Berkeley Symp. Math. Statist and Prob.*, **1**, 187.

HOGG, R. V. and CRAIG, A. T. (1958). On the decomposition of certain variables. *Ann. Math. Statist.*, **29**, 608.

HOJO, T. (1931, 1933). Distribution of the median, quartiles and interquartile distance in samples from a normal population. *Biometrika*, **23**, 315; and: A further note on the relation between the median and the quartiles in small samples from a normal population. *Biometrika*, **25**, 79.

HOLGATE, P. (1970). The modality of some compound Poisson distributions. *Biometrika*, **57**, 666.

HOLT, D. R. and CROW, E. L. (1973). Tables and graphs of the stable probability density functions. *J. Research. Nat. Bureau Standards*, **77B**, 143.

HOTELLING, H. (1953). New light on the correlation coefficient and its transforms. *J. R. Statist. Soc.*, **B**, **15**, 193.

HOTELLING, H., and SOLOMONS, L. M. (1932). The limits of a measure of skewness. *Ann. Math. Statist.*, **3**, 141.

HSU, C. T., and LAWLEY, D. N. (1939). The derivation of the fifth and sixth moments of b_2 in samples from a normal population. *Biometrika*, **31**, 238.

IRWIN, J. O. (1927). On the frequency-distribution of the means of samples from a population having any law of frequency with finite moments, etc. *Biometrika*, **19**, 225 ; correction, **21**, 431.

IRWIN, J. O. (1937). The frequency-distribution of the difference between two independent variates following the same Poisson distribution. *J. R. Statist. Soc.*, **100**, 415.

IRWIN, J. O., and KENDALL, M. G. (1944). Sampling moments of moments for a finite population. *Ann. Eugen.*, **12**, 138.

JACKSON, D. (1921). Note on the median of a set of numbers. *Bull. Amer. Math. Soc.*, **27**, 160.

JACOBSON, H. I. (1969). The maximum variance of restricted unimodal distributions. *Ann. Math. Statist.*, **40**, 1746.

JAMES, G. S. (1952). Notes on a theorem of Cochran. *Proc. Camb. Phil. Soc.*, **48**, 443.

JAMES, G. S. (1955). Cumulants of a transformed variate. *Biometrika*, **42**, 529.

JAMES, G. S. (1958). On moments and cumulants of systems of statistics. *Sankhyā*, **20**, **1**.

JAMES, G. S. and MAYNE, A. J. (1962). Cumulants of functions of random variables. *Sankhyā*, **A 24**, 47.

JAMES,, I. R. (1972). Products of independent Beta variables. . . . *J. Amer. Statist. Ass.*, **67**, 910.

JAYACHANDRAN, T. and BARR, D. R. (1970). On the distribution of the difference of two scaled chi-square random variables. *Amer. Statistician*, **24**, **(5)**, 29.

JENKINSON, A. F. (1955). The frequency-distribution of the annual maximum (or minimum) values of meteorological elements. *Quart. J. R. Met. Soc.*, **81**, 158.

JENSEN, D. R. and SOLOMON, H. (1972). A Gaussian approximation to the distribution of a definite quadratic form. *J. Amer. Statist. Ass.*, **67**, 898.

JOHNSON, N. L. (1949a). Systems of frequency curves generated by methods of translation. *Biometrika*, **36**, 149.

JOHNSON, N. L. (1949b). Bivariate distributions based on simple translation systems. *Biometrika*, **36**, 297.

JOHNSON, N. L. (1957). A note on the mean deviation of the binomial distribution. *Biometrika*, **44**, 532.

JOHNSON, N. L. (1965). Tables to facilitate fitting S_U frequency curves. *Biometrika*, **52**, 547; extensions and corrections, **61**, 203.

JOHNSON, N. L. and KITCHEN, J. O. (1971a, b). Tables to facilitate fitting S_B curves: some notes on. *Biometrika*, **58**, 223; II, Both terminals known, *Biometrika*, **58**, 657.

JOHNSON, N. L. and KOTZ, S. (1967a, b, 1968). Tables of distributions of quadratic forms in central normal variables, I, II. *Institute of Statistics Mimeo Series Nos. 543, 547, University of N. Carolina.* Tables of distributions of positive definite quadratic forms in central normal variables. *Sankhyā*, **B**, **30**, 303.

JOHNSON, N. L. and KOTZ, S. (1969, 1970). *Discrete Distributions*; *Continuous Distributions* (2 volumes). Houghton Mifflin, New York.

JOHNSON, N. L. and ROGERS, C. A. (1951). The moment problem for unimodal distributions. *Ann. Math. Statist.*, **22**, 433.

JOHNSON, N. L., NIXON, E., AMOS, D. E., and PEARSON, E. S. (1963). Table of percentage points of Pearson curves, for given $\sqrt{\beta_1}$ and β_2, expressed in standard measure. *Biometrika*, **50**, 459.

JOSHI, P. C. (1969). Bounds and approximations for the moments of order statistics. *J. Amer. Statist. Ass.*, **64**, 1617.

JOSHI, P. C. (1973). Two identities involving order statistics. *Biometrika*, **60**, 428.

KAGAN, A. M., LINNIK, Yu. V., and RAO, C. R. (1973). *Characterization Problems in Mathematical Statistics.* Wiley, New York.

KAMAT, A. R. (1959). Contributions to the theory of Gini's mean difference. *Volume in onore di Corrado Gini* (Istituto di Statistica, Roma), **1**, 285.

KAMAT, A. R. (1965). A property of the mean deviation for a class of continuous distributions. *Biometrika*, **52**, 288.

KAPLAN, E. L. (1952). Tensor notation and the sampling cumulants of k-statistics. *Biometrika*, **39**, 319.

KAPLANSKY, I. (1945). A common error concerning kurtosis. *J. Amer. Statist. Ass.*, **40**, 259.

KATSNELSON, J. and KOTZ, S. (1957). On the upper limits of some measures of variability. *Archiv. f. Meteor., Geophys. u. Bioklimat.*, **(B)**, **8**, 103.

KAWATA, T. and SAKAMOTO, H. (1949). On the characterisation of the normal population by the independence of the sample mean and the sample variance. *J. Math. Soc. Japan*, **1**, 111.

KENDALL, D. G., and RAO, K. S. (1950). On the generalized second limit theorem in the theory of probabilities. *Biometrika*, **37**, 224.

KENDALL, M. G. (1938). The conditions under which Sheppard's corrections are valid. *J. R. Statist. Soc.*, **101**, 592.

KENDALL, M. G. (1940a, b, c). Some properties of k-statistics. *Ann. Eugen.*, **10**, 106. Proof of Fisher's rules for ascertaining the sampling semi-invariants of k-statistics. *Ibid.*, **10**, 215. The derivation of multivariate sampling formulae from univariate formulae by symbolic operation. *Ibid.*, **10**, 392.

KENDALL, M. G. (1941). Relations connected with the tetrachoric series and its generalisation. *Biometrika*, **32**, 196.

KENDALL, M. G. (1942). On seminvariant statistics. *Ann. Eugen.*, **11**, 300.

KENDALL, M. G. (1949a). Reconciliation of theories of probability. *Biometrika*, **36**, 101.

KENDALL, M. G. (1949b). Rank and product-moment correlation. *Biometrika*, **36**, 177.

KENDALL, M. G. (1952). Moment-statistics in samples from a finite population. *Biometrika*, **39**, 14.

KENDALL, M. G. (1953). Discussion of Hotelling (1953).

KENDALL, M. G., and SMITH, B. B. (1938–9). Randomness and random sampling numbers. *J. R. Statist. Soc.*, **101**, 147, and *Supp. J. R. Statist. Soc.*, **6**, 51.

KERRIDGE, D. F. and COOK, G. W. (1976). Yet another series for the normal integral. *Biometrika*, **63**, 401.

KHAMIS, S. H. and RUDERT, W. (1965). *Tables of the Incomplete Gamma Function Ratio: Chi-square Integral, Poisson Distribution.* Justus von Liebig, Darmstadt, Germany.

KOTLARSKI, I. (1962). On groups of n independent random variables whose product follows the beta distribution. *Colloq. Math.*, **9**, 325.

KOTLARSKI, I. (1964). On bivariate random variables where the quotient of their coordinates follows some known distribution. *Ann. Math. Statist.*, **35**, 1673.

KOTLARSKI, I. (1965). On pairs of independent random variables whose product follows the Gamma distribution. *Biometrika*, **52**, 289.

KOTZ, S., JOHNSON, N. L. and BOYD, D. W. (1967a). Series representations of distributions of quadratic forms in normal variables. I. Central Case. *Ann. Math. Statist.* **38**, 823.

KOTZ, S., JOHNSON, N. L. and BOYD, D. W. (1967b). Series representations of distributions of quadratic forms in normal variables. II. Non-central case. *Ann. Math. Statist.*, **38**, 838.

KRAEMER, H. C. (1973). Improved approximation to the non-null distribution of the correlation coefficient. *J. Amer. Statist. Ass.*, **68**, 1004.

KRATKY, REINFELDS, J., HUTCHESON and SHENTON, L. R. (1972). *Tables of Crude Moments Expressed in Terms of Cumulants.* Computer Center Report No. 7201, University of Georgia, Athens, Georgia.

KRISHNA IYER, P. V. and SIMHA, P. S. (1967). *Tables of Bivariate Random Normal Deviates.* Defence Science Laboratory, Delhi, India.

KULLBACK, S. (1934). An application of characteristic functions to the distribution problem of statistics. *Ann. Math. Statist.*, **5**, 264.

LAHA, R. G. (1954). On a characterisation of the gamma distribution. *Ann. Math. Statist.*, **25**, 784.

LAHA, R. G. (1959). On the laws of Cauchy and Gauss. *Ann. Math. Statist.*, **30**, 1165.

LANCASTER, H. O. (1954). Traces and cumulants of quadratic forms in normal variables. *J. R. Statist. Soc.*, **B**, **16**, 247.

LANCASTER, H. O. (1960). The characterisation of the normal distribution. *J. Austral. Math. Soc.*, **1**, 368.

LEE, Y. S. (1972). Tables of upper percentage points of the multiple correlation coefficient. *Biometrika*, **59**, 175.

LEMPERS, F. B. and LOUTER, A. S. (1971). An extension of the table of the Student distribution. *J. Amer. Statist. Ass.*, **66**, 503.

LENTNER, M. M. and BUEHLER, R. J. (1963). Some inferences about gamma parameters with an application to a reliability problem. *J. Amer. Statist. Ass.*, **58**, 670.

LÉVY, P. (1925). *Calcul des probabilités.* Gauthier–Villars, Paris.

LIAPOUNOFF, A. (1901). Nouvelle forme du théorème sur la limite de probabilité. *Mem. Acad. Sci. St. Pét.*, (8), **12**, No. 5.

LIEBERMAN, G. J. and OWEN, D. B. (1961). *Tables of the hypergeometric probability distribution.* Stanford U.P.

LOMNICKI, Z. A. (1952). The standard error of Gini's mean difference. *Ann. Math. Statist.*, **23**, 635.

LOMNICKI, Z. A. (1967). On the distribution of products of random variables. *J. R. Statist. Soc.*, **B**, **29**, 513.

LUKACS, E. (1942). A characterization of the normal distribution. *Ann. Math. Statist.*, **13**, 91.

LUKACS, E. (1952). An essential property of the Fourier transforms of distribution functions. *Proc. Amer. Math. Soc.*, **3**, 508.

LUKACS, E. (1956). Characterization of populations by properties of suitable statistics. *Proc. 3rd Berkeley Symp. Math. Statist. and Prob.*, **2**, 195.

LUKACS, E. (1970). *Characteristic functions*, 2nd edn. Griffin, London.

LUKACS, E. and LAHA, R. G. (1964). *Applications of characteristic functions.* Griffin, London.

LUKACS, E. and SZÁSZ, O. (1952). On analytic characteristic functions. *Pacific J. Math.*, **2**, 615.

MCCORD, J. R. (1964). On asymptotic moments of extreme statistics. *Ann. Math. Statist.*, **35**, 1738.

McKay, A. T. (1935). The distribution of the difference between the extreme observation and the sample mean in samples of n from a normal universe. *Biometrika*, **27**, 466.

Majindar, K. N. (1962). Improved bounds on a measure of skewness. *Ann. Math. Statist.*, **33**, 1192.

Marcinkiewicz, J. (1938). Sur une propriété de la loi de Gauss. *Math. Zeitschr.*, **44**, 612.

Mardia, K. V. (1965). Tippett's formulas and other results on sample range and extremes. *Ann. Inst. Statist. Math.*, **17**, 85.

Mardia, K. V. (1967). Some contributions to contingency-type bivariate distributions. *Biometrika*, **54**, 235 ; corrections, **55**, 597.

Mardia, K. V. (1970a). *Families of Bivariate Distributions*. Griffin, London.

Mardia, K. V. (1970b). Some problems of fitting for contingency-type bivariate distributions. *J. R. Statist. Soc.*, **B**, **32**, 254.

Mardia, K. V. (1972). *Statistics of Directional Data*. Academic Press, London.

Marsaglia, G. (1965). Ratios of normal variables and ratios of sums of uniform variables. *J. Amer. Statist. Ass.*, **60**, 193.

Martin, E. S. (1934). On the correction for the moment coefficients of frequency-distributions when the start of the frequency is one of the characteristics to be determined. *Biometrika*, **26**, 12.

Matérn, B. (1949). Independence of non-negative quadratic forms in normally correlated variables. *Ann. Math. Statist.*, **20**, 119.

Mathur, R. K. (1961). A note on Wilson–Hilferty transformation of χ^2. *Bull. Calcutta Statist. Ass.*, **10**, 102.

Mehta, J. S. and Gurland, J. (1969). Some properties and an application of a statistic arising in testing correlation. *Ann. Math. Statist.*, **40**, 1736.

Mikhail, N. N. (1968). Multivariate tables of symmetric function. *Egypt. Statist. J.*, **12**, 17.

Miller, J. C. P. (1954). Tables of binomial coefficients. *Roy. Soc. Math. Tables*, vol. 3.

Molina, E. C. (1942). *Poisson's Exponential Binomial Limit*. Van Nostrand, New York.

Moran, P. A. P. (1948). Rank correlation and product moment correlation. *Biometrika*, **35**, 203.

Moran, P. A. P. (1956). The numerical evaluation of a class of integrals. *Proc. Camb. Phil. Soc.*, **52**, 230.

Moriguti, S. (1951). Extremal properties of extreme value distributions. *Ann. Math. Statist.*, **22**, 523.

Moriguti, S. (1952). A lower bound for a probability moment of any absolutely continuous distribution with finite variance. *Ann. Math. Statist.*, **23**, 286.

Nagambal, P. N. and Tracy, D. S. (1970). Products of two polykays when one has weight 5. *Ann. Math. Statist.*, **41**, 1114.

Nair, U. S. (1936). The standard error of Gini's mean difference. *Biometrika*, **28**, 428.

Newman, T. G. and Odell, P. L. (1971). *The Generation of Random Variates*. Griffin, London.

Neyman, J. (1939).* On a new class of "contagious" distributions, applicable in entomology and bacteriology. *Ann. Math. Statist.*, **10**, 35.

Ogasawara, T. and Takahashi, M. (1951). Independence of quadratic quantities in a normal system. *J. Sci. Hiroshima University*, **A**, **15**, 1.

Ord, J. K. (1967a). Graphical methods for a class of discrete distributions. *J. R. Statist. Soc.*, **A**, **130**, 232.

Ord, J. K. (1967b). On a system of discrete distributions. *Biometrika*, **54**, 649.

Ord, J. K. (1968a). Approximations to distribution functions which are hypergeometric series, *Biometrika*, **55**, 243.

Ord, J. K. (1968b). The discrete Student's t distribution. *Ann. Math. Statist.*, **39**, 1513.

Ord, J. K. (1972). *Families of Frequency Distributions*. Griffin, London.

OWEN, D. B. (1956). Tables for computing bivariate normal probabilities. *Ann. Math. Statist.*, **27**, 1075.
OWEN, D. B. (1957). The bivariate normal probability integral. Research Report, Sandia Corp., Washington.
OWEN, D. B. and STECK, G. P. (1962). Moments of order statistics from the equicorrelated multivariate normal distribution. *Ann. Math. Statist.*, **33**, 1286.

PAIRMAN, E., and PEARSON, K. (1919). On the corrections for moment coefficients of limited-range frequency-distributions when there are finite or infinite ordinates and any slopes at the terminals of the range. *Biometrika*, **12**, 231.
PATIL, G. P. and SESHADRI, V. (1964). Characterization theorems for some univariate probability distributions. *J. R. Statist. Soc.*, **B**, **26**, 286.
PATIL, G. P., KAMAT, A. R. and WANI, J. K. (1964). *Certain studies on the structure and statistics of the logarithmic series distribution and related tables.* U.S. Dept of Commerce, Washington.
PAULSON, E. (1942). An approximate normalisation of the analysis of variance distribution. *Ann. Math. Statist.*, **13**, 233.
PEARSON, E. S. (1926). A further note on the distribution of range in samples taken from a normal population. *Biometrika*, **18**, 173.
PEARSON, E. S. (1930). A further development of tests for normality. *Biometrika*, **22**, 239.
PEARSON, E. S. (1932). The percentage limits for the distribution of range in samples from a normal population. *Biometrika*, **24**, 404.
PEARSON, E. S., and HARTLEY, H. O. (1942). The probability integral of the range in samples of n observations from a normal population. *Biometrika*, **32**, 301.
PEARSON, E. S., and HARTLEY, H. O. (1950). Tables of the χ^2 integral and of the cumulative Poisson distribution. *Biometrika*, **37**, 313.
PEARSON, K. (1900).* On a criterion that a given system of deviations from the probable in the case of a correlated system of variables is such that it can be reasonably supposed to have arisen in random sampling. *Phil. Mag.*, (5), **50**, 157.
PEARSON, K. (1919). On generalised Tchebycheff theorems in the mathematical theory of statistics. *Biometrika*, **12**, 284.
PEARSON, K. (1923). On non-skew frequency surfaces. *Biometrika*, **15**, 231.
PEARSON, K. (1924a). On the moments of the hypergeometrical series. *Biometrika*, **16**, 157.
PEARSON, K. (1924b). On a certain double hypergeometrical series and its representation by continuous frequency surfaces. *Biometrika*, **16**, 172.
PEARSON, K. (1931). Appendix to a paper by Professor Tokishige Hojo. On the standard error of the median to a third approximation, etc. *Biometrika*, **23**, 361.
PEARSON, K., STOUFFER, S. A., and DAVID, F. N. (1932). Further applications in statistics of the $T_m(x)$ Bessel function. *Biometrika*, **24**, 293.
PEIZER, D. B. and PRATT, J. W. (1968). A normal approximation for binomial, F, beta, and other common, related tail probabilities, I. *J. Amer. Statist. Ass.*, **63**, 1416.
PITMAN, E. J. G. (1937a). Significance tests which may be applied to samples from any population. II. The correlation coefficient test. *Supp. J. R. Statist. Soc.*, **4**, 225.
PITMAN, E. J. G. (1937b). The "closest" estimates of statistical parameters. *Proc. Camb. Phil. Soc.*, **33**, 212.
PITMAN, E. J. G. (1956). On the derivatives of a characteristic function at the origin. *Ann. Math. Statist.*, **27**, 1156.
PLACKETT, R. L. (1947). Limits of the ratio of mean range to standard deviation. *Biometrika*, **34**, 120.
PLACKETT, R. L. (1965). A class of bivariate distributions. *J. Amer. Statist. Ass.*, **60**, 516.
PLACKETT, R. L. (1968). Random permutations. *J. R. Statist. Soc.*, **B**, **30**, 517.
POLYÀ, G. (1945). Remarks on computing the probability integral in one and two dimensions. *Proc. 1st Berkeley Symp. Math. Statist. and Prob.*, 63.

POSNER, E. C., RODEMICH, E. R., ASHLOCK, J. C. and LUBIN, S. (1969). Application of an estimator of high efficiency in bivariate extreme value theory. *J. Amer. Statist. Ass.*, **64,** 1403.

PRATT, J. W. (1968). A normal approximation for binomial, *F*, beta and other common, related tail probabilities, II. *J. Amer. Statist. Ass.*, **63,** 1457.

PRESCOTT, P. (1974). Variances and covariances of order statistics from the gamma distribution. *Biometrika*, **61,** 607.

PRETORIUS, S. J. (1930). Skew bivariate frequency surfaces, examined in the light of numerical illustrations. *Biometrika*, **22,** 109.

PURI, P. S. and RUBIN, H. (1970). A characterization based on the absolute difference of two i.i.d. random variables. *Ann. Math. Statist.*, **41,** 2113.

PYKE, R. (1965). Spacings. *J. R. Statist. Soc.*, **B, 27,** 395.

QUENOUILLE, M. H. (1949). A relation between the logarithmic, Poisson, and negative binomial series. *Biometrics*, **5,** 162.

QUENOUILLE, M. H. (1959). Tables of random observations from standard distributions. *Biometrika*, **46,** 178.

RABINOWITZ, P. (1969). New Chebyshev polynomial approximations to Mills' ratio. *J. Amer. Statist. Ass.*, **64,** 647.

RAFF, M. S. (1956). On approximating the point binomial. *J. Amer. Statist. Ass.*, **51,** 293.

RAMASUBBAN, T. A. (1958). The mean difference and the mean deviation of some discontinuous distributions. *Biometrika*, **45,** 549.

RAY, W. D. and PITMAN, A. E. N. T. (1963). Chebyshev polynomial and other new approximations to Mills' Ratio. *Ann. Math. Statist.*, **34,** 892.

RÉNYI, A. (1953). On the theory of order statistics. *Acta Math. Acad. Sci. Hung.*, **4,** 191.

ROBERTSON, W. H. (1960). *Tables of the binomial distribution function for small values of P.* Sandia Corp., Albuquerque, New Mexico, U.S.A.

ROMANOVSKY, V. (1923). Note on the moments of a binomial $(p+q)^n$ about its mean. *Biometrika*, **15,** 410.

ROMANOVSKY, V. (1925). On the moments of the hypergeometrical series. *Biometrika*, **17,** 57.

ROMIG, H. G. (1953). *50–100 Binomial tables.* Wiley, New York.

RUBEN, H. (1954). On the moments of order statistics in samples from normal populations. *Biometrika*, **41,** 200.

RUBEN, H. (1960). On the distribution of the weighted difference of two independent Student variables. *J. R. Statist. Soc.*, **B, 22,** 188.

RUBEN, H. (1962). A new asymptotic expansion for the normal probability integral and Mills' ratio. *J. R. Statist. Soc.*, **B, 24,** 177.

RUBEN, H. (1963). A convergent asymptotic expansion for Mills' ratio and the normal probability integral in terms of rational functions. *Math. Ann.*, **151,** 355.

RUBEN, H. (1964). Irrational fraction approximations to Mills' ratio. *Biometrika*, **51,** 339.

RUBEN, H. (1966). Some new results on the distribution of the sample correlation coefficient. *J. R. Statist. Soc.*, **B, 28,** 513.

SAMIUDDIN, M. (1970). On a test for an assigned value of correlation in a bivariate normal distribution. *Biometrika*, **57,** 461.

SAMPFORD, M. (1953). Some inequalities on Mill's ratio and related functions. *Ann. Math. Statist.*, **24,** 130.

SANDIFORD, P. J. (1960). A new binomial approximation for use in sampling from finite populations. *J. Amer. Statist. Ass.*, **55,** 718.

SANDON, F. (1924). Note on the simplification of the calculation of abruptness coefficients to correct crude moments. *Biometrika*, **16,** 193.

SANKARAN, M. (1959). On the non-central chi-square distribution. *Biometrika*, **46,** 235.

SARHAN, A. E., and GREENBERG, B. G. (1956, 1958). Estimation of location and scale parameters by order statistics from singly and doubly censored samples. I. *Ann. Math. Statist.*, **27**, 427. II. Same *Ann.*, **29**, 79.

SARHAN, A. E. and GREENBERG, B. G., eds. (1962). *Contributions to order statistics.* Wiley, New York.

SAVAGE, L. J. (1954). *The foundations of statistics.* Wiley, New York.

SAVAGE, L. J. (1961). The foundations of statistics reconsidered. *Proc. 4th Berkeley Symp. Math. Statist. and Prob.*, **1**, 575.

SAVAGE, L. J (1962). *The foundations of statistical inference: a discussion . . . at a meeting of the Joint Statistics Seminar, Birkbeck and Imperial Colleges, in the University of London.* Methuen, London.

SEN, P. K. (1961). On some properties of the asymptotic variance of the sample quantiles and mid-ranges. *J. R. Statist. Soc.*, **B, 23**, 453.

SHENTON L. R. (1954). Inequalities for the normal integral, including a new continued fraction. *Biometrika*, **41**, 177.

SHEPPARD, W. F. (1898). On the application of the theory of error to cases of normal distributions and normal correlations. *Phil. Trans*, **A, 192**, 101, and *Proc. Roy. Soc.*, **62**, 170.

SHEPPARD, W. F. (1939). *The Probability Integral.* British Ass. Math. Tables, vol. 7. Cambridge Univ. Press.

SHOHAT, J. A. and TAMARKIN, J. D. (1943). *The problem of moments.* Amer. Math. Soc., New York.

SHORACK, G. R. (1969, 1972). Asymptotic normality of linear combinations of functions of order statistics ; *and* Functions of order statistics. *Ann. Math. Statist.*, **40**, 2041, and **43**, 412.

SICHEL, H. S. (1949). The method of frequency moments and its application to Type VII distributions. *Biometrika*, **36**, 404.

SILLITTO, G. P. (1951). Interrelations between certain linear systematic statistics of samples from any continuous population. *Biometrika*, **38**, 377.

SILLITTO, G. P. (1969). Derivation of approximants to the inverse distribution function of a continuous univariate population from the order statistics of a sample. *Biometrika*, **56**, 641.

SINGH, C. (1967). On the extreme values and range of samples from non-normal populations. *Biometrika*, **54**, 541.

SINGH, C. (1970). On the distribution of range of samples from nonnormal populations. *Biometrika*, **57**, 451.

SINGH, C. (1972). Order statistics from nonnormal populations. *Biometrika*, **59**, 229.

SLUTSKY, E. E. (1950). *Tables for the calculation of the Incomplete Gamma Function and the χ^2 probability function.* Moscow, Isdatelstvo Akad. Nauk S.S.S.R.

SMIRNOV, N. V. (1949). Limit distribution for the terms of a variational series. *Amer. Math. Soc. Transl.*, 67.

SMIRNOV, N. V. (1961). *Tables for the distribution and density functions of t-distribution.* Pergamon, Oxford.

SMIRNOV, N. V. (1965). *Tables of the normal probability integral, the normal density and its normalized derivatives.* Transl. from Russian by D. E. Brown. Pergamon, Oxford.

SOLOMON, H. (1960). Distribution of quadratic forms—tables and applications. *Appl. Math. and Statist. Labs., Tech. Report 45, Stanford University.*

SOWEY, E. R. (1972). A chronological and classified bibliography on random number generation and testing. *Int. Statist. Rev.*, **40**, 355.

STECK, G. P. (1958a). A uniqueness property not enjoyed by the normal distribution. *Ann. Math. Statist.*, **29**, 604.

STECK, G. P. (1958b). A table for computing trivariate normal probabilities. *Ann. Math. Statist.*, **29**, 780.

STECK, G. P. (1962). Orthant probabilities for the equicorrelated multivariate normal distribution. *Biometrika*, **49**, 433.

STECK, G. P. (1968). A note on contingency-type bivariate distributions. *Biometrika*, **55**, 262.

STEPHENS, M. A. (1966). Statistics connected with the uniform distribution : percentage points and applications to testing for randomness of directions. *Biometrika*, **53**, 235.

STIELTJES, T. J. (1918). Recherches sur les fractions continues. *Œuvres*, Groningen.

STUART, A. (1958). Equally correlated variates and the multinormal integral. *J. R. Statist. Soc.*, **B**, **20**, 373.

STUART, A. (1962). Gamma-distributed products of independent random variables. *Biometrika*, **49**, 564.

" STUDENT " (1908a).* On the probable error of a mean. *Biometrika*, **6**, 1.

" STUDENT " (1908b).* On the probable error of a correlation coefficient. *Biometrika*, **6**, 302.

STYAN, G. P. H. (1970). Notes on the distribution of quadratic forms in singular normal variables. *Biometrika*, **57**, 567.

SUBRAHMANIAM, K. (1966). Some contributions to the theory of non-normality—I (Univariate case). *Sankyhā*, **A**, **28**, 389.

SUBRAHMANIAM, K. (1969). Order statistics from a class of non-normal distributions. *Biometrika*, **56**, 415.

SUKHATME, P. V. (1937). Tests of significance for samples of the χ^2-population with two degrees of freedom. *Ann. Eugen.*, **8**, 52.

TANIS, E. A. (1964). Linear forms in the order statistics from an exponential distribution. *Ann. Math. Statist.*, **35**, 270.

TEICHROEW, D. (1956). Tables of expected values of order statistics and products of order statistics. *Ann. Math. Statist.*, **27**, 410.

TEICHROEW, D. (1965). A history of distribution sampling prior to the era of the computer and its relevance to simulation. *J. Amer. Statist. Ass.*, **60**, 27.

THIELE, T. N. (1903). *Theory of Observations.* Reprint in *Ann. Math. Statist.*, **2**, 165, of the English version published in 1903.

THOMPSON, C. M. (1941). Tables of percentage points of the χ^2-distribution. *Biometrika*, **32**, 187.

THOMPSON, C. M., PEARSON, E. S., COMRIE, L. J., and HARTLEY, H. O. (1941). Tables of percentage points of the incomplete beta-function. *Biometrika*, **32**, 151.

THOMSON, W. E. (1959). ERNIE—a mathematical and statistical analysis. *J. R. Statist. Soc.*, **A**, **122**, 301.

TIAGO DE OLIVEIRA, J. (1962–3). Structure theory of bivariate extremes : extensions. *Estudos de Matematica, Estatistica e Econometria*, **7**, 165.

TIPPETT, L. H. C. (1925). On the extreme individuals and the range of samples taken from a normal population. *Biometrika*, **17**, 364.

TRACY, D. S. (1968). Some rules for a combinatorial method for multiple products of generalized k-statistics. *Ann. Math. Statist.*, **39**, 983.

TRACY, D. S. (1969). Some multiple products of polykays. *Ann. Math. Statist.*, **40**, 1297.

TRACY, D. S. (1972). *Symmetric Functions in Statistics.* Proceedings of a Symposium in Honour of Professor Paul S. Dwyer. University of Windsor, Ontario.

TRACY, D. S. and GUPTA, B. C. (1973). Multiple products of polykays using ordered partitions. *Ann. Statist.*, **1**, 913.

TRACY, D. S. and GUPTA, B. C. (1974). Generalized h-statistics and other symmetric functions. *Ann. Statist.*, **2**, 837.

TUKEY, J. W. (1950). Some sampling simplified. *J. Amer. Statist. Ass.*, **45**, 501.

VAUGHAN, R. J. and VENABLES, W. N. (1972). Permanent expressions for order statistic densities. *J. R. Statist. Soc.*, **B**, **34**, 308.

VOGLER, L. E. (1964). *Percentage points of the beta distribution.* U.S. Govt. Printing Office, Washington.

VON MISES, R. (1936). La distribution de la plus grande de n valeurs. *Revue de l'Union Interbalkanique*, **1**, 1.

WALKER, A. M. (1968). A note on the asymptotic distribution of sample quantiles. *J. R. Statist. Soc.*, **B**, **30**, 570.

WALLACE, D. L. (1958). Asymptotic approximations to distributions. *Ann. Math. Statist.*, **29**, 635.

WEINTRAUB, S. (1963). *Tables of the cumulative binomial probability distribution for small values of p*. Free Press of Glencoe, New York.

WHITE, J. S. (1970). Tables of normal percentile points. *J. Amer. Statist. Ass.*, **65**, 635.

WICKSELL, S. D. (1917). On logarithmic correlation with an application to the distribution of ages at first marriage. *Medd. Lunds Astr. Obs.*, No. 84.

WIDDER, D. V. (1941). *The Laplace Transform*. Princeton U. P.

WILLIAMS, J. D. (1946). An approximation to the probability integral. *Ann. Math. Statist.*, **17**, 363.

WILLIAMS, J. S. (1966). An example of the misapplication of conditional densities. *Sankhyā*, **A**, **28**, 297.

WILLIAMS, P. (1935). Note on the sampling distribution of $\sqrt{\beta_1}$ when the population is normal. *Biometrika*, **27**, 269.

WILLIAMSON, E. and BRETHERTON, M. H. (1963). *Tables of the Negative Binomial Probability Distribution*. Wiley, London.

WILLIAMSON, E. and BRETHERTON, M. H. (1964). Tables of the logarithmic series distribution. *Ann. Math. Statist.*, **35**, 284.

WILSON, E. B., and HILFERTY, M. M. (1931). The distribution of chi-square. *Proc. Nat. Acad. Sci., U.S.A.*, **17**, 684.

WISE, M. E. (1954). A quickly convergent expansion for cumulative hypergeometric probabilities, direct and inverse. *Biometrika*, **41**, 317.

WISE, M. E. (1960). On normalizing the incomplete beta-function for fitting to dose-response curves. *Biometrika*, **47**, 173.

WISHART, J. (1929). The correlation between product moments of any order in samples from a normal population. *Proc. Roy. Soc. Edin.*, **49**, 78.

WISHART, J. (1949). Cumulants of multivariate multinomial distributions. *Biometrika*, **36**, 47.

WISHART, J. (1952). Moment coefficients of the k-statistics in samples from a finite population. *Biometrika*, **39**, 1.

WOLD, H. (1934a). Sulla correzione di Sheppard. *Giorn. Ist. Ital. Attuari*, **5**, 304.

WOLD, H. (1934b). Sheppard's correction formulæ in several variables. *Skand. Aktuartidskr.*, **17**, 248.

WOODWARD, W. A. (1976). Approximation of Pearson Type IV tail probabilities. *J. Amer. Statist. Ass.*, **71**, 513.

YASUKAWA, K. (1926). On the probable error of the mode of frequency-distributions. *Biometrika*, **18**, 263.

YATES, F. (1935). Some examples of biassed sampling. *Ann. Eugen.*, **6**, 202.

YOUNG, D. H. (1970). The order statistics of the negative binomial distribution. *Biometrika*, **57**, 181.

YUAN, P. T. (1933). On the logarithmic frequency distribution. *Ann. Math. Statist.*, **4**, 30.

YULE, G. U. (1910).* On the distribution of deaths with age when the causes of death act cumulatively. *J. R. Statist. Soc.*, **73**, 26.

YULE, G. U. (1927). On reading a scale. *J. R. Statist. Soc.*, **90**, 570.

ZELEN, M. and SEVERO, N. C. (1960). Graphs for bivariate normal probabilities. *Ann. Math. Statist.*, **31**, 619.

ZIA UD-DIN, M. (1954). Expression of the k-statistics, k_9 and k_{10}, in terms of power sums and sample moments. *Ann. Math. Statist.*, **25**, 800.

ZIA UD-DIN, M. (1959). The expression of k-statistic k_{11} in terms of power sums and sample moments. *Ann. Math. Statist.*, **30**, 825.

INDEX

(References are to pages)